Six Sigma for Students

Fatma Pakdil

Six Sigma for Students

A Problem-Solving Methodology

palgrave
macmillan

Fatma Pakdil
Eastern Connecticut State University
Willimantic, CT
USA

ISBN 978-3-030-40708-7 ISBN 978-3-030-40709-4 (eBook)
https://doi.org/10.1007/978-3-030-40709-4

This Palgrave Macmillan imprint is published by the registered company Springer Nature Switzerland AG
The registered company address is: Gewerbestrasse 11, 6330 Cham, Switzerland

To my husband, Semih Pakdil, M.D., for his endless lifelong support....

Preface

The goal of this book is to present how Six Sigma methodology is used in solving problems and issues that affect the variability and quality of processes and outcomes. This book does not intend to make statisticians out of readers and students, rather it aims to teach readers and students how to integrate statistical perspective into problem solving processes using the Six Sigma approach. This book also aims to provide graduate and undergraduate students with a textbook that helps them learn Six Sigma. The book may be used for the following purposes: (1) to provide an effective understanding of Six Sigma, (2) to guide undergraduate students in Six Sigma journey, and (3) to provide students with a comprehensive text for use in higher education in various majors. What makes this book different from other books in this discipline is that it has a "student-oriented approach" focusing on understanding the fundamental mechanisms of the Six Sigma philosophy.

Students and instructors will benefit from the unique structure of the textbook which allows readers to learn both the theoretical and practical aspects of the topics presented in each chapter: first, the theoretical background of the topic is presented, then examples and questions are given and solved in each section and chapter, and finally discussion questions are offered at the end of each chapter. In each chapter, the practice questions allow the readers to have a better grasp of the topics analyzed in each section. In addition, the tools and methods are accompanied with statistical software.

Fatma Pakdil
Mansfield, CT, USA

Acknowledgments

I have many people to thank to for their support while I was preparing this textbook for publication. First, I would like to thank my parents, Embiye and Enver Besiktepe, who helped me prepare for life. Second, I thank my husband, Dr. Semih Pakdil, who has been extremely understanding throughout the entirety of my academic career and while I was writing this textbook. Third, I thank my children, Ece and Yigit, for their understanding and patience during the preparations for this book. I will never forget how my son drew a picture for my office and wrote on it, "my mom, the best professor ever…" Fourth, I thank my mother-in-law Sema Pakdil and father-in-law Ihsan Pakdil for their endless support and care for my family throughout my academic career… One more sincere appreciation goes to all of my teachers and professors who introduced me new perspectives. Lastly, I have to thank Dr. Karen Moustafa Leonard and Dr. Timothy N. Harwood many times. Dr. Leonard tremendously helped me during the proofreading stages of this textbook and supported the whole project with her ideas and suggestions. I wouldn't have been able to finish this book without her incredible support. I am also grateful for her relentless support and encouragement all through my career since 2005. Dr. Timothy N. Harwood, accepting me as a visiting scholar to Wake Forest University, opened the door to a new world for me and my career in 2001 after I visited the USA for the first time. My colleagues Dr. Nasibeh Azadeh-Fard, Dr. Burçin Çakır Erdener, and Dr. Aysun Kapucugil Ikiz also receive my wholehearted appreciation for their contributions to my textbook. *Six Sigma for Students: A Problem-Solving Methodology* wouldn't have been completed without their invaluable contributions.

Contents

Abbreviations

AHP	Analytic Hierarchy Process
ANOVA	Analysis of Variance
ANP	Analytic Network Process
ANSI	American National Standards Institute
ASME	The American Society of Mechanical Engineers
ASQ	American Society for Quality
ASQC	American Society for Quality Control
ASTM	American Society for Testing and Materials
CDF	Cumulative Distribution Function
CI	Confidence Interval
CL	Central Line
CQI	Continuous Quality Improvement
CTQ	Critical-to-Quality
DEA	Data Envelopment Analysis
DEMATEL	Decision-Making Trial and Evaluation Laboratory
DES	Discrete Event Simulation
DFSS	Design for Six Sigma
DMAIC	Define, Measure, Analyze, Improve, Control
DOE	Design of Experiment
DPMO	Defects per Million Opportunities
DPO	Defects per Opportunity
DPU	Defects per Unit
FDA	Food and Drug Administration
FMEA	Failure Modes and Effects Analysis
FMECA	Failure Mode Effects and Criticality Analysis
HOQ	House of Quality
IMVP	International Motor Vehicle Program
IoT	Internet of Things
ISO	International Organization for Standardization
JIPM	Japan Institute of Plant Maintenance
JIT	Just-in-Time
JUSE	The Union of Japanese Scientists and Engineers
L	Lower Confidence Limit
LCL	Lower Control Limit
LNTL	Lower Natural Tolerance Limit
LSL	Lower Specification Limit
LTL	Lower Tolerance Limit
MAD	Mean Absolute Deviation
MCDM	Multi-Criteria Decision-Making
MR	Moving Range
MSE	Mean Square Error

NIST	National Institute of Standards and Technology
NVA	Non-Value-Added
NVAR	Non-Value-Added But Required
PAF	Prevention-Appraisal-Failure
PCR	Process Capability Ratio
PDCA	Plan, Do, Control, Act
PDF	Probability Density Function
PDPC	Process Decision Program Chart
PMF	Probability Mass Function
PPM	Product per Millions
QFD	Quality Function Deployment
QI	Quality Improvement
RCA	Root Cause Analysis
RPN	Risk Priority Number
RSM	Response Surface Method
SIPOC	Supplier, Input, Process, Output, Customer
SMED	Single Minute Exchange of Dies
SoPK	The System of Profound Knowledge
SPC	Statistical Process Control
SQC	Statistical Quality Control
SSA	Sum of Squares of Factor
SSE	Sum of Squares of Error
SSMI	Six Sigma Management Institute
SST	Total Sum of Squares
SSW	Sum of Squares Within Groups
STM	Scientific Thinking Mechanism
TOC	Theory of Constraints
TOPSIS	Technique for Order of Preference by Similarity to Ideal Solution
TPM	Total Productive Maintenance
TPS	Toyota Production System
TQM	Total Quality Management
U	Upper Confidence Limit
UCL	Upper Control Limit
UNTL	Upper Natural Tolerance Limit
USL	Upper Specification Limit
UTL	Upper Tolerance Limit
VA	Value Added
VOC	Voice of Customer
VSM	Value Stream Map

List of Figures

List of Images

List of Tables

Organization of Six Sigma

Contents

Overview of Quality and Six Sigma

Contents

F. Pakdil, *Six Sigma for Students*, https://doi.org/10.1007/978-3-030-40709-4_1

1

Learning Objectives

After careful study of this chapter, you should be able to:

- Define characteristics of Six Sigma philosophy
- Define and discuss quality definitions using four approaches
- Identify and discuss six dimensions of quality definition
- Identify "value" term in quality definition
- Describe the quality management philosophies of quality gurus and identify their contributions to quality and Six Sigma literature
- Identify historical background of Six Sigma
- Describe international standards used in Six Sigma
- Describe quality cost term
- Identify quality cost categories
- Calculate and interpret quality costs and performance metrics.

1.1 Introduction

As an introductory, this chapter first discusses basic characteristics of Six Sigma philosophy, quality definitions using four approaches, six dimensions of quality definition, and "value" term in quality definition. Quality definitions are categorized into four managerial perspectives: (1) the product-based approach, (2) the manufacturing-based approach, (3) the value-based approach, and (4) the user-based approach. Then, the contributions of quality gurus to quality and Six Sigma literature are discussed in detail. These quality gurus are Walter Shewhart, W. Edwards Deming, Joseph M. Juran, Armand V. Feigenbaum, Kaoru Ishikawa, Taiichi Ohno, Shigeo Shingo, Genichi Taguchi, Philip Crosby, David Garvin, and Douglas Montgomery. Historical background of Six Sigma, standards developed by the International Organization for Standardization (ISO) on Six Sigma, types of costs of quality, and performance metrics used in quality cost calculations are detailed in this chapter. Costs of quality are grouped into two main categories: costs of good quality and costs of poor quality. Costs of good quality comprises of prevention costs and appraisal costs, while costs of poor quality contain internal failure costs and external failure costs.

1.2 The Six Sigma Philosophy

In today's business world, organizations rely on various competitive advantages to make a difference in the marketplace. Drucker (1999) defines the purpose of a business as the creation and retention of satisfied and loyal customers. Considering that loyal customers are gained through flawless service and product, Nasir and Nasir (2005) state that decreasing defects and nonconformities has become a requirement in order to sustain a competitive advantage for business organizations. In *Total Quality Control*, Feigenbaum (1983) explicitly states that "quality is a crucial hinge for business success or failure in quality-performance-oriented markets" (17). Feigenbaum perceives quality as a major factor allowing firms to gain a major competitive advantage, business profitability, and positive cash flow. According to Reichheld (1996), learning about the causes of defects provides an opportunity for the companies to revise and improve their processes. Six Sigma is considered one of the most effective remedies in management and industrial engineering disciplines, not only for improving process performance but also for enhancing organizational performance in today's highly competitive business world.

Sigma is a statistical term used to measure variability in processes. Blakeslee Jr (1999) calls Six Sigma a "high-performance, data-driven approach to analyzing the root causes of business problems and solving them" (78). According to Wedgwood (2015: 56):

> Six Sigma is a systematic methodology to home in on the key factors (known as Xs) that drive variation in the performance of a process (the measures of this are known as Ys), set them at the best levels, and hold them there, thus ensuring a significantly increased performance level for all time. Six

> Sigma is an excellent method to drive much higher reliability in a process by increasing accuracy and decreasing defects.

Therefore, Six Sigma focuses on prioritizing specific topics impacting all relevant stakeholders and aims to solve problems by using data-driven cross-functional teams. Specifically, being customer-focused is considered a key point of Six Sigma. Effectively managed Six Sigma projects are expected to identify organizational issues and define problems (Pyzdek and Keller 2018). "Six Sigma is a rigorous, focused, and highly effective implementation of proven quality principles and techniques. Six Sigma aims for virtually error-free business performance" (Pyzdek and Keller 2018: 3).

According to Hsieh, Lin, and Manduca (2007), Six Sigma is a quality management system adopted by at least 25% of the Fortune 200 firms. In the 1990s, most Fortune 500 companies adopted Six Sigma (Goh 2002). Six Sigma proposes a well-structured problem-solving methodology that focuses on how to increase quality and decrease the probability of nonconformities and nonconforming (defects and defectives) products per million (ppm) unit, rather than using percent of defectives as presented in traditional quality philosophy. Instead, it is based on the perfectionist quality expectations of all stakeholders. It is considered a way of thinking and doing things in business processes. Critical thinking is also a basis of Six Sigma. Quality improvement and control in the manufacturing and service industry is a necessity for responding timely to increasing customer needs and sustainability expectations and requires critical thinking about problems and solutions.

To decrease the variance in functions, business processes, and products, Six Sigma and other related methodologies, such as lean production, have been implemented in business processes and related manufacturing processes. Six Sigma was developed by Motorola as a company-wide strategy against Japanese rivals in the 1980s in the USA. It was developed as a remedy to reduce the reliability issues and increase the quality of the final products at Motorola (Barney 2002; Folaron and Morgan 2003). Former General Electric CEO Jack Welch identified Six Sigma as the most important initiative GE has ever undertaken. In GE's 1998 annual report, Jack Welch stated that (George 2002: ix):

» We have tended to use all our energy and Six Sigma science to "move the mean to … reduce order-to-delivery time to … 12 days… The problem is, as has been said, "the mean never happens," and the customer … is still seeing variances in when the deliveries actually occur – a heroic 4-day delivery time on one order, with an awful 20-day delay on another, and no real consistency… Variation is evil.

As a data-driven approach, Six Sigma projects focus on how to reduce variation, waste, defects, and defective products or services. Each Critical-to-Quality (CTQ) characteristic's specifications should be met by components, products, or services. Otherwise, a defect occurs. Each time a component, product, or service fails to meet a CTQ characteristic's specifications, the product or service is considered defective. Six Sigma is a process to reduce defect rates per million opportunities. The name of Six Sigma was generated from an expectation of achieving a performance level at 3.4 ppm, namely, 3.4 products per million. According to Pyzdek and Keller (2018: 9), 3.4 ppm demonstrates a very conservative upper bound on the nonconformance rate. It is also considered a statistic that measures how close the team, department, section, or entire organization are to the goal. Six Sigma essentially focuses on how to sustain continuous improvement in the organization.

Quantitatively, Six Sigma aims to keep the defect rate at 3.4 defects per million products, and qualitatively it aims to minimize quality and reliability issues throughout the organization. The defect rate is considered a continuous variable in relation to time. That is, the calculated defect rate reflects the performance for a limited and identified period. Defect rate can be calculated for a certain period such as hourly, daily, weekly, and so forth. Therefore, this rate should be systematically and periodically monitored and measured.

Any abnormalities discovered in the process should be analyzed to identify common and assignable causes and, eventually, root causes of abnormalities and variation. Defect rates were once calculated as *percentage of defects* in other management approaches. However, defect rates are presented as defects *per million opportunities* in Six Sigma.

Even though total quality management and continuous quality improvement in the manufacturing industry have been seen as a necessity for responding timely to rapidly changing customer needs and sustainability expectations, some traditional quality improvement approaches have involved uncoordinated team efforts, wasting invaluable resources. These efforts may not be enough to generate expected solutions to solve the problems. However, the Six Sigma-based problem-solving approach has a potential to identify root causes as well as discover appropriate solutions and remedies for performance-related issues and problems. Six Sigma projects are also recognized as a way of *hearing* the voice of process. Six Sigma is a structured, data-driven methodology for eliminating waste from processes, products, and other business activities while having a positive impact on financial performance (Summers 2007). Similarly, Six Sigma also functions as a systematic way to *hear* the voice of the customer to gain a competitive advantage, as detailed in the next sections.

One of the distinctive factors of Six Sigma is that project completion time varies between 3 to 6 months (Levine 2006). Project completion time is limited to increase effectiveness in problem-solving processes. As the allocated time period of project gets longer, the probability of solving the problems may decrease. The other distinctive factor is that the Six Sigma projects focus on how to increase the sigma level and decrease the probability of nonconformities and nonconforming products and services in the system. Six Sigma is a process improvement methodology, not a practice, symptom, function, or activity improvement methodology. This is a key distinction in framing projects, and it is one that newcomers frequently get wrong during project identification, scoping, and selection (Wedgwood 2015: 33). According to Duncan et al. (2011), "the goal of statistical thinking in quality improvement is to make the available statistical tools as simple and useful as possible in meeting the primary goal, which is not mathematical correctness, but improvement in both the processes and outcomes." The frequency of quality improvement (QI) initiatives has been steadily increasing over the last decades (Varkey et al. 2007). The results of some QI initiatives do not meet the goals because of implementation failure of performance measurement and insufficient control systems (Nembhard et al. 2009). Thus, Six Sigma methodology may significantly help business organizations improve the outcomes of QI activities.

Therefore, Six Sigma focuses on customer needs and expectations, preventing defects, as well as minimizing cycle time and costs (Pyzdek and Keller 2018: 5). The ability to respond to customer needs and expectations is a key component of Six Sigma projects (Pyzdek and Keller 2018: 13).

1.3 Quality Definitions

Juran and Gryna (1980: xv) state that "our lives are increasingly dependent on the quality of products and services. Quality failures can and do result in serious human inconvenience, economic waste, and sometimes loss of life." Theoretically and practically, the quality term has been identified and detailed over several decades in many ways and using different aspects. Some researchers and practitioners focused on the qualitative aspect of quality, while others analyzed the measurable aspect of quality. Essentially, *quality* has been analyzed based on four managerial perspectives: (1) the product-based approach, (2) the manufacturing-based approach, (3) the value-based approach, and (4) the user-based approach. These approaches view quality as a *technical*, *operational*, and *manageable* matter. Managerial perspectives present quality as a self-contained process that can be planned, managed, and controlled by relevant functions and departments (Kelemen 2003: 7). In this section, quality term will be analyzed based on these four approaches as follows.

1.3.1 The Product-Based Approach

Feigenbaum (1983: 77) states that "quality must be designed and built into a product, it cannot be exhorted or inspected into it." Quality is perceived as a *measurable* and *precise* variable in the product-based approach (Kelemen 2003). The measurable and countable dimensions of product/service are determined as to assign the quality level of the outcome in the product-based approach. Garvin's (1987) model explaining these dimensions is utilized in measuring the quality term, especially for manufacturing industry, in the literature and practice. These dimensions, *performance, features, reliability, conformance, durability, serviceability, aesthetics, and perceived quality*, cover most of the measurable and countable quality dimensions of products.

Performance refers to a product's primary operating characteristics and attributes. For instance, performance means speed and accuracy in the service industry, while it refers to the ability of processing information and data in computers. As a basic example, the performance dimension of a pencil is its ability to write on paper. Automobiles are a more complex product; therefore, the quietness, speed, gas mileage, and acceleration to 100 km in seconds are some of the operating characteristics and performance attributes. Essentially, measurable and countable characteristics and attributes are used to evaluate the performance dimension in manufacturing and service processes and outcomes.

Features are considered the secondary aspects of performance, as the characteristics that supplement the basic functions of performance. Features are expected to add value to the performance of the product/service. For example, heated seats in automobiles or food and beverage services in low-cost carriers in airline services may be considered features. Additionally, a customer expectation represented in a feature may transfer to *performance* dimension after the product or service is used in the market over time. It may not be very clear and easy to differentiate feature from performance for some products and market segments. Instead, while a customer expectation is explained by a *performance*-related dimension in a certain market segment, the same expectation may be analyzed in *features* in another segment of the market. For instance, food and beverage services are perceived *features* in low-cost carriers, while these services are significant components of *performance* dimension in full-service carriers in airline industry.

Reliability refers to the probability that a product fails to perform at least one of the performance dimensions as designed in a period of time. The customers expect products not to fail while the product is in use. The mean time to first failure and the mean time between failures are the two reliability-related metrics to analyze the reliability dimension in quality. Considering the degree of technological complexity and advancement of the products/services, it is more likely to see highly reliable services and products in relevant marketplaces in the 2000s. Automobile, communication, and healthcare industries are good examples where the *reliability* dimension of quality is considered crucial. Along with this trend, customers' reliability-related expectations are on the rise in many different marketplaces over the last decades. Motorola's former Vice President and Director of External Quality, Paul Noakes, stated that *reliability* was the major root of cause of losing customers of Motorola to Japanese rivals in 1950s and 1960s (▶ https://youtu.be/Ze3TWEXXqUU).

Conformance reflects the degree to which a product or service meets the standards that have been established internally or externally (Garvin 1987). The *design specifications*, as well as industry or product-specific standards, represent the requirements for the products/services. These specifications set the target values that need to be met in inputs, processes, and outcomes, while the *tolerances* determine the maximum and/or minimum limits allowed, based on the standards' requirements. *Specifications* are a set of conditions that provide a detailed description of a quality characteristic for a procedure, process, product, or service. Acceptable bounds on individual quality characteristics are known as *specification limits*. *Standards* are a set of conditions of general or broad application established by some authority to be satisfied by a process, product, or service.

In some industries where deviance from the target value is not accepted, tolerances are considered zero, which means the process only includes target values as the acceptable levels. Aerospace and defense industries are good examples of zero tolerance. Genichi Taguchi's loss function represents how high tolerances and deviance from the target value may be costly for all relevant stakeholders of the final product. According to Garvin (1987), the most common measures of *conformance* are defect rates and service calls. In service industry, timeliness and accuracy of the service are taken into consideration, as well as the number of mistakes and customer calls. Both literature and practice assert that quality is directly related with developing an effective competitive advantage for business organizations. Overconcern in quality turned into excessive bureaucratic implementation of standards in many industries over the last three decades (Kelemen 2003).

Under the *conformance* dimension, Juran and Gryna (1980) emphasize that quality parameters are grouped into two critical sections such as "quality of design" and "quality of conformance." *Quality of design* refers to the degree to which a product or service design process reflects the customer needs and expectations on design outcomes that will be used by the next steps of the entire process. In other words, quality of design focuses on how successfully design processes perceive and utilize customer needs and expectations to develop design outcomes, such as engineering drawings and standards. *Quality of conformance* is evaluated on the basis of differences between what was designed and what was produced in the system. The lower difference between design outcomes and products reflects on higher *conformance* quality. In other words, quality of conformance shows the ability of production processes to produce the outcomes of the design processes. The conformance quality shows how well the product conforms to design specifications. There are various factors affecting the quality of conformance such as types of manufacturing processes, process controls, and inspections, the competency and motivation of employees, and unexpected conditions (Montgomery 2009). Juran (1992: 11) emphasizes that describing quality as "conformance to specification" or "conformance to standards" may be dangerous at managerial levels since "conformance" refers to static standards and specifications. On the other hand, quality is a dynamic concept because of constantly changing customer needs and expectations.

Durability is defined as "the amount of use one gets from a product before it deteriorates or the amount of use one gets from a product before it breaks down and replacement is preferable to continued repair" (Garvin 1987: 106). Essentially, *reliability* and *durability* appear as two critical, hand-in-hand dimensions. A product that fails frequently gives significant alerts that the product has reliability issues. Additionally, product mix changes so quickly in some industries that the durability concept may not as critical for those firms as it was in the past. For example, new smartphones are pushed into the marketplace so rapidly that customers may prefer to replace their current phones with the ones that have the most current technology. However, that does not mean that the current phone is not capable of functioning or meeting the customer's needs and expectations as designed. Other subjective factors, such as not being behind the technology or desire for having the most advanced technology, may affect customers' decisions about that replacement. As another factor, the *durability* expectation may vary in different marketplaces for different products. For instance, durability or life expectancy for automobiles may be longer than cell phones in the 2000s. But, the durability of a T-shirt may not be as long as an automobile for some groups of customers.

Serviceability refers to effectiveness and expansiveness of after-sales services. According to Garvin (1987), the *serviceability* dimension also contains speed, courtesy, competence, and ease of repair. The customers make their purchasing decisions based on various factors, and one of those factors is the likelihood of receiving high-quality and timely services after sales. Timeliness, quality, and speed of after-sales services play a critical role in decisions of customers in many industries. For example, the power service provider companies are expected to respond to calls of power outages, especially in seasons where increment weather affects life. Furthermore, the way the companies handle customer complaints gives customers a general sense about the *serviceability* dimension, which is a differentiating factor in some industries. At the bottom line, customer satisfaction is strongly affected by their perceptions of the *serviceability* of the product/service.

Aesthetics, as a subjective dimension, is related to the appearance, smell, taste, or even sound of the product in Garvin's classification. Due to the subjectivity in this dimension, a certain set of expectations for the same product may not even possible to be identified. For example, some prefer the taste of Starbuck's products, but others don't see the desirability of the brand. Others might prefer the shape of a piece of furniture, while others might consider its comfort.

Perceived quality or *reputation* reflects what previous customers think about the product or service they experienced or heard about before. Word of mouth can have a pivotal role in improving or destroying the reputation of a brand or company. Customers perceive the quality of the product based on the previous interactions with products and/or manufacturers. For example, they may have bought one product from a manufacturer that influences whether they will buy a different product from the same manufacturer. They also rely heavily on previous customers' feedback, thoughts, and opinions. Therefore, *perceived quality* or *reputation* determines the potential and current customers' expectations and satisfaction level.

1.3.2 The Manufacturing-Based Approach

In this approach, quality refers to the degree to which a specific product/service conforms to design and product specifications (Gilmore 1974; Kelemen 2003). Many tools, techniques, methodologies, and philosophies in quality fall into the manufacturing-based approach. For example, statistical process control, design of experiment, and acceptance sampling are manufacturing-based approaches used primarily in manufacturing settings.

The main expectation in this approach is to achieve zero defects level by minimizing variation in manufacturing processes. Design drawings, design quality, and engineering standards, specifications, and requirements result from this approach. ISO, NIST (National Institute of Standards and Technology), and ASME (the American Society of Mechanical Engineers) are the three global and national institutions that focus on creating, developing, and improving standards for a great variety of products and services. Shewhart, Crosby, Deming, Juran, Feigenbaum, Taguchi, and Ohno are the quality gurus supporting the manufacturing-based approach in defining quality between 1920s and 1990s.

Under the manufacturing-based approach, Juran and Godfrey (1999) identify quality as freedom from deficiencies. According to their definition, quality also means freedom from errors that require doing processes over again, which is also known as *rework*, or that result in field failures, customer dissatisfaction, and customer claims. The cost-oriented quality definition results in higher quality and lower cost. Juran and Gryna (1980: xv) also assert that the control of quality is not primarily a matter of statistical techniques. Instead, the quality concept requires a viewpoint that includes managerial, technological, and statistical approaches for all major functions in the organization.

Crosby (1979) identifies quality as "conformance to requirements," while Juran and Gryna (1980) describe quality as "fitness for use" in a general perspective. The American Society for Quality (ASQ) defines quality as the totality of features and characteristics that satisfy needs. Feigenbaum (1983: 9) states that *quality* does not mean "the best" in any abstract sense. Instead, quality means that the product or service is best for satisfying certain customer conditions and expectations. Feigenbaum also points out ten product or service conditions identifying quality:

1. The operating characteristics and specifications of the dimensions
2. The life and reliability objectives
3. The safety requirements
4. The standards relevant to the product and service
5. Cost of the product or service
6. Manufacturing conditions
7. Maintenance and service objectives
8. Energy and material-related factors
9. Environmental and side effects
10. The costs of customer operations and use.

The quality perception of the manufacturer is related to the field performance of the product, and "fitness for use" depends on field performance. As an anecdote, Juran and Gryna (1980: 478) mention the importance of specifically focusing on field performance as follows:

> ... In a company, making rubber products, there was a drop in the sales of a major product type. The sales force blamed the drop on poor quality, whereupon the president ordered an improvement in quality. The factory complied by tightening up on quality standards, resulting in higher shop rejections (and higher costs) for dirt, discoloration, and other factory defects. Actually, the filed problems were different... The factory "improvement" program was raising costs without solving the field problems.

Finally, business organizations still have a widespread belief that quality problems rise solely from manufacturing functions, excluding other relevant functions. Much research indicates that quality problems can arise from any area of the firm, including human resource management, finance, marketing, and firm management (c.f., Pakdil and Leonard 2014).

Montgomery (2013: 6) states that "quality is inversely proportional to variability." If the variability decreases, the quality of product/service increases. The inverse relationship between quality and variability should be kept in mind in all design and production processes. Speaking at micro level, lower variation on a particular CTQ characteristic for a product refers to lower deviation from the target values and tolerances and lower rejection and rework rates. At macro level, lower variation refers to lower customer complaints, returns, and warranty costs. As seen in ◘ Image 1.1, system 1 (Standard deviation = 2) produces lower variation as opposed to system 2 (Standard deviation = 4.315) that results in higher quality.

1.3.3 The Value-Based Approach

Value is used to relate the price paid for a product/service to the benefits provided through that product/service. To optimize the value, the price is expected to be minimized and the benefits maximized. The value-based approach takes into consideration worth, price, value, benefit, cost, and quality of the product/service. In a rational consumer's mind, value is generally calculated as a ratio of worth to cost. Maximizing the ratio depends on increasing worth and decreasing cost simultaneously. All relevant tools and techniques developed in the quality area focus on how to decrease cost, depending on constantly increasing variation.

The complex relationship between cost and quality is taken into account as well in this approach. Kelemen (2003: 11) states that the value-based approach "considers quality as being the degree of excellence at an acceptable price or the control of variability at an acceptable cost." Kelemen (2003) compares Japanese and Western manufacturing paradigms. While Japanese management understanding perceives quality and cost as two negatively correlated terms, Western management presents quality and cost as positively correlated to each other. Furthermore, regardless of value, excessive price is considered a quality indication in some irrational

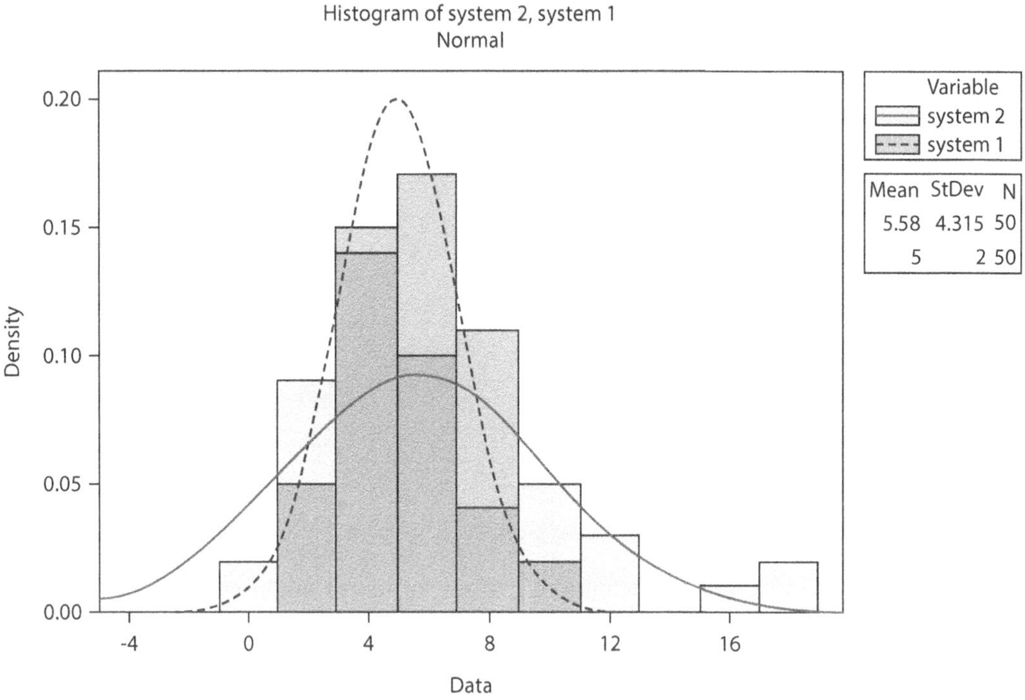

Image 1.1 The variations of systems 1 and 2. (Source: Author's creation in Minitab)

market segments, particularly luxury items. Identifying customer value is the core principle of Six Sigma (Pyzdek and Keller 2018). Hearing the voice of the customer is the first step in defining value in Six Sigma-oriented organizations.

Value refers to outcomes that are gained for every dollar spent. Value depends on outcomes, not on inputs. As the main outcomes of the process, quality, cost, and access to product/service are monitored and optimized to measure and increase value. Value engineering and value analysis methods (Fallon 1980) analyze value by focusing on the economic burden of the item to the organization and the benefits received out of the product/service. Fallon (1980: 1) identifies value analysis as "a method that provides the means for industrial disciplines to cross departmental lines in a joint study of a product's function in relation to its cost." Juran (1992: 192) describes value analysis as "a process for evaluating the interrelationships among (a) the functions performed by product features, and (b) the associated costs." The main goals of value analysis are to identify the essential product functions and product features that can be eliminated and optimize the cost of essential functions. Value is also known as the ratio of worth to cost. Researchers identify the value concept in a broader perspective, presented in the equations in Table 1.1. For instance, Bryant (1986) focuses on customers' wants and needs in the numerator part of the ratio, while resources are used in the denominator. Kaufman and Becker (1981) and Wasserman (1977) see the value as a ratio of functionality of the product/service to the cost.

In definitions where quality is identified as "conformance to specifications," manufacturing-based approach appears more effective. It refers to how well a product/service meet targets and tolerances defined by its designers. Where quality is articulated as "fitness for use," it focuses on evaluating the performance of the product/service for intended use. Value-based definitions concentrate on evaluation of usefulness versus price paid.

Six Sigma helps companies increase profit by improving customer value, efficiency, and effectiveness. Pyzdek and Keller (2018) link Six Sigma with quality term using "poten-

Table 1.1 Value definitions and ratios

Definitions	Definitions	Definitions
Fallon (1980) $V = \frac{\text{Objectives}}{\text{Cost}}$	Bryant (1986) $V = \frac{\text{Wants \& needs}}{\text{Resources}}$	Kaufman and Becker (1981) $V = \frac{\text{Functionality}}{\text{Cost}}$
Wasserman (1977) $V = \frac{\text{Functionality}}{\text{Cost}}$	Bryant (1986) $V = \frac{\text{Worth}}{\text{Effort}}$	Fallon (1980) $V = \frac{\text{Utility}}{\text{Cost}}$
Fallon (1980) $V = \frac{\text{Performance}}{\text{Cost}}$		

Source: Author's creation

tial quality" and "actual quality" definitions. Potential quality refers to maximum possible value added per unit, whereas actual quality means current value added per unit. The difference between potential and actual quality is *waste*.

Defining value should focus on "for whom?". The value definition may vary from stakeholder to stakeholder or from function to function. Although the ultimate goal is to create value for end customers, functions and processes may have different value definitions and expectations. For example, the way the engineering function identifies value may focus on internal (production function) and external customer's needs and expectations, while the CEO also takes into consideration the value perception of the investors and shareholders. The value term may vary based on the stakeholders' needs and expectations. Additionally, "who are the customers?" and "what are the needs and expectations of customers?" help identify value for each stakeholder in the value chain process.

1.3.4 The Customer-Based Approach

Quality has different meanings for different audiences, depending on the expectations and focus of stakeholders. For instance, quality means meeting customer needs and expectations for marketing associates. For engineers, it means that the outcomes of production functions, such as production and assembly lines, flawlessly meet technical specifications and design outcomes. From the customers' point of view, quality means purchasing and receiving better products/services with lower costs. In our daily lives, quality also has many different meanings for different users.

Customer needs and expectations should be fully met to simultaneously increase customer satisfaction and organizational performance. In the literature and practice, quality starts with identifying customers' needs and expectations. Some quality definitions such as "conformance to use" and "fitness for use" focus on the customer-based approach. In examining fitness for use, Juran and Gryna (1980: 443) focus on multiple uses and users. A customer is anyone who is impacted by the product or process, externally and internally.

Critical perspectives perceive quality to conceptualize dimensions of the product/service that matter for internal and external customers. External customers are the ones who purchase a product or service and are impacted by the product or service. Internal customers are the ones whose outcome is accepted and used by the next process within the value chain in the organization. In other words, internal customers are the ones who utilize the outcomes of the previous step or process. The next process or work station in the company is considered the previous step's customer. The customer may be identified as another manufacturer, end customer, supplier, or even internal customer. Juran and Gryna (1980) define user as someone who receives the ultimate benefit of the product, such as an employee or a consumer. Juran and Godfrey (1999) identify quality as "features of products which meet customer needs and thereby provide customer satisfaction. In this sense, the meaning of quality is oriented to income."

A thorough and accurate understanding of customer demands and expectations, called "the voice of the customer," is presented as the key to possessing competitive advantage (Pakdil et al. 2012: 1397). This approach essentially focuses on meeting and exceeding the customers' voiced demands and expectations. Meeting and exceeding the customers' needs and expectations and being a customer-focused organization is considered one of the requirements of the TQM philosophy. The entire organization, from design to after-sales services, strives to meet customers' demand, with management understanding that has "customer needs and expectations-the voice of customer" as the main motto. In this chain, each function and department fulfills different responsibilities, meets different expectations, and focuses on different CTQ characteristics, internally and externally. For example, design functions, processes, and outcomes should first be optimized on the basis of the customer demands to improve design outcomes in manufacturing and service industries.

The ASQ defines quality as the totality of features and characteristics of a product or service that bear on its ability to satisfy given needs. Juran and Gryna (1980: 445) analyze quality from users' and manufacturers' perspectives. What is important for differentiating quality perspective among users and manufacturers is that while the users mostly focus on the degree to which the product/service fits for use, manufacturers put more emphasis on conformance to specifications.

1.4 Quality Gurus and Thinkers

The quality world has had many significant contributors who created, developed, and improved quality terminology and discipline since Industrial Revolution started. These contributors are known as quality gurus, who were engineers, statisticians, physicist, and businessman. Each quality guru contributed to the literature of quality in a great variety of ways. Kelemen (2003) and Bendell (1991) categorize quality gurus into three groups such as (1) early American writers, (2) Japanese writers, and (3) new Western writers. The first category covers Walter Shewhart, W. Edwards Deming, Joseph M. Juran, and Armand V. Feigenbaum. Kaoru Ishikawa, Taiichi Ohno, Shigeo Shingo, and Genichi Taguchi are Japanese quality writers. The last group embraces Philip Crosby. We will also add David Garvin and Douglas Montgomery in the last group. The following sections present their ideas and ideologies about the quality concept and detail their contribution to quality literature and practice.

1.4.1 Walter Shewhart

Shewhart is known as the founder of statistical quality control understanding and specifically control charts. Shewhart, as the first honorary member of ASQ, integrated statistics, economics, and engineering perspectives into quality control activities in the early 1900s. The *Economic Control of Quality of Manufactured Product*, published in 1931, is the first book on basic principles of quality control in the twentieth century. This book allowed American and British manufacturers to integrate statistical approach into quality control (Finison et al. 1993). The *Statistical Method from the Viewpoint of Quality Control* published in 1939, also by Shewhart, is another book that led to developing quality control as an international discipline.

His mathematics and physics knowledge supported him in his quality control and applied statistics journey. Shewhart developed and improved basics of quality control methods at Western Electric between 1918 and 1924 and later at Bell Telephone Laboratories between 1925 and 1956. At Bell, Shewhart and his colleagues worked on the reliability of transmission systems, focusing on the importance of reducing variation and on the relationship between increased variation and decreased quality level. Shewhart developed the control chart framework and categorized out-of-control points as assignable and common (chance) causes. He advised using control charts as a device to detect assignable and/or common causes in the process and supported the idea of minimizing the variance and assigning types of causes by using them. This allowed him to develop "tolerance intervals."

Shewhart globally served as a fellow in many different not-for-profit and government-funded institutions such as the International Statistical Institute, the United Nations, and The New York Academy of Sciences. The concept of Six Sigma developed and matured by Motorola has the roots in Shewhart's philosophy. Deming's Plan-Do-Control-Act (PDCA) cycle is built upon Shewhart's integration of management thinking with statistical analysis. His basic teachings can be categorized into:

- Control charts
- PDCA cycle (Shewhart cycle)
- Assignable and common causes.

Walter Shewhart stated the importance of statistical thinking as follows: "the long-range contribution of statistics depends not so much upon getting a lot of highly trained statisticians into industry as it does in creating a statistically minded generation of physicists, chemists, engineers, and others who will in any way have a hand in developing and directing the production processes tomorrow" (Shewhart 1939: 49).

1.4.2 W. Edwards Deming

Dr. Deming, along with his contemporaries, developed such a comprehensive understanding of quality management that quality science is established mostly upon his teachings and contributions. His collaboration with JUSE (the Union of Japanese Scientists and Engineers) produced many contributions to the Japanese economy after 1950. He worked for Western Electric as a statistician and became an honorary member of ASQ in 1970. He is known as the founder of the third wave of the Industrial Revolution.

Dr. Deming insistently focused on product and service improvement and decreasing variation in processes in order to increase quality. After Shewhart classified the causes of the variation into two groups, common causes and assignable causes, Deming also focused on the causes of variation. Deming (1986) states that 95% of defects are caused by system-related reasons while only 5% are due to human-related reasons. In other words, if the organizations design their systems in a way that do not let the employees make mistakes or fail, the systems will produce lower variation and defects. As the systems improve, the defectives and defect rates will decrease. Dr. Deming's main contributions to quality literature are:

- Deming's Chain Reaction
- PDCA cycle
- The Deming System of Profound Knowledge®
- Deming's 14 points
- Responsibility for quality assignments
- The Red Bead Experiment
- The Funnel Experiments
- Seven deadly sins
- The classification of common and assignable causes.

The Deming's Chain Reaction focuses on the relationships between and among quality, productivity, cost, and sustainability in business organizations (◘ Fig. 1.1). Deming asserts that the more the variation decreases, the more jobs the company can create in the long run. Quality improvement in any process results in higher productivity because of the increasing number of acceptable product/service. Any improvement in the productivity causes cost reduction because of less rework, fewer mistakes and delays, as well as better use of machine time and materials. Furthermore, overall improvement in quality, productivity, and cost leads to a significant increase in market share because of sales price advantage

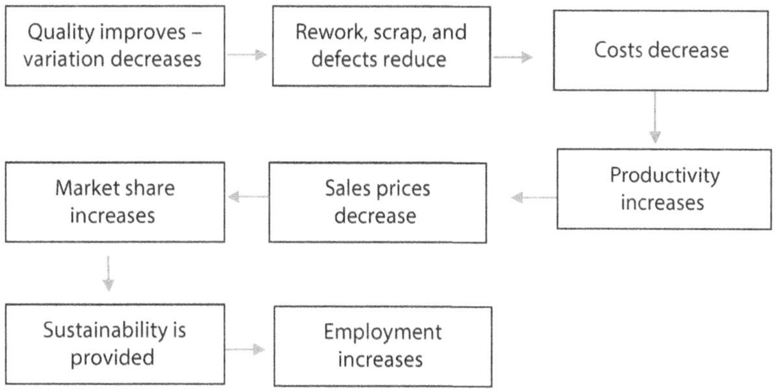

Fig. 1.1 Deming's Chain Reaction. (Source: Adapted from Deming 1986)

and flexibility. Eventually, increasing market share improves the organization's position in the entire market, nationally or globally, and enhances the sustainability of the organization. Afterward, organizations can implement organizational sustainability concept in corporate strategic priorities and planning. As seen from Fig. 1.1, improved quality, as a micro level effort, is an effective strategy of increasing the organization's sustainability capacity.

The seven deadly sins are articulated in Deming's *Out of Crisis* (1986: 97–98) book. All these points address where companies make mistakes in quality journey. These points also demonstrate how companies need to take actions, set goals, and formalize strategies to improve quality and finally organizational performance:

1. Lack of constancy of purpose to plan products and services that will have a market and keep the company in business and provide jobs
2. Emphasis on short-term profits; short-term thinking (just the opposite from constancy of purpose to stay in business), fed by fear of unfriendly takeover and by the push from bankers and owners for dividends
3. Evaluation of performance, merit rating, or annual review
4. Mobility of management; job hopping
5. Management using only of visible data, with little or no consideration of data that are unknown or unknowable
6. Excessive medical costs
7. Excessive costs of liability swelled by lawyers that work on contingency fees.

The Deming System of Profound Knowledge (SoPK) is known as a comprehensive set of his lifelong work. His all teachings and theories are accumulated under SoPK to provide all quality practitioners with a theory of management that covers a framework of quality management. The Deming Institute asserts that "…by management appropriately applying the principles and practices of SoPK), a business can simultaneously reduce costs through reducing waste, rework, staff attrition and litigation, while increasing quality, customer loyalty, worker satisfaction and, ultimately, profitability" (▶ https://deming.org/management-system/sopk). "Dr. Deming's System of Profound Knowledge provides a highly integrated framework of thought and action for any leader wishing to transform an organization operating under the prevailing system of management into a thriving, systemically focused organization." (▶ https://deming.org/explore/so-pk). Deming (1986) states that business organizations go through the transformation process by implementing System of Profound Knowledge in the management processes.

Deming addressed the importance of top management support for quality transformation. He believed that the top executives' and managers' responsibility is to improve overall quality in the organization. Deming believed that, in the long run, the most effective competitive strategy is constructed based on *quality*. He also points out the necessity of using effective statistical techniques in quality improvement activities. Deming's 14 points refer to philosophical aspects of quality improvement and address vital points of quality transformation for organizations. They represent a management philosophy that focuses on continuously improving quality and overall performance. These points have been considered key management principles by quality management-oriented manufacturing and service organizations since 1950s:

1. Create constancy of purpose toward improvement of product and service, with the aim to become competitive and to stay in business, and to provide jobs.
2. Adopt the new philosophy. We are in a new economic age. Western management must awaken to the challenge, must learn their responsibilities, and must take on leadership for change.
3. Cease dependence on inspection to achieve quality. Eliminate the need for inspection on a mass basis by building quality into the product in the first place.
4. End the practice of awarding business on the basis of price tag. Instead, minimize total cost. Move toward a single supplier for any one item, on a long-term relationship of loyalty and trust.
5. Improve constantly and forever the system of production and service, to improve quality and productivity, and thus constantly decrease costs.
6. Institute training on the job.
7. Institute leadership. The aim of supervision should be to help people and machines and gadgets to do a better job. Supervision of management is in need of overhaul, as well as supervision of production workers.
8. Drive out fear, so that everyone may work effectively for the company.
9. Break down barriers between departments. People in research, design, sales, and production must work as a team, to foresee problems of production and in use that may be encountered with the product or service.
10. Eliminate slogans, exhortations, and targets for the workforce asking for zero defects and new levels of productivity. Such exhortations only create adversarial relationships, as the bulk of the causes of low quality and low productivity belong to the system and thus lie beyond the power of the workforce.

    Eliminate work standards (quotas) on the factory floor. Substitute leadership.

    Eliminate management by objective. Eliminate management by numbers, numerical goals. Substitute leadership.
11. Remove barriers that rob the hourly worker of his right to pride of workmanship. The responsibility of supervisors must be changed from sheer numbers to quality.
12. Remove barriers that rob people in management and in engineering of their right to pride of workmanship. This means, inter alia, abolishment of the annual or merit rating and of management by objective.
13. Institute a vigorous program of education and self-improvement.
14. Put everybody in the company to work to accomplish the transformation. The transformation is everybody's job.

The PDCA (Plan-Do-Control-Act) cycle is a vital part of the continuous improvement philosophy and is also known as the Deming Wheel, Deming cycle, and Shewhart cycle. PDCA is located at the heart of continuous quality improvement activities. Deming emphasizes that PDCA should be a way of implementing continuous improvement in all business processes, including operations, as well as all aspects of one's daily life. PDCA is a four repetitive-step process referring to components that need to be followed in continuous quality improvement (◘ Fig. 1.2). The first step, PLAN, focuses on how to analyze the current situation and design a plan for increasing the performance of the systems or processes. It also details ideas and tasks for improvement. The second step, DO, is the step in which plans are executed and activities are performed sequentially on the basis of the plans done in the previous step. The third step, CHECK, is the step to make sure that every activity determined in PLAN has been executed as planned. Any deviations or changes are documented in this step. The main idea in control step is to identify where activities do not meet plans. The fourth step, ACT, is for the activities that will be done if

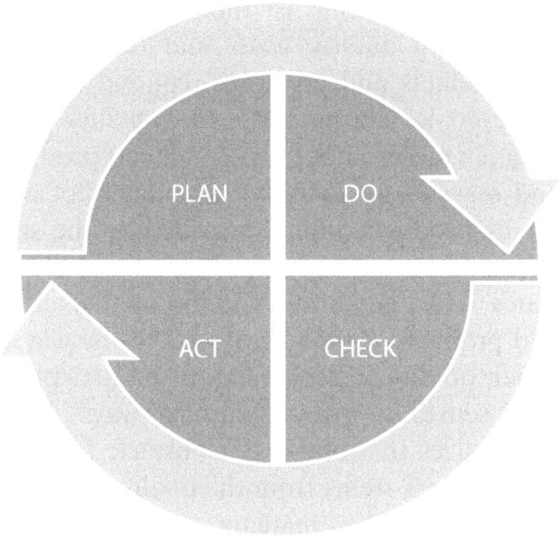

Fig. 1.2 Deming cycle. (Source: Adapted from Deming 1986)

there are any unintended gaps between PLAN and DO steps. Corrective and preventive actions are usually implemented in this step in the practice.

Deming emphasizes the variation concept in his Red Bead Experiment. Using this famous experiment, he explains how impactful a variation may be on production and business processes and that common causes of the variation are unavoidable. There are always various significant factors that affect the product/service and the processes that cannot be eliminated from the system. Because of the variation that is inherent to the common causes, products, services, and process outcomes may vary. The experiment also highlights the critical role of the management function in business organizations. He emphasizes that variation in selecting red and other color beads is a result of management deficiency. The video of Red Bead Experiment can be watched in the following link: (▶ https://youtu.be/ckBfbvOXDvU). Further Deming's teachings and quality philosophy are in Deming's book, *Out of the Crisis*.

1.4.3 Joseph M. Juran

Juran is known as the father of quality management in quality literature. Juran held degrees in engineering and law and pursued a great variety of careers in management as engineer, industrial executive, government administrator, professor, labor arbitrator, corporate director, and management consultant. He received 30 medals, fellowships, and honorary memberships from 12 countries (Juran and Gryna 1980: v).

Juran worked for Western Electric as an industrial engineer and was invited to Japan by JUSE in the 1950s to share his knowledge with Japanese business organizations, as was his colleague Dr. Deming. He also developed the *Western Electric Statistical Quality Control Handbook*. According to Juran, quality basically means "fitness for use." While Deming was focusing on top management's impact on quality transformation, Juran emphasized on middle management's responsibilities for quality. He developed the philosophy that quality should be a responsibility for everyone, every position, and every department, since the quality improvement is everyone's responsibility in the organization. Juran and Gryna (1980: 323) stated that:

> When the situation becomes serious enough, it gains attention because the question is raised: "Who is responsible for quality?" Discussion of this question is generally futile and a waste of time because the question covers too broad a scope. It is like asking "Who is responsible for cost" or "Who is responsible for safety?" Such broad questions result in a haphazard discussion of many issues or conclude with a panacea that "everyone is responsible for quality.

Juran states that there are three different languages in the organizations: (1) upper management speaks dollars, (2) middle management speaks issues and dollars, and (3) lower management or employees speak issues. His contribution to the knowledge of quality management focused on the managerial aspects and requirements of quality management. Juran emphasized the human side of quality management. He became well-known for quality management after his visit to Japan.

Juran integrated Pareto principle into seven old problem-solving tools in quality literature to differentiate the "vital fews" from the "trivial many." During World War II, he worked for the Lend-Lease Administration in Washington, D.C., where he focused on how to improve purchasing and budgeting activities. Afterward, Juran started teaching industrial engineering at New York University. He stated that all quality-oriented problems relate to resistance to change (Edmund 2010). Juran states that about a third of what we do consists of redoing work previously done (Juran 1992) in the USA. He is also the author of the *Quality Control Handbook*, first published in 1951 (▶ http://asq.org/juran/institute-statement.html).

Juran's basic teachings include:

- Juran's trilogy
- PAF (prevention-appraisal-failure) costs of quality system
- Juran's ten steps to quality improvement.

Juran proposed his trilogy including (1) *quality planning*, (2) *quality control*, and (3) *quality improvement* steps in his teachings. Using a holistic structure, Juran focused on how to integrate planning, control, and improvement activities in quality management in one unique system. Juran's trilogy shows that while conducting quality control, the cost of poor quality may increase for a while. As the quality control activities continue and quality improvement activities increase, the cost of poor quality decreases.

Juran and Godfrey (1999) assert that quality is attained, first, by determining the vision of the organization and setting up policies and goals. The three components of Juran's trilogy are associated with each other in a way to maintain and improve quality level. Juran's trilogy starts with quality planning that includes the following requirements:

- Establishing quality goals
- Identifying customers
- Determining customers' needs and expectations
- Developing product features that respond to customers' needs
- Developing processes that are able to produce the product features
- Establishing process controls.

"Quality planning is the activity of (a) establishing quality goals and (b) developing products and processes required to meet those goals" (Juran 1992: 13). Quality planning is required not only for all end products and services for external customers but also for by-products, inputs provided by the suppliers, and internal processes. Juran (1992: 3) states that "numerous specific quality crises and problems have been traced to the way in which quality was planned in the first place. In a sense, we planned it that way." Juran emphasizes that one major obstacle faced by the planners stems from the quality planning activities done by amateurs.

In the next step of Juran's trilogy, product/service features and the details of design are identified to meet and exceed customers' needs and expectations. Next, production/service delivery processes are designed to produce product/service as designed in production processes. Then, process control details are identified, and the second stage, *quality control*, takes place in Juran's trilogy. This step includes:

- Evaluating actual performance
- Comparing actual performance with quality goals
- Acting on the difference.

Chronic waste is identified in the quality control step because of the deficiencies in quality planning activities. All chronic waste is driven down to quality improvement activities, where waste is seen as an opportunity for quality improvement. Overall, Juran's trilogy functions to identify chronic wastes and potential quality improvement opportunities in the processes. The cost of poor quality is minimized in the quality improvement step, whereas it is maximized in quality control activities. Quality improvement step of Juran's trilogy includes:

- Proving the improvement need
- Establishing improvement infrastructure
- Identifying improvement project
- Establishing project teams
- Providing the team with the resources needed
- Establishing control methods.

Juran focused on categorizing costs of quality in the PAF (prevention-appraisal-failure) system. *Cost of quality* is a term referring to the combination of (1) *cost of good quality* and (2) *cost of poor quality*. In other words, it is comprised of conformance and nonconformance costs, where cost of conformance contains the cost of prevention of poor quality and cost of nonconformance includes the cost of poor quality caused by failures (Schiffauerova and Thomson 2006). Cost of poor quality consists of the costs that would disappear if the product or service is flawless (Juran 1992). Cost of quality covers all aspects of production or service delivery processes, good and poor quality, starting from the very first step of the design process to the last step of after-sales services. Juran categorizes the cost of quality into three sections: prevention costs, appraisal costs, and failure costs. In the *prevention costs* category of PAF, the costs of the activities that are oriented toward preventing errors, mistakes, external and internal failures, nonconformities, and defectives are measured and managed. The cost of training programs, quality audits, preventive maintenance, and design functions are some examples of prevention costs. *Appraisal costs* refer to activities that are focused on measuring the items covered in the transformation process: inputs, processes, and outputs. The costs of incoming inspections, process controls, and outgoing inspections are appraisal costs. *Failure costs* refer to the cost of the actions taken for fixing internal and external failures. For example, warranty and recall costs are external failures, while rework, scrap, and delays are internal failure costs. The details of quality costs are presented in the ▶ Sect. 1.7 in this chapter.

Juran states that, if organizations implement ten points of quality improvement into their management systems, they will be able to succeed in quality improvements in the long run. Juran's ten points of quality improvement are:

1. Build awareness of opportunity to improve.
2. Set goals for improvement.
3. Organize to reach goals.
4. Provide training.
5. Carry out projects to solve problems.
6. Report progress.
7. Give recognition.
8. Communicate results.
9. Keep score.
10. Maintain momentum by making annual improvement part of the regular systems and processes of the company.

1.4.4 Armand V. Feigenbaum

Armand V. Feigenbaum is known as the father of total quality control philosophy. He was the company-wide manager of Manufacturing Operations and Quality Control at the General Electric Company in New York City between 1958 and 1968. Feigenbaum was the founder and president of General Systems Company, Inc., in Pittsfield, MA, from 1968 to 2013, the founding chairman of the International Academy for Quality, and the former president of ASQ. Feigenbaum (1983: xxi) stated that "quality is the basic customer decision factor for an explosively growing number of products and services today-whether the buyer is a housewife, an industrial corporation, a government agency, a department store chain, or a military defense program." Feigenbaum articulated quality as the single most important force leading to organizational success and growth in national and international markets. He also states that quality is in its essence a way of managing the organization (Feigenbaum 1983).

Feigenbaum (1983: xxii) claimed that "total quality control is the foundation for the successful management of quality to assure customer satisfaction." According to Feigenbaum (1983), company-wide total quality control comprises the managerial and technical implementation of customer-oriented quality activities that covers marketing, engineering, production, industrial relations, finance, and service as well as of the quality control function itself.

Feigenbaum (1983: 5) identified "the goal of the competitive industry as providing a product or service into which quality is designed, built, marketed, and maintained at the most economical costs which allow for full customer satisfaction." Feigenbaum described total quality control as an effective system for integrating the quality development, quality maintenance, and quality improvement efforts of the various groups in an organization to achieve full customer satisfaction at the most economical level. "The hallmark of modern total quality systems is their effectiveness in providing a strong foundation for the economic control of this complexity, to the benefit of both improved customer quality satisfaction and lower quality costs" (Feigenbaum 1983: 77).

Feigenbaum addressed productivity-related issues in his *hidden plant* concept, where he suggests that there is a part of the organization where 15 percent to 40 percent of productive capacity relates to reworking, reinspecting, and retesting rejected parts and dealing with recalls. Feigenbaum indicates that "there is no better way to improve productivity than to convert this hidden plant to productive use (1983: 47)." Feigenbaum indicates that the three major contributions of quality are (1) promotion of quality ethics, (2) development of total quality control, and (3) quality cost system. Similar to Juran's approach, Feigenbaum categorized quality costs into three sections such as prevention, approval, and internal/external failure.

Feigenbaum stated that Japanese employees would make their decisions concerning the operations if they detect any abnormalities in the manufacturing processes. However, American employees are not allowed to follow the same path in American manufacturing industry.

1.4.5 Kaoru Ishikawa

Ishikawa is known as the father of the fishbone diagram, quality control circles, and the quality movement in Japan. He developed the Japanese quality strategy and focused on how to use principal quality tools, such as fishbone diagrams, in problem-solving and continuous quality improvement processes. He is one of the Japanese quality gurus who significantly contributed to Japanese economic development and quality literature. He shows standardization and quality control as the two principal tools, using the fishbone diagram, which is also known as Ishikawa diagram, as a true representation of organizational operations. He collaborated with Dr. Deming and Juran to contribute to Japanese economic development. Ishikawa developed and proposed the quality control circles methodology in a nation-wide continuous quality improvement movement in Japan. He focused on the importance of integrating employees and their suggestions in problem-solving and continuous quality improvement efforts. Increasing job satisfaction, recognizing and utilizing employees' capabilities, and valuing human relations in the workplace were the driving forces of quality control circles. He also emphasized that quality control and quality improvement are inseparable components of continuous quality improvement.

He wrote several books on quality such as *Guide to Quality Control*, *Introduction to Quality Control*, and *What is Total Quality Control?* His books helped many quality practitioners understand the details of quality improvement tools. The *Guide to Quality Control* was used as a training manual and guide for quality control circles. Ishikawa identified ten aspects of quality:

1. Study quality improvement ahead of anyone else.
2. Establish policies toward promoting quality improvement.
3. Specify priorities for implementing quality improvement and short- and long-term goals.
4. Assume a leadership role in making quality improvement happen.
5. Provide a means for educating people.
6. Check to see if quality improvement is implemented as planned.
7. Make clear the responsibility of top management.
8. Establish a system of cross-functional management.
9. Drive home the notion that outputs from processes are inputs to customers.
10. Provide leadership toward making "breakthrough" happen.

1.4.6 Taiichi Ohno

Ohno is known as the father of lean production, namely, Just-in-Time, philosophy, and practice, and had a tremendous contribution in the lean literature and in practice at Toyota Motor Co. For example, the pull production system, process improvement, kaizen, standardization, teamwork, value stream map, economically producing large variety in small volumes, and reducing changeover time are some of Ohno's contributions to lean production literature.

Ohno joined Toyota Spinning and Weaving Co. in 1932 as a mechanical engineer and transferred to Toyota Motor Co. in 1943, where he developed the Just-in-Time (JIT) philosophy. After analyzing Western manufacturing systems, Ohno stated that producing with large batch sizes and failing to satisfy customer preferences in the final product were the biggest mistakes made by Western manufacturers. He suggested abandoning the idea of large batch production, which results in large inventories. Ohno gradually implemented small-lot production, *jidoka*, and autonomous control systems to eliminate waste and minimize cost at Toyota in the 1940s.

As an engineer working for Toyota Motor Co.., Ohno visited the USA in the 1950s and developed the pull production concept while visiting supermarkets. He also sought to discover the root causes of a problem through "5 whys." Finding root causes of problems is most effective countermeasure in Six Sigma and lean practices. Ohno (1988) stated that "having no problems is the biggest problem of all." Ohno also stated that waste, *muda*, is anything that adds cost, but not value, to a product (Ohno 1988). Ohno identified seven types of waste in lean philosophy: (1) transportation, (2) inventory, (3) motion, (4) waiting, (5) overprocessing, (6) overproduction, and (7) defects. However, Ohno (2013) states that:

> I don't know who came up with it, but people often talk about "'the seven types of waste." This might have started when the book came out, but waste is not limited to seven types. There's an old expression: "He without bad habits has seven," meaning even if you think there's no waster you will find at least seven types. So, I came up with overproduction, waiting, etc., but that doesn't mean there are only seven types. So, don't bother thinking about "what type of waste is this?" Just get on with it and do kaizen.

Ohno's philosophy is combined in a holistic way and presented in the lean house figure (◘ Fig. 1.3) based on two pillars of lean production: autonomation and JIT. As stated by Liker (2004), Ohno learned from walking through the shop floor, specifically about non-value-added waste. His visits to the shop floor enabled him to identify value-added and non-value-added activities. He also mapped the entire value stream in shop floors and systems. Mapping the value stream was a different approach from the traditional mass production systems.

Ohno was charged by Eiji Toyoda, the president and chairman of Toyota Motor Manufacturing, to analyze Ford manufacturing system's productivity. After visiting Ford and reading Ford's book, *Today and Tomorrow*, Ohno came up with several ideas for the Toyota Production System (TPS), such as continuous material and process flow and moving assembly lines. Ohno also pointed out that Ford's mass production system had several flaws, including batch production methods and large work-in-process inventories. He implemented the pull system, one-piece flow system, and principles of *jidoka* in the TPS. Supermarket system that was utilized in daily life in the USA in the 1950s influenced Ohno and his team and inspired him to implement pull system that includes "supermarket" inventory storage areas into TPS. To minimize the size of work-in-process inventory, he implemented the *kanban* system that visually shows the replenishment needs to users and operators.

As another contribution of Ohno's, *Genchi Genbutsu* was implemented in TPS. Ohno says "observe the production floor without preconceptions and with a blank mind. Repeat 'why' five times to every matter" (Liker 2004: 223). *Genchi Genbutsu* means going to the place and seeing the actual situation to understand the

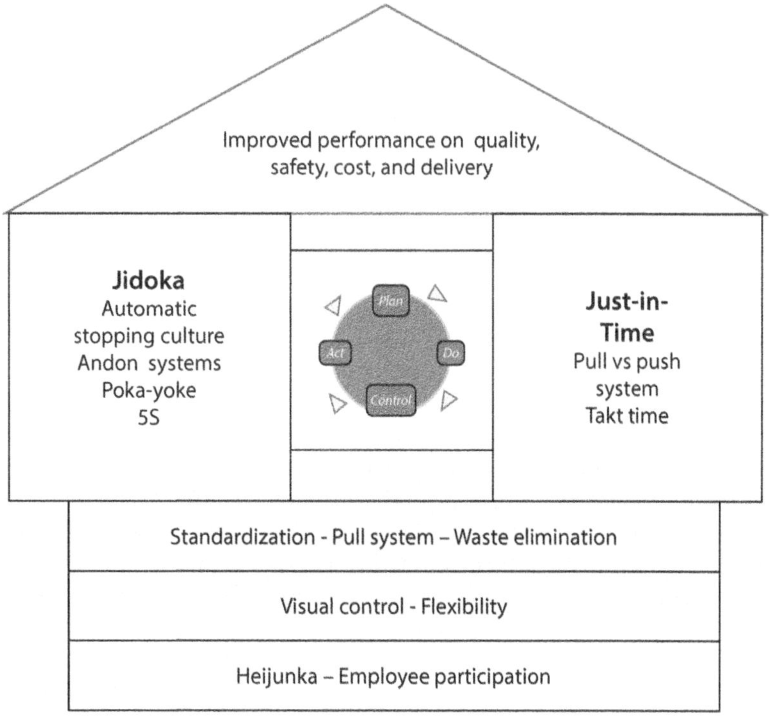

Fig. 1.3 Ohno's lean house. (Source: Adapted from Ohno 1988)

problems. Using the Ohno Circle on the shop floor, employees, engineers, and managers monitor the entire process to discover problems and issues. Standing on the Ohno Circle for a while gives the workforce opportunities to monitor the process, see the problems, and realize the obstacles within the processes.

Six Sigma projects aim at solving problems, issues, and bottlenecks; therefore, using the Ohno Circle may be helpful to visually analyze the process and problems. Using data sets and computers may generate many pivotal and critical information and knowledge. However, analyzing actual areas where the problems and obstacles occur may be very helpful to find out the most effective solutions in Six Sigma projects. Including the Ohno Circle, Ohno implemented several tools to aid in deep thinking at TPS. Ohno says "The Toyota style is not to create results by working hard. It is a system that says there is no limit to people's creativity. People don't go to Toyota to work, they go there to think" (Liker and Ross 2016). In his book titled *Toyota Production System: Beyond Large-Scale Production* (1988, xv), Ohno stated that:

> The Toyota Production System, however, is not just a production system. I am confident it will reveal its strength as a management system adapted to today's era of global markets and high-level computerized information systems.

1.4.7 Dr. Shigeo Shingo

Dr. Shigeo Shingo is one of the Japanese industrial engineers who contributed to creating lean knowledge and practices. Defeo (2017) names Dr. Shingo as the most influential Japanese quality professional. After graduating from Yamanashi Technical College, he worked for Taipei Railway Factory where he deeply explored Frederick Taylor's scientific management principles and the compatibility of scientific management with Japanese culture. Shingo discovered that Japanese management system was different from Western management approach. By adopting scientific management principles with the other employees for 5 years, he decreased repair time to 10 days, increased productivity, and improved operations.

Shingo and his team studied Frank and Lillian Gilbreth's time and motion studies and implemented their methods to improve their processes. Taylor's and Gilbreth's thoughts helped Shingo's future discoveries. In his next position, he doubled production in 6 months by implementing his process improvement ideas. Then, at the Mitsubishi Heavy Industries' Nagasaki shipyards, he decreased hull construction time from 4 months to 2 months in 1956. He also worked for the Japan Management Association for more than 10 years and founded the Institute for Management Improvement.

He developed source inspection, *poka-yoke* system, and zero quality control system. He was hired as a consultant at Toyota in 1955 and had a chance to extensively work with Taiichi Ohno at Toyota, where he developed the JIT system with Ohno. He published 17 books and many articles explaining his thoughts and practices about eliminating waste and implementing lean techniques. His ideas included creating quality at the source, creating and flowing value to customers, ensuring zero inventories, single-minute exchange of dies (SMED), and going to *gemba*.

Shingo is known as the Thomas Edison of Japan. He proposed and implemented SMED to minimize long setup or changeover times to less than 10 minutes. He categorized setup time into two critical parts: external setup and internal setup, converting internal setup time into external setup at Toyota. By transferring internal setup from the manufacturing processes to the suppliers through external setup, he succeeded in minimizing setup time at Toyota. He also cut the setup time on a 1,000 ton press at Toyota from 4 hours to 3 minutes. Shingo helped develop and integrate the *poka-yoke* methodology at Toyota. Dr. Shingo's efforts contributed to eliminating waste and improving overall performance throughout the organization. Based on his experience and knowledge, Shingo developed Scientific Thinking Mechanism (STM) methodology to improve business processes. In addition to his technical contributions to the manufacturing processes, he also emphasized that motivating employees and setting up clear objectives for them would be necessary to improve processes.

In 1988, Shingo received his honorary Doctor of Management from Utah State University. Afterward, the Shingo Prize was developed by the Shingo Institute, established by the School of Business at Utah State University, to recognize his contribution to lean production and thinking in the USA. The Shingo Prize is awarded for excellence in manufacturing by the Shingo Institute. The Model is comprised of four sections: Results, Enterprise Alignment, Continuous Improvement, and Cultural Enablers. Shingo Guiding Principles in the Shingo Prize Model are:

1. Respect Every Individual
2. Lead with Humility
3. Seek Perfection
4. Embrace Scientific Thinking
5. Focus on Process
6. Assure Quality at the Source
7. Improve Flow & Pull
8. Think Systemically
9. Create Constancy of Purpose
10. Create Value for the Customer.

1.4.8 Genichi Taguchi

Genichi Taguchi is a Japanese engineer and quality guru who is well-known in the area of product design optimization. Taguchi (1986) stated that "quality is the loss a product causes to society after being shipped, other than losses caused by its intrinsic functions." Taguchi's quality definition has two categories: (1) loss caused by variability of function and (2) loss caused by harmful side effects. Society, in Taguchi's definition, includes all relevant stakeholders such as producers, customers, employees, related government agencies, and unions. According to Taguchi's philosophy, quality is designed into a product, not inspected into it. System design, parameter design, and process design are utilized to increase design quality. Using these techniques, Taguchi developed robust design and classified variables in a process as control variables (controllable) and noise variables (uncontrollable). Then, he identified controllable and uncontrollable variables that minimize the variability in the response variable. His techniques used fractionated factorial

1

designs and other fractional designs obtained from orthogonal arrays (Montgomery 2013). The details of his techniques will be presented in ▶ Sect. 10.2 in ▶ Chap. 10.

Taguchi implemented the design on experiment method based on robust design in product design steps in quality engineering discipline. He estimated that 80% of defective products are associated with poor product design (Reid and Sanders 2016). Taguchi emphasized the significance of high product design quality in the entire value chain in manufacturing processes to minimize the total cost of the products. Taguchi started his first experiments at Electric Communication Laboratories where he focused on improving and enhancing the quality and productivity of telephone systems. He developed the quality loss function and Taguchi methods in the quality engineering discipline. Taguchi's methods primarily focus on off-line quality control, online quality control, product design quality, and a system of experimental design to improve quality and reduce costs (▶ https://asq.org/about-asq/honorary-members/taguchi). Online quality control involves production, service, and delivery of finished goods, while off-line quality control contains activities such as learning customer needs and expectations through market research, identifying product specifications, and product and process development (◘ Fig. 1.4). Taguchi suggests using system design, parameter design, and tolerance design as three important components of off-line quality system.

The quality loss function is considered a process to measure the quality level of a product or service. It also functions as a way to measure financial aspect of customer dissatisfaction and any deviation from target value of the CTQ characteristics. The greater the deviation from the target value, the greater the loss of quality to the society. Any factor that results in deviating from the target value is "noise" in Taguchi's loss function. These noises can be caused by a great variety of reasons such as production environment, raw material, energy, and employees. The details of noise and variation will be presented in ▶ Sect. 7.5 in ▶ Chap. 7 and ▶ Sect. 10.2 in ▶ Chap. 10.

Taguchi studied on statistics in Japan and worked on industrial experiments for the Institute of Statistical Mathematics from 1948 to 1950. He transferred to the Electrical Communication Laboratory (ECL) in 1950, where he worked on experimental design, data analysis, and life tests analysis. He was awarded Japan's Deming Prize for his contributions in quality engineering. He also developed Quality Research Group at Japanese Standards Association and helped Aoyama Gakuin University in Japan to establish engineering department. Then, he became the executive director of the Ford Supplier Institute, Inc., in

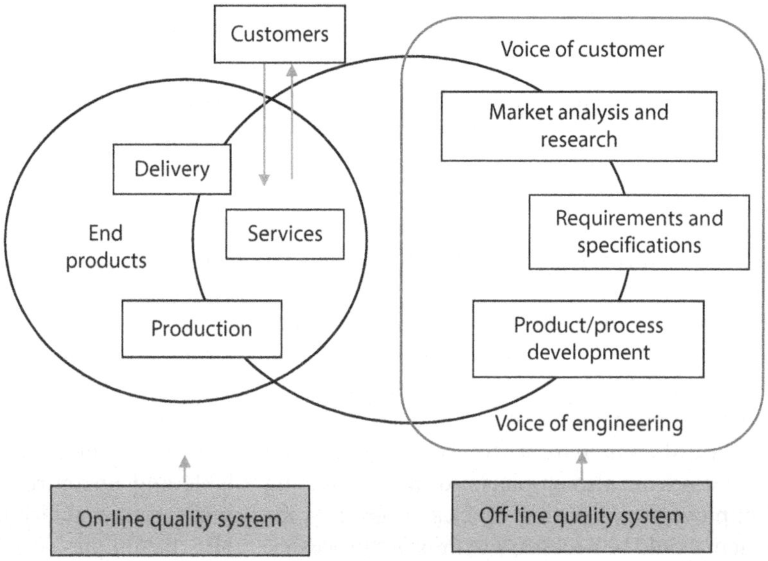

◘ **Fig. 1.4** Taguchi's online and off-line quality systems. (Source: Adapted from Taguchi 1986)

the USA. Taguchi is a recipient of the Indigo Ribbon given by the Emperor of Japan in 1986 for his contributions to Japanese economics and of the International Technology Institute's Willard F. Rockwell Medal for his product design optimization studies. He was also an honorary member of the Japanese Society of Quality Control and the executive director of the American Supplier Institute.

1.4.9 Philip B. Crosby

Crosby (1979) defines quality as "conformance to requirements" and addresses the critical role of "zero defects" in quality literature, suggesting benefits in "doing it right the first time." Crosby was the vice president of Quality at International Telephone & Telegraph (ITT). He focused on "absolutes" of quality and the basic elements of improvement. Some of his books are *Quality Is Free*, *Quality is Still Free*, *The Absolutes of Leadership*, and *Quality Without Tears: The Art of Hassle-Free Management*. Crosby states that organizations must view quality in (1) function, (2) process, and (3) ideology. Crosby identified "five absolutes of quality":

1. Quality means conformance to requirements, not elegance.
2. There is no such thing as a "quality problem."
3. There is no such thing as the economics of quality; it is always cheaper to do the job right the first time.
4. The only performance measurement is the cost of quality.
5. The only performance standard is "zero defects."

Crosby emphasizes that quality of products or services improves if the companies focus on 14-point quality improvement program. The components of this program are:

1. Management commitment.
2. Quality improvement teams.
3. Quality measurements.
4. Cost of quality evaluation.
5. Quality awareness.
6. Corrective action.
7. Ad hoc committees for zero defects.
8. Supervisor training.
9. Zero-defects day.
10. Goal setting.
11. Error cause removal.
12. Recognition.
13. Quality councils.
14. Do it over again.

Crosby identifies four activities to minimize the cycle time in business processes by focusing on the relationship between "zero-defects" philosophy and benefits of reduced cycle time. He states that, first, cycle time should be identified in the "diagnosis of cycle time" step, and then the process should be analyzed in depth in the "diagnosis of the process" step. Afterward, the major factors affecting cycle time of the process are diagnosed. In the last step, potential remedies to reduce cycle time are identified and implemented. Similarly, Crosby also proposes a five-step methodology to minimize the cycle time as follows:

- Define a process.
- List all activities of the process.
- Flowchart the process.
- List the elapsed time for each activity.
- Identify non-value-adding tasks.
- Eliminate all possible non-value-adding tasks.

Crosby proposes a Quality Management Maturity Grid to categorize the level of quality management implementation in organizations. The grid starts with "uncertainty" level and ends at "certainty" level. According to Crosby, the idea of quality management practices is to move companies to "certainty" level going through other steps and improve quality management practices in the organization. At uncertainty level, organizations do not have enough knowledge about quality management. As they go through the grid, they come to certainty level where they learn and successfully implement the requirements of quality management.

1.4.10 David Garvin

David Garvin was a professor of business administration at Harvard Business School. As a researcher and author, he significantly contributed to the literature of management, quality, learning organizations, and leadership concepts. His articles published in *Harvard Business Review* helped students and researchers understand how to identify and improve quality in the American manufacturing industry. He focused on how to minimize American manufacturers' high failure and defect rates. A notable accomplishment in his career was answering "How Japanese automakers could make higher-quality, more-reliable cars than Americans, while charging less for them." His studies also focused on "quality" as a strategic competitive advantage as well as the presentation of consumers' needs and preferences as the paramount in the competition. One of Garvin's landmark studies is the book titled *Managing Quality: The Strategic and Competitive Edge*. His eight critical dimensions of product quality have been perceived as a framework by quality researchers and practitioners. Garvin served on the board of directors of the Malcolm Baldrige National Quality Award. Garvin (1987) stated that "managers have to stop thinking about quality merely as a narrow effort to gain control of the production process, and start thinking more rigorously about consumers' needs and preferences. Quality is not simply a problem to be solved: it is a competitive opportunity."

1.4.11 Douglas Montgomery

Douglas Montgomery is a regents professor of industrial engineering and statistics at Arizona State University. He is known as one of the significant quality thinkers over the last five decades. Montgomery is the author of 16 books in industrial engineering and industrial statistics; *Introduction to Statistical Quality Control*, *Response Surface Methodology*, *Introduction to Time Series Analysis and Forecasting*, and *Design and Analysis of Experiments* are some of his books.

Montgomery's research has contributed to statistical quality control, design of experiment, regression analysis, engineering statistics, optimization, and response surface. In return for his invaluable contributions, Montgomery has received various medals and awards such as the Shewhart Medal, the Brumbaugh Award, the Hunter Award, the Lloyd S. Nelson Award, the Shewell Award, the Ellis R. Ott Award, and the George Box Medal from ENBIS (European Network for Business and Industrial Statistics). Montgomery is also a fellow of many professional associations and societies such as the American Statistical Association, the ASQ, the Royal Statistical Society, the Institute of Industrial Engineers, the International Statistical Institute, and Academician of the International Academy for Quality.

1.5 The Historical Background of Six Sigma

It is not easy to understand what the concept enables for practice and academia today without understanding the historical background of the concept. Several continuous quality improvement methodologies were developed in practice and academia over the last decades, and one of those methodologies is Six Sigma. Quality, and specifically the Six Sigma concept, became a scientific discipline with its own vocabulary, terminology, techniques, rules, and norms. The Six Sigma concept was constructed from the advances of quality concept and total quality management.

The history and origin of quality started at the beginning of human civilization. The quality term has a long historical background that goes to the Code of Hammurabi in early 2000 BC and is considered to be the first structured quality standard. According to the Code, "If a builder has built a house for a man, and his work is not strong, and the house falls in and kills the householder, that builder shall be slain."

Quality still has the same meaning today. What is perhaps different today is the focus on a structured quality approach, that is, a quality plan that is well articulated, included in operational strategies, and integrated in a decision-making framework to attain predefined goals. Keeping this in mind, quality has become a powerful tool through which organizations can achieve competitiveness and survival in the marketplace. This structured approach to quality has helped many leading corporations such as Xerox, Motorola, IBM, Harley-Davidson, and a host of others to survive and thrive in their competitive environment (Madu 1998).

Some researchers (c.f. Benedetto 2003) asserted that the root of Six Sigma could be credited to the early nineteenth century, when Gaussian/normal distribution and sigma was developed as a measurement standard. Prior to the mass production era in the late 1800s, quality heavily relied on craftsmen and operators' skills, abilities, and knowledge and was known as *operator quality control*. In the early 1900s, *foreman quality control* emerged in various industries, when the foreman was responsible for determining quality. During the Industrial Revolution, the quantity of the product, rather than quality, was pivotal for manufacturers to meet constantly increasing demand in the marketplace.

In the 1920s, the importance of quality was realized and upgraded to a statistical direction, which revolutionized the manufacturing industry. The expectations and standards of mass production systems revitalized the quality concept in the following years, especially after the 1920s when statistical process control started its first steps in the USA. Beginning with Shewhart, and followed by many others, statistical sciences were embedded into Six Sigma (Çelikoğlu 2008).

During World War I, manufacturing systems became more complex, and full-time quality inspectors were hired, called the *inspection quality control* approach. World War II and massive production needs and requirements required *statistical process control* practices. The needs and expectations that emerged during and after World War II in both military and non-military areas resulted in increasing quality concerns all over the world. National and global quality movements were accelerated in industrialized countries. JUSE in Japan, ASQC in the USA, and ISO in Europe were contemporary institutions established in the 1930s.

In 1935, Kiichiro Toyoda founded Toyota Motor Corporation in Japan and, in 1945, started producing cars, building 3,000 cars that year. They seeded the first practices of lean production in Japan (Holweg 2007). When companies realized that all departments and functions have an impact on quality of the product and service, organization-wide *total quality control* was seen pivotal for increasing overall performance by a group of quality gurus led by Armand Feigenbaum in the 1960s. Overall, quality gurus such as Shewhart, Deming, Juran, Feigenbaum, Ishikawa, Taguchi, Ohno, and Shingo contributed to quality control, quality assurance, TQM, and lean management between 1920 and 1980. The Six Sigma approach emerged in the 1980s and developed further during the following years.

After the Japanese manufacturing industry became a powerful rival to American manufacturing, Motorola was at risk of losing its market to Japanese rivals in the 1980s. In an interview, Robert Waterman, Jack Welsh, and Paul Noakes, top executives of Motorola, asserted that poor product quality caused Motorola to lose a significant percentage of market share to the Japanese manufacturers (▶ https://youtu.be/Ze3TWEXXqUU). At that time, top executives of Motorola, Robert Galvin and Mikel Harry, developed plans to increase overall performance by decreasing early life failure and the overall failure rate of the products to enable the firm to remain viable against their rivals. The plan was to reduce failure rate to zero by analyzing variation in the processes. They won the Malcolm Baldrige National Quality Award in 1988 and became a quality and profit leader by implementing the award model's requirements based on Six Sigma (Pyzdek and Keller 2018).

In 1994, Larry Bossidy, CEO of AlliedSignal, introduced Six Sigma as a business initiative to improve work processes and performance. General Electric and AlliedSignal implemented Six Sigma as a corporate initiative in 1995. Six Sigma methodology was welcomed by the engineers and managers in General Electric and AlliedSignal and resulted in saving millions of dollars in a company-wide effort. After large organizations showed successful practices, medium- and small-size organizations implemented Six Sigma methodology. Beginning in the 2000s, lean and Six Sigma concepts were combined by many successful companies. Afterward, *Lean Six Sigma* and *Design for Six Sigma* were published, and the synergy created by the combination of lean manufacturing and Six Sigma helped organizations minimize defects and nonconformities and improve sustainability in a great variety of industries.

Advancements of Six Sigma in the academic world stayed behind the impact of Six Sigma practices in industry (i.e., Linderman et al. 2003). In light of this fact, academia focused on this question: from practice to research, or from research to practice? That question has been asked by researchers and practitioners for a long time when analyzing newly emerged research areas and disciplines. Six Sigma is one of those areas that seek to answer this question by looking from several perspectives. Six Sigma was criticized it had a background in practice in industry but not an academic background for a while because it was developed by a firm to minimize its percentage of defectives and defects. The concept of lean Six Sigma as an approach to process improvement has yet to fully mature into a specific area of academic research. In a similar vein, the Toyota Production System (TPS) was criticized by academics because it was created from actual practices. Monden (1998: xiii) states that "it has a strong feature of emphasizing practical effects, and actual practice and implementation over theoretical analysis." As the quality concept gained importance, academic interest on the subject grew (Celikoglu, 2008).

To accept an area as an academic discipline, several requirements are expected to be met. These requirements are (1) being supported by at least one scientific discipline; (2) having at least one undergraduate educational program; (3) having at least one graduate program; (4) books and articles published in scholarly and professional journals; and (5) professional societies or non-profit organizations established that are related to new discipline.

In terms of scientific disciplines that support Six Sigma, Schroeder (2008) and several researchers pointed out that literature started analyzing theoretical background of Six Sigma in the 2000s. Additionally, there are several areas that relate to quality and Six Sigma concepts, such as industrial engineering, statistics, mechanical engineering, and operations management. After the Ford manufacturing system and Taylor's scientific management approaches developed, Six Sigma started its first steps by focusing on increasing productivity, efficiency, and effectiveness. With regard to educational programs, the first department in higher education in industrial engineering was established at Pennsylvania State University in 1909. The first doctoral degree in industrial engineering was awarded in 1933 by Cornell University. Since then, industrial engineering departments have taught efficiency, effectiveness, and productivity improvements, which are the focal points of Six Sigma.

The statistics discipline has supported Six Sigma with its older background. Mikel Harry, an engineer in Motorola, obtained his Ph.D. degree in Six Sigma at Arizona State University in 1984. Helper obtained his Ph.D. degree with a dissertation titled *Supplier Relations and Technical Change: Theory and Application to the U.S. Automobile Industry* in 1987 from the Department of Economics, Harvard University. Krafcik, MacDuffie, and Sherriff followed Helper at MIT and Fujimoto at Harvard. The first book published about quality control was *The Control of Quality in Manufacturing* by G.S. Radford in 1922. Books by Shewhart, Juran, Crosby, Shingo, Ishikawa, and Deming followed. Scholarly journals published the first articles about Six Sigma in the 1990s. Niñerola et al.'s (2019) study reports the numbers of the publications in Six Sigma methodology between 1990 and 2017. The study shows that publications analyzing Six Sigma have been on the rise since 1990.

For the last requirement (establishing organizations), several countries made a promising progress by building organizations that develop and support quality and Six Sigma efforts, such as JUSE, Japan Industrial Management Association, Japan Institute of Plant Maintenance (JIPM), Japan Auto Manufacturers Association, International Motor Vehicle Program (IMVP) at MIT, American Society for Quality Control

(ASQC), the Deming Institute, the Juran Institute, the Motorola Six Sigma Technical Institute, the Motorola Six Sigma Research Institute, Six Sigma Academy, and Six Sigma Management Institute (SSMI) in Japan and the USA.

Based on all these developments, the quality and Six Sigma concepts were systematized and became a pivotal part of corporate performance management systems in business organizations throughout the world. Six Sigma has been recognized as an academic field by researchers and universities.

It was reported that Motorola saved $16 billion from Six Sigma practices between 1986 and 2001, and other giant firms such as Ford, Caterpillar, Boeing (Alimohamadi and Khorshidi 2009), 3M, Honeywell, and GE benefited from Six Sigma in similar ways (Zhang et al. 2009). All Western and Japanese quality gurus' contributions to quality literature and practices supported the base of Six Sigma at Motorola. Maguad (2006) and Goh (1999) stated that the necessity of making sure that the products meet the customer requirements supported the origin of quality engineering that integrates the discipline of statistics into quality improvement activities.

1.6 Standards in Six Sigma

Six Sigma is one of the subjects standardized by the ISO. The ISO's Technical Committee 69 (TC 69) has developed and improved statistical and related techniques for the implementation of Six Sigma under the subcommittee SC 7. There are 11 published active standards that have been developed by SC 7 as of February 2019. The SC 7 has been developing four new standards. The standards developed by SC7 as of January 10, 2019 are as follows:

- ISO/TR 12845:2010: Selected illustrations of fractional factorial screening experiments
- ISO/TR 12888:2011: Selected illustrations of gauge repeatability and reproducibility studies
- ISO 13053-1:2011: Quantitative methods in process improvement – Six Sigma – Part 1: DMAIC methodology
- ISO 13053-2:2011: Quantitative methods in process improvement – Six Sigma – Part 2: Tools and techniques
- ISO/TR 13195:2015: Selected illustrations of response surface method – Central composite design
- ISO/TR 14468:2010: Selected illustrations of attribute agreement analysis
- ISO/TR 16705:2016: Statistical methods for implementation of Six Sigma – Selected illustrations of contingency table analysis
- ISO 17258:2015: Statistical methods – Six Sigma – Basic criteria underlying benchmarking for Six Sigma in organisations
- ISO 18404:2015: Quantitative methods in process improvement – Six Sigma – Competencies for key personnel and their organizations in relation to Six Sigma and Lean implementation
- ISO/TR 29901:2007: Selected illustrations of full factorial experiments with four factors
- ISO/NP 13053-1 [Under development]: Systems for process improvement – Lean & Six Sigma – Part 1: Methodology
- ISO/NP 13053-2 [Under development]: Systems for process improvement – Lean & Six Sigma – Part 2: tools and techniques
- ISO/PRF TR 20693 [Under development]: Statistical methods for implementation of Six Sigma – Selected illustrations of distribution identification studies
- ISO/NP TR 22914 [Under development]: Statistical methods for implementation of Six Sigma – Selected illustration for ANOVA.

ISO/TR 12845:2010 details the requirements of two-level fractional factorial experimental designs and demonstrates six distinct applications of this methodology (▶ https://www.iso.org/standard/51963.html). ISO/TR 12888:2011 describes the measurement process for continuous variables. Measurement processes that include an attribute (i.e., pass/fail) are not analyzed in this document. ISO/TR 12888:2011 presents examples of simple measurement systems and analyzes two major factors: (1) variation between operators or appraisers and (2) within operators or appraisers (▶ https://www.iso.org/standard/52899.html).

ISO 13053-1:2011 describes DMAIC phases that include define, measure, analyze, improve, and control (DMAIC) as a methodology for Six Sigma. The standard recommends the preferred or best practice for each of the phases of the DMAIC methodology. The standard also addresses how Six Sigma projects can successfully be managed and details the expected roles, expertise, and training needs, whereas ISO 13053-2:2011 presents tools and techniques used in each phase (▶ https://www.iso.org/standard/52901.html).

ISO/TR 13195:2015 describes the steps of response surface methodology (RSM) and data analysis using central composite designs (CCD) through four distinct applications of this methodology. RSM analyzes the relationship between the response and the set of quantitative predictor variables (▶ https://www.iso.org/standard/53479.html).

ISO/TR 14468:2010 assesses a measurement process when the CTQ is an attribute variable. ISO/TR 14468:2010 details attribute agreement analysis (AAA) including agreement within appraisers, agreement between appraisers, agreement of each appraiser vs a standard, and agreement of all appraisers vs a standard (▶ https://www.iso.org/standard/52900.html).

ISO/TR 16705:2016 presents the details of contingency table analysis and the method to analyze the relation between categorical variables (including nominal variables and ordinal variables) using different software applications (▶ https://www.iso.org/standard/57453.html).

ISO 17258:2015 presents a methodology for establishing the level of quality, performance, and productivity of processes, products, and services according to Six Sigma principles, regardless of the type of industry. The methodology is presented as a convenient way to start a Six Sigma program in any type of organization and focuses on criteria, measures, measurement process, and comparison process (▶ https://www.iso.org/standard/59489.html).

ISO 18404:2015 identifies the competencies for the positions included in Six Sigma organizations such as black belt, green belt, and lean practitioners. Yellow belt is not included in ISO 18404:2015. ISO 18404:2015 excludes Design for Six Sigma (▶ https://www.iso.org/standard/62405.html).

ISO/TR 29901:2007 describes the steps of full factorial experimental designs with four factors through illustration, with five distinct applications of this methodology (▶ https://www.iso.org/standard/45731.html). There are also some Six Sigma-related draft standards that are still in the approval process at ISO such as ISO/NP 13053-1, ISO/NP 13053-2, ISO/PRF TR 20693, and 3ISO/NP TR 22914.

1.7 Quality Costs

1.7.1 Quality Cost Definition

Quality guru Philip Crosby states that the only performance measurement would be "the cost of quality" for many stakeholders such as customers, manufacturers, service providers, and suppliers. Quality affects not only the manufacturers' quality costs; it also affects the users' costs and satisfaction (Juran and Gryna 1980: 443). Feigenbaum (1983) states that there was a widespread belief in the past that quality could not be practically measured in cost terms. However, quality literature and practice already showed the measurability of quality costs and how quality costs could help develop strategic plans, improve decision-making processes, and identify areas of improvement in continuous quality improvement efforts. ASQ identifies cost of quality as follows:

> Cost of quality is a methodology that allows an organization to determine the extent to which its resources are used for activities that prevent poor quality, that appraise the quality of the organization's products or services, and that result from internal and external failures. Having such information allows an organization to determine the potential savings to be gained by implementing process improvements (▶ http://asq.org/learn-about-quality/cost-of-quality/overview/overview.html).

Feigenbaum (1983) claims that investment on quality programs may be evaluated through quality costs as the foundation of quality-systems economics. As presented in Deming's Chain Reaction previously, in ◘ Fig. 1.1, cost containment takes a pivotal role in improving quality and overall performance. To reduce total costs, including costs of good and poor quality, a system classifying and measuring components of good and poor quality is needed first in business organizations:

> There are many situations in which it is useful to summarize the effect of all deficiencies through a single unit of measure. One way doing so is to convert the effect of all deficiencies into money. The end result of such conversion goes by various names: Cost of Quality, Quality Costs, Cost of Poor Quality Juran (1992: 119).

Research shows that companies that implemented cost of quality systems achieve cost reduction and quality improvement (Schiffauerova and Thomson 2006). In practice, classifying and measuring components of cost of quality may be a challenging topic for quality practitioners and managers. Along with this need, there is no consensus on the definition of quality costs in the literature (Dale and Plunkett, 1993).

1.7.2 Quality Cost Categories

Essentially, cost of quality is considered the total costs of *conformities* and *nonconformities* in practice. Abdul-Rahman et al. (1996) classified the cost of doing the things right at the first time as the *cost of conformity* and the cost of fixing performance-related issues as the *cost of nonconformity*. The cost of conformity includes costs of activities planned for preventing errors, mistakes, and failures throughout the organization. Planning and design activities, quality audits, trainings, and preventive maintenance are some of the items of cost of conformity. The cost of nonconformity is the cost of all failures, errors, and mistakes identified inside and outside the organization. Rework, scrap, returns, customer complaints, warranty reimbursements, and recalls are considered cost of nonconformity.

Feigenbaum (1983) classifies quality costs into two groups such as (1) *costs of control* and (2) *costs of failure of control*. He categorizes prevention and appraisal costs under the costs of control. Internal and external failure costs are grouped into the costs of failure of control. Schiffauerova and Thomson (2006) asserted that most commonly implemented quality of cost model in organizations is the PAF model. Juran developed and suggested PAF model to measure, analyze, and manage the cost of quality. PAF model is comprised of:

1. *Prevention*
2. *Appraisal*
3. *Internal failure*
4. *External failure costs.*

Similarly, Pyzdek and Keller (2018: 76) identify quality costs as the total of the cost occurred by (1) investing in the *prevention of nonconformance* to requirements, (2) *appraising* product or service for conformance to requirements, and (3) *failure* to meet requirements. Quality costs are generated throughout the total life cycle of the product and service, from design to after-sales services. Quality costs may occur in the entire process where producers, consumers, and merchants are involved in production and consumption activities (Feigenbaum 1983).

Prevention costs refer to the costs of activities that are planned and implemented to eliminate failures in the system as a proactive approach. The proactive approach shapes prevention cost groups, since the idea behind prevention costs is to prevent quality problems. The costs of quality planning, supplier selection, marketing research, product design and planning, design validation and verification, new product reviews, training, data analysis, quality audits, quality reporting, operations planning and validation, and improvement projects are listed under prevention costs. According to Juran and Gryna (1980), prevention costs occur to keep failure and appraisal costs to a minimum. The lowest cost of quality occurs when errors and mistakes are prevented from happening (Pyzdek and Keller 2018).

Appraisal costs are related to monitoring and measuring activities and contain the cost of activities that align with incoming inspection and test, supplier audits, certifications, process control, outgoing inspections, and maintaining the accuracy of test equipment. These costs occur while analyzing the condition of the products and processes in inspection, control, and testing activities. As Feigenbaum (1983) stated, in traditional quality control activities, failures and appraisal expenses increase together. It may even be difficult to get them down when one of them begins rising. When the defects or nonconformities occur, the failure costs increase. The traditional solution when failures increased was to do more inspections in the past, which was a contribution to higher appraisal cost (Feigenbaum 1983).

Failure costs are grouped into two sections: *internal failure* and *external failure* costs. *Internal failure costs* include the costs of failures identified and detected in the organization, prior to delivering products to the marketplace. These costs disappear when there is no defect of the product/service before delivering it to the customer. For example, rework, scrap, retest, downtime, overtime, defects, nonconformities, corrective actions, reinspection and retests, and dispositions identified in the processes are categorized as internal failure costs. *External failure costs* are associated with the activities that aim to find a solution for the quality problems discovered by the external customers. Theoretically, these costs disappear if there is no defect or nonconformity in the product and service after delivering it to the customer. The most expensive quality cost stems from nonconformities detected by external customers (Pyzdek and Keller 2018: 76) and other undetected failures and nonconformities that take place in the marketplace. At this point, the *hidden cost* concept comes into play. Hidden costs cannot be computed quantitatively, since the impact of those costs cannot entirely be projected and calculated in the short or long run. Some hidden costs are negative word of mouth, unheard customer complaints and dissatisfaction, unknown lost sales, lost customers, and market share. Warranty claims, returns, recalls, product liability costs, customer complaints, allowances, lost sales, and bad reputation costs are some external failure costs.

To minimize internal and external failure costs, appropriate preventive strategies should be effectively implemented in the system. Appraisal costs contribute to minimize internal and external failure costs. Although quality cost measurement does not have to be accurate to the penny (Pyzdek and Keller 2018), implementing and running an effective measurement system identifies improvement needs in the organization. The outcomes of quality cost measurements are used by Six Sigma practitioners to identify improvement areas and bottlenecks within the system.

A process that operates at the 6 sigma level is expected to fail to meet requirements 3.4 times per million opportunities. Another process that operates at 4 sigma is expected to produce 6,210 times per million opportunities, that is, 6,210 defects (errors) per million opportunities. The costs of poor quality are high for companies that do not implement Six Sigma methodology, and approximately 25 to 40% of total revenues are allocated on fixing problems in companies. For example, General Electric estimated that being at 3 sigma instead of 6 sigma costs them between $8 billion and $12 billion a year (Pyzdek and Keller 2018: 5).

To classify and calculate costs of quality, let's work on three examples as given below.

Exercise 1

In Table 1.2, four scenarios are presented with various quality cost categories used in the PAF system. Analyze and identify the details of each scenario. The values given in scenarios are in dollar ($1,000).

Table 1.2 Quality cost levels for four scenarios

Quality cost categories	Scenario 1	Scenario 2	Scenario 3	Scenario 4
	Low appraisal and prevention cost model	High appraisal cost model	Higher appraisal cost model	Higher prevention cost model
Prevention cost	1	1	1	3
Appraisal cost	3	6	10	3
Internal cost	2	15	8	5
External cost	25	3	2	1
Total quality cost	31	25	21	12

Source: Author's creation

- **Solution**

In the first scenario, the low prevention and appraisal cost model results in higher external costs in the marketplace. This scenario is not desirable for organizations that aim to minimize total quality costs, because having high external cost means that there is a bigger hidden cost that is not easily calculated. In the second scenario, appraisal-related activities are doubled, and external costs significantly decrease, whereas internal costs increase due to the nonconformities identified within the company. Compared to the first scenario, the second scenario shows better performance in terms of the total quality cost. It is a good indicator of effective appraisal activities implemented in the system, since defects and nonconformities are detected and categorized in internal failure cost. In the third scenario, appraisal activities increase, resulting in lower internal and external costs. Eventually, total cost reduces because of decreased defects and nonconformities. In the last scenario, prevention activities are three times those of the previous scenarios, and this increase produces the positive highest impact on total quality costs. As shown in this example, increasing prevention activities and costs can significantly minimize total quality costs as both appraisal and failure costs decrease. Therefore, increasing preventive activities may result in minimizing the total quality costs.

Exercise 2

Sun Legend Inc. aims at measuring, categorizing, and analyzing quality costs of the manufacturing system for the year of 2019. The components of quality costs (percentages) for three phases are shown in ◘ Table 1.3. Analyze each phase and decide which phase shows a better performance compared to the others with regard to quality costs distribution.

◘ **Table 1.3** Quality costs at Sun Legend Inc.

Quality costs	Phases		
	Phase A	Phase B	Phase C
External failure (%)	41.2	20.3	20
Internal failure (%)	45.1	25.1	45
Appraisal (%)	13.1	51.5	30
Prevention (%)	0.6	3.1	5
Total	100	100	100

Source: Author's creation

▪ **Solution**

Considering that one of the goals in quality costs is minimizing both external and internal failure costs, Phase B demonstrates a better performance among three phases, with 45.4% of total quality costs. Looking from another perspective that supports increasing preventive actions, Phase C may be considered a better option as the prevention costs dramatically increase from 0.6% to 5%. In Phase C, external failure costs significantly decrease from 41.2% to 20%. However, internal failure costs in Phase C account for 45% of total cost as it is in Phase A. This may be due to decreased appraisal costs, especially when Phase C is compared to Phase B (from 51.5% to 30%). The inverse relationship between internal failure cost and appraisal cost is presented in ◘ Table 1.3. It shows that appraisal activities are likely to decrease internal failures.

Juran and Gryna state that (1980: 23) "the typical quality cost study will show that the previously underemphasized failure costs are several times the appraisal costs. This comes as a surprise to most managers and forces them to reconsider their emphasis." In the ideal scenario, prevention costs are expected to be higher than any other quality cost category. By increasing preventive activities in the system, internal and external failures are expected to decrease, which is an indication of improved quality system. Furthermore, the relationship between internal failure and external failure is critical, since internal failure costs refer to production planning improvement needs, while external failure costs point to product design-related improvement needs. Internal and external failure costs reflect "quality of conformance" and "quality of design," respectively, in the total quality cost system. The estimated ideal quality cost ratios are (1) prevention costs (0.5–5%), (2) appraisal costs (10–50%), (3) internal failure costs (25–40%), and (4) external failure costs (20–40%).

1.7.3 Performance Metrics in Quality Costs

Total quality costs are related to several performance metrics: (1) labor cost, (2) production cost, (3) total sales, and (4) total manufacturing costs. Some potential metrics that can be used to monitor the performance of quality cost systems are labor index, cost index, sales index, and production index as presented in Eqs. 1.1, 1.2, 1.3, and 1.4.

$$\text{Labor index} = \frac{\text{total quality cost}}{\text{total direct labor cost}} \quad (1.1)$$

$$\text{Cost index} = \frac{\text{total quality cost}}{\text{total production cost}} \quad (1.2)$$

$$\text{Sales index} = \frac{\text{total quality cost}}{\text{total sales}} \quad (1.3)$$

$$\text{Production index} = \frac{\text{quality cost per unit}}{\text{value per unit}} \quad (1.4)$$

As well as industry-specific differences matter, Juran proposes the following optimal quality cost indexes:

$$\frac{\text{total quality costs}}{\text{total production costs}} = 20-30\%$$

$$\frac{\text{prevention costs}}{\text{total production costs}} = 5-8\%$$

$$\frac{\text{appraisal costs}}{\text{total production costs}} = 10-14\%$$

$$\frac{\text{failure costs}}{\text{total production costs}} = 5-8\%$$

Exercise 3

Using the data presented in ◘ Table 1.4, compute (1) the company's cost of good quality and (2) cost of poor quality as a percentage of total quality costs for each of 4 years. Discuss how the company's quality strategy should be formalized based on current quality costs performance. Using the same data, (3) calculate labor index, (4) cost index, (5) sales index, and (6) production index for each year. Analyze the performance of the company.

◘ **Table 1.4** Data used in the example

Quality costs ($ in thousands)	**2015**	**2016**	**2017**	**2018**
Prevention	215	310	560	820
Appraisal	1,155	905	835	800
Internal failure	2,880	3,500	2,599	1,630
External failure	1,885	1,460	770	790
Variables				
Sales price/unit	$2.15	$2.19	$2.25	$2.30
Total sales	33,950	34,700	39,300	48,400
Total manufacturing cost	13,700	14,110	14,600	14,750
Total labor cost	2,130	3,090	3,120	3,250

Source: Author's creation

1

Solution

First, let's combine (1) prevention and appraisal costs under "cost of good quality" category and (2) internal and external failure costs under "cost of poor quality" category (Table 1.5). By combining quality cost items under these categories, we will be able to monitor and group quality costs such as "cost of good quality" and "cost of poor quality," which are two common categories used in theory and practice. As seen in Table 1.5, total cost of good quality in 2015 is \$1,370 and is the sum of \$215 prevention and \$1,155 appraisal costs. Similarly, total cost of poor quality is \$4,765 and is the sum of \$2,880 internal failure and \$1,885 external failure costs in 2015. Then, we calculate each performance metric as detailed in Table 1.5. The calculations for each performance metric for 2015 are presented in detail in the 2015 column in Table 1.5. To illustrate the first ratio, which is % of good quality,

$$\frac{\text{total cost of good quality}}{\text{total quality costs}} = \frac{1{,}370}{6{,}135} = 22\%$$

For the second ratio, the percentage of poor quality in 2015 is

$$\frac{\text{total cost of poor quality}}{\text{total quality costs}} = \frac{4{,}765}{6{,}135} = 78\%$$

As shown in Table 1.5, within 4-year period, the percentage of good quality is on the rise from 20% to 40%, which results in a significant decrease in total quality costs from \$6,135,000 to \$4,040,000. Increasing good quality costs caused in decreasing total quality costs. It seems that the quality strategy implemented in this company makes a positive contribution in minimizing poor quality costs and, ultimately, total quality costs. A deeper analysis reveals that the company can decrease total quality costs by increasing prevention costs from \$215,000 to \$820,000 (384%) and decreasing appraisal costs from \$1,155,000 to \$800,000 (30.73%), which is the ideal scenario for increasing the quality level in production processes. Investing on preventive activities paid off on lowering internal

Table 1.5 Quality cost results

Quality costs (\$ in thousands)	2015	2016	2017	2018
Total cost of good quality	215 + 1,155 = 1,370	1,215	1,395	1,620
Total cost of poor quality	2,880 + 1,885 = 4,765	4,960	3,369	2,420
Total quality costs	1,370 + 4,765 = 6,135	6,175	4,764	4,040
Ratios				
% of good quality	1,370/6,135 = 22%	20%	29%	40%
% of poor quality	4,765/6,135 = 78%	80%	71%	60%
Quality cost/sales	6,135/33,950 = 18.1%	17.8%	12.1%	8.3%
Quality cost/total manufacturing cost	6,135/13,700 = 44.8%	43.8%	32.6%	27.4%
Indexes				
Labor index	6,135/2,130 = 2.88	2.00	1.53	1.24
Cost index	6,135/(13,700 + 2,130) = 0.39	0.36	0.27	0.22
Sales index	6,135/33,950 = 0.18	0.18	0.12	0.08
Product index	(6,135/33,950)/2.15 = 0.08	0.08	0.05	0.04

Source: Author's creation

failure costs from \$2,880,000 to \$1,630,000 (43.40%) and external failure from \$1,885,000 to \$790,000 (58.9%).

For labor index, cost index, sales index, and production index in 2015, we can use the following ratios:

$$3)\ \text{Labor index} = \frac{\text{total quality cost}}{\text{total direct labor cost}} = \frac{6{,}135}{2{,}130} = 2.88$$

$$4)\ \text{Cost index} = \frac{\text{total quality cost}}{\text{total production cost}} = \frac{6{,}135}{13{,}700 + 2{,}130} = 0.39$$

$$5)\ \text{Sales index} = \frac{\text{total quality cost}}{\text{total sales}} = \frac{6{,}135}{33{,}950} = 0.18$$

$$6)\ \text{Production index} = \frac{\text{quality cost per unit}}{\text{value per unit}} = \frac{6{,}135}{33{,}950} * \frac{1}{2.15} = 0.08$$

The proportion of total quality costs, in relation to total sales and in relation to manufacturing costs, decreases each year. These two ratios show that increasing cost of good quality enables the proportion of total quality cost in total sales and total manufacturing cost to be reduced. Similarly, labor index, cost index, sales index, and product index show a better performance year by year due to decreasing total quality cost in each year.

Optimum total quality cost level is located at the point where total cost is minimum (Juran and Gryna 1980). In the classical approach, total cost function is comprised of (1) failure cost and (2) prevention and appraisal costs. As the quality level increases on *X*-axis, the sum of prevention and appraisal costs increases on *Y*-axis as well. Along with this interrelationship, the total failure costs function starts decreasing until a designated defect level and is located at minimum value when the defective rate is zero. After that designated defect rate, the total failure costs function starts increasing in this classical view mechanism. As opposed to this classical approach, it is expected that total cost of quality is located at zero when the defect rate is zero in modern approach.

Feigenbaum (1983: 145) states that "quality control and quality economics must become two of the principal elements of a company's strategic planning for gaining a competitive advantage globally. Quality control activities and quality costs must be directed in a way that provides the firm major added business value."

Key Concepts

Six Sigma, DMAIC, Critical-to-Quality (CTQ) characteristic, quality definitions, six dimensions of quality, value, internal and external customers, specifications, variability, quality thinkers, Walter Shewhart, W. Edwards Deming, Joseph M. Juran, Armand V. Feigenbaum, Kaoru Ishikawa, Taiichi Ohno, Shigeo Shingo, Genichi Taguchi, Philip Crosby, Plan-Do-Control-Act (PDCA), Deming's Chain Reaction, Deming System of Profound Knowledge, Deming's 14 points, Red Bead Experiment, Juran's trilogy, PAF (prevention-appraisal-failure) costs of quality system, Juran's ten steps to quality improvement, Taguchi's quality loss function, standards of Six Sigma, quality costs, costs of good quality, costs of poor quality, prevention costs, appraisal costs, internal failure costs, external failure costs, and performance metrics in quality costs.

Summary

This chapter aims to create a base for Six Sigma. Basic characteristics of Six Sigma philosophy, quality definitions, six dimensions of quality definition, contributions of quality gurus to quality and Six Sigma literature, historical background of Six Sigma, standards developed by International Organization for Standardization (ISO) on Six Sigma, types of costs of quality, and performance metrics used in quality cost calculations are detailed in this chapter.

Practice and Discussion Questions

1. Discuss the details of Six Sigma philosophy.
2. What does 3.4 ppm mean in Six Sigma?
3. What are the ultimate goals of Six Sigma in continuous quality improvement?
4. What are the four approaches used for defining quality term in the literature? Discuss the details of each approach.
5. What are the dimensions of Garvin's quality definition? Analyze and exemplify the details of each dimension.
6. Who are the quality gurus in the literature? Categorize quality gurus into groups, and briefly list those quality gurus in each group.
7. What are the main thoughts of Walter Shewhart? Discuss his main ideas and approaches.
8. What are the main thoughts of Dr. W. Edwards Deming? Discuss his main ideas and approaches.
9. Explain the mechanism of Deming's Chain Reaction.
10. What is the mechanism of PDCA cycle? Explain what each step means in PDCA cycle.
11. What does the System of Profound Knowledge represent in Deming quality approach?
12. Briefly explain what is represented in Deming's 14 points.
13. What is explained and modeled in the Read Bead Experiment in Deming's philosophy?
14. Explain the main difference between common causes of variation and assignable causes of variation.
15. Discuss the details of Juran's trilogy.
16. What are the activities that need to be done in each phase of Juran's trilogy?
17. What are the ten steps of quality improvement in Juran's quality approach?
18. What does PAF stand for in Juran's quality approach?
19. What does total quality control mean in Feigenbaum's quality approach?
20. What are the contributions of Ishikawa in quality literature?
21. What are the contributions of Ohno in lean production and quality literature?
22. What are the two pillars of lean house?
23. What concepts create the base of lean house?
24. What are the contributions of Shingo in lean and quality literature?
25. What are the contributions of Taguchi in quality literature?
26. Discuss Taguchi's off-line and online quality system.
27. What does Taguchi's loss function represent in quality literature?
28. What are the contributions of Crosby in quality literature?
29. What does "zero defects" represent in quality literature?
30. What methodology is suggested by Crosby to minimize the cycle time in manufacturing setting?
31. Discuss the quality understanding in prehistoric age.
32. Discuss the quality understanding before and right after the Industrial Revolution.
33. Discuss the quality concepts that dominated quality improvement efforts between 1900 and 2000.
34. What terms in quality literature emerged right after World War II in manufacturing world.
35. How did Six Sigma journey start at Motorola? What were the driving forces behind Six Sigma efforts at first?

36. What are the requirements for an area to be considered an "academic discipline"?
37. Analyze and discuss those requirements for Six Sigma methodology.
38. Discuss the impact of ISO on developing standards about Six Sigma concept.
39. Define what "quality cost" concept mean in quality literature.
40. Categorize quality costs into groups and define what each group represents in PAF system.
41. Give examples for each category of PAF system.
42. What does "hidden cost" term mean in quality costs?
43. What are the ideal percentages of each quality cost item in PAF system?
44. List performance metrics of quality costs.
45. Discuss the alternative ways to find the minimum level of total quality cost in PAF system.

References

Abdul-Rahman, H., Thompson, P. A., & Whyte, I. L. (1996). Capturing the cost of non-conformance on construction sites. *International Journal of Quality & Reliability Management*.

Alimohamadi, B., & Khorshidi, N. (2009). Six Sigma Perspectives: In Academic World and Business Environment.

Barney, M. (2002). Motorola's second generation. *Six Sigma Forum Magazine, 1*(3), 13–16.

Bendell, T. (1991). *The quality gurus: What can they do for your company?* London: Department of Trade and Industry.

Benedetto, A. R. (2003). Adapting manufacturing-based six sigma methodology to the service environment of a radiology film library. *Journal of Healthcare Management, 48*(4), 263.

Blakeslee, J. A., Jr. (1999). Implementing the six sigma solution. *Quality Progress, 32*(7), 77.

Bryant, J. W. (1986). Customer Oriented Value Engineering (COVE). *Value World, 9*(1).

Çelikoğlu, Y. (2008). A literature review on six sigma (Doctoral dissertation, DEÜ Fen Bilimleri Enstitüsü).

Crosby, P. B. (1979). *Quality is free: The art of making quality certain* (Vol. 94). New York: McGraw-Hill.

Dale, B., Plunkett, J. J. (1993). Quality Costing. London: Chapman & Hall.

Defeo, J. (2017). *Juran's quality management and analysis*. Singapore: McGraw-Hill Higher Education.

Deming, W. E. (1986). *Out of the crisis*. Cambridge, Mass: Massachusetts Institute of Technology, Center for Advanced Engineering Study.

Drucker, P. (1999). *Management challenges for the 21st century*. New York: HarperCollins.

Duncan, G. T., Elliot, M., & Salazar-González, J. J. (2011). Why statistical confidentiality? In *Statistical confidentiality* (pp. 1–26). New York: Springer.

Edmund, M. (2010). Joseph Juran 1904–2008. *Quality Progress, 43*(11), 20–21.

Fallon, C. (1980). *Value analysis*. Washington: Miles Value Foundation.

Feigenbaum, A. V. (1983). *Total quality control*. New York: McGraw-Hill.

Finison, L. J., Finison, K. S., & Bliersbach, C. M. (1993). The use of control charts to improve healthcare quality. *Journal for Healthcare Quality, 15*(1), 9–23.

Folaron, J., & Morgan, J. P. (2003). The evolution of six sigma. *Lean & Six Sigma Review, 2*(4), 38.

Garvin, D. (1987). Competing on the eight dimensions of quality. *Harvard Business Review*, 101–109.

George, M. L. (2002). *Lean six sigma: Combining six sigma quality with lean speed*. New York: Mc-Graw Hill.

Gilmore, H. L. (1974). Product conformance cost. *Quality Progress, 7*(5), 16–19.

Goh, T. N. (1999). "Perspectives on statistical quality engineering". *The TQM Magazine, 11*(6), 461–466.

Goh, T. N. (2002). A strategic assessment of six sigma. *Quality and Reliability Engineering International, 18*(5), 403–410.

Holweg, M. (2007). The genealogy of lean production. *Journal of operations management, 25*(2), 420–437.

Hsieh, C. T., Lin, B., & Manduca, B. (2007). Information technology and six sigma implementation. *Journal of Computer Information Systems, 47*(4), 1–10.

Juran, J. M., & Godfrey, A. B. (1999). *Quality handbook*. New York: Republished McGraw-Hill.

Juran, J. M., & Gryna, F. M. (1980). *Quality planning and analysis*. New York: McGraw-Hill.

Juran, J. M. (1992). *Juran on quality by design: The new steps for planning quality into goods and services*. New York: The Free Press.

Kaufman, J. J., & Becker, R. F. (1981). *Value engineering: An executive overview*.

Kelemen, M. L. (2003). *Managing Quality*. London: Sage.

Levine, D. M. (2006). *Statistics for six sigma green belts with Minitab and JMP*. Upper Saddle River: Prentice-Hall.

Linderman, K., Schroeder, R. G., Zaheer, S., & Choo, A. S. (2003). Six Sigma: a goal-theoretic perspective. *Journal of Operations management, 21*(2), 193–203.

Liker, J. K. (2004). *The Toyota way – 14 management principles from the World's greatest manufacturer*. New York: McGraw-Hill.

Liker, J. K., & Ross, K. (2016). *The Toyota way to service excellence: Lean transformation in service organiza-*

1

tions: Lean transformation in service organizations. New York: McGraw Hill Professional.

Madu, C. N. (1998). *Handbook of Total Quality Management.* Boston, MA: Springer.

Maguad, B. A. (2006). The modern quality movement: Origins, development and trends. *Total Quality Management & Business Excellence, 17*(2), 179–203.

Monden, Y. (1998). Toyota production system: An integrated approach to just-in-time engineering and management press. Norcross, GA: IEE.

Montgomery, D. C. (2009). *Statistical quality control* (Vol. 7). New York: Wiley.

Montgomery, D. C. (2013). *Introduction to statistical quality.* New York: Wiley.

Nasir, S., & Nasir, V. (2005). Analyzing the role of customer-base differences in developing customer relationship management strategies. *Journal of American Academy of Business, 7*(2), 32–38.

Nembhard, I., Alexander, J., Hoff, T., & Ramanujam, R. (2009). Why does the quality of health care continue to lag? Insights from Management Research. *The Academy of Management Perspectives, 23*(1), 24–42.

Niñerola, A., Sánchez-Rebull, M. V., & Hernández-Lara, A. B. (2019). Six sigma literature: A bibliometric analysis. *Total Quality Management & Business Excellence*, 1–22.

Ohno, T. (1988). *Toyota production system: Beyond large-scale production.* Cambridge, MA: Productivity Press.

Ohno, T. (2013). *Taiichi Ohno's Workplace Management: Special 100th Birthday Edition.* McGraw-Hill.

Pakdil, F., Işın, F. B., & Genç, H. (2012). A quality function deployment application using qualitative and quantitative analysis in after sales services. *Total Quality Management & Business Excellence, 23*(11–12), 1397–1411.

Pakdil, F., & Leonard, K. M. (2014). Criteria for a lean organisation: Development of a lean assessment tool. *International Journal of Production Research, 52*(15), 4587–4607.

Pyzdek, T., & Keller, P. A. (2018). *The six sigma handbook* (5th ed.). New York: Mc-Graw Hill.

Reichheld, F. (1996). Learning from customer defections. *Harvard Business Review, 74*(2), 56–69.

Reid, D., & Sanders, N. (2016). *Operations management: An integrated approach* (6th ed.). New York: Wiley.

Schiffauerova, A., & Thomson, V. (2006). Managing cost of quality: Insight into industry practice. *The TQM Magazine, 18*(5), 542–550.

Schroeder, R. G., Linderman, K., Liedtke, C., & Choo, A. S. (2008). Six Sigma: Definition and underlying theory. *Journal of Operations Management, 26*(4), 536–554.

Shewhart, W. (1939). *The statistical method from the viewpoint of quality control.* New York: Dover Publications.

Summers, D. C. S. (2007). *Six sigma: Basic tools and techniques.* Upper Saddle River, New Jersey: Pearson & Prentice Hall.

Taguchi, G. (1986). *Introduction to quality engineering: Designing quality into products and processes* (No. 658.562 T3).

Varkey, P. F., Reller, M. K., Resar, R., & Resar, R. K. (2007). Basics of quality improvement in health care. *Mayo Clinic Proceedings, 82*(6), 735–739.

Wasserman, M. (1977). Sales market Value. *Proceedings, Society of American Value Engineers, 12*, 105–114.

Wedgwood, I. (2015). *Lean sigma: Rebuilding capability in healthcare.* New York: Pearson Education.

Zhang, W., Hill, A. V., & Gilbreath, G. H. (2009). Six Sigma: a retrospective and prospective study. In POMS 20th Annual Conference, Orlando, Florida USA, May (Vol. 1).

Organization for Six Sigma

Contents

F. Pakdil, *Six Sigma for Students*, https://doi.org/10.1007/978-3-030-40709-4_2

Learning Objectives

After careful study of this chapter, you should be able to:

- Describe and understand effective Six Sigma leaders' roles and responsibilities
- Describe and understand effective requirements of Six Sigma organizational structure
- Describe and understand all Six Sigma positions' roles and responsibilities.

2.1 Introduction

Constantly changing managerial and organizational needs of businesses affect the organizational performance at the bottom line. Prior to implementing new practices in business organizations, decision-makers need to make sure that the organizational modifications are successfully done and that the organizational structure is ready to initiate new practices. Without preparing the organizational structure to meet the needs and requirements of the new practices, immediately starting new practices may bring unexpected obstacles and issues, as were experienced in Western organizations in the past. For example, even though the significance of quality control circles was discovered by the Western practitioners, the results of the practice have not been as effective as expected in Western organizations. Practitioners and academicians developed several theories as to why the practice didn't meet the expectations. One of the important reasons identified by the researchers was the lack of an appropriate organizational structure needed prior to starting quality control activities.

Before starting the Six Sigma journey, the top executives need to make sure that the organizational structure is ready for Six Sigma practices. Therefore, this chapter focuses on the needed organizational structure for Six Sigma that should increase success.

Employees did not have a responsibility to improve quality in Taylor's scientific management approach that dominated a majority of the Western production world in the late nineteenth and early twentieth centuries. This lack delegating the responsibility for quality to the employees became a regular behavior and norm in Western management understanding. In these times, the low education level of supervisors and employees appeared as a limitation to implement his scientific management principles, although some were skilled craftsman. As a result, Taylor suggested creating separate planning departments in organizations, separating planning from execution. His practices resulted in planning and inspection departments staffed with engineers, separated from other areas, which eventually increased overall productivity. In the long run, these practices were broadly spread out to manufacturing and service industry. Along with Adam Smith's thoughts, "division of labor" and "specialization" became a norm in many organizations.

A unionized work environment and the structure of management in some industries was another contributing factor in the failure to delegate quality responsibilities to the workers. Excessive division of labor and standardization were other pivotal factors for employees to accept "just do the work" approach. This approach limited employees' potential contributions to company-wide quality improvement efforts. In addition, all quality improvement activities were done by the organizational teams but excluded knowledge and experience of the employees who work in *gemba*. Afterward, it was found that specific problems and issues arose in departments and divisions that could not be solved by supervisors and upper-level managers alone, but also needed employee participation. Japanese management demonstrated that employee participation is necessary and that improvements were best made by relying on quality control circles involving all levels of the organizational workforce. Rather than expecting upper-level decision-makers to solve problems and improve quality, employee involvement in problem-solving processes and quality improvement efforts was found to be a more effective approach. In the 1980s, Western companies integrated this approach into their practices, and many gained significant benefits, such as Motorola.

2.2 Six Sigma Leaders' Approaches and Organizational Vision

As seen in all business settings, if management function of organizing is not conducted effectively and efficiently, Six Sigma project efforts will fail, resulting in low performance. To increase the success of Six Sigma projects, first and foremost, the organizational structure of Six Sigma should be constructed, developed, and improved. The entire organization expects many positive and value-added outcomes out of Six Sigma projects, so creating and establishing organizational structures that are needed in Six Sigma practices should be the highest priority in the entire Six Sigma project plans, *before* starting Six Sigma projects. In other words, without establishing the organizational structure of Six Sigma, the next steps should not be followed in Six Sigma journey; otherwise the likelihood of failure is high.

According to Feigenbaum (1983), quality can be achieved by focusing on a total quality system that includes a company-wide system that aims to increase customer satisfaction and to improve the economic cost of quality. Looking from a total quality control point of view, appropriate organization for Six Sigma requires a company-wide system in which all departments and functions are interconnected and coordinated. Six Sigma efforts should not be isolated or devoted to a few departments in the organization. Ignoring the other departments' roles and responsibilities to achieve the goals of Six Sigma efforts will lead to failure of the project(s). Instead, Six Sigma should effectively disseminate into each function and department in the organizational structure. Leaders must make sure that each and every function and department is included in Six Sigma organization.

Six Sigma efforts also require a well-designed management system in which processes are operated by satisfied and happy employees, and customers and owners are satisfied (Pyzdek and Keller 2018: 8). Six Sigma benefits managers, employees, customers, owners, and investors by creating a comprehensive chain reaction that includes more productive and satisfied employees, shortened product design cycles, and minimized costs. Essentially, Six Sigma practices benefit all related stakeholders. Successful Six Sigma implementation is perceived as a comprehensive transformation and a competitive advantage to the organization. Therefore, leaders need to make sure that each stakeholder equally benefits from Six Sigma efforts and that each stakeholder has opportunities to participate in Six Sigma efforts.

"The philosophy and tools/techniques of Six Sigma are strikingly similar to prior quality management approaches. However, the way Six Sigma is practiced represents a new organizational structural approach to improvement" (Schroeder et al. 2008: 537). Schroeder et al. (2008) link Six Sigma with contingency theory and perceive Six Sigma as a *switching mechanism* in ambidextrous organizational structures, where organic structure is created from mechanistic structure when new ideas are needed. Burns and Stalker's (1961) theory of mechanistic and organic organizations identifies certain structural features based on various organizational dimensions: standardization, formalization, specialization, authority, and task complexity. Successful Six Sigma practices require leaders to implement more organic organizational structure, since organic organizations cope best with uncertainty and unstable work environments. But it does not mean that Six Sigma organizations do not need to have mechanical organizational structure features. For example, high level of standardization and formalization helps Six Sigma teams analyze problems and identify root causes in a more stable and well-defined environment. "Six Sigma should be understood to be a philosophy as well as a scientific approach and this has growing acceptance" (Keller 2001: 100). Therefore, leaders need to make sure that relevant mechanistic and organic organizational structure requirements are implemented as needed prior to starting the journey in Six Sigma.

All practical and scientific findings show that successful Six Sigma deployment requires organizations to implement several factors, systems, and activities prior to starting Six

2

Sigma projects. Pyzdek and Keller (2018) focus on the following items for successful Six Sigma deployment, and it is top management's responsibility to make sure that all these requirements are met in the organization:

1. Leadership
2. Infrastructure
3. Communication and awareness
4. Stakeholder feedback systems
5. Process feedback systems
6. Project selection
7. Project deployment.

Training is a key part of leadership and infrastructure requirements. For example, successful Six Sigma practices require leaders and everyone in the organization to be "data-driven" decision-makers. Therefore, Six Sigma also requires a high degree of training for staff who will apply and use the methodology. If this is not done, the company should not start the Six Sigma journey with widespread implementation. The timing of training is critical to the success of Six Sigma. The material should be taught to the members of the organization whenever they need to learn it. Providing training too early or too late may cause serious problems as well as a lack of retention.

Moreover, an effective organizational structure for Six Sigma requires a systematic hierarchical composition including several positions. As Pzydek and Keller (2018: 21) stated, "successful Six Sigma deployment is always a top-down affair." The positions of the hierarchical system in Six Sigma organizations are:

1. Executive committee
2. Project champions
3. Deployment managers
4. Process owners
5. Master black belts
6. Black belts
7. Green belts
8. Finance representatives
9. Team members.

The roles and positions in Six Sigma organization are presented in ◘ Fig. 2.1 (ISO 13053-1:2011 standard). As seen in ◘ Fig. 2.1, the Six Sigma Steering Committee is at the top level of the hierarchical structure, and the Six Sigma Deployment Manager and Six Sigma Champion are authorized by the Steering Committee to run Six Sigma projects in the organization. The Committee decides upon which projects should be developed and in which order. At the next levels of the organizational structure, project sponsors, master black belts, black belts, green belts, and yellow belts (team members) take different responsibilities in Six Sigma projects.

The degree of coordination and cooperation among the hierarchical levels deter-

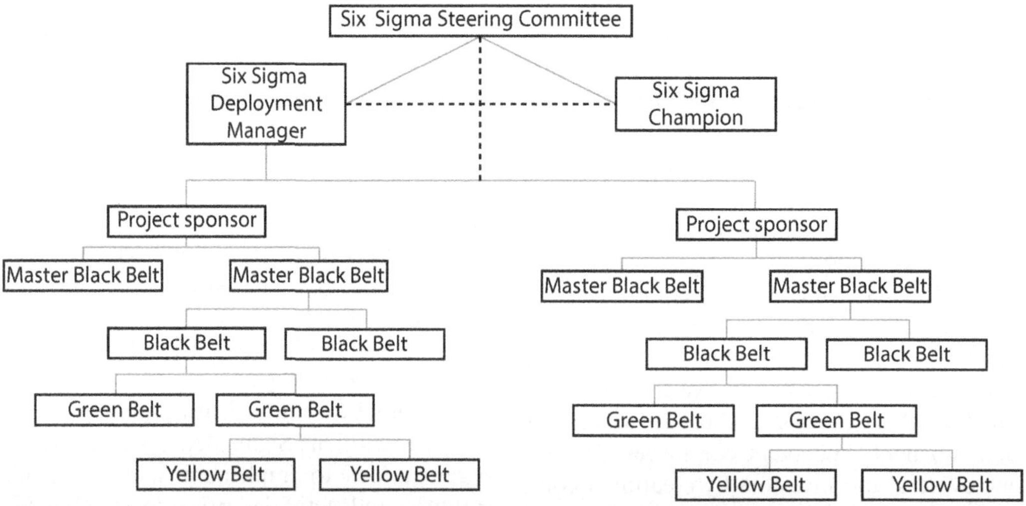

◘ **Fig. 2.1** Positions in Six Sigma organization. (Source: Author's creation based on ISO 13053-1:2011)

mines the quality of project outcomes. The more coordination and cooperation among the relevant actors and functions is created, the higher the performance is observed in Six Sigma projects. This is top management's responsibility to make sure that these organizational requirements are met in the organization. The responsibilities and roles of each position are presented below.

2.3 Roles and Responsibilities in Six Sigma Organization

2.3.1 Executive Committee

As reported by Collins and Porras in *Built to Last: Successful Habits of Visionary Companies*, when the top management of a company seriously focuses on continuous quality improvement efforts, they outperform their counterparts that do not involve in the same effort. George (2002: 63) states that:

> When a CEO shows passion and support, I have never seen Lean Six Sigma fail. If, however, the CEO does not show this passion, I have never seen it succeed. If he or she leaves the initiative up to the divisions to decide to use Lean Six Sigma, it will generally fail to produce breakthrough results. If he or she fails to enforce the commitment of fulltime champion and black belt resources, it will fail.

The CEO makes it clear to everyone in the organization how seriously Six Sigma efforts are regarded by seeing the resources committed to the efforts. The CEO should show the link between corporate strategy and Six Sigma initiatives. The employees should be able to see Six Sigma commitment by the top management and that it is implemented, not only in a few functions but also in the overall organization. The senior committee, senior executive committee, or executive committee are authorized by the top management team to provide direction for Six Sigma projects in the organization to succeed. This committee is called the Six Sigma Steering Committee in the ISO 13053 standard. Successful Six Sigma projects receive full support and commitment from the top management. Clear direction, full engagement, and support in Six Sigma projects are the main roles of executive committees.

Even though the awareness and initiatives of Six Sigma may start at lower levels in some cases, when these initiatives receive full support, commitment, and leadership from top management, the outcomes of the projects are more effective and efficient. The members of the executive committee play a critical role in Six Sigma projects. The degree of support received from the executive committees significantly determines the sustainability of Six Sigma projects. The executive committee approves project selections and suggests potential impactful projects. The lack of support or interest in executive committee members may result in lack of motivation among employees in the short term and devastate Six Sigma efforts in the long term.

Commitment also includes members' participation in Six Sigma training sessions. To support training and efforts of employees, top executives and the executive committee members should systematically be trained in Six Sigma methodology. Executives and members are expected to cascade knowledge and information to lower hierarchical levels of the organization through effective training sessions managed by the executive committee members.

The executive committee members' job description includes the following tasks and responsibilities:

- Learn Six Sigma methodology and teach the other people involved in Six Sigma in the organization.
- Deploy Six Sigma philosophy and manage entire transformation process in the organization.
- Lead project champions, master black belts, black belts, and green belts.
- Set the general direction of Six Sigma projects that will be organized in the organization.
- Remove barriers and obstacles to Six Sigma projects.
- Provide teams with the required resources.
- Develop and improve Six Sigma understanding in the organization.

- Analyze the reports of the Six Sigma teams and accept solutions suggested by the teams.
- Review performance of Six Sigma projects throughout the organization.

2.3.2 Project Champions

Project champions are the leaders in conducting and managing Six Sigma projects. Champions are trained business leaders who oversee the deployment of Six Sigma throughout the organization. Prior to taking this role, champions should take intensive training courses in Six Sigma methodology, techniques, and tools. Champions are executive committee members and function as a link between the committee and individual projects. Project champions help Six Sigma teams in training the members and removing barriers preventing the team from effectively and efficiently functioning. The project champions are more active and more dominant in the Define phase of the DMAIC process. "In large organizations Six Sigma is led by a full-time, high level champion, such as an executive vice president" (Pyzdek and Keller 2018: 27). The project champions' job description includes the following tasks:

- Identify and direct Six Sigma projects with Six Sigma teams.
- Identify and eliminate organizational barriers and obstacles faced by Six Sigma teams.
- Help Six Sigma team develop project goals, objectives, and charter.
- Review project objectives and project charters with the executive committee.
- Assign black/green belts to lead and manage Six Sigma projects.
- Remove barriers, find a remedy, or bring help to Six Sigma team.
- Help team and the executive committee communicate.
- Review the performance of the Six Sigma team.
- Coach and help the Six Sigma team.

According to the ISO 13053-1:2011 standard, project champions should be a senior member of the organization and influential within the organization. Project champions are charge of determining strategies for Six Sigma initiatives and implementation. They are also responsible for setting and promoting Six Sigma-related business objectives.

2.3.3 Deployment Manager

ISO 13053-1:2011 contains a position called deployment manager that is essential in Six Sigma organizations. The roles attained to this position in the standard are:

- Promote the Six Sigma initiative.
- Determine, along with senior management, the nature of the expansion of Six Sigma within the company; the size of the populations of master black belts, black belts, green belts, etc.; and the duration of the secondments for these personnel.
- Liaise with and report to senior management about the progress of any Six Sigma initiative.
- Involve new project sponsors and recruit new master black belts and black belt candidates for the purpose of Six Sigma.
- Negotiate with the different areas of the company for the secondment, and later the redeployment, of the candidate black belts.
- Manage any facility that is provided for the pursuit of Six Sigma, e.g., a Six Sigma center, for the use of the master black belts and the black belts.
- Seek potential projects.
- Participate in "major" gate reviews, as necessary.

2.3.4 Process Owners

Process owners lead and manage the processes of resolving problems or issues. They are also known as project sponsors in the ISO 13053-1:2011 standard. Process owners are responsible for improving the processes and making sure that the processes are error-free. Collaborating with project champions, master black belts, black belts, and green belts, process owners work in Six Sigma projects to create value for their own processes. They are responsible for implementing the solutions suggested by Six Sigma teams. The process owners are more active and dominant in the Control phase of the DMAIC process. The process owners' job description includes the following tasks:

- Develop project goals, objectives, and charter with Six Sigma team members.
- Monitor and analyze the process and its outputs with team members.
- Empower team members to monitor and analyze the process and collect and analyze data.
- Analyze data with team members.
- Help the team follow project charter.
- Go through DMAIC process with the team members.

In addition to these, the ISO 13053-1:2011 (p.10) standard identifies process owners' roles and responsibilities as follows:

- Champion the Six Sigma methodology with peers and others higher in the organization.
- Support the nominated Six Sigma project.
- Provide resources requested by the black belt and required for the Six Sigma project.
- Remove any "road blocks" encountered by the black belt in discharging the project.
- Participate in all gate reviews directly and sign off on the phase when the work has been done properly.
- Ensure the full implementation of all recommendations of the Six Sigma project.
- Ensure that improvements identified within the nominated projects are realized and maintained.
- Ensure that completed projects are evaluated for potential application across other businesses or elsewhere within the same business.

2.3.5 Master Black Belts

Master black belts are highly trained and proficient leaders and technical experts who develop, improve, and deploy Six Sigma methodology in the organization. They are responsible for advising senior executives, black belts, process owners, and green belts. They also train, educate, and mentor black and green belts in the organization, functioning as a change agent throughout the organization. They report the progress of organization-wide Six Sigma projects to the executive committee and upper levels of the hierarchy. The master black belts are active and dominant in all of the DMAIC processes and are trained in all aspects of Six Sigma. Black belts and green belts can train the other members in Six Sigma projects under the supervision of master black belts. Master black belt's job description includes the following tasks:

- Follow DMAIC process phase by phase.
- Advise and consult executive committee as well as other positions in Six Sigma organization.
- Mentor and train black belts and green belts.
- Improve and support all Six Sigma activities and organization.
- Help executive committee prioritize potential Six Sigma projects.
- Help improve awareness of Six Sigma throughout the organization.

In addition to these, the ISO 13053-1:2011 (p.10) standard identifies master black belts' roles and responsibilities as follows:

- Coach and mentor the black belts in the application of the DMAIC methodology and the selection and use of the tools and techniques required.
- Provide "internal" consultancy in advanced statistics.
- Assist in the identification of suitable improvement projects.
- Assist in the determination of the scope of the selected improvement project.
- Assist in periodic reviews of the improvement projects.
- Provide training in the tools and techniques associated with Six Sigma to black and green belts as required.
- Determine if any training activities are appropriate and effective.
- Lead improvement projects as required.

2

2.3.6 Black Belts

The black belts are quality professionals who are mentored by a master black belt. Black belts are also considered change agents, taking an improvement leadership role throughout the project period. They implement proven advanced methods and train and mentor green belts. They actively work in project teams, and can come from all disciplines that require college-level mathematics knowledge in the organization, but they are not supposed to be just statisticians. Black belts are trained on 3- to 6-week curricula in Six Sigma. They are trained on methods and through real projects to apply the knowledge they gain through training sessions. Since learning is a function of time, the material should be taught to black belt candidates over a time frame where they can easily digest the material taught each week. Black belts are active and dominant in all of the DMAIC processes. Black belts lead Six Sigma projects and ensure that the team's efforts are likely to result in beneficial outcomes for the organization and function as a consultant for Six Sigma teams. They are expected to answer Six Sigma methodology questions in the team. Black belts should have good soft skills, such as oral and written communication skills. Six Sigma projects and teams require a high level of communication skills among black belts. Additionally, black belts are expected to have high levels of conceptual and technical skills. A black belt's job description includes the following tasks:

- Follow DMAIC process phase by phase.
- Help Six Sigma team prepare project charter.
- Lead the project and team.
- Communicate with all Six Sigma positions and inform them of the progress of the team.
- Schedule meetings.
- Train Six Sigma team members.
- Provide Six Sigma team with technical knowledge.

The ISO 13053-1:2011 (p.10) standard identifies black belts' roles and responsibilities as follows:

- Work with others to identify and quantify opportunities for improvement.
- Organize multidisciplinary teams (process organization), where necessary, and manage improvement projects.
- Lead improvement projects or facilitate Green Belt Projects using the DMAIC methodology.
- Train, coach, and mentor green belts on DMAIC methodology and associated process improvement techniques.
- Participate in all gate reviews directly through prepared presentations of the work accomplished to date with an emphasis on the accomplishments in the phase being reviewed.

2.3.7 Green Belts

Green belts are Six Sigma project leaders and work in projects part-time. They may be a team member or a project leader. Levine (2006) states that most managers working in organizations that implement mature Six Sigma projects are green belts. Green belts are also fully trained employees who are capable of applying Six Sigma skills to projects in their work areas. The green belts are more active and dominant in the Measure, Analyze, and Improve phases in the DMAIC process. They communicate with the other teams to learn what they work on in Six Sigma projects. The essential responsibilities of green belts are (Levine 2006):

- Follow DMAIC process phase by phase.
- Finalize project charter.
- Review and improve the project charter with project champions.
- Select the members of the team.
- Communicate with other members of the organization such as master back belt, black belts, and process owners to run the project.
- Facilitate and help the team throughout the project period.
- Help team members analyze the data.
- Train team members on Six Sigma tools and methodology.

Training for green belts includes such diverse topics as project management, problem-solving tools, and statistical data analysis tools. Similar to black belts, they also function as a change agents in the organization. Throughout the project term, green belts are supported by black belts when they need help, since they are not necessarily trained in all Six Sigma tools. The roles and responsibilities assigned to green belts in the ISO 13053-1:2011 standard are:

- Work with the local "line management" to identify and quantify opportunities for improvement within the local environment.
- Work under the direction of a black belt as a member of a larger Six Sigma project led by the black belt.
- Lead a smaller Six Sigma project under the direction of a black belt.
- Possibly coach process operators (yellow belts) on process improvement methods and activities.

2.3.8 Finance Representatives

Finance representatives are responsible for developing the financial portrait of Six Sigma projects. Cost-benefit analysis is one tool that finance representatives heavily rely upon in Six Sigma projects. The teams expect finance representatives to estimate potential benefits and costs of the projects, as well as outline resource requirements, including human resource needs. As one of the crucial outcomes of Six Sigma projects, financial benefits (savings) and potential losses of the projects should be estimated prior to kicking off the projects. In other words, financial representatives are responsible for financially validating the potential projects. The financial representatives are more active and dominant in the Improve and Control phases in the DMAIC process. Although calculating the cost of the potential project might be simpler, calculating all expected benefits may not be.

2.3.9 Team Members

Team members are employees who experience issues and problems in the relevant processes under consideration. They operate in Six Sigma projects with all other actors by following the Six Sigma methodology. They are trained in Six Sigma methodology and tools. In the entire project session, they are the ones capable of analyzing the processes where the problems develop that affect quality. Team members are more active and dominant in the Measure and Analyze phases in the DMAIC process, expected to analyze and identify root causes of the problem. Team members' job description includes the following tasks:

- Follow DMAIC process phase by phase.
- Identify root causes of the problem with green belt and black belt.
- Follow project charter and develop potential solutions for the problem analyzed in Six Sigma team.

The ISO 13053-1:2011 standard uses the term *yellow belt* for team members. The roles and responsibilities assigned to team members in this standard are:

- Work with the local green belt to identify and quantify opportunities for improvement within the local environment.
- Work under the direction of a black belt or a green belt as a member of a larger Six Sigma project led by the black belt.
- Participate in a smaller Six Sigma project under the direction of a green belt.

◘ Table 2.1 demonstrates where all these positions take relatively more active roles and responsibilities in DMAIC process. Each position in a Six Sigma organization requires different skill sets. Katz's (1974) managerial skills model can be used to categorize these skills in a structural way. Furthermore, Katz's managerial skills model contains three core skills: technical skills, human skills, and conceptual skills. Technical skills refer to specialized skills and knowledge about a profession and include specialized knowledge of using particular tools, methods, and techniques used in the proficiency. Technical skills in Six Sigma are associated with the technical details of DMAIC process. For example, the knowledge of hypothesis testing, project charter, or regression analysis implies technical skills for Six Sigma practitioners. Human skills refer to the ability of effectively and efficiently working

2

Table 2.1 DMAIC process

Define	Measure	Analyze	Improve	Control
Executive committee Project champions Master black belts Black belts Team members	Black belts Green belts Team members	Black belts Green belts Process owners Team members	Black belts Green belts Process owners Financial representatives Team members	Process owners Financial representatives

Source: Author's creation

with subordinates, superiors, and groups. Each position is expected to have a high degree of human skill sets to perform successfully in Six Sigma projects. For example, the ways the project champions communicate with master black belts and team members explicitly show the quality of human skills of the project champions. Conceptual skills imply the ability of seeing the entire system as a whole. It also refers to seeing the interrelationships between functions and departments and predicting the effects of changes on the components of the system. In a Six Sigma organization, higher-level positions such as executive committee, project champions, deployment managers, and process owners need to have higher conceptual skills.

Key Concepts

Mechanistic and organic organizational structure, executive committee, project champions, process owners, master black belts, black belts, green belts, finance representatives, team members, and DMAIC.

Summary

Quality discipline has traveled from a time period in which employees would not have any responsibility to improve quality to another time period in which employees have an authority to stop manufacturing processes in case they detect any abnormalities in the processes. In this transforming journey, Six Sigma projects require business organizations to establish a well-defined organizational structure that includes various positions. This chapter represents the responsibilities of these positions.

Practice and Discussion Questions

1. Discuss the responsibilities assigned to employees in Taylor's scientific management philosophy.
2. What is expected from employees in quality control circles?
3. Compare Taylor's scientific management approach and quality control circles in terms of employee participation and empowerment.
4. What are the roles and responsibilities of top executives in Six Sigma?
5. What are features of mechanistic organizations?
6. What are features of organic organizations?
7. Analyze Six Sigma philosophy based on mechanistic and organic organizations.
8. Discuss Pyzdek and Keller's (2018) items for successful Six Sigma deployment.
9. What are the positions of Six Sigma organization?
10. Describe the roles and responsibilities of executive committee in Six Sigma.

11. Describe the roles and responsibilities of project champions in Six Sigma.
12. Describe the roles and responsibilities of deployment managers in Six Sigma.
13. Describe the roles and responsibilities of process owners in Six Sigma.
14. Describe the roles and responsibilities of master black belts in Six Sigma.
15. Describe the roles and responsibilities of black belts in Six Sigma.
16. Describe the roles and responsibilities of green belts in Six Sigma.
17. Describe the roles and responsibilities of finance representatives in Six Sigma.
18. Describe the roles and responsibilities of team members in Six Sigma.
19. What kind of managerial skills are expected for each position that takes place in Six Sigma organization? Discuss these skills using Katz's managerial skills model.

References

Burns, T., & Stalker, G. M. (1961). *The management of innovation*. London: Tavistock.

Feigenbaum, A. V. (1983). *Total quality control*. New York: McGraw-Hill.

George, M. L. (2002). *Lean six sigma: Combining six sigma quality with lean speed*. New York: Mc-Graw Hill.

ISO 13053-1:2011: Quantitative methods in process improvement — Six sigma — Part 1: DMAIC methodology. ISO Standard.

Katz, R. L. (1974). Skills of an effective manager. *Harvard Business Review, 52*, 90–102.

Keller, P. A. (2001). Recent trends in six sigma. *Annual Quality Congress Proceedings*, 98–102.

Levine, D. M. (2006). *Statistics for six sigma green belts with Minitab and JMP*. Upper Saddle River: Prentice-Hall.

Pyzdek, T., & Keller, P. A. (2018). *The six sigma handbook* (5th ed.). New York: Mc-Graw Hill.

Schroeder, R. G., Linderman, K., Liedtke, C., & Choo, A. S. (2008). Six sigma: Definition and underlying theory. *Journal of Operations Management, 26*(4), 536–554.

Cultural Considerations for Effective Six Sigma Teams

Karen Moustafa Leonard

Contents

F. Pakdil, *Six Sigma for Students*, https://doi.org/10.1007/978-3-030-40709-4_3

3

Learning Objectives

After careful study of this chapter, you should be able to:

- Compare two Six Sigma teams working in different organizational, professional, or societal cultures
- Evaluate the management of two Six Sigma teams working in different cultures
- Identify the cultural changes needed for effective Six Sigma teams.

3.1 Introduction

An excellent definition of culture is by Schein (2010), who said that culture is simply the way that people have agreed to do things in a group. The groups discussed in this chapter are organizational, professional, and societal. It is essential that these cultures be considered before starting Six Sigma implementation and when working in Six Sigma teams. Cross-cultural problems arise because the differences and resulting complications are not acknowledged, accepted, and solved. The final section of this chapter is a discussion of methods to make changes that may be needed in culture before a Six Sigma team can be effective.

3.2 Different Faces of Culture

Culture is a concept, which means that it cannot be seen or touched; it operates on the unconscious level within societies, groups, and individuals (Bate 1984; Hofstede 1990). Culture has been evident since the dawn of time, when people first noticed that other groups did things in different ways, based on the types of people, skills, food, climate, and other parts of the individual and environmental settings around them. Many who write about culture call it, simply, "how we do things around here." Culture results from an adaptation to an environment, in this case, to an organization, profession, or society. It is "a system of [a.] shared values defining what is important, and [b.] norms, defining appropriate attitudes and behaviors, that guide members' attitudes and behaviors" (O'Reilly and Chatman 1996, p. 166). Culture allows predictable, shared valued, norms, and beliefs (Swidler 1986) which affect social, professional, and organizational behavior (Steel and Taras 2010; Taras et al. 2010). It is stable, becoming a part of the history that has allowed the group to survive (Hofstede et al. 2010), but at the same time dynamic, changing as needed through individual exchanges (Beck 2000). Although it is dynamic, new behaviors or structures are not easily changed (Cameron and Quinn 1999; Hofstede et al. 2010; Swidler 1986). Similarly, any new, organization-wide changes are not easily implemented, as seen in Six Sigma practices. If the societal, professional, or organizational culture is not ready, change will not occur. It is difficult to import new ideas into culture, particularly if those ideas are not consistent with existing social culture (Westney 1987; White et al. 1999; Wittrock 2015). This is clear in the attempts to import Japanese management into US organizations (Pakdil and Leonard 2015, 2016).

Culture is found in workgroups, teams, departments, organizations, professions, and societies. Each of these different levels of culture has been studied by various prominent researchers, although there is still considerable work to be done in the area of professional culture. Culture can help us understand what effects it can have on Six Sigma implementation, Six Sigma teams, on their formation, choices of problems, processes, and outcomes.

3.3 Organizational Culture

Aspects of organizational culture can enable or block Six Sigma team formation, decision-making, process realignment, and implementation programs (Crandall and Crandall 2011). Implementation of Six Sigma changes cannot be imposed as though introducing a new machine because support for these efforts must be supported throughout the organization; it is a precondition of successful Six Sigma implementation (Dahlgaard and Dahlgaard-Park 1999). This is true for any successful effort requiring collaboration of the entire organization (Cameron and Quinn 1999). Before initiating any effort, the organizational culture must be examined to ensure that it can sustain the changes (Zu et al. 2010). If the culture does not support change, then the culture must be altered (Shook 2010).

Organizational culture is the accepted way to do things in the organization where you work. It affects how individuals behave at work (Calori and Samin 1991; Yin-Cheong 1989). Cultures can be very different from industry to industry and from organization to organization. If you consider the retail clothing industry, it is a very different feel to walk into a discount store (such as Walmart, Lojas Americanas, or Reitangruppen AS) than to walk into an upmarket store (Louis Vuitton, Maison Margiela, or Prada). Discount stores have different strategies and objectives than do upmarket ones. Each has a mission to serve customers in a designated niche, gender, or socioeconomic group. This means that their expected behaviors, that is, their cultures, are different.

"How we do things around here" is one of the first things that new employees are taught when they begin work in an organization (Schein 2010). Organizational culture is defined by rituals, processes, and stories that people tell in the workplace (Beyer and Trice 1987; Meyer 1995; Schein 2010). Sometimes this is given to them through an orientation, but it is further solidified as they work with others and become socialized. However, the socialization may not be what is intended by the manager or top management team. For example, if a new employee works with a sour, cynical employee, then they might adapt the behaviors of that individual.

Socialization has a huge effect on workgroups, teams, departments, and organizations (Bate 1984; Calori and Samin 1991; Jablin 2001), but it is often one of the least considered concepts when identifying effects on behaviors in organizations. However, shared values lead to efficiency, in the long run, because it helps employees determine which behaviors are acceptable and which ones they should expend in different workplace situations so that they fit into the organization or subgroup (Calori and Samin 1991; Denison 1984; Glisson and James 2002; Jehn et al. 1997). Those who are poorly socialized may not fit into the organization and are considered outsiders, which is detrimental to proper organizational or subgroup operations (Myers and McPhee 2006). Cultures within organizations are controlled by those who engage in activities in the particular culture, so a culture can be defined at the total organization level, at the department level, or even at the workgroup or team level.

Multiple researchers have categorized organizational culture, and many provide valuable insights, such as Kets de Vries and Miller (1986), Martin (1992), and Quinn and Rohrbaugh (1983). However, Quinn and Spritzer (Quinn and Spreitzer 1991) suggested the Competing Values Framework, which is currently the most common way to visualize culture and, if needed, to change it. Cameron and Quinn (1999, 2011) visualized four types of organizational culture from spatial analysis of effectiveness measures, each valuable within a certain type of organization, but all needed in some amount within an organization for stability. Each quadrant in their framework has similarity with more complex representations of culture (Cameron and Quinn 1999), allowing a more complete, although simplified, view of culture in an organization (Zammuto & Krakower, Quantitative and qualitative studies of organizational culture, 1991). Effectiveness is listed along a continuum, with control and stability at one end and discretion and flexibility at the other. Focus is listed along an intersecting continuum, with external focus, differentiation, and independence at one end and internal focus, consistency, and cohesion at the other (Cameron and Quinn 1999). This creates the four quadrants of the Competing Values Framework: group (also called clan), developmental, hierarchical, and rational (market). By looking at each quadrant, it is easy to identify the core values of each type of culture. This Framework is illustrated in ◘ Fig. 3.1.

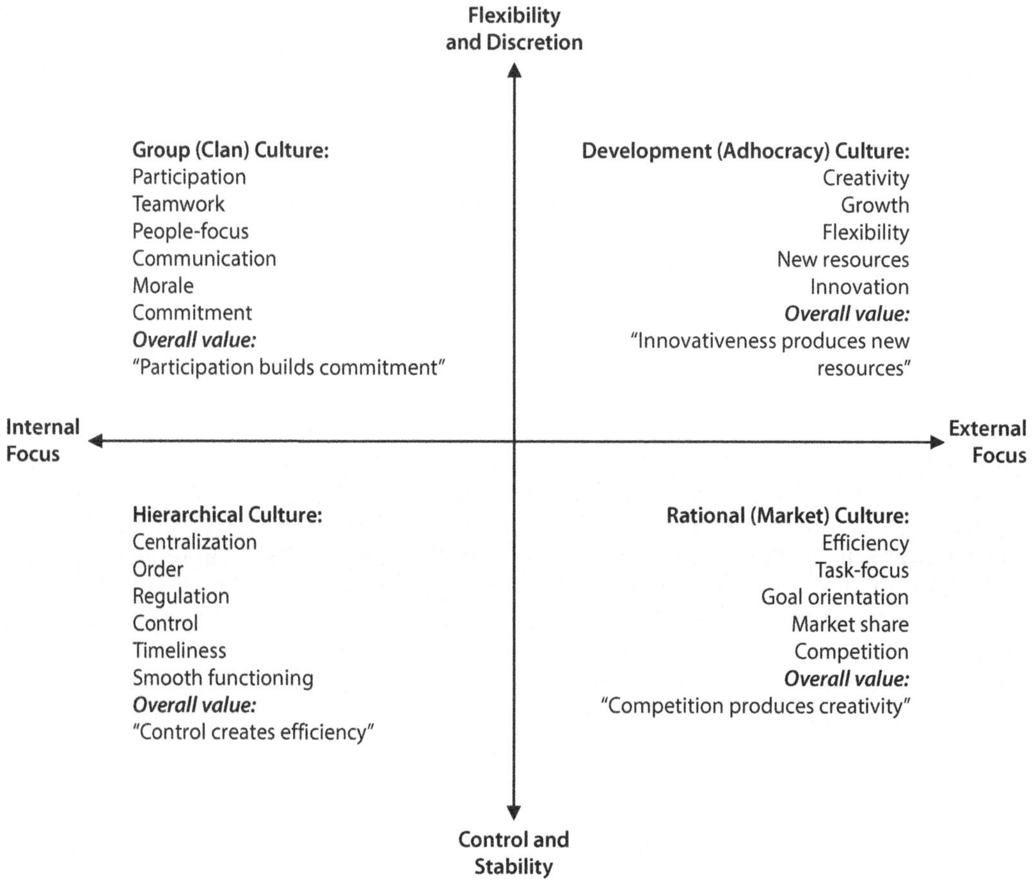

Fig. 3.1 Competing values framework. (Source: Adapted from Quinn and Spreitzer 1991)

Group (Clan) Culture The family-type values and structure in group culture liken it to a clan (Cameron and Quinn 2011). In the model, there is a focus on the employee, flexibility, and teamwork. Participation and collaboration drive the firm forward in organizations with this culture (Ouchi 1980; Ouchi and Johnson 1978). Generally, quality circles and other teamwork-based efforts are used to address problems, employees are given voice within the processes, and rewards are based on teamwork rather than individual work. In group cultures, organizational goals are defined by needs of the firm, customer, and employees, so all are considered partners (Cameron and Quinn 2011).

Maintaining group culture requires high internal agreement on organizational goals, and the firms often have unclear performance expectations, due to the nature of the firm (Ouchi 1980). Within this culture, there is high morale, job rotation, employee participation, knowledge sharing, and flatter hierarchies than in other cultures, making it most effective for firms needing high levels of participation and morale (Cameron and Freeman 1991; Cameron and Quinn 2011).

Six Sigma implementations often fail when transferred to firms without the respect for people that is a hallmark of the group culture (Liker & Franz, The Toyota Way to continuous improvement, 2011). This is because implementing and sustaining changes required by Six Sigma decisions needs continuous quality improvement embedded within the culture as well as the ability among groups to build consensus (Liker et al. 1999). It is the employees who actually improve systems through collaboration and group decision-making. However, as we will discuss further, a firm with **only** group culture may not be optimum for implementation and sustainability of Six Sigma processes. Therefore, a firm characterized by group culture will have a high degree of employee involvement and teamwork (Pakdil and Leonard 2015), which creates an effective base for successful Six Sigma implementation.

Developmental Culture Sometimes, developmental culture is called adhocracy or entrepreneurial culture because it is the most responsive to competitive forces because of its focus on the external environment and its high flexibility (Cameron and Quinn 2011). This culture emphasizes employees, future preparation, and risk. It is useful for high-risk organizations, such as technology firms, because of the vibrant nature of their environment. Such high-risk organizations require every employee to be empowered, that is, developed and trained, because a change in project needs a change in structure, resulting in flexible (or no) hierarchical charts, offices, or job roles (Quinn and Cameron 1983). Developmental cultures, with entrepreneurial leadership, flexibility, creativity, and innovation, allow high-risk organizations to succeed (Cameron and Quinn 1999).

Six Sigma processes are encouraged in developmental cultures, because managers are responsible for employee development as well as for performance (Liker and Franz 2011). Decentralization from the flatter hierarchical structures allows front-line employees to feel more trusted, resulting in empowerment and freedom within the structure (Naor et al. 2010). For example, in developmental cultures, employees feel more empowered to approach managers about work problems without the usual reserve found in more centralized environments. When the manager is a facilitator rather than a micromanager, empowerment leads to higher performance (Naor et al. 2010) because employees are given authority and have the training and education to be involved in various planning and decision processes (Dahlgaard and Dahlgaard-Park 1999). Ideas and improvement suggestions can move freely within the hierarchy when the organizational culture is developmental (Dennis 2002). Therefore, a developmental culture will encourage decentralization, creativity, and effective problem-solving processes (Pakdil and Leonard 2015). The components of developmental culture support successful Six Sigma implementation efforts in business organizations.

Hierarchical Culture This was the first culture defined in scientific management (Weber 1947), with strict rules, specialized training and tasks, impersonality, and meritocracy – in other words, bureaucracy. It was the main form of organizational culture through the late 1960s (Cameron and Quinn 1999). There is low flexibility and little internal focus, but instead, high uniformity, strict rules, and consistent output. Elimination of waste and redundancy results from effectiveness and predictability (Cameron and Quinn 1999). Employees are socialized into strict behaviors, with written and unwritten rules of behavior (Mehri 2006).

In hierarchical cultures, quality strategies involve rules, with measurement and error detection key, along with controls of the process to prevent errors and abnormalities (Cameron and Quinn 2011; Dennis 2002). There is no equality and everyone knows their rank in the hierarchy; decision-making is left with the leadership (Hall and Hall 1987; Naor et al. 2010). Most governments are hierarchical, with decisions made along the decision tree. There is little opportunity to deviate from accepted, routinized decisions. Therefore, the hierarchical culture will encourage control and standardization. Using a standardized DMAIC process in Six Sigma effort may be aligned with the hierarchical culture. For example, developing a project charter, identifying improvement goals, building a well-defined organizational structure, and following the phases of DMAIC in Six Sigma show that Six Sigma efforts may support a hierarchical culture.

Rational Culture Also called market culture, rational culture is based on concepts Ouchi (1981) and Williamson (1975) developed; they identified this culture as supporting activities creating effectiveness and targeting selective customers, with low flexibility and an external focus (Cameron and Quinn 2011). The focus on a hostile environment means that customers, suppliers, and regulators are the primary consideration in its decisions (Cameron and Freeman 1991; Cameron and Quinn 2011), making them very efficient when there is a great deal of conflict about goals, but performance is relatively well understood (Ouchi 1980).

The rational culture core values of competitiveness and productivity aid in their focus on profit, allowing them to concentrate on customers and on improving their competitive advantage (Cameron and Quinn 2011). Their leaders focus on winning and achievement. This culture uses strategies to improve productivity, measure preferences of customers, and involving customers and suppliers (Cameron and Quinn 2011). The business organization culture of most restaurants is usually rational, as the focus on efficiency and productivity are necessary to be successful in such a competitive industry. Rational cultures are most supportive of efficiency, continuous quality improvement, and profit. Considering the core values of rational culture and main goals of Six Sigma methodology, it is obvious that Six Sigma is also aligned with rational culture. Focusing on customer needs and expectations, efficiency, productivity, and continuous quality improvement shows that rational culture is associated with Six Sigma philosophy.

Balanced Culture Few groups or organizations operate at extremes. Using Six Sigma effectively requires a balanced culture, as each culture type described by Cameron and Quinn have some elements necessary to manage the process appropriately. A balanced culture is one where values within each of the four quadrants is strong; that is, uniformity is valued, but so is flexibility, where appropriate within the organizational processes. Six Sigma processes have a requirement to consider both internal and external environments for successful implementation of change. Cooperation is a hallmark of Six Sigma, but all organizations require control (Mann 2005). There will be an emphasis on one (or maybe two) of the cultural types in almost every organization, based on the mission, vision, and strategy of the organization (Cameron and Quinn 2011).

An example of a balanced culture is found at Toyota, which promotes this organizational culture type (Liker and Franz 2011; Liker et al. 1999; Mehri 2006). Because of the balance within each quadrant, true organizational effectiveness is possible (Cameron and Freeman 1991; Zammuto and Krakower 1991; Zu et al. 2010). Balance improves productivity because "too much of anything can have negative consequences" (Quinn and Spreitzer 1991, p. 128). The authors noted that weaker quadrants should be improved through resource allocation, to bring organizational culture back into balance, suggesting that a balanced culture will have the most effective processes, when compared with single quadrant cultures. The more balanced the culture within the four quadrants, the more likely that Six Sigma processes will be successful and sustained (Pakdil and Leonard 2016). An example of a culture with a weak hierarchical culture and a relatively weak rational culture is shown in ◘ Fig. 3.2, while a balanced culture is demonstrated in ◘ Fig. 3.3.

Exercise 1

After surveying an organization, you obtain the following results: (1) hierarchical culture 20/100; (2) market culture 60/100; (3) developmental culture 50/100; and (4) group culture 50/100. Draw the culture diagram. Is the culture balanced or not? Which culture needs more resources so that the organizational culture is balanced?

Answer

1. Start by using the graph to draw the points on each culture (◘ Fig. 3.4).
2. By observation, we can see that the hierarchical culture needs more resources to create a balanced culture.

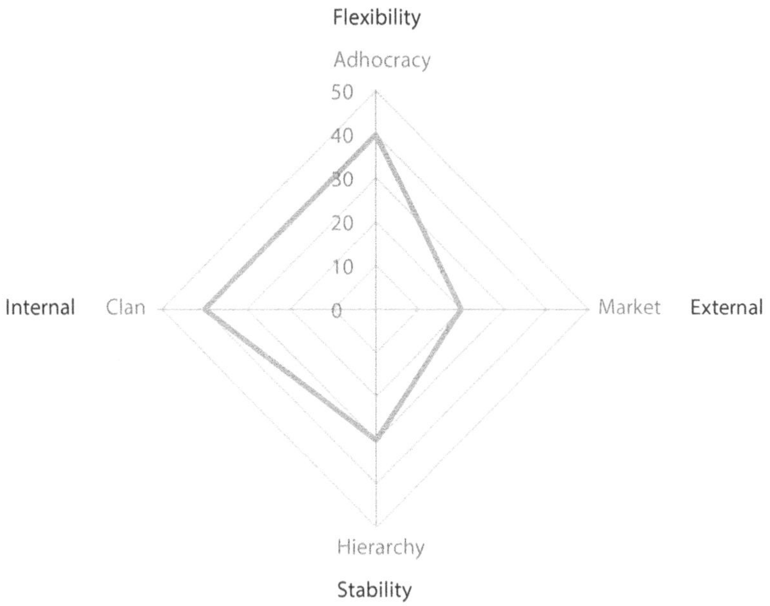

Fig. 3.2 An example of weak hierarchical and relatively weak market cultures. (Source: Adapted from Demir et al. 2011). Notes: In this figure, group (clan) culture, 40; development (adhocracy) culture, 40; rational culture, 20; hierarchical culture, 30

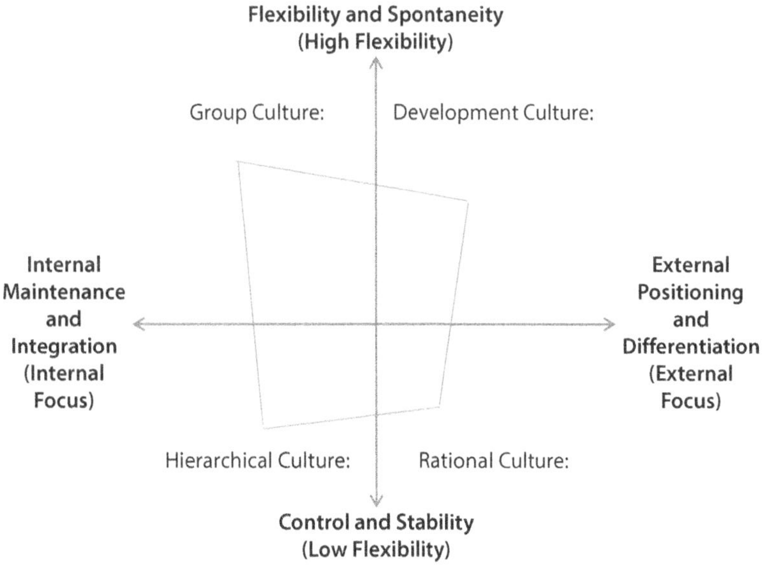

Fig. 3.3 An example of a more balanced culture. (Source: Adapted from Cameron and Quinn 1999)

3.4 Professional Culture

Professional culture is, as it states, how things are expected to be done in a profession. Some professions have strong cultures with strict rules about what is appropriate and inappropriate, developing from shared education and experiences of those in the same occupation or profession (Brief and Nord 1990; Greenwood 1957). For example, nursing and accounting professions have very strict rules about behavior. These rules often supersede the organizational culture's rules, causing conflict (Trice 1993). For example, if the hospital's culture is all about making money, the nurse will still behave in an appropriate manner toward the patient and his/her care or leave the hospital. If an accountant is asked to change input on a spreadsheet in a manner that is contrary to good practice, then he is likely to refuse and be fired or leave the firm, as was seen in the high turnover in Enron (McLean and Elkind 2003).

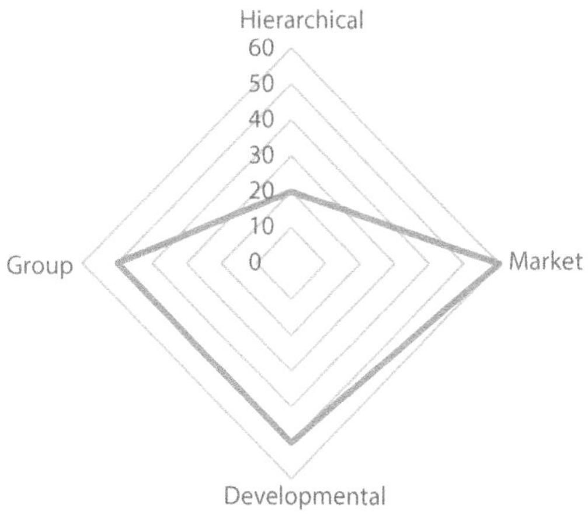

Fig. 3.4 An organizational culture radar chart. (Source: Author's creation)

Professional socialization is similar to organizational socialization; newcomers observe others to obtain cues about behavior (Bandura 1977; Olson-Buchanan and Boswell 2008). In general, they behave as they see others behave, both positive and negative, because these patterns are generally well established (Halbesleben 2009; Olson-Buchanan and Boswell 2008).

Professional culture is still under investigation among academics. Some believe that there is a professional culture set up in organizations by workers performing the same tasks, such as housekeeping, and that this culture may persist across different organizations in the same industry, even though the workers do not correspond with each other. It is a professional culture based on the needs of the tasks within that industry (Greenwood 1957; Meyer 1995).

Some work on cultural dimensions has been done on various professions, such as nursing, fire fighters, and police (Fullerton et al. 1992; Glisson and James 2002; Myers and McPhee 2006; Rothwell and Baldwin 2006). However, all professions have a culture, and these may be strong or weak, based on the type of rules, the enforcement mechanisms, severity of penalties for violation, and the cohesion of the profession's members. The more cohesive and educated the professionals, the more likely that they have a strong professional culture. One example is the medical profession. It has strict rules of behavior, enforcement mechanisms, severe penalties if the rules are broken, and high cohesion among members.

Unions play a role in professional culture among their members. Since culture is the way that things are done, it is easy to see that a strong union can create a strong professional culture within its members. One problem may exist where there are members and non-members working in the same organization. There could be an exchange of culture from the unionized members to the non-members, creating a multiorganizational professional culture within the industry. Once again, the more training and partnership between new and current employees, the more likely that the culture will persist.

Exercise 2

As an administrator and part of the hospital's Six Sigma team, you are asked to work with the nurses on revising the process of giving medications to patients to aim for greater efficiency. What professional cultural differences might you expect to encounter? List three and give ways to negotiate through them.

1. What are the professional culture rules and processes that administrators follow? What may be some differences between administrative and nursing cultures?
   Some differences will include jargon, definitions of efficiency, rules about patient care, financial emphasis, and nursing routines established over time.

For each difference:

Jargon – Be clear about the information given and received. If nursing jargon, such as h.s. (each night) and p.o. (by mouth) are not clear, then ask for explanations. If you are using ROI (return on investment), make sure you use the words and not the initials.

Efficiency – Be clear about efficiency in terms of administration and nursing cultures. Administrative paperwork is unimportant to nurses when faced with a patient who needs care. You must decide how to balance your needs as an administrator with those of nurses attending to patients.

Rules about Patient Care – There are specific nursing rules about patient care, just as there are administrative rules about admission, finance, and other matters. Patient care supersedes the requirements of any administrative dictate, in nursing culture.

Financial Emphasis – For most hospital administrators, the meaning of financial efficiency is to enable the hospital to continue to operate. This may have little meaning for those in the nursing profession, as care of patients overrides all else. However, keeping the hospital viable does mean continuing to care for patients. When administrators limit the number of caregivers for patients in the name of efficiency, nurses have concerns for their patients. The wise administrator can negotiate these financial needs in terms of patient care.

Nursing Routines – Over time, the ways that patients are cared for by nurses has become a standard. Alterations of these routines will require careful thought and reasoning to be successful, but they can be done.

One way to bridge the gap between cultures is to find the best rationale that suits both cultures, but it takes a lot of time, effort, and critical thinking.

3.5 Societal Culture

Researchers use other mechanisms to represent culture, such as whether a person has their own goals or uses the goals designated by the group, which is the individualism dimension discussed by Hall and Hall (1987), Triandis (1989, 1995), and Hofstede (1990). Another is to determine how much structure a person needs in his or her organization, profession, or environment or how much inequality is acceptable. These are uncertainty avoidance and power distance dimensions, respectively (Hofstede et al. 2010; Triandis 2004). In business research and textbooks, these are three aspects of culture that are most often used because they are convenient and fairly easy to study. However, they are not the only ones. There are many different variations of cultural dimensions, but we will use these three in our discussion of societal culture.

Hofstede (1990; 2010) clearly identifies cultural dimensions as being concepts, that is, they do not exist in any solid form. They are merely ways to think about differences between groups of people. These dimensions and their opposites are seen as being on a continuum, where most societies fall along the continuum line and not at either extreme. His work is described in great detail at ▶ https://geerthofstede.com. Here we will only mention the ideas most relative to the discussion of Six Sigma teams: individualism (and its opposite, collectivism), uncertainty avoidance, and power distance (Li et al. 2006).

Individualism (and its opposite, collectivism) Triandis (2002b) stated that the meaning of work varies among social cultures. Hofstede (n.d.) gives dimension maps of the world explaining which countries are more individualistic. It is clear that those countries considered Western, including Australia and New Zealand, are more individualistic, while the most collectivistic countries include China, Ecuador, Colombia, Venezuela, Peru, Indonesia, Pakistan, and Afghanistan, along with some African countries.

In more collectivist countries, pleasing oneself is less important than pleasing the group (Triandis 1995). In individualistic cultures, personal, short-term goals are more important than those of the group. This is because the way that individuals consider themselves in relation to the group is different in individualistic and collectivistic cultures. In individualistic cultures, people most generally see themselves as being independent of any particular group, that is, their identity does not depend only upon the group (Markus and Kitayama 2003). However, most of those in collectivistic societies see themselves as being interdependent with other group members, that is, the majority of their identity is imbedded within the group (Markus and Kitayama 2003).

Teamwork is generally viewed as easier to develop within more collectivistic societies, as members are more used to working within groups (Hofstede et al. 2010). However, in these societies, the groups may be very narrowly defined. For example, if you put someone from an engineering team with another from an accounting team working in the same firm, problems may arise because each is from a different group. They might see each other as from an *outgroup*, where rules are different than when dealing with their *ingroup* members. In some collectivistic societies, lying and corruption are allowed between outgroup members but are highly sanctioned between ingroup members (Li et al. 2006). More individualistic societies do not generally encourage teamwork, but unfair dealings between individuals (whether in the in- or outgroup) are habitually sanctioned with high penalties, which is how the society manages to exist without the tight ingroup bonds of collectivists (Triandis 1995).

Uncertainty Avoidance Avoiding uncertainty is necessary within some cultures that are high on this dimension; therefore, the culture has created structure to reduce uncertainty in interpersonal behaviors. Peru, Argentina, Spain, France, Central Europe, Russia, Japan, Turkey, and Iran are some of the countries that rank usually high on this dimension and cope with uncertainty through ritual and strict rule of behavior (Hofstede, n.d.; Hofstede et al. 2010). For example, the meanings of colors are very clear to members of these societies – giving someone a white carnation as a gift would be considered rude or an insult as they represent purity and grief in Japan. However, a white carnation means innocence, purity, and luck in the USA, if it means anything at all. Sometimes, it is just the color available at the florist, and little thought is given to the actual meaning of the color.

Strict rules were put into place in high uncertainty avoidance cultures to ensure that everyone is in harmony and that there is a structure to society. But in low uncertainty avoidance cultures, it is sometimes difficult to determine the correct rule for behavior (Triandis 2004). Often, the rules are flexible and can even change over time. This can be problematic in Six Sigma teams in lower uncertainty cultures. Leaders are required to create rules within teams so that the work can be done; this may be less necessary in teams within high uncertainty cultures because the rules are already there. However, the rules in these high uncertainty societies may not be the ones that promote good Six Sigma teamwork (Hofstede 1990). For example, if the society rule prohibits discussion of a leader's decision, then discussion may not be an option for the Six Sigma team.

Power Distance Hofstede (1990) and Singelis, Triandis, Bhawuk, and Gelfand (Singelis et al. 1995) defined power distance as the amount of inequality permitted by a society. In his work, Hofstede (The 6-D model of national culture, n.d.) identified Central Europe, Iraq, Saudi Arabia, Indonesia, Malaysia, Brunei, and Russia as being among the highest ranked in power distance. In contrast, he ranked Scandinavia, the UK, Australia, New Zealand, Canada, and the USA among the lowest power distance countries.

In high power distance cultures, inequality is the norm (Hofstede et al. 2010; Triandis 2002a, b). People are born into a hierarchy and often are unable to move out of this birth level, in more extreme high power distance cultures. Generally, tribes are used as examples for the extreme of the high power distance continuum. In low power distance cultures, there is little inequality, with the aim being total equality. Where you were born in the hierarchy or who your parents were means less than in high power distance cultures. An example of low power distance cultures is Scandinavian cultures, where society makes attempts to create equality (Hofstede et al. 2010; Triandis 2004); in Norway, even the King rides the tram.

In societies with high power distance, the head or leader of the group generally determines the goals of the group. Contradicting the leader is seen as an insult; therefore, free speech is usually limited in high power distance. In societies with low power distance, the leader usually expects that there might be some debate over decisions and may even build in the opportunity for discussion of options with other group members before decisions are made (Hofstede 1998; Khatri 2009). Therefore, it is easy to see how power distance might affect a Six Sigma team. It may be that, in higher power distance countries, the Six Sigma team in the Define phase of DMAIC may be told to work on a problem that is considered important to those of higher status. In lower power distance countries, the potential that the Six Sigma group will discuss potential issues and project topics in the Define phase of DMAIC and arrive at a consensus is more likely.

Much of the work surrounding current practices in lean management were developed in Japan. However, its culture is significantly different from other countries, even those nearby: China, Russia, and South Asia (Lowe 1998; Onishi and Bliss 2006). There is a continuing concern that they might not be effective elsewhere because the cultural values are not the same (Kull and Wacker 2009; Lozeau et al. 2002; Zhao et al. 2004), although they were originally supported by American quality thinkers, W. Edwards Deming, Joseph Juran, and Armand Feigenbaum. Deming designed systems around the work of Walter Shewhart and his Plan-Do-Control-Act (PDCA) cycle. However, it is his work in Japan that is most noted around the world, along with his book, *Out of the Crisis* (1986).

3.6 Cultural Change

Cultures are both stable and dynamic; that is, they are consistent, but they do change over time. They are reinforced through every process and procedure performed in the organization by its members as well as by the interaction of members through the workplace. If procedures and processes are performed to standard throughout the organization, then that will reflect its culture; however, if they are not performed to standard, that will also reflect its organizational, professional, and societal cultures. All are integrated. Professional and societal cultures are difficult to change, but they evolve through time.

Professional culture changes are generally initiated by those who are foremost in those professions. New technologies or techniques can change the culture if they are supported by the leaders of the profession. Generally, a single organization will not have an effect on professional culture until others in the profession begin to make those changes and they are accepted by the leaders of the profession. New leaders can also change the culture of the profession by bringing new ideas into it. The most striking case is that of Florence Nightingale, who changed the culture of nursing over many decades by demonstrating to individual nurses and doctors how much better sanitized conditions were for patients – just one of her many changes to the nursing culture.

Societal culture changes very gradually, unless some environmental shift necessitates a change. Single organizations can rarely change a society, but one large event can do so. You can easily see the changes that occurred in society following the 9/11 attack on the World Trade Center, not only in the USA, but around the world. Travel became more controlled, particularly air travel. In the USA, many restrictions on freedom were instituted by the government, some that are debated to this day. China's technology has increased its restrictions on populations, instituting a system of monitoring individuals by computer as they go about their daily lives. Africa's culture has been changed by violent tribal and other wars throughout the centuries. But rarely can one organization create an organizational change. The next major societal changes are likely to occur because of climate change or the event of a massive solar flare or a large meteoroid or asteroid strike. Changing organizational culture is much easier than changing professional or societal culture.

3.6.1 Changing Organizational Culture

Organizational culture change may be necessary to encourage appropriate Six Sigma processes and implementation (Liker 2004). Without an appropriate change, most change initiatives, such as total quality management and Six Sigma initiatives, will fail (Cameron and Quinn 2011). Decentralization and flatter structures are critical for Six Sigma processes to work, but this is not appropriate in many organizational or societal cultures, creating difficulties for changing processes. Child and Keiser (1979) and Ralston et al. (1997) found that cultural values may not change even with urbanization or when foreign enterprises become a normal part of the environment (Child and Keiser 1979; Ralston et al. 1997). Therefore, an active cultural change may be needed.

Schein (2010) has done work in the area of organizational change for many years. He states that equilibrium and stability is the norm for all human systems. In groups, departments, organizations, professions, and societies, the set of assumptions shared by the members provides both stability and meaning. But the environment changes, necessitating changes within the system. Changing organizational culture is to aid in its evolution to meet environmental changes.

Lewin (1947) proposed a model of unfreezing, change, and refreezing, that Schein has adapted for his model. In the first instance, there must be enough people who see that a change is needed (disconfirming data) that can be done (identity/self-image), but if not done will cause problems (anxiety or guilt). Disconfirming data can be anything that shows that the organization is losing ground in some way, including increased absenteeism or turnover, quality problems, increased customer problems, and decreased demand for the product or service. Seeing these problems can create anxiety or guilt through the realization that if change does not occur, then something negative will occur, either to the individual, group, department, or organization. The organization faces up to these problems when it has enough confidence in its underlying identity to make changes to solve the problems and, often, modify its self-image.

The next stage in Schein's model is the actual change or cognitive restructuring, as he terms it. This is the response to the unfreezing that allows the organization to continue to thrive, although not with its present set culture. Often it involves some rephrasing or reconceptualizing of values that the organization has held as a core part of the culture. The example that Schein (2010) gives is a common one: economic necessity to reduce payroll costs. If the organization has a core value of keeping employees and never laying anyone off, then they will have to redefine layoffs as "transitions," "early retirements," or "restructuring" to preserve the core value of treating employees well. Often this change is made incrementally to individual beliefs, but in a rapidly changing environment, these changes may need to be made more quickly through dramatic restructuring of ideas and values. Although Schein acknowledges that change may be essential, he does not believe that coercion works well; he cites the example of Communism, where people living in Communist countries did not automatically become Communists although living under the rule for over 50 years.

The final step in Schein's model is refreezing or reinforcing the changes that have been made through data that confirms that the change created positive results. With this data, the changes become part of the value and belief system, become stabilized and internalized, and continue until new environmental disconfirmations cause the cycle all over again. If the leader models the change, it will refreeze more quickly, depending on the leader's relationship with the group, department, or organizational members.

In discussing the change model, Schein (2010) states that psychological safety is a necessary condition for the model's success and lists eight steps to achieve this (p. 332):

1. Compelling vision: Organizational members must believe that they will be better off if they make the change.
2. Training: Some changes require reskilling or additional training, and members must be assured that they will receive it.
3. Learner involvement: Each learner has a slightly different way of learning and of incorporating the learning into his/her own work, so they must help design training processes and outcomes.
4. Group and team training: Cultural values and beliefs are rooted within groups, so the entire group must have training to make the changes needed.

5. Time, resources, teaching, and feedback: These are all a part of learning, and if they are not provided, then learning will not occur to the extent needed by the change.
6. Models: Role models are needed so that organizational members can pattern their behavior upon the model. They must see the new values and beliefs in others that they can identify with.
7. Sharing learning problems: Individuals should feel comfortable in airing difficulties and problems in learning the new change, and groups and coaches provide this venue. Groups are very valuable because they can teach each other the skills needed and reinforce learning.
8. Reward and structures: Reward systems and organizational structures must support the change. For example, to encourage teamwork, it is necessary to have team rewards, and individually selfish behaviors must be punished. Otherwise, the change will not refreeze.

Schein's model has been used in many firms, both nationally and internationally, with significant success. Schein's model can be implemented in the cultural change process in business organizations prior to beginning Six Sigma initiatives, if necessary.

Cameron and Quinn (2011) discuss culture change at some length, stating that individual behavior changes are necessary to institute an overall cultural change, be it organizational, professional, or societal. They use a different but valid approach to change. Individuals must identify the change that they wish to see and identify activities that will produce that change. Many change processes will depend upon the skills and abilities of the manager and whether they can reinforce the changes needed. In their book, Cameron and Quinn (2011) identify the different competencies needed by managers, depending on where change is needed. If the clan culture needs strengthening, then interpersonal relationships, teams, and employee development skills are needed. If the development culture needs improvement, then continuous improvement and innovation management skills are needed, along with the ability to envision, communicate, and move toward a clear vision. Competitive capabilities need to be upgraded for improvements in the market culture, as well as enhancing incentivization and inspiration for employees and exceeding customer expectations. Changing the hierarchical culture means that managers must ensure that controls are in place and that employees understand what is expected of them and able to coordinate their activities. In balanced cultures, all of these different segments of the culture are important, but the mission, vision, and strategy of the organization determines which segment(s) should be emphasized over others.

Cameron and Quinn suggest organizational culture change steps that would be helpful for most organizations (Cameron and Quinn 2011, p. 160), and these are given below:

1. Complete the evaluation of the organizational culture and compute a cultural profile based on the present culture.
2. Specify what you want the organizational culture to be.
3. Pinpoint ways in which the culture might be changed and which changes might be problematic because of either positive or negative consequences.
4. Identify a few key artifacts (key stories and incidents) that already occur in the organization that you can use to reinforce the culture that is desired.
5. Design strategies that will help achieve the culture change desired – what should you do more of, less of, or eliminate totally in the culture.
6. Construct some small wins that can be made quickly to reinforce the culture change.
7. Identify the managerial skills and abilities needed for the change and to manage the transformed culture.
8. Develop measures, metrics, and milestones to determine whether the change is occurring and at a steady rate.
9. Communicate the changes, using a well-developed strategy for this.

Cameron and Quinn (2011) emphasize that changing culture is difficult and requires commitment and dedication on the part of the top management team to implement and sustain these changes. To improve the success chance of Six Sigma in business organizations, organizations need to be ready from top to bottom for a cultural change in case it is seen as a necessity prior to starting the Six Sigma journey.

3.6.2 Diagnosing Potential Organizational Culture to Implement Six Sigma

You do not necessarily need to change your organization's culture to implement Six Sigma, but you need to analyze which structure you have to determine whether it needs changing or where you might encounter problems in getting Six Sigma started or the decisions implemented. To start, consider the stories and gossip occurring in the organization. A saying circulating around management now is from an unknown author: "If you change the way you talk about something, you will probably change the way you think about it. If you change the way you think about it, you are likely to change the way you act around it [behavior and culture]."

Cameron and Quinn (2011) outline one way to measure the culture in your organization, and it has been very effective in those organizations where we have used it (Pakdil and Leonard 2015; Pakdil et al. 2018). The survey and measurement from our instrument were different from theirs in some ways, to make it more relevant for the firms that we studied, although we used their information to compile our survey. It is shown in ◘ Table 3.1.

◘ **Table 3.1** Survey of organizational culture types

Please *circle* one answer in each line across after each question, indicating how strongly you agree or disagree with each of the statements about your company as it is now							
1. The organization is a very personal place. It is like an extended family. People seem to share a lot of themselves	Strongly disagree	1	2	3	4	5	Strongly agree
2. The organization is a very dynamic and entrepreneurial place. People are willing to stick their necks out and take risks	Strongly disagree	1	2	3	4	5	Strongly agree
3. The organization is very results oriented. A major concern is with getting the job done. People are very competitive and achievement oriented	Strongly disagree	1	2	3	4	5	Strongly agree
4. The organization is a very controlled and structured place. Formal procedures generally govern what people do	Strongly disagree	1	2	3	4	5	Strongly agree
5. The leadership in the organization is generally considered to be mentoring, facilitating, or nurturing	Strongly disagree	1	2	3	4	5	Strongly agree
6. The leadership in the organization is generally considered to be entrepreneurship, innovating, or risk taking	Strongly disagree	1	2	3	4	5	Strongly agree
7. The leadership in the organization is generally considered to have a no-nonsense, aggressive, results-oriented focus	Strongly disagree	1	2	3	4	5	Strongly agree
8. The leadership in the organization is generally considered to be coordinating, organizing, or smooth-running and efficient	Strongly disagree	1	2	3	4	5	Strongly agree
9. The management style in the organization is characterized by teamwork, consensus, and participation	Strongly disagree	1	2	3	4	5	Strongly agree

Table 3.1 (continued)

Please *circle* one answer in each line across after each question, indicating how strongly you agree or disagree with each of the statements about your company as it is now							
10. The management style in the organization is characterized by individual risk-taking, innovation, freedom, and uniqueness	Strongly disagree	1	2	3	4	5	Strongly agree
11. The management style in the organization is characterized by hard-driving competitiveness, high demands, and achievement	Strongly disagree	1	2	3	4	5	Strongly agree
12. The management style in the organization is characterized by security of employment, conformability, predictability, and stability in relationships	Strongly disagree	1	2	3	4	5	Strongly agree
13. The glue that holds the organization together is loyalty and mutual trust. Commitment to this organization runs high	Strongly disagree	1	2	3	4	5	Strongly agree
14. The glue that holds the organization together is commitment to innovation and development. There is an emphasis on being innovative and progressive	Strongly disagree	1	2	3	4	5	Strongly agree
15. The glue that holds the organization together is the emphasis on achievement and accomplishment. Aggressiveness and winning are common themes	Strongly disagree	1	2	3	4	5	Strongly agree
16. The glue that holds the organization together is formal rules and policies. Maintaining a smooth-running organization is important	Strongly disagree	1	2	3	4	5	Strongly agree
17. The organization emphasizes human development. High trust, openness, and participation are important	Strongly disagree	1	2	3	4	5	Strongly agree
18. The organization emphasizes acquiring new resources and creating new challenges. Trying new things and looking for opportunities are valued	Strongly disagree	1	2	3	4	5	Strongly agree
19. The organization emphasizes competitive actions and achievement. Hitting stretch targets and winning in the marketplace are dominant	Strongly disagree	1	2	3	4	5	Strongly agree
20. The organization emphasizes permanence and stability. Efficiency, control, and smooth operations are important	Strongly disagree	1	2	3	4	5	Strongly agree
21. The organization defines success on the basis of the development of human resources, teamwork, employee commitment, and concern for people	Strongly disagree	1	2	3	4	5	Strongly agree
22. The organization defines success on the basis of having the most unique or newest products. It is a product leader and innovator	Strongly disagree	1	2	3	4	5	Strongly agree
23. The organization defines success on the basis of winning in the marketplace and outpacing the competition. Competitive market leadership is most important	Strongly disagree	1	2	3	4	5	Strongly agree
24. The organization defines success on the basis of efficiency. Dependable delivery, smooth scheduling, and low-cost production are critical	Strongly disagree	1	2	3	4	5	Strongly agree

Source: Adapted from Cameron and Quinn (1999), in Pakdil et al. (2018)

3

Table 3.2 Types of cultures and question numbers

Type of culture examined	Question #
Group (clan) culture	1, 5, 9 13, 17, 21
Developmental culture	2, 6, 10, 14, 18, 22
Hierarchical culture	3, 7, 11, 15, 19, 23
Rational (market) culture	4, 8, 12, 16, 20, 24

Source: Author's creation based on Pakdil and Leonard (2015)

Answer Key The following sets of questions measure the strength of the type of culture that exists (Table 3.2)

Once the results are obtained, you map them on the radar map, as shown in Figs. 3.2 and 3.3.

One key consideration is "change fatigue." A survey by the Center for Creative Leadership found that it is one of the top leader concerns (Ditch change fatigue and embrace continual evolution 2019). To prevent this fatigue, make sure that the change is clearly communicated and that need for change is clearly stated. Even positive changes can create havoc if not properly done. The authors' talk emphasizes that communication, collaboration, and commitment are necessary in any change process.

Key Concepts
- Culture is the way things are done within an organization.
- There are three types of culture that affect Six Sigma teams: organizational, professional, and societal.
- Each type of culture must be taken into consideration when designing Six Sigma solutions.

Summary
In concluding this chapter, we must note that all three cultures, organizational, professional, and societal culture, are influences on organizations and on Six Sigma processes. The one most likely to change over a short period of time is organizational culture, although there are processes to follow to get the results that you want. If you can understand the changes that are necessary, you can design methods to change the organizational culture to suit the need for change. Otherwise, systems will not work well because of a lack of support for those changes.

Practice and Discussion Questions
1. Consider the effect of a collectivistic culture with high power distance on Six Sigma teamwork. How might Six Sigma teamwork be different in this culture than in an individualistic culture with low power distance? You might consider the difference between Six Sigma teamwork in Sweden and China, for example.
2. Consider the effect of an organization with hierarchical culture and one with clan culture on Six Sigma teamwork. Discuss the differences and similarities.
3. What problems would you expect in a Six Sigma team with individuals from different professions? Discuss the issues that might arise and why if the following professionals were included in the team. How would you manage the team, given the cultural differences?
   - Finance/accounting professionals
   - Engineering professionals
   - Management professionals
   - Marketing professionals
   - Human resource professionals.

References

Bandura, A. (1977). *Social learning theory*. New York: Prentice Hall.

Bate, P. (1984). The impact of organizational culture on approaches to organizational problem solving. *Organization Studies, 5*(1), 43–66.

Beck, U. (2000). The cosmopolitan perspective: Sociology of the second age of modernity. *British Journal of Sociology, 51*(1), 79–105.

Beyer, J. M., & Trice, H. M. (1987). How an organization's rites reveal its culture. *Organizational Dynamics, 15*(4), 5–24.

Brief, A., & Nord, W. (1990). *Meanings of occupational work*. Lanham: Lexington Books.

Calori, R., & Samin, P. (1991). Corporate culture and economic performance: A French study. *Organization Studies, 12*(1), 49–74.

Cameron, K. S., & Freeman, S. (1991). Cultural congruence, strength, and type: Relationships to effectiveness. *Research in Organizational Change and Development, 5*(1), 23–58.

Cameron, K. S., & Quinn, R. E. (1999). *Diagnosing and changing organizational culture* (1st ed.). Reading: Addison-Wesley.

Cameron, K. S., & Quinn, R. E. (2011). *Diagnosing and changing organizational culture: Based on the competing values framework* (3rd ed.). San Francisco: Jossey-Bass.

Child, J., & Keiser, A. (1979). Organizational and managerial roles in British and West German companies: An examination of the culture-free thesis. In C. J. Lammers & D. J. Hickson (Eds.), *Organizations alike and unlike: International and interinstitutional studies in the sociology of organizations* (pp. 251–271). London: Routledge/Kegan Paul.

Crandall, R. E., & Crandall, W. E. (2011). Three little words. *Industrial Engineering, 43*(6), 42–47.

Dahlgaard, J. J., & Dahlgaard-Park, S. M. (1999). Integrating business excellence and innovation management: Developing a culture for innovation, creativity, and learning. *Total Quality Management, 10*(4/5), S465–S472.

Deming, W. E. (1986). *Out of the crisis*. Cambridge, MA: MIT Press.

Demir, C., Unnu, N. A. A., & Erturk, E. (2011). Diagnosing the organizational culture of a Turkish pharmaceutical company based on the Competing Values Framework. *Journal of Business Economics and Management, 12*(1), 197–217.

Denison, D. R. (1984). Bring corporate culture to the bottom line. *Organizational Dynamics, 13*(2), 59–76.

Dennis, P. (2002). *Lean production simplified*. New York: Productivity Press.

Ditch change fatigue and embrace continual evolution. (2019, September 24). Retrieved from Center for Creative Leadership: https://www.ccl.org/blog/change-fatigue-continual-evolution/?utm_campaign=1-LeadersAtAllLevels%2C2-LE_Article%2C3-Custom&utm_medium=social&utm_source=linkedin&utm_content=1572102120

Fullerton, C. S., McCarroll, J. E., Ursano, R. J., & Wright, K. M. (1992). Psychological responses of rescue workers: Fire fighters and trauma. *American Journal of Orthopsychiatry, 62*(3), 371–378.

Glisson, C., & James, L. R. (2002). The cross-level effects of culture and climate in human service teams. *Journal of Organizational Behavior, 23*(6), 767–794.

Greenwood, E. (1957). Attributes of a profession. *Social Work, 2*, 45–55.

Halbesleben, J. R. (2009). The role of pluralistic ignorance in the reporting of sexual harassment. *Basic and Applied Social Psychology, 31*(3), 210–217.

Hall, E. T., & Hall, M. R. (1987). *Hidden differences: Doing business with the Japanese*. New York: Anchor Press.

Hofstede, G. (1990). *Cultures and organizations: Software of the mind*. New York: McGraw-Hill.

Hofstede, G. (1998). Attitudes, values, and organizational culture: Disentangling the concepts. *Organization Studies, 19*(3), 477–493.

Hofstede, G. (n.d.). *The 6-D model of national culture*. Retrieved 11 2019, from Geert Hofstede: https://geerthofstede.com/culture-geert-hofstede-gert-jan-hofstede/6d-model-of-national-culture/

Hofstede, G., Hofstede, G. J., & Minkov, M. (2010). *Cultures and organizations: Software of the mind: Intercultural cooperation and its importance for survival* (3rd ed.). New York: McGraw-Hill.

Jablin, F. M. (2001). Organizational entry, assimilation, and disengagement/exit. In F. M. Jablin & L. L. Putnam (Eds.), *The new handbook of organizational communication: Advances in theory, research, and methods* (pp. 732–818). Newbury Park: Sage.

Jehn, K. A., Chadwick, C., & Thatcher, S. M. (1997). To agree or not to agree: The effects of value congruence, individual demographic dissimilarity, and conflict on workgroup outcomes. *International Journal of Conflict Management, 8*(4), 287–305.

Kets de Vries, M. R., & Miller, D. (1986). Personality, culture, and organization. *Academy of Management Review, 11*(2), 266–279.

Khatri, N. (2009). Consequences of power distance orientation in organizations. *Vision – Journal of Business Perspective, 13*(1), 2–9.

Kull, T. J., & Wacker, J. G. (2009). Quality management effectiveness in Asia: The influence of culture. *Journal of Operations Management, 28*(3), 223–239.

Lewin, K. (1947). Group decision and social change. In T. N. Newcomb & E. L. Hartley (Eds.), *Readings in social psychology*. New York: Holt, Rinehart, and Winson.

Li, S., Triandis, H. C., & Yu, Y. (2006). Cultural orientation and corruption. *Ethics and Behavior, 16*(3), 199–215.

Liker, J. K. (2004). *The Toyota way: 14 management principles from the World's Greatest Manufacturer*. New York: McGraw-Hill.

Liker, J. K., & Franz, J. K. (2011). *The Toyota way to continuous improvement*. New York: Mc-Graw Hill.

Liker, J. K., Fruin, W. M., & Adler, P. S. (1999). *Remade in America: Transplanting and transforming Japanese management systems*. New York: Oxford University Press.

Lowe, S. (1998). Culture and network institutions in Hong Kong: A hierarchy of perspectives. A response to Wilkinson: Culture, institutions, and business in East Asia. *Organization Studies, 19*(2), 321–344.

Lozeau, D., Langley, A., & Denis, J. L. (2002). The corruption of managerial techniques by organizations. *Human Relations, 55*(5), 537–564.

Mann, D. (2005). *Creating a lean culture*. New York: Productivity Press.

Markus, H. R., & Kitayama, S. (2003). Culture, self, and the reality of the social. *Psychological Inquiry, 14*(3&4), 277–283.

Martin, J. (1992). *Cultures in organizations: Three perspectives*. New York: Oxford University Press.

McLean, B., & Elkind, P. (2003). *The smartest guys in the room: The amazing rise and scandalous fall of Enron*. New York: Portfolio Hardcover.

Mehri, D. (2006). The darker side of lean: An insider's perspective on the realities of the Toyota production system. *Academy of Management Perspectives, 20*(2), 21–42.

Meyer, J. C. (1995). Tell me a story: Eliciting organizational values from narratives. *Communication Quarterly, 43*(2), 210–224.

Myers, K. K., & McPhee, R. D. (2006). Influences on member assimilation in workgroups in high-reliability organizations: A multilevel analysis. *Human Communication Research, 32*(4), 440–468.

Naor, M., Linderman, K., & Schroeder, R. (2010). The globalization of operations in Eastern and Western countries: Unpacking the relationship between national and organizational culture and its impact on manufacturing performance. *Journal of Operations Management, 28*(3), 194–205.

O'Reilly, C. A., & Chatman, J. A. (1996). Culture as social control: Corporations, cults, and commitment. In B. Staw & L. L. Cummings (Eds.), *Research in organizational behavior* (Vol. 18, pp. 157–200). Greenwich: JAI Press.

Olson-Buchanan, J. B., & Boswell, W. R. (2008). An integrative model of experiencing and responding to mistreatment at work. *Academy of Management Review, 33*(1), 76–96.

Onishi, J., & Bliss, R. E. (2006). In search of Asian ways of managing conflict: A comparative study of Japan, Hong Kong, Thailand, and Vietnam. *International Journal of Conflict Management, 17*(3), 203–225.

Ouchi, W. G. (1980). Markets, bureaucracies, and clans. *Administrative Science Quarterly, 25*(1), 129–141.

Ouchi, W. J. (1981). *Theory Z: How American business can meet the Japanese challenge*. Reading: Addison-Wesley.

Ouchi, W. G., & Johnson, J. (1978). Types of organizational control and their relationship to emotional well-being. *Administrative Science Quarterly, 23*, 293–317.

Pakdil, F., & Leonard, K. M. (2015). The effect of organizational culture on implementing and sustaining lean processes. *Journal of Manufacturing Technology Management, 26*(5), 725–743.

Pakdil, F., & Leonard, K. M. (2016). Implementing and sustaining lean processes: The dilemma of societal culture effects. *International Journal of Production Research, 55*(3), 700–717.

Pakdil, F., Toktas, P., & Leonard, K. M. (2018). Validation of qualitative aspects of the lean assessment tool (LAT). *Journal of Manufacturing Technology Management, 29*(7), 1094–1114.

Quinn, R. E., & Cameron, K. S. (1983). Organizational life cycles and shifting criteria of effectiveness. *Management Science, 29*(1), 33–51.

Quinn, R. E., & Rohrbaugh, J. (1983). A spatial model of effectiveness criteria: Toward a competing values approach to organizational analysis. *Management Science, 29*(3), 363–377.

Quinn, R. E., & Spreitzer, G. M. (1991). The psychometrics of the competing values culture instrument and an analysis of the impact of organizational culture on quality of life. *Research in Organizational Change and Development, 5*, 115–142.

Ralston, D. A., Holt, D. H., Terpstra, R. H., & Kai-Cheng, Y. (1997). The impact of national culture and economic ideology on managerial work values: A study in the United States, Russia, Japan, and China. *Journal of International Business Studies, 28*(1), 177–207.

Rothwell, G. R., & Baldwin, N. J. (2006). Ethical climate theory, whistle-blowing, and the code of silence in police agencies in the state of Georgia. *Journal of Business Ethics, 60*, 341–361.

Schein, E. H. (2010). *Organizational culture and leadership*. (4 ed. New York: Wiley.

Shook, J. (2010). How to change a culture: Lessons from NUMMI. *MIT Sloan Management Review, 51*(2), 63–68.

Singelis, T. M., Triandis, H. C., Bhawuk, D. P., & Gelfand, M. J. (1995). Horizontal and vertical dimensions of individualism and collectivism: A theoretical and measurement refinement. *Cross-Cultural Research, 29*(3), 240–275.

Steel, P., & Taras, V. (2010). Culture as a consequence: A multi-level multivariate meta-analysis of the effects of individual and country characteristics on work-related values. *Journal of International Management, 16*(3), 211–233.

Swidler, A. (1986). Culture in action: Symbols and strategies. *American Sociological Review, 51*(2), 273–286.

Taras, V., Steel, P., & Kirkman, B. L. (2010). Negative practice-value correlations in the GLOBA data: Unexpected findings, questionnaire limitations, and research directions. *Journal of International Business Studies, 41*(8), 1330–1338.

Triandis, H. C. (1989). The self and social behavior in differing cultural contexts. *Psychological Review, 96*(3), 506–520.

Triandis, H. C. (1995). *Individualism and collectivism*. Boulder: Westview Press.

Triandis, H. C. (2002a). Motivation to work in cross-cultural perspective. In J. M. Brett & F. Drasgow (Eds.), *The psychology of work: Theoretically based empirical work* (pp. 101–117). Mahwah: Lawrence Erlbaum Associates.

Triandis, H. C. (2002b). Subjective culture. *Online Readings in Psychology and Culture, 2*(2). https://doi.org/10.9707/2307-0919.1021.

Triandis, H. C. (2004). The many dimensions of culture. *Academy of Management Executive, 18*, 88–93.

Trice, H. C. (1993). *Occupational subcultures in the workplace*. Ithaca: International Labor Relations Press, Cornell University.

Weber, M. (1947). *The theory of social and economic reform*. (Henderson, & Parsons, Trans.). New York: Free Press.

Westney, E. (1987). *Innovation and imitation: The transfer of Western Organizational patterns to Meiji, Japan*. Cambridge, MA: Harvard University Press.

White, R. E., Pearson, J. N., & Wilson, J. R. (1999). JIT manufacturing: A survey of implementation in small and large US manufacturers. *Management Science, 45*, 1–15.

Williamson, O. (1975). *Markets and hierarchies*. New York: Free Press.

Wittrock, C. (2015). Reembedding lean: The Japanese cultural and religious context of a world changing management concept. *International Journal of Sociology, 45*(2), 95–111.

Yin-Cheong, C. (1989). Organizational culture: Development of a theoretical framework for organizational research. *City University of Hong Kong Education Journal, 17*(2), 128–147.

Zammuto, R. F., & Krakower, J. Y. (1991). Quantitative and qualitative studies of organizational culture. *Research in Organizational Change and Development, 5*, 83–114.

Zhao, X. D., Leung, A. L., & Lee, T. S. (2004). Quality management and organizational context in selected service industries of China. *Journal of Operations Management, 22*(6), 575–587.

Zu, X., Robbins, T. L., & Fredendall, L. D. (2010). Mapping the critical links between organizational culture and TQM/six sigma practices. *International Journal of Production Economics, 123*(1), 86–106.

Six Sigma Process: DMAIC

Fundamentally, all decisions and actions should rely on measurement, data, and statistical evidence in modern management approaches, including quality management, lean management, and Six Sigma. Rather than implementing subjective approaches in decision-making processes, Six Sigma practitioners make their decisions based on facts, data, knowledge, and statistical interferences. When Six Sigma methodology is combined with other management philosophies and methodologies, it functions as a powerful management tool to improve organizational performance and solving performance problems (Kawaguchi et al. 2015). As Pyzdek and Keller (2018: 3) stated, "Six Sigma relies on tried and true methods that have been used for decades."

Quality and performance improvement efforts are most effectively constructed on a project basis in Six Sigma. DMAIC (Define-Measure-Analyze-Improve-Control) has been accepted as the most frequently used and best structured project base for quality improvement and Six Sigma projects. DMAIC has been utilized by practitioners and academicians in the past in a great variety of industries and settings. It enables team members to focus on a systematic problem-solving methodology, concentrates on improving processes in which the problem appeared, and helps organizations maintain and sustain improvement and solution methods. DMAIC is a problem-solving methodology that focuses on:

1. Identifying problem accurately
2. Analyzing problem with all details to discover root cause(s)
3. Deciding the best solution(s) to solve the problem identified in the first phase of DMAIC.

Wedgwood (2015) states that DMAIC is supported by multiple data-driven tools and techniques to allow teams to understand quality problems and develop solutions and remedies for the problems. It is not just associated with Six Sigma-based projects. Quality management-oriented problem-solving processes, quality improvement projects, and lean projects aiming at improving various performance indicators can be developed and conducted based on DMAIC process.

The DMAIC process encourages creative thinking about the problem and its potential solutions. In each phase of DMAIC, creative thinking brings various approaches to define, measure, analyze, improve, and control the problems and processes (Montgomery 2009). Creative thinking focuses on the details of the problems and creative and effective solutions for the problems. When it is determined necessary to start over and redesign the process from the beginning, the Improve phase of DMAIC becomes a Design phase. In this case, the regular DMAIC process turns into a Design for Six Sigma (DFSS) structure.

Even if projects are a base for quality improvement in Six Sigma methodology, Juran (1992) emphasized the pitfalls of project-by-project quality improvement efforts. He suggests not depending on project-based improvements. Juran and other quality gurus emphasized that quality improvement must be integrated into daily quality improvement activities through *kaizen*. Daily improvements throughout the organization are pivotal in continuous quality improvement efforts. If Six Sigma strategy is appropriately integrated into continuous quality improvement activities, Six Sigma projects elevate and support quality improvement activities.

There are several questions that need to be asked in each phase to make the process more understandable when using DMAIC, including:

- *Define*: Who is the customer and what does the customer care about? Is this the right project now?
- *Measure*: How is the performance of products, problems, or factors identified in the Define phase measured? How is the process performed? How efficient and effective is the process?
- *Analyze*: Are the data collected in Measure phase analyzed? If so, how? If not, why not? Is it necessary that the process be performed this way? Why?
- *Improvement*: How are the causes of defectives, defects, and problems identified? How are the causes eliminated? How should the process be improved?
- *Control*: How can the organization maintain the improvement level?

The strength of DMAIC process is identified as the sequence of tasks, tools, and critical thinking implemented in the process. Each phase of DMAIC includes a group of tasks that need to be done in DMAIC flow. Major steps of DMAIC roadmap are listed below:

Define:

- Analyze the process and map the current state.
- Analyze customer needs and expectations.
- Identify stakeholders.
- Project prioritization and selection.
- Build up Six Sigma team.
- Develop project charter.
- Develop project plan.

Measure:

- Understand and measure the process.
- Develop data collection plan.
- Data collection.
- Measurement system analysis.
- Identify sigma level.
- Measure poor and good cost of quality.

Analyze:

- Analyze the problem.
- Analyze the process and data.
- Identify the root causes.
- Prioritize the root causes.

Improve:
- Develop alternative solutions to improve sigma level.
- Select the optimal solution(s).
- Map future state.
- Analyze the potential risks in the new process.
- Pilot test the new process.
- Implement the new process.

Control:
- Implement ongoing measurements.
- Standardize the improvement.
- Monitor the improvement.
- Project closure.

The second part of the book analyzes the details of DMAIC phases and tools used in each phase. The following chapters discuss and analyze each phase of the DMAIC process. Each chapter in this part highlights examples and solutions after presenting the theoretical background of the phases.

References

Kawaguchi, M., Tanaka, Y., & Furuya, H. (2015). Improving the safety and quality of surgical patient care: what can we learn from quality management of industries? *Journal of Anesthesia, 29*(4), 485–486.

Pyzdek T & Keller, PA. (2018). *The Six Sigma Handbook*, 5th edition. New York: Mc-Graw Hill.

Wedgwood, I. (2015). *Lean Sigma: Rebuilding Capability in Healthcare*. New York: Pearson Education.

Montgomery, D. C. (2009). *Introduction to Statistical Quality Control*. New York: Wiley.

Juran, J. M. (1992). *Juran on Quality by Design: The New Steps for Planning Quality into Goods and Services*. New York: The Free Press.

Contents

Define Phase: D Is for Define

Contents

F. Pakdil, *Six Sigma for Students*, https://doi.org/10.1007/978-3-030-40709-4_4

Learning Objectives

After careful study of this chapter, you should be able to:

- Explain Define phase of DMAIC
- Apply process analysis and documentation
- Develop transformation process, value stream map, flow chart, SIPOC diagram, swim lane, and spaghetti diagram
- Define boundaries and bottlenecks in the process
- Perform stakeholder analysis
- Prioritize and select Six Sigma projects
- Define and develop project charter, problem statement, goal statement, project scope, deliverables, metrics, and Critical-to-Quality (CTQ) characteristics
- Construct House of Quality (HOQ)
- Identify relatively more important voice of customer and voice of engineering.

4.1 Introduction

As Bruhl (2018: 1) states, "every investigation starts-and ends-with a purpose: I want (or need) to understand something about something." Therefore, Six Sigma projects start with the Define phase. This phase concentrates on selecting and identifying the research question or problem that will be analyzed in the Six Sigma project. In this phase, the details of the problem are identified along with the costs, expected benefits, and financial portrait of the problem. The details of the project are discovered, and the problem that will be analyzed during the project is detailed in the Define phase.

Projects should include an improvement opportunity within an existing process, product, or problem. Problems may emerge internally or externally in the environment of the organization. That is, internal problems can be detected in the organizational structure, before these problems affect the external environment of the organization. External problems include customer complaints and dissatisfactions, returned products, warranty claims, recalls, negative word-of-mouth, and boycotts. The voice of the customer is taken into consideration to identify what kind of topics should be analyzed in the Define phase.

In the Define phase, potential value opportunities of the project are anticipated and identified. If possible, the potential value of the project should be linked with the organizational goals and objectives. The financial impact of the project is analyzed at the beginning and end of the process, which makes project evaluation more rigorous and meaningful. It also presents the potential and real costs and outcomes of the project. The financial assessment helps decision-makers evaluate and compare potential and current financial contribution of all efforts. The CTQ characteristics and performance metrics/indicators related to the project are then identified and linked with the estimated benefits of the project. In this phase, various performance indicators, such as defect rate, rework rate, scrap rate, number of customer complaints, cost reduction, lead time, return on equity, return on invested capital, earnings, growth in operating income, and market share, can be measured and utilized in decision-making processes (Montgomery, 2013).

The questions that need to be answered in the Define phase include:

- Who are my customers?
- What matters to my customers?
- How is our performance in terms of delivery, cost, quality, and quantity, in the view of customer?
- What is the problem?
- What is the project scope?
- What is my process about?
- What does my process produce?
- Can I map my process?
- What is the "defect" in my processes?
- What is my baseline and entitlement?
- Are there financial benefits out of this project?

As a comprehensive phase, the Define phase contains six main deliverables: (1) process analysis outcomes, (2) stakeholder analysis outcomes, (3) customer needs and expectations, (4) project prioritization and selection, (5) project charter, and (6) project plans. The following sections will present how these deliverables are developed in Six Sigma projects.

One of the visual benefits of DMAIC lies in ease of use of the tools in the entire process. ◘ Table 4.1 represents a compre-

Table 4.1 The tools used in DMAIC phases

	Quality costs	Process analysis tools	Stakeholder analysis	Project charter	House of quality	Data collection plan	Data collection	Six Sigma measurements	Gage R&R	Descriptive statistics	Inferential statistics	Statistical distributions	Hypothesis testing	Correlation analysis	Regression analysis	Analysis of Variance	Process capability analysis	Taguchi's Loss Function	Seven old tools	Seven new tools	Brainstorming	5 Whys analysis	Gantt chart	Dot plot	Run chart	Bar chart	Box-and-whisker plot	Probability plot	Design of Experiments	Simulation	Lean	Failure Modes and Effects Analysis (FMEA)	Control plans	Statistical Process Control (SPC)
DEFINE																																		
Analyze the process and map the current state																																		
Analyze customer needs and expectations																																		
Identify stakeholders																																		
Project prioritization and selection																																		
Build up Six Sigma team																																		
Develop project charter																																		
Develop project plan																																		
MEASURE																																		
Understand and measure the process																																		
Develop data collection plan																																		
Data collection																																		
Measurement system analysis																																		
Identify Sigma level																																		
Measure poor and good cost of quality																																		

(continued)

Table 4.1 (continued)

	Quality costs	Process analysis tools	Stakeholder analysis	Project charter	House of quality	Data collection plan	Data collection	Six Sigma measurements	Gage R&R	Descriptive statistics	Inferential statistics	Statistical distributions	Hypothesis testing	Correlation analysis	Regression analysis	Analysis of Variance	Process capability analysis	Taguchi's Loss Function	Seven old tools	Seven new tools	Brainstorming	5 Whys analysis	Gantt chart	Dot plot	Run chart	Bar chart	Box-and-whisker plot	Probability plot	Design of Experiments	Simulation	Lean	Failure Modes and Effects Analysis (FMEA)	Control plans	Statistical Process Control (SPC)
ANALYZE																																		
Analyze the problem																																		
Analyze the process and data																																		
Identify the root causes																																		
Prioritize the root causes																																		
IMPROVE																																		
Develop alternative solutions to improve Sigma level																																		
Select the optimal solution/s																																		
Map future state																																		
Analyze the potential risks in the new process																																		
Pilot the test new process																																		
Implement the new process																																		
CONTROL																																		
Implement ongoing measurements																																		
Standardize the improvement																																		
Monitor the improvement																																		
Project closure																																		

Source: Author's creation

hensive list of the tools used in each phase. These tools are detailed in the relevant chapters.

4.2 Process Analysis and Documentation Tools

The Define phase, first and foremost, focuses on analyzing the processes and identifying problems that result in low performance in any components of the system. Therefore, the Define phase starts with analyzing and mapping the processes in detail. The way the processes are designed and operated give a hint about the root causes of the problem. In the Define phase, business processes are analyzed and documented through using tools, such as the transformation process, value stream map (VSM), flow charts, SIPOC diagrams, swim lanes, and spaghetti diagrams. The following sections detail how these tools can be employed in the Define phase of DMAIC process, respectively.

During process analysis in the Define phase, some questions are answered by Six Sigma project team. The answers for these questions shape the next phases of process analysis. These questions are:

- What position *owns* the process?
- What position is *responsible* for the improvement of the process?
- What are the *boundaries* and *bottlenecks* of the process?
- What are the *objectives* of the process?
- What *measurements* are followed in this process?

4.2.1 Transformation Process

In the Define phase, the components of the transformation process can be analyzed to discover where potential issues are more likely to occur. Transformation processes that include inputs, processes, and outputs of the system represent a generalized overall picture of the process, product, service, or organization. It shows what inputs are transferred into processes to produce designated outputs. There is also a feedback loop that links inputs with outputs to give feedback about the performance of outputs. The feedback is designed to hear the voice of the customers, internally and externally, and to improve the performance of the transformation process. Inputs demonstrate what kinds of resources are consumed in the system (◘ Fig. 4.1).

A process is a sequence of activities that aim to achieve a set of goals and objectives. Juran (1992: 219) identifies process as "a systematic series of actions to the achievement of a goal." Processes display the systems and activities that are built into the organization to produce the outputs. The processes are the key to error-free performance.

Outputs demonstrate the expected or planned outcomes of the system. The performance-related issues and problems may be related to components of the transformation process. Therefore, the details of the transformation process are analyzed in the Define phase of DMAIC process while developing problem statements and project charter. The transformation process also establishes the base of a SIPOC diagram.

According to Pyzdek and Keller (2018: 40), "process is the act of creating value." Being process-focused means that stakeholder values are determined and activities classified as value-added and non-value-added activities. In this sense, processes are analyzed to determine whether they *effectively* and *efficiently* create value for the stakeholders. *Effectiveness* is to deliver what customers demand on time, whereas *efficiency* means doing the things right or using fewer resources in finalizing planned activities. The following example shows how transformation process can be mapped and drawn in a fast-food restaurant.

► Example 1

Identify inputs, processes, and outputs of a fast-food restaurant using the transformation process. Draw the transformation process that includes inputs, processes, and outputs. ◄

▪ Solution

As seen in ◘ Fig. 4.1, *inputs* use a great variety of resources to run and feed the system, including employees, buildings, equipment,

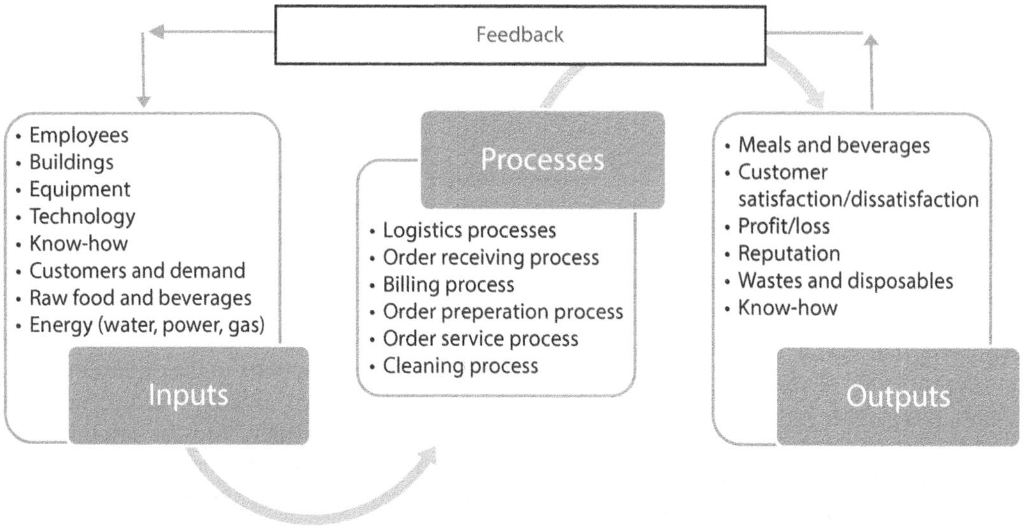

Fig. 4.1 Transformation process in a fast-food restaurant. (Source: Author's creation)

technology, know-how, customers and demand, raw food and beverages, and energy. Each input has a different function and task in the transformation process. *Processes* can be broken down into several segments starting from the first step of the value chain as presented in Fig. 4.1. Additionally, each sub-process can be detailed based on the structure of the sub-processes. *Outputs* vary based on the processes, such as meals and beverages, customer satisfaction or dissatisfaction, profit or loss, reputation, wastes and disposables, and know-how. Possible components of the transformation process are represented in Fig. 4.1.

4.2.2 Value Stream Analysis and Map

The VSM is considered one of the critical and most beneficial process analysis tools used in Six Sigma and Kaizen projects. VSM is a good way to picture how the process flows, how long each piece of process takes (cycle time), and where the obstacles and bottlenecks occur in a process or system. According to Womack and Jones (1996), VSM is the most important tool that Six Sigma project teams need to make sustainable progress in the war against *muda*, which means waste.

Ohno (1988) states that "[a]ll we are doing is looking at the time line (defined as lead time), from the moment the customer gives us an order to the point when we collect the cash. And we are reducing the time line by reducing the non-value adding wastes." From Ohno's point of view, VSM is one of the tools that can be used to picture the entire process to identify value-added and non-value-added activities. The activities placed on VSM are evaluated in terms of the potential value created for the final customer. Value added (VA), non-value-added but required (NVAR), and non-value-added (NVA) activities are identified on VSMs. Additionally, CTQ characteristics and performance indicators are shown on VSM. As a lean tool and an advanced version of traditional flow charts, VSM also demonstrates the flow of inventory, material, people, design outcomes, energy, and information throughout the system. VSM is also used to calculate value-added and non-value-added process time to shed some light on developing a problem statement.

VSM is categorized based on two decision-making criteria: (1) *level of VSM* and (2) *state of VSM*. In the level of VSM, there are two types of VSM: *high-level VSM* and *detailed VSM*. The high-level VSM contains major steps with cycle times of the process and is utilized in the Define phase to see the general structure of the system. The detailed VSM is used in the Measure phase to understand the details of the system and processes by breaking each process down into detailed steps (Wedgwood, 2007). In terms of state of VSM, *current-state* and *future-state VSMs* are generated for the systems. The current-state VSM shows how the system and process flow in the current system design, whereas the future-state VSM demonstrates where the process is targeted to be following the quality improvement activities.

To draw a VSM, a Six Sigma team can follow these steps:
- Step 1: Build the team.
- Step 2: Select the process.
- Step 3: Go to gemba.
- Step 4: Collect required data.
- Step 5: Draw and analyze current-state map.
- Step 6: Identify value added and non-value-added steps.
- Step 7: Construct future-state map.
- Step 8: Develop action plan on future-state map.
- Step 9: Measure improvements.

If the entire flow of the organization is mapped in VSM, a cross-functional team will definitely be required. As Six Sigma team follows this process step-by-step, the first process that will be analyzed is determined by the team by being in *gemba* to physically analyze the environment where the process takes place. Being in *gemba*, walking through the flow and process as a team, gives the team an opportunity to see the process and identify problematic points that create waste and bottlenecks. The team can also diagnose value-added and non-value-added activities that occur in the process during *gemba*. Required data, such as cycle time and waiting times between sub-processes, are collected and analyzed to determine the performance of CTQ characteristics on VSM. Cycle time and wait time are identified through time-and-motion studies performed by experts. Afterward, the current-state VSM is collaboratively drawn and analyzed by the team. Based on CTQ characteristics' values, NVA and VA activities are determined and marked on the VSM. The team, then, formalizes how the process should be performed. This analysis is collaboratively done to generate the future-state VSM. To shape the future-state VSM, additional inputs are required, such as target values of the CTQ characteristics determined at the upper hierarchy levels of the process. Before starting to draw the current-state VSM, several questions should be answered by the team, such as:
- What kind of bottlenecks and wastes will be targeted to be eliminated?
- What is the takt time in the process?
- What is the cycle time in each sub-process?
- How can we improve the flow of the process?
- Where can we reduce inventory, safety stock, and WIP?
- What are target values of CTQ characteristics?

Based upon the answers, the team can recognize how to shape the future-state VSM. After finalizing the future-state VSM, details of the action plan to achieve the future-state VSM are framed and formalized by the team. Upon implementing action plan, performance of the CTQ characteristics are measured and analyzed. Each step in the process flow of VSM takes careful consideration.

There are a great variety of symbols used in the VSM drawing process. Each symbol represents a particular meaning, action, activity, and information. Prior to engaging in the VSM drawing process, team members engage in training sessions where these symbols and their importance are highlighted.

► Example 2

The chair of an emergency room (ER) at a hospital receives complaints about the excessive wait times over the last months. Prior to analyzing the issue, the chair wants to see the system flows in the department. A Six Sigma team formed in the department analyzes the system and collects data to see the cycle time of each sub-process and activity. The team and chair draw a VSM to calculate the proportion of VA and NVA activities in the system. Cycle time and wait time in each sub-process are as follows, respectively:

- Triage: 5 and 3 minutes
- Registration: 4 and 2 minutes
- Initial exam: 5 and 23 minutes
- Physician exam: 8 and 10 minutes
- Lab test orders: 20 and 10 minutes
- Physician decision: 4 and 25 minutes.

Using the data, draw the VSM and calculate the percentages of VA and NVA activities in the ER. ◂

▪ **Solution**

The team analyzes the system comprehensively and draws VSM as presented in ▣ Image 4.1. Total lead time of the process was found to be 119 minutes, whereas VA time is 46 minutes for the system. The difference between total lead time and VA time refers to NVA activities with 73 minutes. After analyzing the flow of the process, it is seen that the longest waste time of 40 minutes occurs at the end of the system in between two last steps of the process. The longest cycle time is 12 minutes in "completion of labs and reqs" in the last sub-process. The department should start root cause analysis on the sub-processes that have the longest cycle time and wait time to decrease the total lead time and patient complaints.

4.2.3 Flow Chart

The flow chart is a graphic means to lay out the steps of a process (Juran, 1992: 46). ASQ defines flow chart as "a picture of the separate steps of a process in sequential order." Essentially, flow charts contain sequence of activities taken place in a process. The process flow charts show the steps of each job in sequence. They depict how the streams of materials flow during processing time and make it easier to understand and manage the system design (Juran and Gryna, 1980: 270). According to Deming (1986), the first step in any organization is to draw a flow diagram to show the relationship between com-

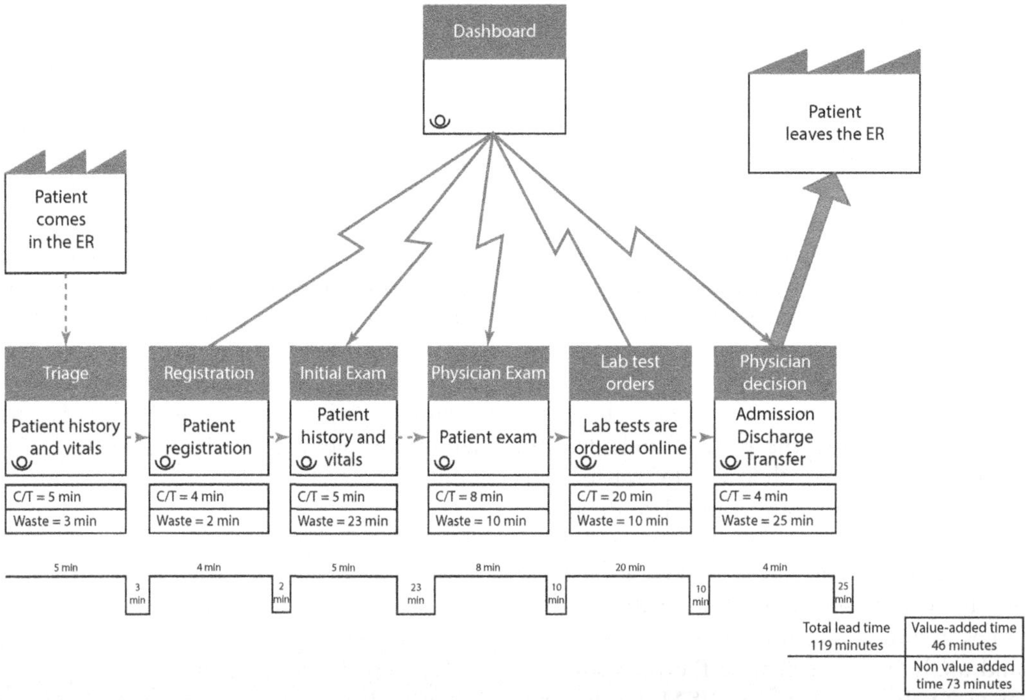

▣ **Image 4.1** A current-state VSM example. (Source: Author's creation)

Table 4.2 The symbols used in flow charts

Symbol	Meaning	Symbol	Meaning
	Process steps or activity symbol		Start and end of the process
	Direction of the process flow		Start and end of the process
	Decision-making points		Documents
	Delay or wait		Data storage
	Input or output		Preparation

Source: Author's creation based on ISO 5807 (1985)

ponents of the process. It also contains inputs and outputs, decisions made in each step, and actors taking a role in each step. Flow chart is also the most basic process mapping tool. To detail the process using advanced process mapping tools such as VSM, a basic flow chart is always drawn first. Flow charts can be used for a great variety of purposes in manufacturing, service, and administrative processes. A flow chart is useful for analyzing and picturing the process, identifying improvement needs, bottlenecks, and obstacles of the process, teaching and training the others on the process flows, documenting the process, and planning an improvement project on the process. A flow chart is drawn by a team of employees working in the relevant process or an internal expert who knows the process well. Drawing the flow chart should *not* be assigned to a group of experts from outside the company. It may be more beneficial for a multifunctional team in the organization to draw the process. When the flow charts are drawn by multifunctional teams, they usually depict the details of the process in a way to understand the system as a whole, taking into consideration customers, and identifying potential waste points. The flow charts can be drawn manually or electronically using software. The flow charts include a group of particular symbols. A list of most frequently used symbols standardized by ANSI (American National Standards Institute) is shown in Table 4.2. To construct a process flow chart, a straightforward flow can be used, as follows:

- Step 1: Define and identify the process.
- Step 2: Analyze the process step-by-step by the team members.
- Step 3: Identify the steps back to back that take place in the process.
- Step 4: Draw the process flow chart by using appropriate symbols.

The flow charts are also used in identifying the points that need to be improved and fixed in the process. An example of flow chart is demonstrated in Fig. 4.2. A general process flow of a fast-food restaurant can be drawn as depicted in Fig. 4.2. The first and last steps are shown as rounded rectangles. Decision-making points are displayed with diamond shapes. The flow of the process is shown by arrows. If necessary, the other symbols listed in Table 4.2 can be used as well.

4.2.4 SIPOC Diagram

The Six Sigma project teams draw and analyze SIPOC diagrams to define the processes. The main idea behind SIPOC diagram is to picture whole relationships graphically, starting from suppliers to relevant customers. SIPOC stands for suppliers, inputs, processes, out-

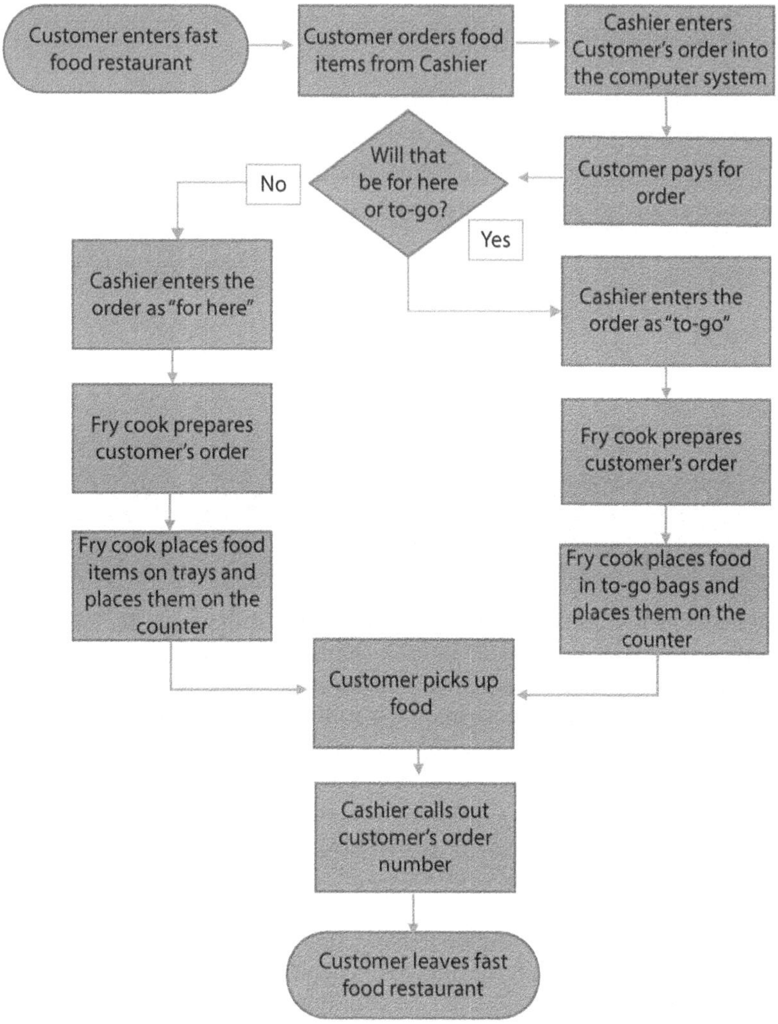

Fig. 4.2 A flow chart of fast-food restaurant. (Source: Author's creation)

puts, and customers. Juran (1992) termed the SIPOC diagram as a TRIPROL diagram in his *Juran on Quality by Design* book. SIPOC diagrams display a broader transformation process, including inputs, processes, and outputs. It identifies the components of the transformation process within a systems approach (Montgomery, 2009). It is considered a high-level flow chart to visualize the process and is a simple map of the process. The SIPOC diagram defines and documents the suppliers, inputs, key processes, expected outputs, CTQ characteristics, and potential customers of the process. It also helps team members create a common understanding about the scope and goal of the process and the project. SIPOC is a good way to visualize the components of the process. Detailed SIPOC diagrams give system analyzers and decision-makers a chance to see the whole picture of the system, identify potential and current bottlenecks, and pinpoint improvements in the organization. A SIPOC diagram is expected to answer the following questions:

- Suppliers – Who are the suppliers of the inputs in the process?
- Inputs – What kind of inputs are used in the process?
- Process – What kind of process/es are run?
- Process – What are the details of the sub-processes?
- Process – How is the process performed?
- Customers – Who are the customers of the process, internally and externally?

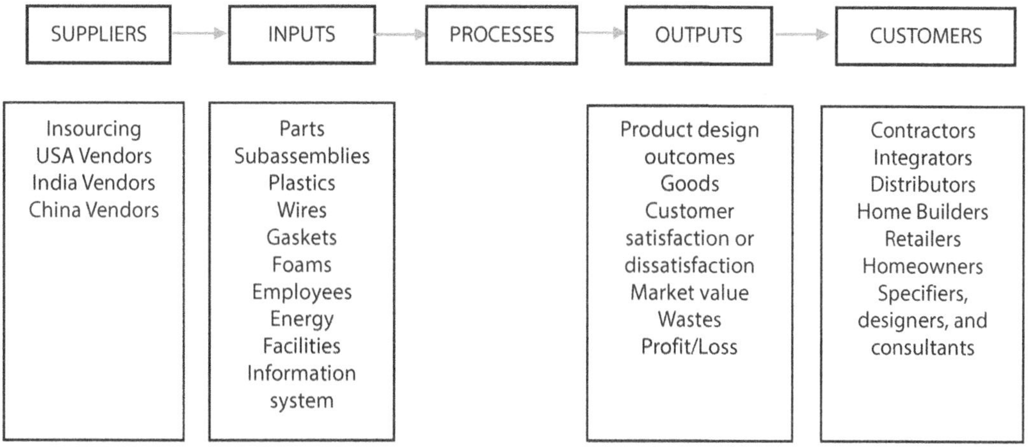

Fig. 4.3 SIPOC diagram of a wire producer. (Source: Author's creation)

A SIPOC diagram is drawn using the following steps:

- Step 1: Define and identify the system or process.
- Step 2: Analyze the system or process step-by-step by the team members.
- Step 3: Identify suppliers, inputs, processes, outputs, and customers of the process or system.
- Step 4: Draw the SIPOC diagram.

In the SIPOC diagram drawing process, the process or system that will be analyzed by the Six Sigma team is first defined and identified. Then, the team analyzes the system or process comprehensively. After a detailed analysis, all related suppliers, inputs, processes, outputs, and customers are placed in SIPOC diagram.

There are several advantages of using the SIPOC diagram for Six Sigma practitioners. First, as a broader version of the classical transformation process, a SIPOC diagram pictures the process or system with all related pre-process and post-process components of the system. If the Six Sigma team is in the Analyze phase of DMAIC, that is, if the team is trying to find out the causes of the problem, a SIPOC diagram displays the relationships between the components. Second, a SIPOC diagram functions as training material for beginners in the organization. Also, a SIPOC diagram is a data collection tool to analyze the process or system. Finally, SIPOC diagrams can be used for determining customer needs, expectations, and requirements. An example of a SIPOC diagram developed in a wire producer is shown in Fig. 4.3.

4.2.5 Swim Lane

Similar to previous process analysis tools, swim lanes help teams visualize and understand the flow of the process. Unlike the other tools, swim lanes also include functions and positions where the activities take place. Using swim lane as a process analysis tool gives team members a chance to see what is done by each department and position. Swim lanes explicitly show the responsibilities shared by the functions located in the process. In analyzing wastes and improvement needs, swim lanes help diagnose bottlenecks and obstacles within the process. Swim lanes also give teams an opportunity to effectively identify VA and NVA activities, redundancies, and synchronization issues between departments and functions.

Swim lane can also be drawn using a computer or manually, if necessary. A regular swim lane drawing process is detailed below.

- Step 1: Identify the lanes based on functions or departments that take a role in the process.
- Step 2: Identify the start point of the process.
- Step 3: Sequence the activities back to back starting from the first one.

- Step 4: Locate each activity in designated lanes depending on the functions and departments involved in the process.
- Step 5: Connect activities to each other and finalize the swim lane.

Swim lanes demonstrate concurrent activities and are drawn after VSM is complete. Swim lanes help other departments and functions understand the roles that others have in the entire process. If any duplication occurs anywhere in the process, the team will have an opportunity to diagnose them. Swim lanes can also be used to model the future and needed changes in the process.

▶ Example 3

Draw a swim lane for a fast-food restaurant that you have visited before. ◀

Solution

Using the steps given above, first, we split the positions and roles into three groups: customers, cashiers, and fry cook at a fast-food restaurant. Then, we add each piece of the process to relevant positions. Image 4.2 demonstrates an example of swim lane drawn for a fast-food restaurant.

4.2.6 Spaghetti Diagram

Spaghetti diagrams portray how long the inputs and outputs of the process travel in the work environment, process, or job shop. The movement of any kind of input, such as employees, raw material, or information, as well as outputs can be tracked on spaghetti diagrams. Spaghetti diagrams address efficiency and effectiveness needs emerging in the process, allowing resources to be used more effectively and efficiently.

Tracing the path of the inputs is a visual aid. Since the main idea of Six Sigma projects is to improve the process or system and reduce the waste and defectives, spaghetti diagrams generate useful information to determine where the improvements are needed. Spaghetti diagrams are used to identify where current bottlenecks and wastes occur and where to focus to eliminate waste and bottlenecks. Wastes and excessive transportation of the inputs are visualized using spaghetti diagrams, and further decisions can be made based on the information generated through spaghetti diagrams. In this sense, spaghetti diagrams are used to identify the problems in the processes. The idea behind the spaghetti diagrams is to (1) represent travel distance and patterns and (2) minimize the travel distance of the inputs and outputs throughout the system. In addition to team members, whoever works in the process is expected to join the team to help draw the spaghetti diagram.

To draw spaghetti diagrams, the following steps can be followed in a process approach:

- Step 1: Identify the scope and content of the process.

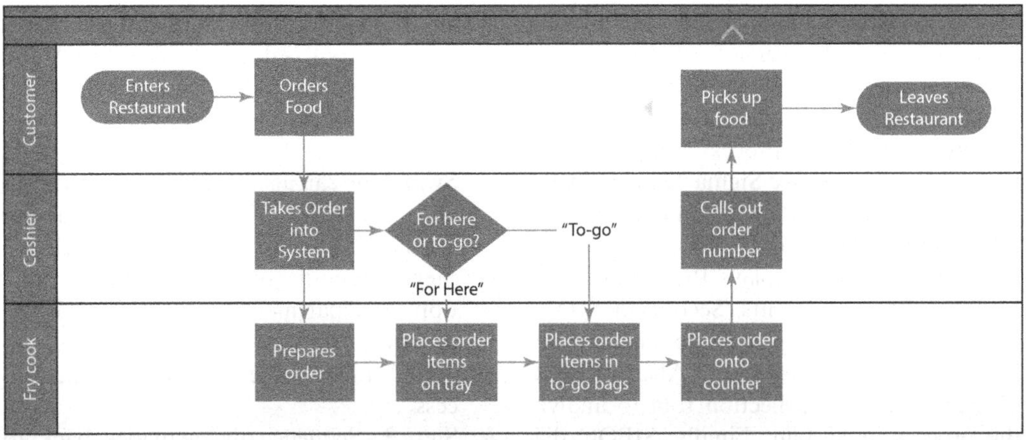

Image 4.2 A swim lane of fast-food restaurants. (Source: Author's creation)

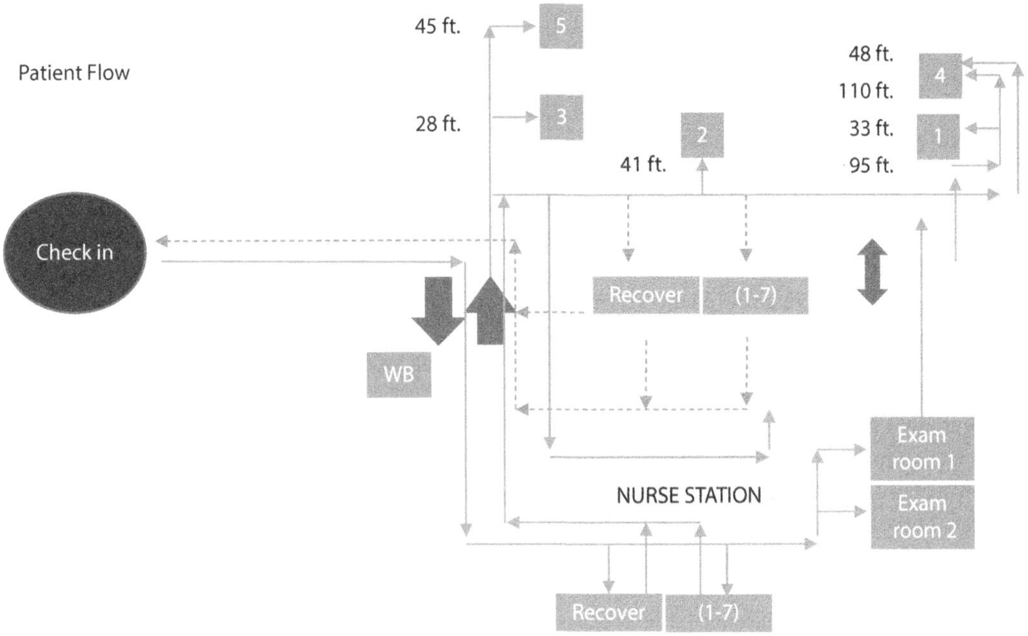

Fig. 4.4 Spaghetti diagram at the clinic. (Source: Author's creation)

- Step 2: Draw the process/facility layout.
- Step 3: Mark the activities, motions, and steps on the process/facility layout.
- Step 4: Show the links between all activities, motions, or steps.
- Step 5: Calculate the travel distance for each input.
- Step 6: Identify where the wastes occur.
- Step 7: Draw current-state spaghetti diagram.

► Example 4

An outpatient clinic that provides endoscopy services aims at minimizing total lead time of the services for patients in a few weeks. The industrial engineers working in the clinic in a Six Sigma team want to see how patients travel in the system during the service delivery process. Draw a spaghetti diagram and demonstrate potential travel distances and points in the clinic. ◄

Solution

Using the steps given above, first, Six Sigma team analyzes the flow of the system and calculates the distances between sub-processes that take place in service delivery process. Figure 4.4 demonstrates the spaghetti diagram drawn by the team. Flat lines show how the patients come in the system and follow the sub-processes. Then, dotted lines demonstrate where patients start leaving the system.

4.3 Stakeholder Analysis

Stakeholders are the groups without whose support the organization would cease to exist (Elias, 2016). They are the individual or institutional entities who have a direct or indirect interest or relationship with the organization. Employees, internal functions, and departments are considered internal stakeholders. External customers include suppliers, investors, shareholders, customers, regulators, unions, educational institutions, research and development institutions, the media, and federal and state government agencies.

As presented in the stakeholder theory (Freeman, 1984, 2010) and experienced in practice, stakeholders are critical, not only for Six Sigma success but also for the entire organization. Previous research (e.g., Glasgow et al., 2010; Laureani et al., 2013; Snee, 2010; Andersson et al., 2014) reveals that communication with stakeholders and understanding their roles and expectations in Six Sigma proj-

4

ects is a key success factor. Given the critical role of stakeholders and their satisfaction in overall organizational performance, Six Sigma projects must focus on stakeholders' needs and expectations in the Define phase in Six Sigma.

After analyzing and documenting the processes, the Six Sigma team conducts a stakeholder analysis. An effective stakeholder analysis should be able to show the Six Sigma team how stakeholders could react to potential improvements at the end of the project. The steps of stakeholder analysis are (1) identify stakeholders of the project, (2) identify and prioritize their interests and stakes in the project, and (3) document how stakeholders can impact the project and how they can be affected by the project. These three steps are detailed as follows:

- Step 1: Identify stakeholders of the project. First, the Six Sigma team identifies internal and external stakeholders of the process in the organization. Stakeholders in Six Sigma are the actors who are likely to affect the progress of the project and be affected by the project outcomes. Stakeholder analysis reveals who will help and challenge the Six Sigma project. In this step, the Six Sigma team draws a stakeholder map. Multiple brainstorming sessions may be helpful for teams to map stakeholders. An example of a stakeholder map is shown in ◘ Table 4.3.
- Step 2: Identify and prioritize their interests and stakes in the project.
  In this step, the Six Sigma team analyzes each stakeholder's interests and stakes. Their interests show the Six Sigma team what factors are important for stakeholders and what factors should be taken into account while identifying the goals, scope, and performance metrics of the project. Identifying stakeholders and understanding their needs and expectations regarding the Six Sigma project helps the Six Sigma team organize tasks, activities, and analyses. The Six Sigma team may begin with analyzing the various responsibilities of the team and company toward stakeholders using a stakeholder/responsibility matrix (◘ Table 4.4). Then, the team identifies and prioritizes the interests and stakes of each. Multiple brainstorming sessions help the Six Sigma team in this step.

◘ Table 4.3 Stakeholder map

Political environment	**Social environment**
Federal government agencies Local government agencies State government agencies	Unions Social activists Retired employees Community Media Educational institutions Special interest groups
Economic and technological environment	**Internal environment**
Actual and potential customers Investors Shareholders Research institutions Suppliers and business partners Rivals	Employees Owners Board of directors

Source: Adapted from Carroll and Buchholtz (2012)

While identifying and prioritizing stakeholders' interests and stakes in the project, several approaches can be taken by Six Sigma teams. One of the approaches is categorizing stakeholders using stakeholder categorization matrix that contains two dimensions: (1) stakeholders' potential for threatening the project and (2) stakeholders' potential to cooperate with the project (◘ Table 4.5). Using the matrix allows the Six Sigma team to differentiate which stakeholders are more likely to collaborate with the team or oppose the project. Identifying the position of stakeholder and quadrant in stakeholder categorization matrix helps the Six Sigma team determine what strategy may be effective for increasing the success of the project.

- Step 3: Document how stakeholders can impact the project and how they can be affected by the project.
  In this step, the Six Sigma team systematically documents the stakeholder map, stakeholder/responsibility matrix, stakeholder categorization matrix, and stakeholders' interests and stakes.

Table 4.4 The stakeholder/responsibility matrix

Stakeholders	Responsibilities			
	Legal	Ethical	Economic	Philanthropic
Political environment				
Federal government	X	X	X	
State government	X	X	X	
Social environment				
Unions	X	X		
Community	X	X		X
Media	X	X		
Economic and technological environment				
Actual customers	X	X	X	X
Potential customers	X	X		
Suppliers	X	X	X	X
Internal environment				
Employees	X	X	X	X
Board of directors	X	X	X	
Owners	X	X	X	

Source: Adapted from Carroll and Buchholtz (2012)

Table 4.5 Stakeholder categorization matrix

		Stakeholders' potential for threat to project	
		High	Low
Stakeholders' potential for cooperation with the project	High	Stakeholder: Mixed blessing Strategy: Collaborate	Stakeholder: Supportive Strategy: Involve
	Low	Stakeholder: Non-supportive Strategy: Defend	Stakeholder: Marginal Strategy: Monitor

Source: Adapted from Carroll and Buchholtz (2012)

4.4 Project Prioritization and Selection

Project prioritization and selection is the most important part of any business improvement process, specifically in Six Sigma efforts (Montgomery 2009). The management of the project selection step will cause long-term rejection or acceptance of Six Sigma approach throughout the organization in the early stages of the implementation. Decision-makers utilize *qualitative* and *quantitative* approaches to prioritize and select projects in Six Sigma. They can also prioritize and select projects based on single or multiple criterions. This section will analyze these qualitative and quantitative approaches that include single or multiple criterions.

4

4.4.1 Qualitative Approaches

Potential project topics are analyzed and evaluated in terms of several factors, considering the expectations of all related stakeholders in Six Sigma projects. For example, financial benefits of alternative projects, process priorities emerging on site, current performance level, customer expectations and complaints in the marketplace, employees' ideas, thoughts, and suggestions, or the reputation of the company are just a few of those factors. The projects that have potentially more influence on overall business performance and stakeholders are prioritized in Six Sigma. Additionally, the potential *value creation opportunity* of alternative projects is assessed to determine which project/s have the highest priority in this step. The financial impact of alternative projects must be demonstrated by Six Sigma teams and the executive committee. Montgomery (2013) suggests that if a project proposes at least a 50% annual return, it should be started. For example, Motorola exceeded 65% improvement rate in the first years of Six Sigma implementation. Choosing the right projects is the responsibility of executive committees in Six Sigma organizations. The executive committees have the authority to select project topics and identify emerging areas and priorities for the teams. More specifically, from the executive committees' perspective, performance on competitive advantage factors such as customer satisfaction, sales growth, new product development, and growth in operating income are some of the performance indicators on which potential impact of projects can be assessed. The potential projects that have relatively more impact on corporate performance should be prioritized. For example, if one potential project impacts customer dissatisfaction, defect, scrap and rework rates, customer returns, and warranty claims, this project should be given a higher priority than others.

Potential projects well aligned with corporate business vision, goals, and objectives are identified in this step as well. The association between (1) potential project scope and (2) strategic business direction and objectives of the company can help determine the importance of potential projects for company. In this approach, current strategic goals and objectives are identified and listed first. Second, the key business processes and key performance metrics associated with the project scope are listed. Then, executive committee members, project champions, and master black belts discuss and analyze the relationships between the two variables to determine the feasibility and impact of the potential projects. The projects with a real impact on key business metrics and strategic direction are selected. An effective and efficient project prioritization and selection improves the quality of Six Sigma project outcomes. In other examples, companies establish project selection committees and organize meeting sessions including various stakeholders to select projects for Six Sigma.

In the project prioritization and selection process, the time frame of the potential project also is considered to analyze the feasibility of the project. If the expected time frame is relatively longer, the priority will be given to alternative projects that have relatively shorter time frames. In some alternative projects, it may not be possible to accurately estimate the time frame. In this case, the best estimation is made for projects and the first priority is given to the projects with shorter projected completion times. Another important decision-making criterion for project prioritization and selection process is the payback period of the potential project. When payback period and payback probability are included in project prioritization and selection process, decision-making processes may generate the best results.

Gitlow et al. (2015) suggest using a project identification matrix, as presented in ◘ Table 4.6, to determine the types of information used for identifying the potential project topics. *Proactive* and *reactive approaches* can be implemented in the project prioritization and selection process. Furthermore, *internal* and *external information* are utilized in this process to broaden the project selection. For example, Juran and Gryna (1980: 477) emphasize customer complaints depend on economic climate, age, affluence, technological skills of users, seriousness of the defect or problem, and

Table 4.6 Project identification matrix

Information resources	Proactive	Reactive
Internal	Strategic and tactical plans Voice of employee Employee focus groups Employee survey results	Employee suggestions and complaints
External	Voice of customers Customer focus groups Customer survey results Regulatory compliance issues	Customer suggestions and complaints Regulatory compliance issues

Source: Adapted from Gitlow et al. (2015: 231)

Table 4.7 An example of weight-based project evaluation

Project topic: analyzing strength of metal gaskets **Master black belt: director of operations** **Black belt: supervisor I** **Members:**		**Evaluation score:** **Evaluation date:**	
Evaluation criterions	**Weight**	**Assigned score**	**Weighted score**
Financial impact	0.20	90	18
Resource availability	0.15	80	12
Time span	0.19	70	13.3
Complexity	0.23	100	23
Scope	0.23	90	20.7
Total weighted score			87

Source: Author's creation

price of the product/service. They also state that low complaint rate does not guarantee high customer satisfaction. Instead, high customer complaint rate is an indication of customer dissatisfaction. Similar to the voice of customers, external information resources are taken into consideration when prioritizing projects. Besides, the voice of employees brings many different points to the attention of Six Sigma teams and considered internal information.

4.4.2 Quantitative Approaches

To evaluate and prioritize potential projects, quantitative approaches can be implemented in decision-making processes by Six Sigma teams and executive committees. Based on the basic quantitative approach, point- and weight-based evaluation systems help decision-makers assign priorities to potential project topics. An example of evaluation mechanism is presented in Table 4.7. In this example, decision-makers consider five evaluation criterions: financial impact, resource availability, time span, complexity, and scope. Each evaluation criterion is assigned a weight, as presented in the second column in Table 4.7. The third column shows the assigned scores (out of 100) specified by the decision-makers for each criterion. The last column represents the weighted score for each criterion as the product of weight and assigned score. For example, the weight of the financial impact criterion is 0.20, and assigned score by the decision-makers is 90 out of 100. Weighted score is 18, as the product of 0.20 and 90. Total weighted score is calculated for each potential project topic as presented in Table 4.7. For the example, total weighted score is 87 out of 100. Other project topics are evaluated using the same approach, and decision-makers give priority to the project topic that has the highest total weighted score.

For project prioritization and selection problems, advanced quantitative approaches can be used by the teams. For example, Banuelas et al. (2006) reported that some quantitative methods, such as cost-benefit analysis, Pareto analysis, analytic hierarchy process (AHP), theory of constraints (TOC), and QFD, were implemented in project prioritization and selection practices in the UK. Given that problems in business world are multi-faced, it may not be easy to pick and choose one problem based on just one decision-making criterion. While problem A may seem a better option from a *cost* point of view, problem B may appear more convenient one based on both *cost* and *time* perspectives. Project prioritization and selection is a challenging process for decision-makers, considering the degree of the complexity of the potential project topics and problems. Several potential project topics can be analyzed, but only one of them is selected based on several criteria to focus on in each cycle. Considering that decisions are made on the basis of multiple criteria, the nature of the decision-making process in project prioritization and selection step is heavily aligned with multi-criteria decision-making (MCDM). Wang et al. (2014) identify project prioritization and selection in Six Sigma methodology as an MCDM problem. Since there is more than one criterion affecting the selection of the project, Six Sigma team members or executive committee may have different ideas and selections on the topics. MCDM is helpful for optimizing decisions on project selection in that regard. The MCDM approach has been used in a great variety of areas over the last several decades. Some of the most common MCDM methods used in project prioritization and selection are multi-attribute utility theory, analytic hierarchy process (AHP), fuzzy set theory, case-based reasoning, data envelopment analysis (DEA), simple multi-attribute rating technique, goal programming, ELECTRE, PROMETHEE, simple additive weighting, fuzzy AHP, fuzzy weighted additive goal programming model, analytic network process (ANP), fuzzy ANP, decision-making trial and evaluation laboratory (DEMATEL), and technique for order of preference by similarity to ideal solution (TOPSIS) (Velasquez & Hester, 2013; Wang et al., 2014).

The literature and practice suggest using multiple methods together to minimize the effects of disadvantages of each method. MCDM methods require a mathematical background to run the algorithms. To improve decision-making processes, particularly while selecting a project for Six Sigma teams, MCDM methods help decision-makers focus on more than one criterion. Also, the subjectivity in decision-making process is minimized due to the number of the decision-makers, the structure of the criteria, and the algorithm implemented in the methods. Since Six Sigma is a project-driven methodology, prior to implementing any MCDM method, Six Sigma teams and decision-makers need to understand the advantages and disadvantages of the methods. Since the main focus of the chapter is not about teaching MCDM, we will not detail tools of MCDM in this section.

4.5 Project Charter

One of the basic and most important deliverables of the Define phase is considered "project charter." The project charter demonstrates *general structure* of the project, *project overview*, and *schedule* of the project. The project charter is built and outlined by the project champion, process owner, master black belt, black belt, green belt, and team members. The project charter is a formal overall summary of the project. It is also considered a contract between the project team and other stakeholders in the organization. If the project team needs to make any changes on the project plans, the project charter is revised accordingly. Project charters are expected to be dynamic documents. An example of a project charter is presented in ◘ Table 4.8. The basic elements of a project charter are:

- Problem statement
- Goal statement
- Project scope
- Project metrics and CTQ characteristics
- Estimated benefits of the project
- External and internal stakeholders
- Project team members
- Project leader and facilitator
- Project champion
- Process owner
- Project deliverables
- Project timetable.

Table 4.8 An example of project charter

Project charter			
Project #:		Project name:	
Project champion:		Process owner:	
Project leader:		Process:	
Start date:		Estimated final date:	
Define:	Measure:	Define:	Measure:
Analyze:	Improve:	Analyze:	Improve:
Project team members			
Name	Department	Position in the team	Position in the dept.
Project overview			
Problem statement:			
Goal statement:			
Project scope:			
Project metrics:			
Estimated benefits:			
Signatures			
Project champion:			Date:
Project lead:			Date:
Process owner:			Date:

Source: Author's creation

4.5.1 Problem Statement

The problem statement presents the problem that will be analyzed in the project. In the project charter, the definition of the problem focuses on the current status of the problem at the beginning of the project, whereas the goal statement represents where the project team plans to be at the end of the project. A good problem statement is clear, concise, specific, understandable, and measurable. It also should not include the reasons and solutions for the problem. A project charter answers the *what, where, who, when,* and *how* questions of a project. While articulating the problem statement, these questions are taken into consideration:

- *What* is the problem?
- Does the problem focus on symptoms, not causes or solutions? (Montgomery, 2009)
- *Where* does this problem occur?
- *When* does this problem occur?
- *What* are the conditions under which this problem occur?
- *Who* are the key stakeholders involved in this problem?
- *When* and *how* did you notice the problem?
- *What* is the extent of the problem?
- *What* are the impacts of problem on processes, products/service, other functions, stakeholders, overall performance, and throughout the organization?
- Has process been mapped using SIPOC, value stream map, etc.?
- *What* is the process that will be analyzed in the project?
- *Who* is the owner of the process?

Six Sigma teams should make sure that the team concentrates on a feasible problem that has been determined in the project prioritization and selection step, using either qualitative or quantitative approaches. The way the problem is identified and articulated in the project charter affects the focus and quality of the project. The quality of the problem statement is linked with the knowledge, skills, and abilities of the team members, leader, and facilitator.

4

► Example 5

As example problem statements in the Define phase, let's focus on these two statements:

1. Since the beginning of 2020, customers have been complaining about the prolonged waiting time in the ordering process.
2. The rate of defective units in assembly line 1 has been on the rise for the last 6 months and increased to 7.5% since last month. ◄

▪ Solution

Each problem statement answers *where, what, how,* and *when* questions. They are *time bound,* include *CTQ characteristic* and *performance level*, and address *where* the problems occur. These examples show how a problem is articulated by Six Sigma teams. Let's answer the following questions for each problem statement as follows:

Statement 1:

- What is the problem? Customers complain about the long wait time.
- Where is it occurring? Ordering process.
- When did you notice it? Since the beginning of 2020.
- What is the extent of the problem? No information about it.
- What is its impact? Customer complaints occur.

Statement 2:

- What is the problem? The rate of defective has been on the rise.
- Where is it occurring? Assembly line 1.
- When did you notice it? Over the last 6 months.
- What is the extent of the problem? No information about it.
- What is its impact? Defectives occur.

► Example 6

Identify problem statement for the scenario given below:

The customers submit complaints about the quality of the products sold in an online shopping web site in the last month. The decision-makers analyze the details of the customer complaints and find that the majority of the customer (87.25%) also returned the products last month. ◄

▪ Solution

Problem statement: The majority (87.25%) of the complaining customers of an online shopping web site returned the products last month.

► Example 7

Identify problem statement for the scenario given below:

The team of a manufacturing cell studies the effects of "temperature" on "tensile strength of metal cable" in a multinational metal wire manufacturing plant in New England area. The team knows that temperature is the most significant factor of the tensile strength in the process and that the ideal range of temperature is between 510 ° F and 540 ° F in the cell. The team discovers that rework rate significantly increases in the cell when the temperature is out of the ideal range and the rework rate has been 56.73% in the last 2-week period. ◄

▪ Solution

Problem statement: The rework rate has been 56.73% in the last two-week period in the cell when the temperature is out of ideal range.

4.5.2 Goal Statement

A goal statement includes what is meant to be accomplished (outcomes) in the chosen Six Sigma project. The goal of the team is articulated to represent what the team plans to accomplish and what CTQ characteristics or performance metrics will be measured and improved throughout the project cycle. Goal statements are SMART: *specific, measurable, achievable, realis-*

tic, and *time oriented.* The following questions help develop an effective goal statement:
- What are the goals of the process?
- What will be improved in the project?
- What is targeted at the end of the project?
- What business metrics are measured in the process and project?
- What business goals and objectives are linked with the project?

The goal statement begins with a verb such as "reduce," "improve," "eliminate," or "increase" (Gitlow et al. 2015). The structure of the goal statement can be constructed as follows:

» Improve (primary metric) from (baseline level) to (target level) by (a certain date)

► Example 8

Let's articulate goal statements for the problem statements in ► Example 5. To recall, the first problem is: "Since the beginning of 2020, customers have been complaining about the prolonged waiting time in ordering process." The second problem is: "the rate of defective units in assembly line 1 has been on the rise for the last 6 months and increased to 7.5% since last month." ◄

■ Solution

To generate a goal statement for the first problem statement, let's focus on the format above:

» Improve (primary metric) from (baseline level) to (target level) by (a certain date).

Then, we can develop the following goal statement based on the targeted values and time periods by the Six Sigma team.

» Reduce (customer complaints) from (25%) to (10%) by (the end of 2020).

The second problem in ► Example 5 is "the rate of defective units in assembly line 1 has been on the rise for the last 6 months and increased to 7.5% since last month." To use the format, let's remember the format of the goal statement above.

» Improve (primary metric) from (baseline level) to (target level) by (a certain date).

Then, now, we can generate our goal statement using the format above:

» Reduce (defective rate in assembly line 1) from (7.5%) to (2.5%) by (the end of this year).

Practice question 1

Recall the problem given in Example 6 in problem statement section above and develop a goal statement using the format. Here is the problem:

The customers submit complaints about the quality of the products sold in an online shopping web site in the last month. The decision-makers analyze the details of the customer complaints and find that the majority of the customer (87.25%) also returned the products last month.

Practice question 2

Recall the problem given in Example 7 in problem statement section above and develop a goal statement using the format. Here is the problem:

The team of a manufacturing cell studies the effects of "temperature" on "tensile strength of metal cable" in a multinational metal wire manufacturing plant in New England area. The team knows that temperature is the most significant factor of the tensile strength in the process and that the ideal range of temperature is between 510 ° F and 540 ° F in the cell. The team discovers that rework rate significantly increases in the cell when the temperature is out of the ideal range and the rework rate has been 56.73% in the last 2-week period.

4.5.3 Project Scope

The project scope identifies the boundaries of the problem in terms of types of processes, resource and time availability, responsibility scope, obstacles, and other constraints. The project scope also may refer to the functions or departments involved in the project. A well-defined project scope helps identify root causes of problems and the potential solutions. Focusing on a broader scope may limit a Six Sigma project team and use resource ineffec-

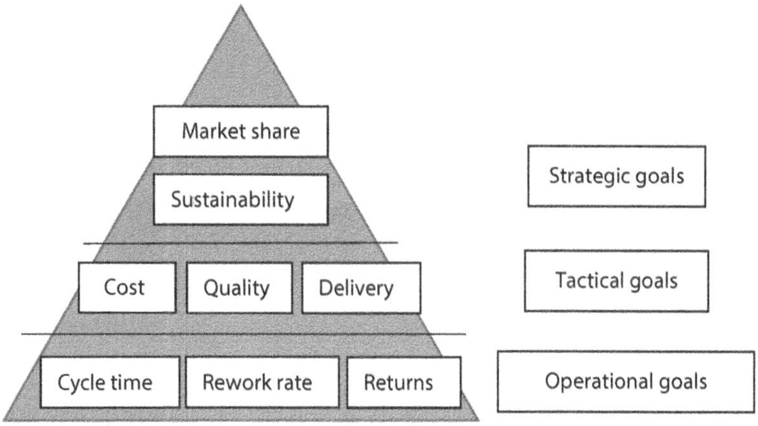

Fig. 4.5 The levels of the metrics. (Source: Author's creation)

tively during the project time. Therefore, a project charter needs to be very specific about the scope of the project. If there is more than one related scope that are covered in the same project, the team needs to clearly identify the boundaries of the project. In this process, identifying external and internal stakeholders will allow the team to understand the problem, identify the scope, and determine related variables and factors. Therefore, analyzing external and internal stakeholders' roles and expectations in the problem will help identify the scope of the project.

4.5.4 Project Metrics

Deciding what to measure is crucial to the success of the organization. Selecting wrong or improper metrics will result in ineffective Measure phase in Six Sigma (Pyzdek and Keller, 2018: 145). Depending on the problem, scope, and goal of Six Sigma project, many different metrics can be measured and analyzed. Joiner (1994) addresses three high-level performance metrics: (1) customer satisfaction, (2) cycle time, and (3) first-pass quality for business organizations. Some metrics are a sole metric, whereas some others are linked with other upper level ones that affect the overall performance of the organization. These metrics are placed in dashboards to communicate the performance of the company to all relevant stakeholders. For example, defect rate, number of defective products, or customer complaints are monitored as metrics. Although these metrics seem to be operational metrics, they are strategically related to some upper metrics, such as quality costs, market share, and reputation of the company. The metrics are directly linked with the long-term planning of the organizations. Although some metrics are directly process- and product-oriented at the micro level, they allow the organization to achieve operational, tactical, and strategic goals and objectives. Therefore, the link between long-term performance and the structure of metrics must be carefully analyzed and mapped. In addition, monitoring and managing the performance of those metrics is communicated throughout the organization. To analyze impact of metrics on the long-term performance, a cascade system that includes multiple levels of goals and objectives, that is, operational, tactical, and strategic, can be used in the organizations (Fig. 4.5).

Metrics must be determined in a participative and collaborative teamwork where each function, product, service, process, or item is equally represented. If the metrics are developed collaboratively, measurement of metrics and making decisions based on these metrics will be more easily embraced by departments and functions. If employees do not have a chance to contribute to the metric identification and development process, they are more likely to react negatively about monitoring these metrics and making decisions. Data collection, data analysis, and decision-making processes are problematic when employees are not involved in the metric identification and development process. The customer needs and expectations, features of the product or service, factors affecting the performance of the product or service, the steps of the process, and standards and regulations related to product, service, or process impact the process of identifying metrics. After deciding which metrics will be monitored, the decision is communicated to the relevant departments and functions in the organization.

Regardless of the type of the metrics, they can be seen as "*Y*" (dependent, response) variables in Six Sigma projects. Since many independent or interdependent factors affect metrics, identifying those factors affecting the performance of metrics allows teams to discover the relationship between all these factors and their metrics.

4.5.5 Project CTQ Characteristics

According to Juran (1992: 7), "A product feature is a property possessed by a product that is intended to meet certain customer needs and thereby provide customer satisfaction... A more technical definition of product feature is quality characteristic." CTQ characteristics also function as operational definition. According to Carey (2003), the aim of an operational definition is to create a clear communication environment between the actors since a good operational definition creates an understandable meaning for everyone and reduces variability. Feigenbaum (1983) states that *reliability, serviceability,* and *maintainability* are some of the individual operational characteristics that make up product and service quality. As well as David Garvin's quality dimensions (► Chap. 1), these dimensions are taken into consideration while formalizing operational definitions of the products and services. Operational definitions in Six Sigma represent the features of the topic that are observed, monitored, and measured. At the very bottom line, quality dimensions help identify operational definitions for products, services, and processes. Operational definitions are finalized prior to the data collection process. Pyzdek and Keller (2018: 412) give a few examples for operational definitions:

- Operational definition of "Intelligence": administer the Stanford-Binet IQ test to a person and score the result. The person's intelligence is the score on the test.
- Operational definition of acids and bases: An acid is any substance that increases the concentration of the H^+ ion when it dissolves in water. A base is any substance that increases the concentration of the OH^+ ion when it dissolves in water.

The CTQ characteristics are derived from the technical requirements and specifications of a product/service or process based on all relevant stakeholders' needs and expectations. Six Sigma project topics consist of measurable or countable CTQ characteristics. In other words, CTQ characteristics are quality dimensions of a product/service or process. There are other equivalent terms of CTQ characteristics, such as key output variables, key performance indicators, or Big Ys. To generate CTQ characteristics, a few critical questions can be asked by Six Sigma teams by comparing expectations and actual performance. For example:

1. Did the process or product/service meet the expectations or fail to meet the expectations?
2. What do the customers expect and want from product/service?
3. What does the voice of customer say?

In Six Sigma projects and other continuous quality improvement efforts, the CTQ characteristics are determined by the relevant actors in the projects and articulated in depth. The first set of actors is external customers because customer needs and expectations identify and shape the characteristics of the final product/service. The second set of actors is internal customers and functional internal stakeholders such as marketing experts, financial specialists, design engineers, and production engineers. In a cross-functional approach, all related departments and functions are involved in this process. In some cases, CTQ characteristics can be easily identified if customers' needs are explicit and understandable. However, in some cases, customers' key metrics are not easy to articulate, define, and measure. If customers' key metrics are not well-defined, internal metrics and project deliverables are used to generate CTQ characteristics. Additionally, the CTQ characteristics are related to the national and global standards.

The CTQ characteristics are categorized into two main groups: continuous and discrete CTQ characteristics. Continuous CTQ characteristics are measurable characteristics of the product/service or process. To exemplify, these can be the diameter of a part of the product, weight and height of a product part, thickness of a part, diameter of a metal cylinder, length of a pencil, and average wait time per customer in queue in a bank per day. Discrete CTQ characteristics are attribute data, such as number of defectives, defects, customer complaints, and returns. For example, the percentage of billing errors in a coffee shop per hour and number of medical errors done per day are good examples. Specifically, in Six Sigma organizations, defects per unit (dpu), defects per million units (dpmu), defects per opportunities (dpo), and defects per million opportunities (dpmo) are discrete CTQ characteristics. These CTQ characteristics, also known as Six Sigma measurements, are analyzed in the ► Sect. 5.7 in ► Chap. 5.

If the CTQ characteristics vary uncontrollably over time, major problems and issues may appear, especially during the data analysis stages. The operational definitions should be unambiguous. Measurable and identifiable terms articulate the CTQ characteristics. For example, if the CTQ characteristic is "temperature of a metal component," things to consider include (1) what that temperature technically means for this metal component, (2) what the target value/s is/are, (3) what the tolerances are, and (4) how the temperature is measured, which will allow data to be identified in depth. Quality plans demonstrate the details of the CTQ characteristics of a product, service, or process. Quality plans include:

1. CTQ characteristics that show the expected performance of the product/service or process
2. Units of measure of CTQ characteristics
3. Target (goal) value
4. Tolerances
5. Quality control methods
6. Frequency of measurement
7. Sample size
8. Decision-making criteria
9. Measurement tool
10. Measurement operator
11. Recording of measurement
12. Responsible decision-makers.

Depending on the structure of the product/process, quality plans are constructed by multifunctional teams and include at least R&D (research and development), design, and production functions. A quality plan example is shown in ◘ Table 4.9 below. If the pencil is the end product, let's assume that (1) length, (2) weight, (3) diameter of the edge, (4) color tone, and (5) strength of the pencil are the CTQ characteristics identified by the relevant functions in light of customer needs and expectations and standards developed for pencils and presented in quality plans. For example, in ◘ Table 4.9, quality control methods give the details of inspection procedures and instructions for each CTQ characteristic. Decision-making criteria is as 1/10, meaning that detecting one pencil out of ten in a batch will cause the batch to be rejected or failing to pass quality control.

4.5.6 Project Deliverables

The project deliverables refer to performance measures or indicators to evaluate the success of the project. The project deliverables are specific outputs of the work activities in a project. Each activity's output identifies the deliverables of the relevant activity. The critical deliverables are related to *cost, quality,* and *time* in the bigger picture of the project. The deliverables reflect the importance of project in terms of organizational performance. The relationships between project outcomes and performance indicators, such as quality, cost, and time, are explicitly presented in project charters. The potential outcomes of the project are stated in a way to allow decision-makers to see the impact of the project on cost, quality, and time.

Project deliverables are categorized based on internal and external stakeholders in traditional project management. The DMAIC flow also helps Six Sigma teams identify key deliverables phase by phase. Each phase of DMAIC

Table 4.9 A sample of quality plan

Product: Pencil			Product group code: A12–56-5			Revision: 0013		Revision date: 1/1/2020		
CTQ characteristic	**Unit of measure**	**Target**	**Tolerances**	**Quality control methods**	**Frequency of measurement**	**Sample size**	**Decision-making criteria**	**Measurement tool**	**Measurement operator**	**Form**
Length	cm	10 cm	±0.5 cm	Instruction 001	Each batch per line	10/batch	If 1/10 is rejected, reject the batch	Digital gauge	Quality control technician 1	**A12–56-5/F1**
Weight	gram	2.9 g	±0.005 gr	Instruction 002	Each batch per line	10/batch	If 1/10 is rejected, reject the batch	Digital gauge	Quality control technician 2	**A12–56-5/F2**
Diameter of the edge	cm	0.5 cm	±0.0004 cm	Instruction 003	Each batch per line	10/batch	If 1/10 is rejected, reject the batch	Digital gauge	Quality control technician 1	**A12–56-5/F3**
Color tone	Color palette	Dark black	0	Instruction 004	Each batch per line	10/batch	If no defects found, passes	Color palette	Quality control technician 3	**A12–56-5/F4**
Strength	Kg	75 kg min	+0.5 kg	Instruction 005	Each batch per line	10/batch	If no defects found, passes	Visual count	Quality control technician 2	**A12–56-5/F5**

Source: Author's creation

process presents different key deliverables as follows:

Define Process analysis outcomes, customer needs and expectations, House of Quality, stakeholder analysis outcomes, Six Sigma team, project charter, Project plan

Measure Data collection plans, data, Gage R&R analysis results, Six Sigma measurements, quality costs

Analyze Descriptive statistics of data, inferential statistics of data, distribution of data, hypothesis testing results, correlation and regression analysis results, ANOVA results, process capability analysis results

Improve Design of experiment results, simulation results, failure modes and effects analysis results

Control Control plans, statistical process control plans, project disclosure

4.6 Project Planning

Prior to proceeding in DMAIC process, Six Sigma teams focus on the details of the project plans. Project management discipline helps organize the details of the project plans. Project plan should answer a group of questions for Six Sigma team. Some of these questions are as follows:

- What activities and tasks will be performed in the project?
- What is the sequence and predecessors of the activities?
- What is the activity duration time of each activity in the project?
- What is the estimated project duration time?
- What position is responsible for what activity in the project?
- Is there any lack of resources to run the project?
- What is the estimated cost of the project?

As one of the main deliverables in the Define phase, the Gantt chart is used to schedule the project and assign the activities to relevant roles in Six Sigma projects (see ▶ Sect. 8.3.3 in ▶ Chap. 8). Since the details of the project management is not in the scope of this chapter, further readings can be used for the details of the project plans.

4.7 Quality Function Deployment

Businesses need to hear the "voice of the customer" in detail by using effective tools and methodologies, to understand the needs and expectations of customers before their rivals. In the Define phase, the voice of the customers is heard and transferred into pool of project topics to consider all potential problems for Six Sigma projects.

As stated by Pakdil et al. (2012: 1397), "a thorough and accurate understanding of customer demands and expectations is the key to possessing competitive advantage in the marketplace." Quality function deployment (QFD) is one of the most effective tools to understand and apply these demands and expectations in production lines. The main idea of QFD is to meet customer needs and expectations by transferring the voice of customer into design specifications and outcomes.

QFD is utilized in interdisciplinary teamwork to increase the performance of business organizations in the marketplace. QFD was developed by Yoji Akao in the 1960s and introduced at the Kobe Shipyard of Mitsubishi Heavy Industries Ltd in the 1970s (Akao 1990). QFD has been implemented by large-size organizations such as Toyota, GM, Ford, HP, and Texas-Instruments since 1980s. Bicknell and Bicknell (1995, p. 28) define QFD as "a systematic approach mapping the customers' needs into definable and measurable product and process parameters, using matrices and other quantitative and qualitative techniques." As a result, QFD addresses the most important customer expectations on customer side and technical requirements on manufacturer's side.

To hear the voice of customer, several questions should be answered by relevant departments, such as marketing, R&D, engi-

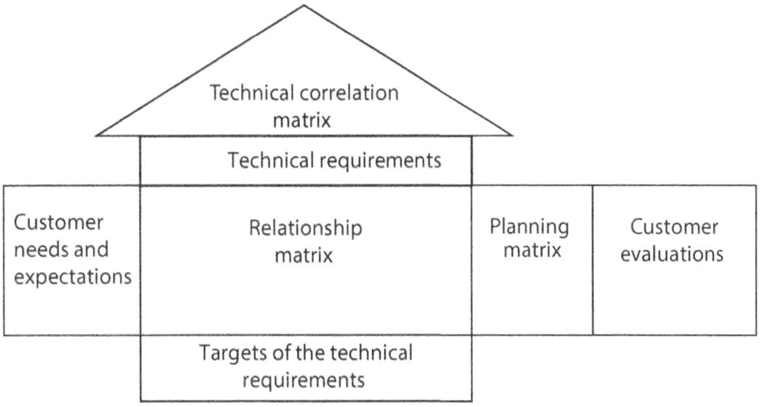

Fig. 4.6 House of Quality (HOQ). (Source: Author's creation)

neering, and production, after contacting the customers. Some of these questions are:
- What do our customers need from us?
- What are our customers' needs and expectations?
- How capable are our products/services and processes to meet my customers' needs and expectations?
- How is our process performance from the customer perspective?
- How do our customers evaluate and view my products/services and processes?
- How does my customer measure my process?
- How can we improve our products/services and processes?
- How would my customer like for our process to perform?
- How can we do better?

The QFD procedure is operated through the House of Quality (HOQ) diagram. An HOQ combines detailed information about the customers' needs (what) and design specifications (how). The regular QFD process is performed in six phases of HOQ:
1. Customer needs and expectations
2. Planning matrix
3. Technical requirements
4. Relationship matrix
5. Technical correlation matrix
6. Customer evaluations (Fig. 4.6).

These phases are interconnected within the HOQ. Generally, QFD is processed in HOQ as follows. First and foremost, the voice of customer is heard and converted out to *customer needs and expectations* section. *The planning matrix* focuses on identifying the most important customer expectations, with several variables. Then, the voice of customer is transformed into *technical requirements* in columns of HOQ to identify design specifications and CTQ characteristics. Each customer expectation is associated with relevant technical requirements in the *relationship matrix*. This step is essential to see how the technical requirements are met to satisfy each customer expectation. *The technical correlation matrix* assigns associations between technical requirements in the roof of the HOQ. Finally, the satisfaction level of the customers on each expectation is compared with the competitor's performance in the *customer evaluations*. The details of this process are presented below, and each step is detailed in the following sections:
- Step 1: Build the QFD team.
- Step 2: Determine customer needs and expectations.
- Step 3: Build the planning matrix.
- Step 4: Convert customer needs and expectations into technical requirements.
- Step 5: Create the relationship matrix.
- Step 6: Construct the technical correlation matrix.
- Step 7: Customer evaluations.
- Step 8: Determining relatively more important customer needs and expectations and technical requirements.

- Step 1: Build the QFD team

Similar to other tools and methodologies used in Six Sigma, QFD requires an effective and systematic teamwork. Prior to running QFD process, the firm needs to make sure that a team including all relevant departments and functions has already been established. Marketing, engineering, operations, manufacturing, R&D, and finance departments are represented in such QFD teams.

4

- Step 2: Determine customer needs and expectations
  One of the competitive advantages for business organizations is to understand current and potential customers' demands and expectations. The firms are required to hear the voice of their customer more accurately and earlier than their rivals in the marketplace (Pakdil & Kurtulmuşoğlu, 2014). Therefore, QFD team starts with identifying customer needs and expectations in detail. Marketing-oriented tools and techniques, such as market research, focus groups, interviews, content analysis, critical incident analysis, and surveys, are effectively utilized in this step. To determine customer needs and expectations and answer the questions given at the beginning of this chapter, companies can also use process mapping tools such as SIPOC diagrams, process flow charts, and value stream maps. While process mapping tools are mostly used by quality professionals, the other tools are effectively used by marketing functions.

  Juran (1992: 73) classifies customer needs into five categories: (1) stated needs, (2) real needs, (3) perceived needs, (4) cultural needs, and (5) needs traceable to unintended use. Customers state their needs in terms of what they would like to buy. However, their real needs may not be as they stated. For example, they may state an automobile as a need; however, their real needs are related to transportation. While they state a need about taking and passing a course at a college, the real need is to complete a degree's requirements. Missing the differences between stated needs and real needs results in complicated problems for manufacturers and service providers. Perceived needs imply the needs felt by the potential customers and are derived from use of a product. They naturally state their needs based on their perceptions and act on their perceptions. Therefore, for the designers and manufacturers, customers' perceptions and their articulation of their needs and expectations are both important. In some industries, customers' particular cultural values and norms influence their needs and expectations. For specific products and services, considering those cultural values and norms in design and manufacturing processes is pivotal.
- Step 3: Build the planning matrix
  In the planning matrix, several variables are calculated to determine the most important customer needs and expectations. These variables are:
  - *The importance rate*
  - *The customer rate* (current and rival firms)
  - *The quality plan score*
  - *The rate of level up*
  - *The priority factor*
  - *The absolute weight*
  - *The relative weight.*

The *importance rate* for each customer need and expectation can be learned from customers using customer surveys, focus groups, or other marketing tools. The *customer rates* in planning matrix for both current and the competitors' customer satisfaction levels are calculated from customer satisfaction scores for each need and expectation. The mean value for each expectation ($\bar{X}_i$) is computed as presented in Eq. 4.1, where i is the item of customer needs and expectations and m is the number of customers.

$$\bar{X}_i = \frac{\sum_{i=1}^{m}\left(Perception_i - Expectation_i\right)}{m} \quad (4.1)$$

The QFD team assigns *quality plan scores* for each customer need and expectation by (1) comparing customer expectation and perception scores and (2) taking into account their views about the quality targets. As the additional input in this step, the goals and strategic priorities of the firm are also considered. General scoring for quality plan varies between 1 (unimportant) and 5 (very important).

Next in planning matrix, *rate of level up score* is calculated as presented in Eq. 4.2.

$$RL_i = QP_i \div CR_i \quad i = 1,2,\ldots,n \qquad (4.2)$$

where

- n = the number of customer expectation
- RL_i = the rate of level up
- QP_i = the quality plan score
- CR_i = the current customer rate.

The rate of level up scores demonstrates whether customer needs and expectations necessitate any improvement compared to customers' satisfaction. The rate of level up variable identifies customer needs and expectations that are relatively more important than the others, considering both customers and managers' points of view. The higher the score in the rate of level up is, the more important the customer need and expectation is. If RL_i is greater than 1, relevant customer need and expectation must be improved since a higher-quality plan score reflects an improvement for meeting customers' needs and expectations.

Akao (1990) points out that *priority factors* correlate customer expectations with expected total sales linked to relevant customer expectation. The priority factors are assigned by the QFD team, considering that 1 point refers to "no improvement possibility in the sales," 1.2 points to "medium improvement possibility," and 1.5 points to "strong improvement possibility" (Akao, 1990).

In the last step of planning matrix, *absolute weights* and *relative weights* for each customer expectation item are calculated using Eqs. 4.3 and 4.4.

$$AW_i = IR_i * RL_i * PF_i \quad i = 1,2,\ldots,n \qquad (4.3)$$

$$RW_i = \left(AW_i \div \sum_{i=1}^{n} AW_i \right) * 100 \quad i = 1,2,\ldots,n \qquad (4.4)$$

where

- n = the number of customer expectation
- AW_i = the absolute weight of ith customer expectation
- IR_i = the importance rate score of ith customer expectation
- RL_i = the rate of level up of ith customer expectation
- PF_i = the priority factor score of ith customer expectation
- RW_i = the relative weight of ith customer expectation.

The main outcome of the planning matrix is to determine "the most important customer expectations" at the end of the process.

- Step 4: Convert customer needs and expectations into technical requirements
  To bring customer demands into the product development process, the voice of engineering is generated through the voice of customer in technical requirements section in HOQ. While customer needs and expectations indicate *what* should be produced and done (what to do), technical requirements inform *how* to meet those needs and expectations (how to do) (Büyüközkan & Berkol, 2011). Utne (2009: 727) stated that "going from stakeholder requirements to technical specifications involves transformation of qualitative requirements to quantitative and measurable characteristics." Technical requirements are engineering characteristics that affects one or more of the customer requirements (Büyüközkan & Berkol, 2011) and determined by product/service design teams that include engineering, R&D, product, and design experts.
- Step 5: Create the relationship matrix
  The relationship matrix in HOQ demonstrates whether there are any relationships between customer needs (in rows) and technical requirements (in columns). The correlation coefficients are assigned by QFD teams and experts who know which technical requirement (hows) is associated with which customer expectation (whats). A quantitative scale is utilized to assign the correlation coefficients developed by Akao

(1990) as follows: 9, a strong relationship; 3, a medium relationship; and 1, a weak relationship. Using Eqs. 4.5 and 4.6, absolute weights and relative weights for each technical requirement are calculated.

$$AW_j = \sum_{i=1}^{n} (r_i * RW_i)$$
$$j = 1,2,\ldots,k \quad i = 1,2,\ldots,n \tag{4.5}$$

$$RW_j = AW_j \div \sum_{j=1}^{k} AW_j \tag{4.6}$$

where

- k = the number of technical requirements
- n = the number of customer expectations
- RW_i = the relative weight of ith customer expectation
- r_i = the correlation coefficient between ith customer expectation and jth technical requirement
- AW_j = the absolute weight of jth technical requirement
- RW_j = the relative weight of jth technical requirement.

The higher relative weights identify relatively more important technical requirements that address which technical requiremensts should be taken into account in all design and production activities.

- Step 6: Construct the technical correlation matrix
  Technical requirements developed in the previous steps are analyzed by the QFD team to discover potential correlations between and among technical requirements in the roof of HOQ. Those relationships help designers understand how the final outcome and other technical requirements are affected by a change in the other technical requirements. Technical correlation matrix is presented by symbols of "+" positive and "-" negative in the roof of HOQ.
- Step 7: Customer evaluations
  The customer evaluation section focuses on how satisfied current customers and competitor company's customers are. A comparative analysis is performed in this section to help QFD team understand what customer need and expectation need to be analyzed in more detailed. The company can see what customer expectations need to be better met, compared to rival company's customer evaluations.
- Step 8: Determine relatively more important customer needs and expectations and technical requirements
  The planning matrix and relationship matrix help identify two important variables in HOQ:
  1. Relatively more important customer needs and expectations
  2. Relatively more important technical requirements after calculating relative and absolute weights for customer needs and technical requirements.

Depending on budget, time, and other limitations, the QFD team and other product-related teams and functions may prefer to focus on particular items with higher relative weights. Product and design optimization efforts are required to take into consideration all these relatively more important customer needs and expectations and technical requirements.

► Example 9

A large-size, 200-store shopping mall located in New England aims to improve its products and services for the customers. To do that, the mall management association wants to implement QFD in their decision-making processes. Using customer needs and expectations as shown below, demonstrate how QFD process can be utilized in this example, step-by-step.

Reliability

- Courtesy of employees
- Experience and knowledge of employees
- Prompt response to customer concerns
- Walking easiness
- Security of mall
- Not being crowded and loud
- Quality of service in food court.

Responsiveness

- Quality of support services
- Playground and entertainment areas
- Activities for kids

- Employees' responsiveness to customer needs
- Rest areas for adults
- Baby changing rooms
- Entertainment alternatives
- Distinctive brands/products
- Services for disabled customers
- Movie theaters and cafes.

Features
- Adequacy of number of stores
- Presence of favorite stores
- The ease of finding products
- Modern shopping environment
- Prestigious mall
- Pleasant leisure time.

Physical features and accessibility
- Escalator/elevator availability
- Pleasant decoration of stores and storefronts
- Mall lighting
- Cleanliness of mall
- Adequacy of signs and information desk
- Ease of access of mall
- Availability of public transportation
- Availability of parking area. ◄

▪ Solution

Let's follow the process flow of QFD step-by-step as follows:

1. Build the QFD team
   As the first step, we need to make sure that the mall management association created a QFD team within the association prior to starting QFD practices.
2. Determine customer needs and expectations
   The 31 customer needs and expectations are already determined within 4 factors as presented in the question. We assume that customer needs and expectations were learned and listed by marketing department in this study.
3. Build the planning matrix
   The QFD team created a planning matrix as follows in this example:

- *Importance rates* and *customer rates* are directly learned from customers' feedback (◘ Table 4.10). Importance rates and customer rates for each customer need and expectation are shown in ◘ Table 4.10. The customer rates show how each customer need and expectation is perceived and assessed by the customers. For example, "the courtesy of employees" has 4.41 out of 5.00 importance rate, whereas the customers are satisfied with "the courtesy of employees" as 3.86 out of 5.00.
- The *quality plan scores* are determined by the QFD team to decide how each customer need and expectation are related with the long-term strategic orientation of the firm (Pakdil & Kurtulmusoglu, 2018). The *rate of level up* scores is computed using Eq. 4.2. To illustrate, since "the courtesy of employees" are ranked as 5 in the quality plan score by QFD team, the rate of level up score is calculated as the ratio of quality plan score to customer rate [5.00/3.86 = 1.29].
- *Priority factors* are assigned by the QFD team to decide which customer need and expectation is more impactful on the sales. For example, priority factor is assigned as 1.5 for "the courtesy of employees." Then, *relative* and *absolute weights* are computed as shown in Eqs. 4.3 and 4.4. For "the courtesy of employees," absolute weight is calculated as follows using Eq. 4.3.

$$AW_i = IR_i * RL_i * PF_i = 4.41 * 1.29 * 1.5$$
$$= 8.56$$

- Then, relative weight of "the courtesy of employees" can be computed as follows using Eq. 4.4.

$$RW_i = \left(AW_i \div \sum_{i=1}^{n} AW_i \right) * 100$$
$$= \left(\frac{8.56}{194.28} \right) * 100 = 4.4\%$$

The last column of planning matrix (◘ Table 4.10), relative weight column, tells us what customer needs and expectations are relatively more important than the other ones. According to planning matrix in this example, "providing services for disabled customers" has the highest relative weight with 5.0%.

4

Table 4.10 The planning matrix

Factors	Customer needs and expectations	Importance rate	Customer rate	Quality plan score	Rate of level up	Priority factor	Absolute weight	Relative weight
Reliability	Courtesy of employees	4.41	3.86	5.00	1.29	1.5	8.56	4.4%
	Experience and knowledge of employees	4.23	3.78	4.00	1.06	1.2	5.38	2.8%
	Prompt response to customer concerns	4.48	3.54	5.00	1.41	1.5	9.48	4.9%
	Walking easiness	4.47	3.99	5.00	1.25	1.5	8.40	4.3%
	Security of mall	4.48	3.70	5.00	1.35	1.5	9.09	4.7%
	Not being crowded and loud	4.49	3.64	5.00	1.37	1.5	9.24	4.8%
	Quality of service in food court	4.32	3.84	4.00	1.04	1.2	5.40	2.8%
Responsiveness	Quality of support services	3.76	3.49	5.00	1.43	1.0	5.40	2.8%
	Playground and entertainment areas	4.06	3.64	4.00	1.10	1.0	4.46	2.3%
	Activities for kids	3.93	3.51	4.00	1.14	1.0	4.47	2.3%
	Employees' responsiveness to customer needs	4.25	3.72	5.00	1.34	1.2	6.85	3.5%
	Rest areas for adults	3.72	3.40	5.00	1.47	1.0	5.48	2.8%
	Baby changing rooms	3.66	3.30	4.00	1.21	1.0	4.44	2.3%
	Entertainment alternatives	4.20	3.67	4.00	1.09	1.2	5.48	2.8%
	Distinctive brands/products	3.99	3.47	5.00	1.44	1.0	5.74	3.0%
	Services for disabled customers	4.47	3.47	5.00	1.44	1.5	9.66	5.0%
	Movie theaters and cafes	4.41	3.88	4.00	1.03	1.5	6.81	3.5%

Table 4.10 (continued)

Factors	Customer needs and expectations	Importance rate	Customer rate	Quality plan score	Rate of level up	Priority factor	Absolute weight	Relative weight
Features	Adequacy of number of stores	4.18	3.74	5.00	1.34	1.2	6.71	3.5%
	Presence of favorite stores	4.32	3.89	5.00	1.29	1.5	8.33	4.3%
	The ease of finding products	4.45	3.80	4.00	1.05	1.5	7.03	3.6%
	Modern shopping environment	4.47	3.97	5.00	1.26	1.5	8.44	4.3%
	Prestigious mall	3.81	3.69	3.00	0.81	1.0	3.10	1.6%
	Pleasant leisure time	4.29	3.84	4.00	1.04	1.2	5.36	2.8%
Physical features and accessibility	Escalator/ elevator availability	4.43	3.89	4.00	1.03	1.5	6.82	3.5%
	Pleasant decoration of stores and storefronts	4.01	3.86	3.00	0.78	1.0	3.12	1.6%
	Mall lightning	4.12	3.88	4.00	1.03	1.2	5.09	2.6%
	Cleanliness of mall	4.51	4.08	5.00	1.23	1.5	8.29	4.3%
	Adequacy of signs and information desk	4.28	3.65	3.00	0.82	1.2	4.23	2.2%
	Ease of access of mall	4.08	3.58	3.00	0.84	1.0	3.42	1.8%
	Availability of public transportation	4.18	3.53	4.00	1.13	1.2	5.68	2.9%
	Availability of parking area	3.80	3.53	4.00	1.13	1.0	4.31	2.2%
					SUM		194.28	100%

Source: Author's creation

4. Convert customer needs and expectations into technical requirements
   In the relationship matrix, the voice of engineering is represented with 21 technical requirements in the columns (Table 4.11). Technical requirements are determined by QFD team and representatives from mall management association. Table 4.11 is fully presented in the data files of the textbook in web site.
5. Create the relationship matrix
   The QFD team also assigns subjective correlation coefficients for the relationships between customer needs and relevant tech-

Table 4.11 Relationship matrix

	The size and quality of food court	Reliable service	Employees' attributes	The variety of tenant mix	The number of well-known stores	The period of time to find a product	The number of security staff
Courtesy of employees			9				
Experience and knowledge of employees		9	9				
Prompt response to customer concerns		9	9				9
Walking easiness						3	
Security of mall		3	3				9
...	...	...	...	...	...	...	...
Sum	239	393	329	260	278	432	197
Relative weight (%)	4%	6%	5%	4%	4%	7%	3%

	The noise level	The size and number of playground	The number of cultural and social entertainments	The size and number of rest area	The size and number of baby changing room	Disabled friendly infrastructure	The number of movie theater, cafe, and restaurant
Courtesy of employees							
Experience and knowledge of employees							
Prompt response to customer concerns							
Walking easiness	3	9		9		9	
Security of mall							
...	...	...	...	...	...	...	...
Sum	291	251	215	313	180	373	276
Relative weight (%)	5%	4%	3%	5%	3%	6%	4%

	The mall size	The number of elevator and escalator	The number of signs	The lightning level	The size of parking area	The number of shuttles	The cleanliness infrastructure of mall
Courtesy of employees							
Experience and knowledge of employees							
Prompt response to customer concerns							
Walking easiness	9	9	9	3	9		3
Security of mall	3	1	3	3	9	3	
...	...	...	...	...	...	...	...
Sum	479	363	257	302	568	205	245
Relative weight (%)	7%	6%	4%	5%	9%	3%	4%

nical requirements as shown in ◘ Table 4.11, using 9, a strong relationship; 3, a medium relationship; and 1, a weak relationship. The last two rows of relationship matrix at the bottom show the absolute weights and relative weights of technical requirements, respectively. For example, for the first technical requirement, absolute weight is first calculated as follows using Eq. 4.5 in "sum" row in ◘ Table 4.11.

$$AW_j = \sum_{i=1}^{n}(r_i * RW_i) = 9(5.40) + 9(5.48) + 3(9.66) + 9(6.81) + 3(8.44) + 3(3.10) + 3(5.36) = 238.89$$

Then, relative weight of the first technical requirement is calculated as follows using Eq. 4.6.

$$RW_j = AW_j \div \sum_{j=1}^{k} AW_j = \frac{238.89}{6,444} = 4$$

The relative weight line at the bottom of relationship matrix in ◘ Table 4.11 tells us which technical requirements are relatively more important than the others. According to ◘ Table 4.11, "employees' attributes" and "the size of parking area" have the highest relative weights and address the areas on which further improvements are expected to be done.

6. Construct the technical correlation matrix
   The technical correlation matrix (◘ Fig. 4.7) demonstrates the relationship between relevant technical requirements. In design activities, it is helpful to see if at least two technical requirements affect each other, positively and/or negatively. Then, further actions are taken by the QFD team to optimize specifications of technical requirements.
7. Customer evaluations
   This step was not detailed in this example.
8. Determine relatively more important customer needs and expectations and technical requirements
   As the main outcomes of QFD:
   1. Relatively more important customer need and expectation is "providing services for disabled customers"
   2. Relatively more important technical requirements are "employees' attributes" and "the size of parking area."

Key Concepts

DMAIC, design for Six Sigma (DFSS), Define phase, process analysis and documentation, transformation process, value stream map, flow chart, SIPOC diagram, swim lane, spaghetti diagram, bottlenecks, project prioritization and selection, project charter, problem statement, goal statement, project scope, project deliverables, metrics, Critical-to-Quality (CTQ) characteristics, House of Quality (HOQ), voice of customer, and voice of engineering.

Summary

The Define phase focuses on analyzing and documenting processes and identifying boundaries and bottlenecks of the processes, using tools such as transformation process, value stream map, flow chart, SIPOC diagram, swim lane, and spaghetti diagram. The Define phase also includes stakeholder analysis, prioritizing and selecting Six Sigma projects, and developing project charters that contain problem statement, goal statement, project scope, deliverables, metrics, and CTQ characteristics. Additionally, HOQ is constructed in the Define phase to identify relatively more important voice of customer and voice of engineering.

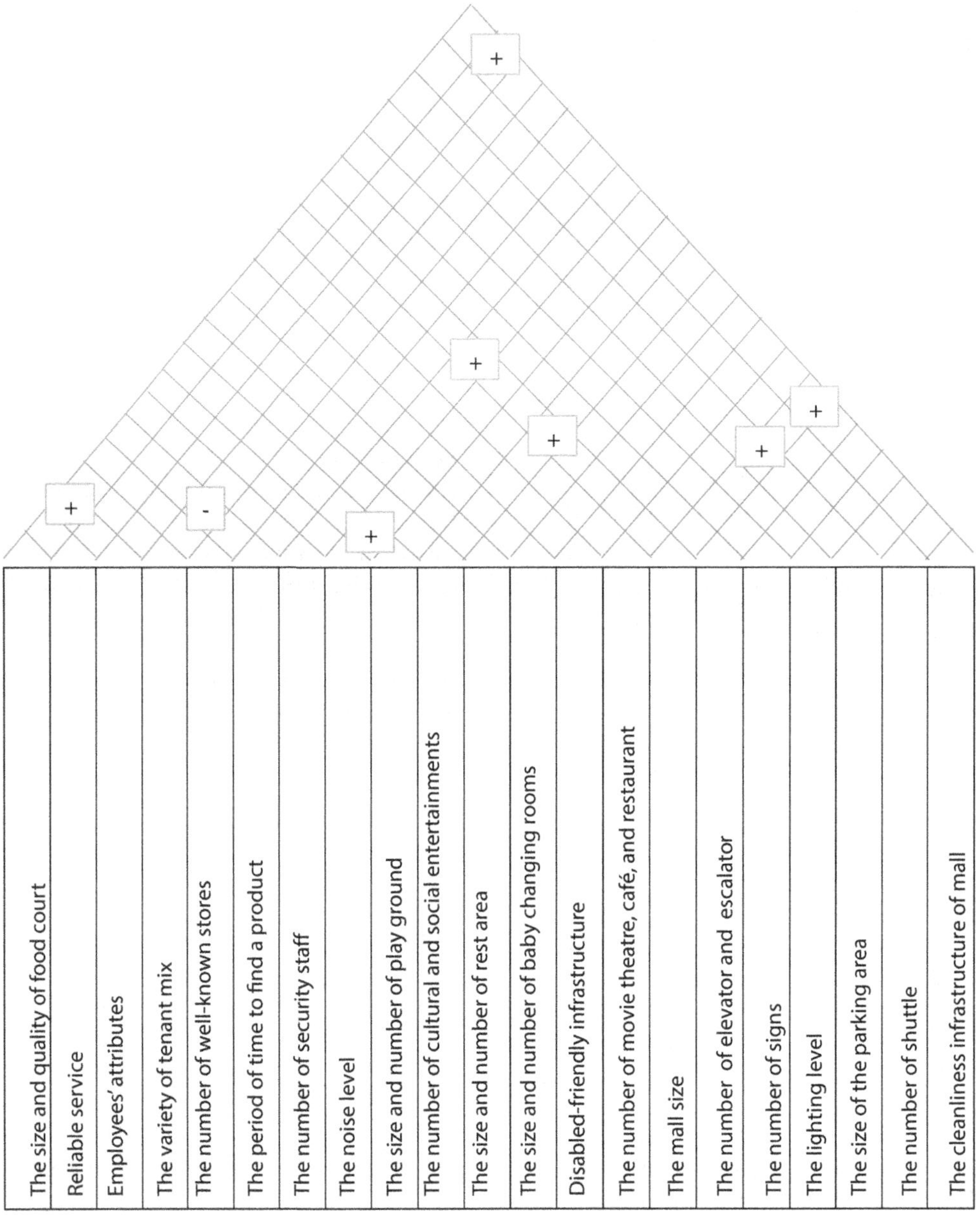

Fig. 4.7 Technical correlation matrix. (Source: Author's creation)

Practice and Discussion Questions

1. Briefly discuss the phases of DMAIC process.
2. What are the general questions that need to be answered by Six Sigma teams in DMAIC process?
3. Draw a transformation process for a hospital and discuss each component.
4. Consider dentistry and draw transformation process of the dentistry including inputs, processes, and outputs.
5. Think about a supermarket and draw transformation process of the system including inputs, processes, and outputs.
6. Think about an automobile factory and draw transformation process of

the system including inputs, processes, and outputs.
7. Discuss each function of value stream map in process analysis and documentation.
8. Discuss value added, non-value added, and non-value added but required activities. Give examples for these activities in a manufacturing and service setting.
9. Categorize value stream maps based on the level.
10. Categorize value stream maps based on the state.
11. What questions should be answered before drawing VSM?
12. List and discuss steps of VSM drawing process.
13. Consider dentistry and draw a hypothetical VSM of the dentistry including inputs, sub-processes, cycle time for each sub-process, outputs, and VA and NVA activities.
14. Think about a supermarket and draw a hypothetical VSM of the system including inputs, sub-processes, cycle time for each sub-process, outputs, and VA and NVA activities.
15. Think about an automobile factory and draw a hypothetical VSM of the factory including inputs, sub-processes, cycle time for each sub-process, outputs, and VA and NVA activities.
16. List and discuss steps of flow chart drawing process.
17. Consider dentistry and draw the flow of the dentistry that includes all relevant activities and work steps.
18. Think about a supermarket and draw the flow of the supermarket that includes all relevant activities and work steps.
19. Think about an automobile factory and draw the flow of the system that includes all relevant activities and work steps.
20. What questions should be answered before drawing SIPOC diagram?
21. List and discuss steps of SIPOC diagram drawing process.
22. Consider dentistry and draw a SIPOC diagram of the system.
23. Think about a supermarket and draw a SIPOC diagram of the system.
24. Think about an automobile factory and draw a SIPOC diagram of the system.
25. List and discuss steps of swim lane drawing process.
26. Consider dentistry and draw a swim lane of the system.
27. Think about a supermarket and draw a swim lane of the system.
28. Think about an automobile factory and draw a swim lane of the system.
29. List and discuss steps of spaghetti diagrams drawing process.
30. Consider dentistry and draw a hypothetical spaghetti diagram of the system.
31. Think about a supermarket and draw a hypothetical spaghetti diagram of the system.
32. Think about an automobile factory and draw a hypothetical spaghetti diagram of the system.
33. Define "stakeholder" term.
34. Discuss the role of stakeholder analysis in Six Sigma projects.
35. List the steps of stakeholder analysis.
36. Discuss how to utilize stakeholder map in Six Sigma projects.
37. Discuss how to utilize stakeholder/responsibility matrix in Six Sigma projects.
38. Discuss how to utilize stakeholder categorization matrix in Six Sigma projects.
39. Discuss qualitative approaches used in project prioritization and selection.
40. Discuss quantitative approaches used in project prioritization and selection.
41. Discuss the function of "the problem statement" in project charter.
42. Discuss SMART feature for goal statement in project charter.
43. Discuss the structure of the goal statement proposed by Gitlow et al. (2015).
44. Discuss the function of "project scope" in project charter.
45. Discuss the function of "project deliverables" in project charter.

46. Discuss the function of "CTQ characteristics" in project charter.
47. Discuss the function of "quality function deployment" in the Define phase in DMAIC process.
48. Discuss why it is so critical for business organizations to hear "the voice of the customer."
49. Discuss the usefulness of Juran's customer need classification for Apple products.
50. Discuss how importance rates are determined in planning matrix in HOQ.
51. Discuss how customer rates are determined in planning matrix in HOQ.
52. Discuss how quality plan scores are determined in planning matrix in HOQ.
53. Discuss how rate of level up scores are determined in planning matrix in HOQ.
54. Discuss how priority factor scores are determined in planning matrix in HOQ.
55. Discuss how absolute weights are determined in planning matrix in HOQ.
56. Discuss how relative weights are determined in planning matrix in HOQ.
57. Discuss the functions of technical correlation matrix in HOQ.
58. What are the most important outputs of QFD for Six Sigma practitioners?

References

Akao, Y. (1990). *Quality function deployment, integrating customer requirements into product design*. New York: Productivity Press.

Andersson, R., Hilletofth, P., Manfredsson, P., & Hilmola, O. P. (2014). Lean six sigma strategy in telecom manufacturing. *Industrial Management & Data Systems, 114*(6), 904–921.

Banuelas, R. C., Tennant, I. T., & Tang, S. (2006). Selection of six sigma projects in the UK. *The TQM Magazine, 18*(5), 514–527.

Bicknell, B. A., & Bicknell, K. D. (1995). *The road map to repeatable success using QFD to implement change*. Boca Raton: CRC Press.

Bruhl, R. (2018). *Understanding statistical analysis and modeling*. Los Angeles: Sage.

Büyüközkan, G., & Berkoll, Ç. (2011). Designing a sustainable supply chain using an integrated analytic network process and goal programming approach in quality function deployment. *Expert Systems with Applications, 38*(11), 13731–13748.

Carey, R. G. (2003). *Improving healthcare with control charts: Basics and advanced SPC methods and case studies*. Milwaukee: ASQ Quality Press.

Carroll, A. B., & Buchholtz, A. K. (2012). *Business and society*. South Western Cengage Learning. Mason.

Deming, W. E. (1986). *Out of the crisis*. Cambridge, MA: Massachusetts Institute of Technology, Center for Advanced Engineering Study.

Elias, A. A. (2016). Stakeholder analysis for lean six sigma project management. *International Journal of Lean Six Sigma, 7*(4), 394–405.

Feigenbaum, A. V. (1983). *Total quality control*. New York: McGraw-Hill.

Freeman, R. E. (1984). *Strategic management: A stakeholder approach*. Boston: Pitman Publishing.

Freeman, R. E. (2010). *Strategic management: A stakeholder approach*. New York: Cambridge University Press.

Gitlow, H. S., Melnyck, R. J., & Levine, D. M. (2015). *A guide to six sigma and process improvement for practitioners and students* (Second ed.). Upper Saddle River, NJ. Pearson.

Glasgow, J. M., Scott-Caziewell, J. R., & Kaboli, P. J. (2010). Guiding inpatient quality improvement: A systematic review of lean and six sigma. *The Joint Commission Journal on Quality and Patient Safety, 36*(12), 533–AP5.

ISO 5807:1985 Information processing—Documentation symbols and conventions for data, program and system flowcharts, program network charts and system resources charts. Retrieved on 26 Oct 2019.

Joiner, B. L. (1994). *Fourth Generation Management: The New Business Consciousness*. New York: Mc-Graw-Hill.

Juran, J. M. (1992). *Juran on quality by design: The new steps for planning quality into goods and services*. New York: Simon and Schuster.

Juran, J. M., & Gryna, F. M. (1980). *Quality planning and analysis*. New York: McGraw-Hill.

Laureani, A., Brady, M., & Antony, J. (2013). Applications of lean six sigma in an Irish hospital. *Leadership in Health Services, 26*(4), 322–337.

Montgomery, D. C. (2009). *Introduction to statistical quality control*. New York: Wiley.

Montgomery, D. C. (2013). *Introduction to statistical quality control*. New York: Wiley.

Ohno, T. (1988). *Toyota production system: Beyond large-scale production*. Cambridge, MA: Productivity Press.

Pakdil, F., & Kurtulmuşoğlu, F. B. (2014). Improving service quality in highway passenger transportation: A case study using quality function deployment. *European Journal of Transport and Infrastructure Research, 14*(4), 376–394.

Pakdil, F., & Kurtulmuşoğlu, F. B. (2018). Quality improvement initiatives based on customer and service provider perspectives in shopping malls.

International Journal of Quality and Service Sciences, 10(2), 176–195.

Pakdil, F., Işın, F. B., & Genç, H. (2012). A quality function deployment application using qualitative and quantitative analysis in after sales services. *Total Quality Management & Business Excellence, 23*(11–12), 1397–1411.

Pyzdek, T., & Keller, P. A. (2018). *The six sigma handbook* (5th ed.). New York: Mc-Graw Hill.

Snee, R. D. (2010). Lean six sigma–getting better all the time. *International Journal of Lean Six Sigma, 1*(1), 9–29.

Utne, I. B. (2009). Improving the environmental performance of the fishing fleet by use of quality function deployment (QFD). *Journal of Cleaner Production, 17*(8), 724–731.

Velasquez, M., & Hester, P. T. (2013). An analysis of multi-criteria decision making methods. *International Journal of Operations Research, 10*(2), 56–66.

Wang, F. K., Hsu, C. H., & Tzeng, G. H. (2014). Applying a hybrid MCDM model for six sigma project selection. *Mathematical Problems in Engineering, 2014*, 730934.

Wedgwood, I. (2007). *Lean sigma a practitioner's guide*. Boston: Prentice Hall, U.S.

Womack, J., & Jones, D. (1996). *Lean thinking*. New York: Simon and Schuster.

Further Reading

Heagney, J. (2016). *Fundamentals of project management*. New York: Amacom.

Kerzner, H. (2017). *Project management: A systems approach to planning, scheduling, and controlling*. Hoboken: Wiley.

Meredith, J. R., Mantel, S. J., Jr., & Shafer, S. M. (2017). *Project management: A managerial approach*. Hoboken: Wiley.

Measure Phase: M Is for Measure

Contents

F. Pakdil, *Six Sigma for Students*, https://doi.org/10.1007/978-3-030-40709-4_5

5

Learning Objectives

After careful study of this chapter, you should be able to:

- Explain Measure phase of DMAIC
- Create data collection plans
- Define and differentiate types of variables
- Define and differentiate types of sampling
- Identify the required sampling method
- Measure limits of the CTQ characteristics
- Measure sigma level of the process and system.

5.1 Introduction

In the second phase of DMAIC process, major components of "Measure" are:

1. Identifying types of variables
2. Identifying types of sampling methods
3. Identifying data collection plans
4. Collecting data
5. Identifying measurement system components
6. Measuring limits of the CTQ characteristics
7. Measuring sigma level of the processes and systems
8. Running measurement system analysis (Gage R&R).

The purpose of the Measure phase is to measure, evaluate, and comprehend the current state of the processes (Montgomery 2005). In this phase, the team collects data on all relevant CTQ (Critical-to-Quality) characteristics and indicators. All CTQ characteristics should be clearly identified, based on the problems analyzed and the project charter developed in the Define phase. Data collection is one of the critical parts of this phase. Data can be collected from either historical data or the relevant processes over a continuous period of time. Data collection can even be done by sampling. Big "Y," the response or dependent variable, is also called CTQ characteristic in Six Sigma projects. X variables are called independent variables, since these variables are likely to be impactful on response variables. The response variable is considered a function of one or more independent variables. While the response variable is also called a "desired outcome," independent variables are known as drivers of the response variable. When assuming that there is more than one variable affecting the response variable, the relationships between and among the variables can be analyzed as well. Decisions should be made based on measurement, data, information, and knowledge in Six Sigma projects, rather than relying on intuitive and personal ideas, opinions, and beliefs. Use of data is expected to reduce the probability of inappropriate or incorrect decisions. The data collected in the Measure phase are used to diagnose the current state of performance. A great variety of tools and diagrams are used in the Measure phase. These tools will be detailed in this chapter.

Berwick et al. (2003: 35) state that "to improve performance, organizations and individuals need the capability to control, improve, and design processes, and then to monitor the effects of this improvement work on the results. Measurement alone will not suffice."

The questions that should be answered in Measure phase are:

- How are we measuring the performance?
- How do we identify and describe CTQ characteristics?
- What data do we need?
- How valid, accurate, and reliable are our data and measurement systems?
- What is the most current performance level of process and product?
- How are the central tendency and variation of the process and product measured?
- What are the measurement system analysis results?
- Are there any assumptions made during the Measure phase?
- Are the data collected enough to picture and understand the process and its performance?

At the end of the Measure phase, the project charter is reviewed, and changes are incorporated, if necessary. In this phase, the goal of the project, the combination of the team, timetable, current or potential issues, or problems emerging in the Measure phase are reviewed in depth. In the Measure phase, Six Sigma teams should make sure that the data collection methods and data sets reflect the process analyzed. Measurement system analysis, sampling, rational subgrouping, and operational definitions are considered in this phase (Mohammed et al. 2008).

5.2 What Are Data?

Businesses rely on data for planning and improving overall performance. Over the last decades, most modern businesses collected data and information about every transaction performed by the organization (Gitlow et al. 2015). Business processes, operations, and transactions are monitored, observed, and recorded for several purposes, such as improving performance, eliminating wastes, and decreasing the probability of potential abnormalities.

Any kind of information collected about a product, service, process, or item is called data. In the past when the computer technology was not an advanced level, the data would be collected manually in many industries. Today, the data are collected, recorded, and stored electronically in *data warehouses*. Because data warehouses enable the storage of data and information in large platforms, the big data concept has been popular among decision-makers, statisticians, and analyzers in a great variety of industry. Some terms that are related to data are given as follows:

- *Data*: collection of facts.
- *Population*: well-defined collection of the objects.
- *Census*: all desired information is collected for all objects in the population.
- *Observation*: data values or individual *observations* include information collected regarding a specific subject based on the "Five Ws": *who*, *what*, *when*, *where*, and *why*. *How* is also added to this list.
- *Sample*: a subset of the population that represents the general features and characteristics of the population.
- *Variable*: a characteristic on which an object in the population is measured and analyzed.
- *Parameter*: a characteristic of a population to describe statistical values such as measures of central tendency and variability.
- *Statistic*: a characteristic or an estimator of a sample.

Data, population, census, observation, sample, variable, parameter, and statistic are some of the terms used in data analysis concepts. *Variable* represents the characteristics of an item representing the features of quality, called CTQ. The quality of a product, service, component, or process is managed based on the CTQ characteristics. To measure and improve the quality level of the relevant subject, CTQ characteristics are determined, based on the voice of customer and voice of engineering. Then, engineering specifications are determined on the basis of CTQ characteristics. Additionally, a *statistic* is used to make inferences on the population parameters that are typically unknown. *Statistic* is also known as a summary assessment of a set of values observed for some property for some set of phenomena (Bruhl 2018: 6). He also states that a *statistic* must answer specific questions about the data sets, either by defining or explaining the data. A *statistic* is an answer to a specific question regarding a set of observations, and any number of different questions might be posed for any set of observations (Bruhl 2018: 1).

▶ Example 1

A 4-year college in New England area would like to estimate the next years' enrollments. The college administration assigns this task to a team that includes vice presidents and directors working at the college. Prior to starting their research, they plan to identify what the following terms mean in this project: data, population, census, observation, sample, variable, parameter, and statistic. Help them describe each term. ◀

■ Solution

In this case, *data* is the collected information about potential number of students who may enroll at a college. The team needs to identify and specify their target students in the population. *Population* represents any students who may consider applying for a college. The team may identify the scope of the population in this study such as demographics of the potential students, towns, cities, states, or high schools targeted. Therefore, the population can be articulated on the basis of the scope of the study at this time period by the team. However, if the time period, team members, and/or scope of the study change, the definition of population may change as well. Since the college is located in New England, the team may limit the geographical scope of the study with the states of New England and with students who will graduate from high school in 2 years and consider individuals within these parameters as the *population* of this study.

Census is the entire population, which in this case is the high school students who will potentially graduate in 2 years in the New England area. If the study does not rely on a sample, that means the entire population is represented in the study, the *census*.

Observation in this study refers to each and every single student who falls into this category as "residing in New England area" and "potentially graduating in 2 years." When the team collects data about the project, based on the scope of the project, they will focus on each and every single student who is categorized in this definition. The data collection step will include all these high school students residing in New England.

To narrow the scope of study, the team may prefer to focus on a few states in New England, such as Maine, Vermont, and New Hampshire. By narrowing the states, they may want to do *sampling* and take into consideration just these states. As another data stratification, they may even think of focusing on high schools located in cities with more than 30,000 residents. In this case, the data represents *sampling* within a group of observations, not the total population.

After identifying the structure of the population and sample, the team focuses on what *variables* will be taken into consideration in data collection and analysis. For example, age, gender, number of siblings of students, annual income of parents, education level of parents, population of cities, the distance between college, and cities are some of the *variables* that can be used in this study. *Parameters* may be mean, mode, and median of the variables of population, that is, central tendency measures, whereas standard deviation, range, and variance of the variables are the variability or dispersion measures. The team members may analyze mean and standard deviation of the age of students in the population. If the team wants to focus on mean, mode, median, standard deviation, range, or variance of the sample (not population), these measures are called *statistics*, not parameter.

Data collection is a very pivotal process in Six Sigma projects, as successful completion of a Six Sigma project relies on data availability. Data collection is a process where observations that represent the CTQ characteristics are accumulated. According to Bruhl (2018: 1–2):

> » data collection, "the process of making observations, is a cognitive activity that requires numerous decisions on the part of the investigator as to how the phenomena of interest are to be represented and how the observations are to be collected. As a cognitive process, we "observe" phenomena by the "properties" that we have "learned" to perceive. In data collection processes, Six Sigma teams need to determine how to represent each phenomena and the properties of that phenomena where the teams can identify the quality and the value of the phenomena as an ordinal or cardinal scale (Bruhl 2018: 2).

Along with the properties of the CTQ characteristics, Six Sigma teams also need to decide how to collect data, organize the project, present properties of CTQ characteristics, and categorize and analyze the data. The data collection process has a priority since the other tasks depend on the quality of data collected. Data collection can be constructed based on two approaches, depending on data availability. First, historical data can be used if required data are available in existing data warehouses. In cases where the required data are not available, the Six Sigma team can consider collecting new data. The details of the data collection process can be shaped based on these two approaches. In the era of Industry 4.0, collecting data is not as difficult as in the past. In this digital age, big data are collected using the Internet of Things (IoT), artificial intelligence, machine learning, digital systems, software-driven devices with wired and wireless networks, and the Internet. Real-time data are automatically collected, processed, analyzed, and delivered to the points where the knowledge will be used in a great variety of industries. Big data enables use of artificial intelligence, cloud services, mobile technology, and machine learning in data collection processes.

5.3 Data Collection Plans

The data collection plans help the Six Sigma team ensure that the data needed in the project are collected appropriately and in a timely manner. The performance of the Six Sigma team relies on the quality of data collected during the project period in DMAIC. Therefore, the team needs to make sure that the data collection plans are ready and effective. Data collection plans picture the current state. It is a detailed document in which Six Sigma practitioners can easily see the step-by-step process of data collection. A detailed data collection plan helps team members collect reliable, valid, and correct data on time while maintaining a standardized approach throughout the organization. Using a standardized data collection plan, Six Sigma team members provide a common language and consistency in the data collection processes and can easily recognize what will be done. Data collection plans should be living documents, and required revisions should be reflected in the plan as soon as the need arises. After the revision, Six Sigma team members should ensure that data collection processes follow the revised plans. An example of a data collection plan is presented in ◘ Table 5.1. The sections of the data collection plans can be listed as follows:

- Purpose of data collection
- What to measure
- Type of data
- CTQ characteristics
- Operational definition
- Specification limits
- Data collection sources
- Data collection method
- Who collects data
- Where to collect data
- How long to collect data
- How to collect data
- The format of the data
- When to collect data
- Sampling method
- Sample size or census
- Frequency
- Reporting format
- Reporting responsibility.

5.4 Types of Variables

As stated by Knapp (2016: 12) "comprehending the types of variables involved in a data set or research design is essential when it comes to properly selecting, running, and documenting the results of statistical tests." Variables are categorized into several groups in various disciplines such as mathematics, statistics, and research methods. We will categorize variables into six main groups:

1. Categorical and numerical variables
2. Univariate, bivariate, and multivariate variables
3. Continuous and discrete variables
4. Nominal, ordinal, interval, and ratio variables
5. Nominal, the smaller, and the larger variables
6. Independent and dependent variables.

Categorical and Numerical Variables Categorical variables take on one of a limited number of possible values, usually categories. Nominal variables are used interchangeably with categorical or attribute ones. Attribute data places an item into categories. For example, the results of incoming inspection can be categorized as *defectives* and *non-defectives* or *conforming* and *nonconforming* ones. Categorical or attribute data consists of the frequencies of a phenomenon. Numerical variables can be a range of values that are actual numbers. Ordinal values

5

Table 5.1 An example of data collection plan

CTQ	Type of data	Operational definition	LSL	USL	Target	Method	Frequency	Sampling method	Sample size	Location	Responsible position	Format of reporting
Cycle time	Continuous	The time between start and complete the process	−0.5 minutes	+0.5 minutes	5.00 minutes	I 1.1	Every other cycle time/day	Simple random	Minimum 25	Assembly line 1	Technician 1	E-Form A1
Diameter	Continuous	The length of the diameter of the cylinder tube	−0.002 mm	+0.002 mm	7.5 mm	I 1.9	Each part produced	No sampling	Census	Job shop 8	Technician 3	E-Form D2
Number of defectives	Discrete	The number of the defective component code 312 detected in incoming inspection	0	2	0	I 1.12	Each batch	Simple random	5/100	Incoming Inspection Lab 2	Incoming inspector of part 312	E-Form D4

Source: Author's creation

are used when the variables are put in an order. For example, students can be ordered based on seniority.

Univariate, Bivariate, and Multivariate Variables Variables can be grouped as *univariate, bivariate,* and *multivariate*. Univariate variables consist of data that have a single variable. For example, the height of students in a class is a univariate variable. Bivariate variables refer to those represented with two different variables for the same characteristic. For example, the height and weight of the students in a class are bivariate variables. Multivariate variables represent the variables that have more than two characteristics.

Continuous and Discrete Variables This categorization is also compatible with probability distributions. Continuous variables are the ones in which the sets of all possible values on the number line are infinite, while discrete variables have the set of all possible values that are finite. Continuous variable can theoretically have any values in the number line, integer or non-integer, positive or negative. These variables are also known as variable quality characteristics. For example, height, weight, length, temperature, pressure, pH level, and lead time can be given as examples for continuous variable. Discrete variables can take only finite and positive integer values. These variables are also known as attribute quality characteristics. Discrete data are collected by counting whereas the continuous ones by measuring. Attribute data are organized in discrete units and counts, such as number of red apples in a box; number of defective or nonconforming products; number of defects, failures, or nonconformities; number of patients waiting in the line more than a certain amount of time; number of students graduating at the end of academic year; and number of new customers in a period of time. The simplest form of attribute data is go/no-go, pass/fail, yes/no, and accept/reject data.

Nominal, Ordinal, Interval, and Ratio Variables Variables can also be categorized on the basis of the level of measurement scale such as *nominal, ordinal, interval,* and *ratio*. Attribute data is also known as *nominal* scale data. Defective and non-defective product classification is an example of nominal scale data. An *ordinal* scale is used for ranking the data based on a number of ranks. For example, customer satisfaction level can be ranked using a 5-point Likert scale varying between 1 and 5, where 1 = very dissatisfied, 2 = dissatisfied, 3 = neutral, 4 = satisfied, and 5 = very satisfied. In *interval* scale, the differences between measurements are meaningful. For example, the customers in a market analysis can be grouped into age intervals such as 0–6 years, 6–12 years, 12–18 years, and so forth. In *ratio* scale, along with the meaningful differences between measurements, the scale starts from zero and has measurable intervals. Weight is a good example for ratio scale. There are no negative values on this scale.

N Type, S Type, and L Type variables In Taguchi's optimization approach, variables are categorized into three groups such as (1) *nominal is best*, (2) *the smaller is best*, and (3) *the larger is best*. Many QTC characteristics can be categorized into one of those groups. "The-nominal-is-best" group, also known as N-type variable, refers to a characteristic with a specific numerical goal or target value. Most quality dimensions fall into this group. Some of the N-type variables are height, length, width, diameter, area, volume, pressure, density, viscosity, time, percent mixture, percent moisture, PH level, and voltage. "The-smaller-is-best" group, also known as S-type variable, refers to a characteristic that targets zero. For example, percent shrinkage, machine downtime, residue, percent contamination, source lines of code, loudness, product deterioration, wasted resources, response time, and power dissipation are S-type variables. "The larger-is-best" group, also known as L-type variable, targets the highest value possible. Strength of a metal piece, ignition temperature, mean time between failures, and melting point are some examples of L-type variables.

Independent and Dependent Variables While analyzing the relationships between and among variables, categorizing variables such as "independent" and "dependent" helps Six Sigma teams in their analyses. The dependent variable is termed because it depends on the level of the independent variable. While the independent variable is considered "cause," the dependent variable is considered "effect." The independent variable corresponds to predictors, and the dependent variable corresponds to results or effects. The independent variable is used to predict the values of dependent variable. The values of the independent variable are assumed to be known. The dependent variable is predicted or estimated using independent variables. The independent variable is on the *X*-axis, whereas the dependent variable is plotted on the *Y*-axis.

5

5.5 Types of Sampling

In many cases, it may be more convenient and feasible to collect data on a sample of the population, rather than using the population. Specifically, if the quality control methods destroy the products, sampling is the only way to design quality control activities. The sampling allows the team to draw conclusions about the population based on the sample taken from the population (Defeo 2017). Although sampling seems more convenient in some data analysis processes, collecting samples from the population may contain a degree of uncertainty, bias, and some degree of error. Hart and Hart (2002) state that sampling generates meaningful results and represents a population if the process is stable. Six Sigma teams may not know if the sample perfectly and adequately represents the population. It is also possible to have bias and errors while calculating and using sample statistics as estimates of population parameters.

The main concern for Six Sigma teams is that they may work on a sample that does not truly represent the population and that statistical inferences made based on the sample may not be correct and adequate for the population. Therefore, sampling methods have a pivotal role in minimizing uncertainty, bias, and sampling error in data collection and analysis processes. From a quantitative analysis perspective, sampling and sampling methods may be related to the generalization of sample statistics to estimate the population parameters. Depending on the research questions and topics analyzed by Six Sigma teams, proper sampling methods are integrated into data collection and sampling. To decide the type of sampling method, Six Sigma teams need to identify the population of the research question, the size of the population, the availability of data, and sampling frame. The sampling frame is a list of elements from which a sample is selected. It includes the elements of the population and can be related to maps, census database, and employee database. If the sampling frame misses some important elements of the population, it is likely to create a difference between sampling statistics and population parameters, called sampling error. Therefore, the sampling frame should be designed to minimize the likelihood of sampling error.

Sampling methods are categorized into two groups: probability sampling methods and non-probability sampling methods. In probability sampling methods, elements of a sampling frame have a known probability of being selected in a sample. These methods are simple random sampling, stratified sampling, systematic sampling, and cluster sampling. In non-probability sampling methods, the probability of being selected for the elements of a sampling frame in a sample is unknown. These methods are quota sampling, purposive sampling, self-selecting sampling, judgmental sampling, snowball sampling, and convenience sampling.

5.5.1 Probability Sampling Methods

Probability sampling methods are designed to produce more accurate sample statistics to estimate population parameters than sample statistics obtained through non-probability methods. These methods also require smaller sample sizes than do the non-probability methods.

5.5.1.1 Simple Random Sampling

Simple random sampling includes a set of n elements in a population of N elements where all are equally likely to be selected. For example, if a quality control technician randomly selects 10 products from a container that includes 50 finished products and each product has equal chance to be selected by the technician, this sampling represents simple random sampling as detailed below. The most distinctive features of simple random sampling are (1) each element of the sampling frame has an equal probability of being selected in the sample and (2) selecting elements is done randomly in the population. It is considered the best method of selecting subjects to form a sample of the population. However, simple random sampling does not guarantee that each population characteristic will be represented in the sample. When the population characteristics are homogenous, the likelihood of representing the population in sampling is greater. When the degree of heterogeneity of the population increases, that likelihood significantly decreases.

Once the population is identified, the sample frame is determined by the researchers. The sample frame shows the individual elements in the population. As long as the elements of the sampling unit have equal chance to be selected, simple random sampling methods will be acceptable and easy to use. It also creates a base for statistical inference. Elements can be selected randomly using with or without replacement methods from the sampling frame. In cases where an element is replaced, this element has another equal probability of being selected. If the element is not replaced again, the element does not have any other probability of being selected in the sampling frame. Random number generators or tables of random numbers can be used for identifying each element of the sampling frame in simple random sampling. Bingo and lottery balls and drawing numbers of elements from a bowl are good examples of simple random sampling. To use simple random sampling, the next steps can be followed:

- Step 1: Define and identify the sampling frame.
- Step 2: List all elements of population.
- Step 3: Assign a sequential number to each element of population.
- Step 4: Decide sample size.
- Step 5: Decide randomness rule or use a random generator to select the sample. For example, if you decide to have 50 elements in your sample size and the population is 500, generate 50 random numbers between 1 and 500.
- Step 6: Randomly select the samples up to sample size.

▶ Example 2

Consider that you are a member of a Six Sigma team collecting data from an assembly line, where the CTQ characteristic is the cycle time of the assembling of the parts. The team wants to sample the assembly line using a simple random sampling method for 7 days. If the volume of the total parts is a big number and half of them are considered to be selected, help them collect data based on simple random sampling. Assume that confidence level is 95% and margin of error is 5%. ◀

■ Solution

Let's follow the steps of simple random sampling given below as follows:

- Step 1: Define and identify the sampling frame.
  Because the sample frame shows the individual elements in the population, each cycle time should be considered an element of the sample frame for each day.
- Step 2: List all elements of population.
  List each part used in this assembly line for 7 days.
- Step 3: Assign a sequential number to each element of population.
  We need to assign sequential numbers to each part starting from 1 (1, 2,..., n). If there are 10,00,000 parts used in this assembly line, the numbers should vary between 1 and 10,00,000.
- Step 4: Decide sample size.
  Several approaches can be taken to decide sample size. Let's focus on Cochran's sample size approach using Eq. 5.1.

$$n_o = \frac{Z^2 pq}{e^2} \tag{5.1}$$

where

- e is the desired level of precision or margin of error,
- p is the estimated proportion of the population,

- Z is the corresponding standard normal distribution value of the probability,
- q is $1-p$.

Let's plug the variables into formula as follows:

$$n_o = \frac{Z^2 pq}{e^2} = \frac{(1.96)^2(0.5)(0.5)}{0.05^2} = 384.16$$

That means 385 cycle times should be sampled.

5

- Steps 5 and 6: Decide randomness rule or use a random generator to select the sample, and randomly select the samples up to sample size.
  Using a random number generator, the team can select 385 different numbers out of 10,00,000 parts. Whatever numbers are selected through random generator, those parts assigned with those numbers are picked from the population.

5.5.1.2 Stratified Random Sampling

If the data collection process and/or sampling method is for a heterogeneous population, stratified sampling method is considered a better sampling method. The sampling frame is divided into layers or subgroups depending on the structure and basic features of the population. Each subgroup is known as "strata" and expected to equally represent important features or characteristics of the sampling frame and population. For example, if the data set includes age, gender, industry, season, or any other similar variables, the data can be grouped based on each of those variables. Then, samples are randomly drawn from each group. Percentages of each subgroup among the data set are taken into consideration to proportionally determine the sample size of each group. Stratified random sampling provides a convenient and systematic way to represent subgroups in the sample. When the population is very large, using simple random sampling may be difficult and unrepresentative of the whole population. In this case, stratified random sampling functions better to represent various stratus of the population, since stratifying the data may create useful subgroups in which the large data is analyzed in more depth. The following steps can be used to implement stratified sampling method in sampling and data collection:

- Step 1: Identify the sampling frame.
- Step 2: Divide the population into subgroups depending on the structure of the population and research question.
- Step 3: Identify the proportion of each subgroup.
- Step 4: Identify sample size for each subgroup.
- Step 5: Identify each member of the subgroups.
- Step 6: Decide randomness rule or use a random number generator to select the sample.
- Step 7: Randomly select the samples from each subgroup.

► Example 3

Consider that you are a member of a Six Sigma team collecting data from an assembly line, where the CTQ characteristic is the cycle time of the assembling of the parts. The team wants to sample the assembly line using stratified sampling method for 7 days to analyze the performance of the assembly line based on "shifts." Help them implement this sampling method for three shifts as follows: (1) 07:00 am–3:00 pm, (2) 3:00 pm–11 pm, and (3) 11 pm–7:00 am. ◄

■ Solution

Let's follow the steps given above to implement stratified sampling method in this example.

- Step 1: Identify the sampling frame.
  The sampling frame is assembly line identified in this question.
- Step 2: Divide the population into subgroups depending on the structure of the population and research question.
  As given in the question, assembly line runs in three shifts. Therefore, cycle times can separately be monitored and recorded in each shift as follows: 300 parts in the first shift, 200 parts in the second shift, and 100 parts in the third shift.
- Steps 3 and 4: Identify the proportion and sample size of each subgroup.
  Let's assume that the proportions were determined as 50%. Therefore, three shifts contain 150, 100, and 50 sample, respectively.

- Step 5: Identify each member of the subgroups.
  Each member of the subgroups is assigned to relevant time period.
- Steps 6 and 7: Decide randomness rule or use a random number generator to select the sample, and randomly select the samples from each subgroup.
  Using a random number generator, the team selects 150 parts from the first shift, 100 parts from the second shift, and 50 parts from the third shift. Whatever numbers are selected through the random number generator, those parts assigned with those numbers are picked from the population.

5.5.1.3 Systematic Sampling

The sample selection in systematic sampling requires each sample to be ordered in a sequence, and selection is done in a fixed interval with a repeating pattern after randomly picking the first element in sampling frame. The steps given below can be followed in systematic sampling. This method is more preferred over the previous methods because of its simplicity.

- Step 1: List all elements of the sampling frame.
- Step 2: Identify the sample size.
- Step 3: Randomly select the first sample.
- Step 4: Determine fixed interval for selecting the next elements.
- Step 5: Add up the fixed interval on the first element's order.
- Step 6: Identify the numbers for the samples.
- Step 7: Select the samples.

▶ Example 4

Consider that you are a member of a Six Sigma team collecting data from an assembly line, where the CTQ characteristic is the cycle time for assembling 100 parts. The team wants to sample the assembly line using systematic sampling method for 7 days to analyze the performance of the assembly line. ◀

▪ Solution

Let's follow the steps one by one below and see what parts will be sampled using systematic sampling method.

- Step 1: List all elements of the sampling frame.
  There are 100 parts to be considered for the sampling. Therefore, all parts will be numbered between 1 and 100.
- Step 2: Identify the sample size.
  Various approaches can be taken to identify sample size. For instance, the Six Sigma team may prefer to sample 10% of the population. Then, ten parts need to be sampled in the sample framing. The Six Sigma team can also use Cochran's sample size method as presented in simple random sampling method above. Let's assume that margin of error is 5%, confidence level is 95%, and only 1% of the sampling frame is considered to be selected. Then, the sample size is 15 or 16 parts.

$$n_o = \frac{Z^2 pq}{e^2} = \frac{(1.96)^2(0.01)(0.99)}{0.05^2} = 15.21$$

- Step 3: Randomly select the first sample.
  Let's say the first item in the list is randomly selected.
- Step 4: Determine fixed interval for selecting the next elements.
  To select 15 or 16 items, let's assume that fixed interval should be every 6th part.
- Step 5: Add up fixed interval on the first element's order and identify the numbers for the samples.
  Starting from the first part, the numbers should be 1, 8, 15, 22, 29, 36, 43, 50, 57, 64, 71, 78, 85, 92, and 99.
- Step 6: Select the samples.
  The parts numbered with the numbers given in the previous step are drawn from the sampling frame.

5.5.1.4 Cluster Sampling

When individual sampling is not convenient, cluster sampling is used. In this method, the individual elements are grouped in "clusters," and individual elements are selected from the clusters, designed as internally heterogeneous groups. Larger population sizes are required in cluster sampling. Compared to previous probability sampling methods, cluster sampling is considered less efficient in terms of population size required in the method. The team needs to make sure that each cluster is given an equal chance of being selected. The cluster sampling method is also less costly and easy to implement and produces more accurate results in data analysis.

When the sample frame that contains individual elements does not seem practical to list all these elements, cluster sampling is preferred, and individual elements can be clustered based on several characteristics in the population. The team lists the clusters and takes samples from that list of clusters. Rather than drawing individual samples from the population, a sample cluster is drawn from the list of clusters. For example, if the team is interested in high school students, they can make a list of high schools in the area and then randomly draw a certain amount of high schools for sampling.

Although cluster sampling and stratified sampling look very similar, there is an important difference between two sampling methods. The sampling unit in cluster sampling method is the clusters, not individual elements. In stratified sampling method, samples are drawn from each strata or subgroup. The cluster sampling method is divided into several categories: single-stage cluster sampling, two-stage cluster sampling, and multi-stage cluster sampling. The next steps can be followed in cluster sampling:

- Step 1: Identify the sampling frame.
- Step 2: Select the cluster factor and group the entire population using clusters.
- Step 3: Sample the clusters.
- Step 4: Sample individual elements from each sampled cluster using simple random sampling or systematic sampling.

▶ Example 5

Consider that you are a member of a Six Sigma team collecting data from an assembly line, where the CTQ characteristic is the cycle time to assemble 100 parts. The team wants to sample the assembly line using cluster sampling method for 7 days to analyze the performance of the assembly line. ◀

▪ **Solution**

Let's follow the steps one by one below and see what parts will be sampled in systematic sampling method.

- Step 1: Identify the sampling frame.
- Step 2: Select the cluster factor and group the entire population using clusters.
  Let's use production hours as the cluster factor in 8 hours in a day. That means we will have eight clusters to categorize the parts produced in an hour.
- Step 3: Sample the clusters.
  The Six Sigma team may use simple random sampling method to select the clusters from eight different clusters. Let's say two clusters were sampled by the team.
- Step 4: Sample individual elements from each sampled cluster using simple random sampling or systematic sampling.
  The Six Sigma team may implement simple random sampling or systematic sampling to sample individual parts from two different clusters.

5.5.2 Non-probability Sampling Methods

In non-probability sampling methods, the probability of being selected for an element is unknown. These methods are easy to use and do not include any predetermined sampling frame. When the teams have limitations to access resources, such as data, time, money, and workforce, they may consider non-probability sampling methods. The availability and accessibility of the elements in the population determines which sample from the sample frame or population is selected in non-probability sampling methods. Since the probability of an element being selected is not known, it is likely to see that the sample may not represent the population. This may affect the generalization of the findings obtained through non-probability sampling methods.

5.5.2.1 Quota Sampling

In quota sampling, a constant number of element is sampled on the basis of the research question and relevant characteristics of the population. Depending on the characteristics and variables of the population, the population is divided into subgroups. Then, the proportions of the subgroups are identified; in other words, the quotas are determined. Finally, the sample size is identified, and samples are drawn from the subgroups. For example, if the team focuses on the age and gender of the respondents in a market research, sample size is determined using percentages of each gender and age group represented in the data set. The team may say that those under the age of 25 and women respondents will be sampled. Then, the team may determine the sample size taking various approaches. Since quota sampling focuses on one or more characteristics of the population, not all characteristics are represented in this sampling method. Although quota sampling is similar to stratified sampling, elements are not selected randomly in quota sampling.

5.5.2.2 Snowball Sampling

Snowball sampling is used when the item is very rare, and it is hard to find samples. Once the initial sample is determined and contacted, the initial sample helps expand the sampling frame and the persons to be contacted in the network. The Six Sigma team may expand the size of the sample by implementing snowball sampling if the team works on a topic where it is challenging to find contact points. Scherbaum and Shockley (2015: 39) state that snowball sampling is not a sampling method, but a process through additional elements are sampled from the population. When the potential current sample is not enough and it is hard to reach out to potential samples, snowball sampling may function as an effective way to increase the size of the sample.

5.5.2.3 Convenience Sampling

Convenience sampling focuses on the availability of the sample. The sample is comprised of whoever or whatever is available at the moment the team needs to collect data. The method is mostly utilized in market research, psychological research, and pools since this method is economic, easy to use, and time-saving. For example, if an organic food producer aims at conducting a market research in Sam's Club on Sunday and asks questions of the customers in the store, the customers are considered the sampling frame for the convenience sampling. This method is *not* a random method. When generalizing findings based on convenience sampling in analyses, convenience sampling method is not recommended because it does not represent an unbiased sample of the population. However, if the size of the population is limited or research budget is tight, convenience sampling may be considered by the Six Sigma teams.

5.5.2.4 Purposive Sampling

Purposive sample is selected non-randomly depending on the structure of the population. Some desirable characteristics of the population may be used to select items in the sample using purposive sampling. This method may function to establish a population. Although purposive method may seem to utilize resources effectively, Six Sigma teams need to make sure that the population generated using the method represents the upper-level population. For example, assume that the team is interested in analyzing products returned by the customers at Walmart. If the team creates another population in the entire returned products, such as only electronics products to analyze the returned reasons just for these products, this is purposive sampling. Purposive sampling may contain bias or error in constituting a sample from the population, depending on the priorities of the researchers.

5.6 Measuring Limits of the CTQ Characteristics

While identifying CTQ characteristics, each characteristic should be supported by several limits in Six Sigma project, including tolerance limits, specification limits, confidence limits, and control limits.

Tolerance limits are determined by design and/or engineering functions, considering customer expectations, the product's expected performance, and relevant CTQ characteristics. Tolerance limits have minimum (*lower tolerance limit* – LTL) and maximum (*upper tolerance limit* – UTL) values for each CTQ characteristic on the two extreme points, where the specification or tolerance limit is two-sided. For example, if a pencil (final product) is considered to be *non-defective* when the length is between 10.30 cm (LTL) and 10.33 cm (UTL), that shows that the pencil has two-sided tolerance limits. For one-sided specification or tolerance limits, either minimum or maximum values are used, and the product/service is expected to remain either higher than minimum value or lower than maximum value. For example, if the pencil is considered to be *defective* when the length of the pencil is greater than 10.33 cm (UTL), that shows that the pencil has a one-sided tolerance limit. To check if the process is statistically in control, individual measurements are compared to tolerance limits (Juran and Gryna 1980: 294). Based on normally distributed data assumption, tolerance limits can be calculated statistically. Table K in Tables A.12 and A.13 represents factors for calculating statistical upper and lower tolerance limits for sample for which sample mean and sample standard deviation are given. The tolerance limits are calculated as Eq. 5.2.

$$\bar{X} \pm Ks \tag{5.2}$$

where the factor K is the function of the confidence interval desired, $\bar{X}$ is the sample mean, and s is the sample standard deviation. Tolerance limits will be detailed in ▶ Chap. 7.

▶ Example 6

Assume that a normally distributed data set includes 12 individual, randomly selected diameter measurements from a process with a sample mean of 14.07 cm, and sample standard deviation of 0.0016 cm. It is assumed that tolerance limits include 99% of the items in the population and confidence level is 95%. Calculate the upper and lower tolerance limits. ◀

■ **Solution**

In Table K in Table A.12, the value of K is 4.150, and upper and lower tolerance limits can be calculated as

$$\bar{X} + Ks = 14.07 + 4.150 * (0.0016) = 14.0766$$

$$\bar{X} - Ks = 14.07 - 4.150 * (0.0016) = 14.0634$$

Upper and lower tolerances show that we are 95 percent confident that 99 percent of the products in the population will have a diameter between 14.0766 cm and 14.0634 cm. Juran and Gryna (1980: 300) state that calculating tolerance limits is more rigorous than natural tolerance limits. It is also noted that range (R) can be used for calculating statistical tolerance limits. As an important assumption in tolerance limit calculations, the sample is accepted as an unbiased best representative of the population.

Natural tolerance limits refer to $\pm 3\sigma$ distance, starting from the mean of data that are normally distributed with mean μ and standard deviation σ. Natural tolerance limits contain actual variation of the process. For example, if the process is centered at 2.90 cm (μ) and standard deviation (σ) is 0.01 cm, natural tolerance limits including upper natural tolerance limit (UNTL) and lower natural tolerance limit (LNTL) can be calculated as follows:

$$\text{UNTL} = \mu + 3\sigma = 2.90 + 3(0.01) = 2.93\,\text{cm}$$

$$\text{LNTL} = \mu - 3\sigma = 2.90 - 3(0.01) = 2.87\,\text{cm}$$

UNTL and LNTL show the interval in which the CTQ characteristic may acceptably vary. If one measurement of relevant CTQ characteristic goes beyond UNTL and LNTL, that measurement refers to a defect or nonconformity. It is assumed that the natural tolerance limits represent 99.73% of the measurements of CTQ characteristic. In other words, 0.27% of the measurements are expected to fall beyond the natural tolerance limits (◘ Fig. 5.1).

According to Pyzdek and Keller (2018: 8):

> The traditional quality paradigm defined a process as capable if the process' natural tolerance spread, plus and minus three sigma, was less than the engineering tolerance. Under the assumption of normality, this three sigma quality level translates to a process yield of 99.73%. A later refinement considered the process location as well as its spread and tightened the minimum acceptance criterion so that the process mean was at least four sigma from the nearest engineering requirement. Six Sigma requires that processes operate such that the nearest engineering requirement is at least Six Sigma from the process mean.

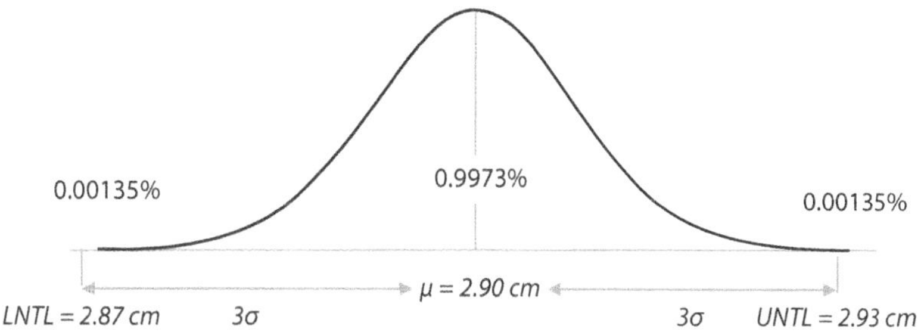

Fig. 5.1 UNTL and LNTL on normally distributed data sets. (Source: Author's creation)

The *specification limits*, upper specification limit (USL) and lower specification limit (LSL), refer to the interval in which the relevant CTQ characteristic is considered acceptable or non-defective by customers or standard setters. First, customers are the most important specification setters. Based on their needs and expectations, they determine where the values of CTQ characteristic should be located in order for products to be considered acceptable. Second, global institutions, such as ASTM (American Society for Testing and Materials) and ISO (International Organization for Standardization), and national institutions, such as ANSI (American National Standards Institute) and NIST (National Institute of Standards and Technology), set and develop global and national standards in some industries. ASTM developed more than 13,000 standards in testing and materials sciences. ISO has a great variety of standards in a very broad spectrum. ANSI and NIST in the USA provide practitioners with standards as well as calibration services, conformity assessment, and laboratory accreditation. As Juran and Gryna (1980: 317) stated, as products become more complex, the definition of process conditions is critical, so using checklists to make sure that the process conditions are well-defined is recommended. Some of the following questions can be asked for evaluating the process conditions (Juran and Gryna 1980: 317):

1. Are there written product specifications? If written in more than one place, are they all the same? Are they conveniently accessible to the operator?
2. Does the specification define the relative importance of different quality characteristics? If control charts or other control techniques are to be used, is it clear how these relate to product specification?
3. Are standards for visual defects displayed in the work area?
4. Are the written specifications given to the operator the same as the criteria used by inspectors?
5. Does the operator know how the product is used?
6. Has the operator been adequately trained to understand the specification and perform the steps needed to meet the specification? Has the operator been evaluated by test or other means to see if he or she is qualified?
7. Does the operator know the effect on future operations and product performance if the specification is not met?
8. Does the operator receive specification changes automatically and promptly?
9. Does the operator know what to do with defective raw material and defective finished product?

Confidence limits include a range of values that includes the true value of a population parameter. Confidence limits refer to the upper and lower boundaries of the confidence intervals. Confidence limits will be detailed in ► Chap. 7.

Control limits are calculated based on the performance of the process depending on the relevant statistical parameter such as mean, range, or standard deviation. The details of the control limits will be presented in ► Chap. 9. These limits cannot be determined as specification or tolerance limits. The magnitude of the standard deviation decides how far control limits are located from the parameter of the process. Generally, in control charts for variables, control limits are calculated as presented in Eq. 5.3.

$$UCL, LCL = \mu \pm 3\frac{\sigma}{\sqrt{n}} \tag{5.3}$$

where

- UCL = upper control limit
- LCL = lower control limit
- μ = the mean of the relevant parameter
- n = the sample size (subgroup size)
- σ = the standard deviation of the parameter.

5

Under the operational definitions, Six Sigma teams also need to define error, defect, and nonconformity, as they apply to each process. Error is identified as "the concept of an individual performing a task that is not completed as intended or an individual making a decision that does not produce the desired results" (Stewart and Grout 2001: 442). Eliminating human and system-related errors causing customer dissatisfaction improves processes (Pakdil et al. 2009). The emphasis on quality improvement efforts shifted away from inspecting to preventing defects and errors at one time (Snell and Atwater 1996), but the importance of eliminating errors, mistakes, and defects has risen over the last decades. Literature and practice proactively focus on how to eliminate and minimize errors in the processes in lean, TQM, and Six Sigma concepts. For instance, in the activation-trigger-schemata system (Norman 1981), when repeating the same actions over a long period of time, they are executed unconsciously and may result in an error. Error is also defined as "a natural consequence of overtaxing the resource or using the stored routines and rules in inappropriate situations" (Stewart and Grout 2001: 442). Defects or nonconformities occur because of variations, errors, and mistakes. In the quality literature, it is believed that errors and defects are linked with system-related problems, not with human-related ones. It is also believed that inadvertent errors occur randomly and are unavoidable (Snell and Atwater 1996).

A product deficiency is a product failure that results in customer dissatisfaction with a product. The major impact of deficiencies is on the costs incurred to redo prior work, to respond to customer complaints, and other related issues. Montgomery (2013) suggests that defects or nonconformities can be categorized based on their importance. It may be beneficial to categorize defects or nonconformities in a systematic way, particularly when the products and services are complex. Not all types of defects are equally important, so a demerit system for attribute data can be implemented to rank or classify the defects or nonconformities. In such a demerit system, the severity of the defects and nonconformities on the final products and services is considered to construct classifications as follows (Montgomery 2013):

- Class A defects: Very serious defects. The product or service will fail in performing as planned. The product life will reduce, and maintenance cost will be high.
- Class B defects: Serious defects. The product or service will generate somewhat less serious operating problem. The product life will reduce, and maintenance cost will be high.
- Class C defects: Moderately serious defects. The product or service will fail and generate somewhat less serious problems.
- Class D defects: Minor defects. The product or service will not fail in performing as planned. Minor defects will be seen in appearance or quality of work.

While minor defects may not have a severe or critical impact on the performance of the final product, very serious defects may result in halting the entire system.

5.7 Six Sigma Measurements

Six Sigma teams use certain terms and concepts to communicate inside and outside the team and measure the performance of the CTQ characteristic analyzed by the team. Six Sigma jargon has been developed over the last decades, as the Six Sigma concept has evolved in academia and practice. Important terms used in Six Sigma projects are explained below.

CTQ Characteristic CTQ characteristic refers to an important/critical measurable or countable feature of the product or service. The importance of the features is analyzed from all stakeholders' perspective such as customers, engineers, regulatory bodies, and standard setters. CTQ characteristics also refer to *Y* or response variables in Six Sigma terminology. CTQ characteristics are identified and detailed in product/service design processes. CTQ characteristics are also articulated by translating customer needs and expectations into technical requirements and specifications. CTQ characteristics are analyzed in ► Sect. 4.4 in ► Chap. 4.

Unit A unit refers to the smallest item or piece that is analyzed to monitor the CTQ characteristic of the product or service.

Defect A defect is when a product, service, and CTQ failed to meet CTQ characteristic's target values or acceptable levels, those standards established in design steps. A defect is a nonconformity. Using number of defects as a measure may help identify the frequency of defectives. Depending on the severity of the defect, a defect can turn into a defective product or service. According to Juran and Gryna (1980: 104), a defect is any state of unfitness for use or nonconformance to specification.

Defective The term defective refers to a particular product/service that fails at least once to meet at least one of the CTQ characteristics. Defective also refers to a nonconforming product or service that is not capable of meeting specifications of CTQ characteristics. In other words, the product or service does not fall into at least one of the CTQ characteristic's specifications. For example, the end product is a blackboard marker, and we have four CTQ characteristics identified for that product:

1. *Length* of the product
2. *Weight* of the product
3. *Diameter* of the cap
4. The *tone* of black color.

If we detect that a particular blackboard marker does not meet the specifications of at least one of these CTQ characteristics, that board marker will be identified as "defective." The same board marker may also fail to meet other CTQ characteristics' specifications at the same time. To be considered "defective," at least one CTQ characteristic should be located out of specifications. A product deficiency refers to a product failure that results in product dissatisfaction. The major impact of product deficiency occurs from the costs of rework and responding to customer complaints and dissatisfactions (Juran 1992: 7).

Defect Opportunity A defect opportunity shows how many times a unit of a product or service may fail to meet CTQ characteristics. The total number of defect opportunities of a product or service is related to the number of CTQ characteristics and the complexity of the product or service. The more complex the product and service, the more defect opportunities are expected. Historical data can also be used to figure out the defect opportunities.

Defects per Unit (DPU) DPU shows the average number of defects observed in a sample. The smaller the DPU is, the higher the level of quality and the lower level of variation is. Equation 5.4 is used to compute DPU for a unit.

$$\text{DPU} = \frac{\sum_{i=1}^{n} D_i}{n} \quad i = 1, 2, \ldots, n \tag{5.4}$$

where D_i = number of defects observed in a unit per batch and n = sample size per batch.

► Example 7

In a large-size dishwasher manufacturer that operates in Europe, the end products are subject to several comprehensive final inspections and tests prior to being delivered to end markets. Quality control department inspects and tests 150 end products per day. Calculate DPU (1) for each day and (2) for the total data given in Table 5.2 (the first two columns). ◄

- **Solution**

Using Eq. 5.4, DPU for each day is calculated as presented in Table 5.2. To exemplify, DPU for the first day is calculated as follows:

$$\text{DPU}_{\text{day1}} = \frac{\sum_{i=1}^{n} D_1}{n} = \frac{\sum_{i=1}^{n=150} 20}{150} = \frac{20}{150} = 0.1334$$

5

Table 5.2 Six Sigma measures

Day	Number of defects (D_i)	DPU per day	Defect opportunities per product	DPO per day	DPMO = DPO * 10^6	Z_{lt}	Z_{st}
1	20	20/150 = 0.1334	10	0.1334/10 = 0.01334 or 20/(150 * 10) = 0.01334	0.01334 * 10^6 = 13,333.34	2.22	3.72
2	22	22/150 = 0.1467	10	0.1467/10 = 0.01467 or 22/(150 * 10) = 0.01467	0.01467 * 10^6 = 14,666.67	2.18	3.68
3	13	13/150 = 0.0867	10	0.0867/10 = 0.00867 or 13/(150 * 10) = 0.00867	0.00867 * 10^6 = 8,666.67	2.38	3.88
4	17	17/150 = 0.1134	10	0.1134/10 = 0.01134 or 17/(150 * 10) = 0.01134	0.01134 * 10^6 = 11,333.33	2.28	3.78
5	20	20/150 = 0.1334	10	0.1334/10 = 0.01334 or 20/(150 * 10) = 0.01334	0.01334 * 10^6 = 13,333.33	2.22	3.72
	Total	92/750 = 0.1227	50	0.1227/10 = 0.01227	0.01227 * 10^6 = 12,227.77	2.25	3.75

Source: Author's creation

DPU for the entire data set is calculated as follows:

$$\text{DPU} = \frac{\sum_{i=1}^{n} D_i}{\sum_{i=1}^{n} n} = \frac{20+22+13+17+20}{150(5)}$$
$$= \frac{92}{750} = 0.1227$$

Defects per Opportunity (DPO) DPO shows the number of defects that is expected per opportunity. In other words, it is the ratio of the number of defects divided by the number of total defect opportunities. DPO is computed as presented in Eq. 5.5.

$$\text{DPO} = \frac{\text{DPU}}{\sum_{i=1}^{n} o_i} = \frac{\text{DPU}}{\text{total number of defect opportunities}}$$
$$= \frac{\sum_{i=1}^{n} D_i}{\sum_{i=1}^{n} n * o_i} \; i = 1,2,\ldots,n \tag{5.5}$$

where

- DPU = defects per unit
- n = sample size per batch
- o_i = number of defect opportunities per unit.

▶ Example 8

Using the question given in ▶ Example 7, calculate DPO (1) for each day and (2) for the total data. It is known that defect opportunities per product are ten each day. ◀

Solution

DPO for the first day can be calculated as follows:

$$\text{DPO}_{\text{Day1}} = \frac{\text{DPU}_i}{\sum_{i=1}^{n} o_i} = \frac{\text{DPU}_1}{\text{total number of defect opportunities}_{\text{Day1}}}$$
$$= \frac{0.1334}{10} = 0.01334 = \frac{20}{150 * 10}$$
$$= 0.01334 = 1.3334\%$$

For the other days, DPOs per day are presented in ◘ Table 5.2.

DPO for the entire data set is calculated as follows:

$$\text{DPO} = \frac{\text{DPU}}{\sum_{i=1}^{n} o_i} = \frac{\text{DPU}}{\text{total number of defect opportunities}}$$
$$= \frac{0.1227}{10} = 0.01227$$
$$= \frac{20+22+13+17+20}{150 * 5 * 10}$$
$$= 0.01227 = 1.227\%$$

Defects per Million Opportunities (DPMO) DPMO is calculated as the product of DPO by one million (Eq. 5.6). The aim of DPMO is to show the Six Sigma team how many defects are expected out of one million products.

$$\text{DPMO} = \text{DPO} * 10^6 \tag{5.6}$$

▶ Example 9

Using the same example that we used previously, calculate DPMO for each day and total 5-day data. ◀

Solution

To exemplify DPMO calculation for the first day,

$$\text{DPMO}_{\text{Day1}} = \text{DPO}_{\text{Day1}} * 10^6 = 0.01334 * 10^6$$
$$= 13{,}333.34$$

For the other days, DPMOs can be calculated as presented in ◘ Table 5.2.

For the total 5-day data, DPMO is calculated as follows:

$$\text{DPMO} = \text{DPU} * 10^6 = 0.01227 * 10^6$$
$$= 12{,}227.77$$

First Time Yield First time yield, in other words, is the ratio of total number of acceptable units to total number of units produced (Eq. 5.7).

$$Y_{\text{ft}} = \frac{\sum_{i=1}^{n} A_i}{\sum_{i=1}^{n} U_i} \; i = 1,2,\ldots,n \tag{5.7}$$

where A is the number of acceptable units in a batch, U is the number of units produced in a batch, and n is size of the batch.

For example, if total number of units produced in August 2018 is 250 and total number of units within specifications is 225, the first time yield would be $Y_{ft} = \frac{225}{250} = 0.90$.

For a process that includes multiple steps, overall process yield is calculated as the product of each step. For example, if a process has two steps that have yields of 85% and 90%, respectively, the overall yield will be 0.85 * 0.90 = 0.765 = 76.5%. The overall process yield is less than the yield of the step that has the lowest yield. In many industries, business and manufacturing processes include multiple sub-processes and work stations that lower overall process yield. The more steps the processes have, the more likely the process will have a lower overall yield level.

Process Sigma Process sigma is divided into two groups: (1) process sigma long term and (2) process sigma short term. Process sigma long term is calculated as the corresponding Z value of DPO that is found out from standard normal distribution table. Process sigma short term is calculated by adding 1.5 sigma on the process sigma long term.

► Example 10

For the same question given previously, calculate process sigma long-term (Z_{lt}) and process sigma short-term (Z_{st}) values. ◄

- **Solution**

For the first day, DPMO was found to be 13,333.34. To calculate process sigma long term, we first need to calculate DPO by dividing DPMO by 10^6. Then, we need to find out the closest associated Z value (Z_{lt}) from the standard normal distribution table (Table A.2).

$$\text{DPO} = \frac{13,333.34}{10,00,000} = 0.01333334$$

The corresponding Z value (Z_{lt}) for 0.01333334 DPO in the standard normal distribution table is approximately Z = 2.22 (Table A.2). Therefore, process sigma long term for the first day (Z_{lt}) is 2.22. For the other days, process sigma long-term (Z_{lt}) values are presented in ◘ Table 5.2. Total process sigma long term in this process is calculated as 2.25 (◘ Table 5.2) since DPO was 0.01227 for the entire process. To calculate Z_{lt} and Z_{st} in MS Excel, NORMSINV function can be used.

To compute process sigma short term (Z_{st}), 1.5 sigma is added to process sigma long term (Z_{lt}). The last column of ◘ Table 5.2 above displays Z_{st} values. The 1.5 sigma adjustment is considered a correction to eliminate the effects of the factors that have not been added to the models. Pyzdek and Keller (2018) states that 1.5 sigma adjustment creates a more helpful model than the traditional models and suggests using the process sigma levels associated with process ppm levels. Essentially, sigma level is associated with the quality performance of the system.

► Example 11

In a fast-food restaurant, the manager wants to calculate process sigma long-term and short-term levels for the order fulfilling process. The process flow chart is presented in ◘ Fig. 5.2. Consider that 105 orders are received daily. The defect opportunities and defective products for each step are presented in ◘ Table 5.3.

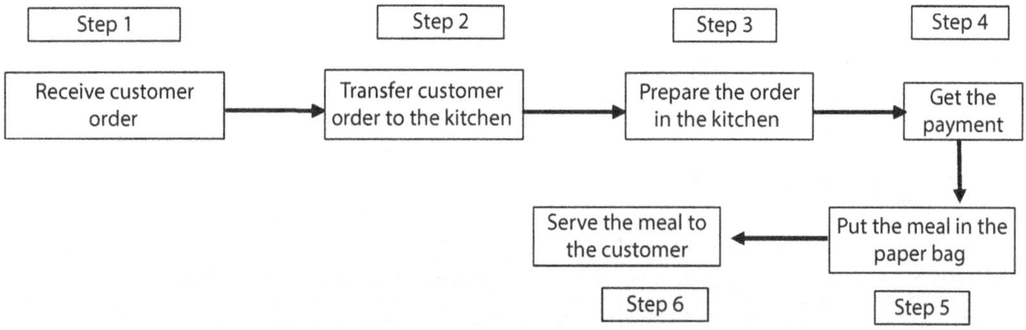

◘ **Fig. 5.2** The process flow of the system. (Source: Author's creation)

Table 5.3 Six Sigma measures for the process

Step	Processes	Defect opportunities	Customer orders (A)	Defective products (B)	Output (C)	Rework (D)	Output for Yft (E)	Yft (F)	Defect Opp (G)	Total Opp (H)	Total defects (I)	DPU (J)	DPO (K)	DPMO (L)	Zlt (M)	Zst (N)
1	Receive customer order	6	105	2	103	2	101	0.962	6	630	4	0.03810	0.00635	6349	2.49	3.99
2	Transfer customer order to the kitchen	1	103	1	102	0	102	0.990	1	103	1	0.00971	0.00971	9709	2.34	3.84
3	Prepare the order in the kitchen	1	101	1	100	1	99	0.980	1	101	2	0.01980	0.01980	19802	2.06	3.56
4	Get the payment	1	101	1	100	1	99	0.980	1	101	2	0.01980	0.01980	19802	2.06	3.56
5	Put the meal in the paper bag	1	100	2	98	1	97	0.970	1	100	3	0.03000	0.03000	30000	4.50	6.00
6	Serve the meal to the customer	1	99	2	97	0	97	0.980	1	99	2	0.02020	0.02020	20202	2.05	3.55
Total		11		9		5		0.870	11	1134	14		0.01235	12346	2.25	3.75

Source: Author's creation

Notes: C = A – B; E = C – D; F = E/A; H = A * G; J = 1 – F; K = J/G; L = K * 10^6; M = |NORMSINV(K)|; N = M + 15

Calculate process sigma long-term and short-term levels, along with the other Six Sigma measures, for this process, and evaluate the overall performance. ◀

Solution

The Y_{ft}, DPU, DPO, DPMO, Z_{lt}, *and* Z_{st} for each process are calculated and presented in ▣ Table 5.3. To exemplify, let's calculate Y_{ft}, DPU, DPO, DPMO, Z_{lt}, *and* Z_{st} for the first process.

$$Y_{\text{ft first process}} = \frac{\sum_{i=1}^{n} A_i}{\sum_{i=1}^{n} U_i} = \frac{105-(2+2)}{105}$$

$$= \frac{101}{105} = 0.962$$

as presented in column F in ▣ Table 5.3.

$$\text{DPU} = \frac{\sum_{i=1}^{n} D_i}{\sum_{i=1}^{n} n} = \frac{2+2}{105} = 0.03810$$

as presented in column J in ▣ Table 5.3.

$$\text{DPO} = \frac{\text{DPU}}{\sum_{i=1}^{n} o_i} = \frac{0.03810}{6} = 0.00635$$

as presented in column K in ▣ Table 5.3.

DPMO = DPO $*$ 10^6 = 0.00635 $*$ 10^6 = 6,349 as presented in column L in ▣ Table 5.3.

Z_{lt} = |NORMSINV(DPO)| = |NORMSINV(0.00635)| = 2.49 as presented in column M in ▣ Table 5.3.

$Z_{st} = Z_{lt} + 1.5 = 2.49 + 1.50 = 3.99$ as presented in column N in ▣ Table 5.3.

Key Concepts

DMAIC; measure phase; data collection plan; types of variables; CTQ (Critical-to-Quality) characteristics; response or dependent variable; independent variable; data; population; census; observation; sample; variable; parameter; statistic; operational definition; Industry 4.0; digital age; big data; the Internet of Things (IoT); artificial intelligence; machine learning; digital systems; software-driven devices; categorical and numerical variables; univariate, bivariate, and multivariate variables; continuous and discrete variables; nominal, ordinal, interval, and ratio variables; nominal, the smaller, and the larger variables; sampling methods; probability sampling methods; simple random sampling; stratified sampling; systematic sampling; cluster sampling; non-probability sampling methods; quota sampling; purposive sampling; self-selecting sampling; judgmental sampling; snowball sampling; convenience sampling; data collection; upper and lower specification limits (USL and LSL), upper and lower tolerance limits (UTL and LTL), upper and lower natural tolerance limits (UNTL and LNTL), confidence limits, and upper and lower control limits (*UCL* and *LCL*); ASTM (American Society for Testing and Materials), ISO (International Organization for Standardization), and national institutions, such as ANSI (American National Standards Institute) and NIST (National Institute of Standards and Technology); sigma level of the process and system; unit; defect; defective; defect opportunity; defects per unit (DPU); defects per opportunity (DPO); defects per million opportunities (DPMO); first time yield; process sigma long term (Z_{lt}) and process sigma short term (Z_{st}); and standard normal distribution.

Summary

This chapter focuses on identifying types of variables and sampling methods, data collection plans, collecting data, identifying measurement system components, measuring limits of the CTQ characteristics, and sigma level of the processes and systems.

Practice and Discussion Questions

1. What are the main questions answered in the "Measure" phase of DMAIC process?
2. Describe what data, population, census, observation, sample, variable, parameter, and statistic terms mean.
3. Describe the main difference between "parameter" and "statistic."
4. Discuss the importance of data collection in Six Sigma methodology.
5. Discuss the importance of collecting reliable, accurate, and appropriate data in Six Sigma methodology.
6. Discuss the functions of data collection plans in Six Sigma.
7. What basic variables should take place in data collection plans?
8. Identify categorical and numerical variables and give examples for each.
9. Identify univariate, bivariate, and multivariate variables and give examples for each.
10. Identify continuous and discrete variables and give examples for each.
11. Identify nominal, ordinal, interval, and ratio variables and give examples for each.
12. Identify nominal, the smaller, and the larger variables, and give examples for each.
13. Identify independent and dependent variables and give examples for each.
14. What are the two basic types of sampling?
15. Classify sampling methods into categories.
16. Briefly describe when and how to use simple random sampling.
17. Briefly describe when and how to use stratified random sampling.
18. Briefly describe when and how to use systematic sampling.
19. Briefly describe when and how to use cluster sampling.
20. Briefly describe when and how to use quota sampling.
21. Briefly describe when and how to use snowball sampling,
22. Briefly describe when and how to use convenience sampling.
23. Briefly describe when and how to use purposive sampling.
24. Discuss what type of limits of CTQ characteristics are evaluated in Six Sigma.
25. What functions and/or stakeholders determine control limits?
26. What functions and/or stakeholders determine tolerance limits?
27. What functions and/or stakeholders determine specification limits?
28. Describe CTQ characteristic, unit, defect, defective, defective opportunity, defects per unit (DPU), defects per opportunity (DPO), defects per million opportunities (DPMO), first time yield, and process sigma long term (Z_{lt}) and short term (Z_{st}).
29. In a production line at an automobile factory, the supervisor of the line wants to calculate process sigma long-term and short-term levels. The process flow chart is presented in ■ Fig. 5.3. Consider that 110 pieces

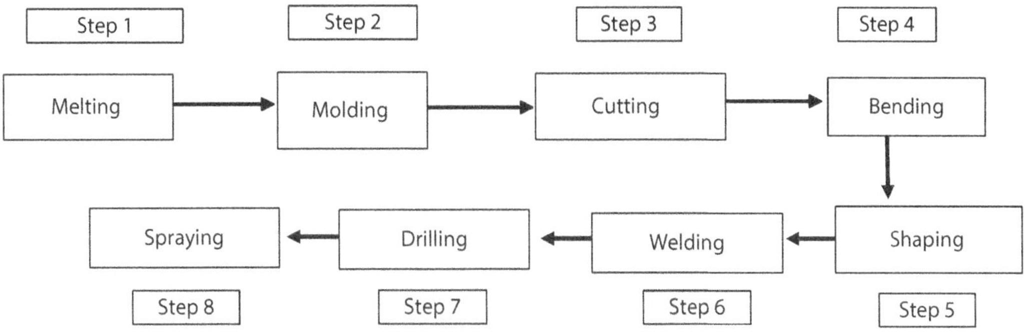

■ **Fig. 5.3** The process flow of the production line at automobile factory. (Source: Author's creation)

Table 5.4 Error opportunities, number of defectives, and rework production in the production line at automobile factory

Step	Activities	Error opportunities	Defective products	Reworked products
1	Melting	1	1	0
2	Molding	2	1	0
3	Cutting	3	1	0
4	Bending	1	1	1
5	Shaping	2	1	1
6	Welding	1	2	1
7	Drilling	1	0	0
8	Spraying	2	0	0
Total		13	7	3

Source: Author's creation

5

are processed daily in the line. The defective products, reworks, and defect opportunities for each step are presented in Table 5.4. Calculate process sigma long-term and short-term levels, along with the other Six Sigma measures, for this process, and evaluate the overall performance.

References

Berwick, D. M., James, B., & Coye, M. J. (2003). Connections between quality measurement and improvement. *Medical Care, 41*(1), I–30.

Bruhl, R. (2018). *Understanding statistical analysis and modeling*. Los Angeles: Sage.

Defeo, J. (2017). *Juran's quality management and analysis*. Singapore: McGraw-Hill Higher Education.

Gitlow, H. S, Melnyck, R. J, & Levine, D. M. (2015). *A guide to six sigma and process improvement for practitioners and students* (2nd ed.). Upper Saddle River, NJ: Pearson.

Hart, M. K., & Hart, R. F. (2002). *Statistical process control for health care*. Duxbury: Thomson Learning.

Juran, J. M. (1992). *Juran on quality by design: The new steps for planning quality into goods and services*. New York: The Free Press. Simon and Schuster.

Juran, J. M., & Gryna, F. M. (1980). *Quality planning and analysis*. New York: McGraw-Hill.

Knapp, H. (2016). *Practical statistics for nursing using SPSS*. Los Angeles: Sage Publications.

Mohammed, M. A., Worthington, P., & Woodall, W. H. (2008). Plotting basic control charts: Tutorial notes for healthcare practitioners. *BMJ Quality and Safety, 17*(2), 137–145.

Montgomery, D. C. (2005). *Introduction to statistical quality control* (5th ed.). New York: John Wiley & Sons.

Montgomery, D. C. (2013). *Introduction to statistical quality control*. New York: Wiley.

Norman, D. (1981). Categorization of action slips. *Psychological Review, 88*(1), 1–15.

Pakdil, F., Özkök, O., Dengiz, B., Kara, I., Selvi, N., & Kargı, A. (2009). A systematic approach to reduce human and system-related errors causing customer dissatisfaction in a production environment. *Total Quality Management & Business Excellence, 20*(1), 129–137.

Pyzdek, T., & Keller, P. A. (2018). *The six sigma handbook* (5th ed.). New York: Mc-Graw Hill.

Scherbaum, C., & Shockley, K. (2015). *Analysing quantitative data for business and management students*. London: Sage.

Snell, T., & Atwater, J. B. (1996). Using poka-yoke concepts to improve a military retail supply system. *Production and Inventory Management Journal, 37*(4), 44.

Stewart, D. M., & Grout, J. R. (2001). The human side of mistake-proofing. *Production and Operations Management, 10*(4), 440–459.

Further Reading

Montgomery, D. C., & Runger, G. C. (2014). *Applied statistics and probability for engineers*. New York: John Wiley and Sons.

Measurement System Analysis: Gage R&R Analysis

Contents

F. Pakdil, *Six Sigma for Students*, https://doi.org/10.1007/978-3-030-40709-4_6

Learning Objectives

After careful study of this chapter, you should be able to:

- Conduct and analyze Gage R&R Analysis in Measure phase of DMAIC
- Calculate measurement system variance, total variance, and process variance
- Interpret Gage R&R Analysis results.

6.1 Introduction

The decisions made in production and assembly lines and service settings heavily rely on measurement systems. The main goal of measurement systems in manufacturing settings is to collect valid and reliable data. The measurement systems contain:

- Operations
- Procedures and instructions
- Measurement instruments or gauge
- Fixtures and other equipment
- Software
- Operators
- Environmental conditions
- Assumptions.

According to Juran and Gryna (1980: 387), the quantification of product or process characteristics involves:

1. Definition of standardized units, called units of measure.
2. Instruments which are calibrated in terms of these standardized units of measure.
3. Use of these instruments to quantify or measure the extent to which the product or process possesses the characteristic under study. This process of quantification is called measurement.

Measurement systems rely heavily on unchanged fundamental units. A unit of measure is defined as a tool that allows the clear, understandable, and precise evaluation in numbers enabling future accurate measurements (Juran 1992). *Hours* to deliver a service or complete an order and the *temperature* of an oven are units of measure. In ancient Egypt, a royal cubit was used as the base unit of measure. The distance between two elbows was considered the standard unit in construction. Similarly, Architect Amenemipt's ruler was employed as a base unit of measure in similar measurement processes (■ Image 6.1). In the modern world, industrialized countries establish national bureau of standards such as ANSI (American National Standards Institute) and NIST (National Institute of Standards and Technology) to construct primary reference standards. Additionally, professional societies such as ASTM (American Society for Testing and Materials) and international organizations such as ISO (International Organization for Standardization) develop standards and rules for standard units and measurement systems.

Because measurements and data have a critical role in Six Sigma projects, teams must make sure that the data collected through the chosen measurement system are dependable. When the measurement error increases, the process capability and behavior will not be understood, and further measurements will not be reliable. Measurement systems may generate wrong measures and readings of characteristics due to problems of *accuracy* and *precision*.

■ **Image 6.1** Architect Amenemipt's ruler, Horemheb B.C.E. 1319–1307. (Source: ▶ https://nistdigitalarchives.contentdm.oclc.org/digital/collection/p15421coll3/id/205/)

> The *accuracy* of an instrument is the extent to which the average of a long series of repeat measurements made by the instrument on a single unit of product differs from the true value. This difference is usually due to a systematic error in the measurement system. The *precision* of a measurement system is the extent to which the instrument repeats its results when making repeat measurements on the same unit of product (Juran and Gryna 1980: 390).

Accuracy and *precision* have three conditions: (1) definition of the test method; (2) definition of systemic of causes of variability; and (3) existence of a statistically controlled measurement process. *Definition of the test method* contains the detailed description of test method such as measurement procedure, measurement equipment, and test conditions used in the measurement process. *Definition of the system of causes of variability* refers to material, operators, apparatus, laboratories, days, etc. *Existence of a statistically controlled measurement process* means that stability of the measurement system should be maintained to provide high accuracy and precision (Juran and Gryna 1980).

Walter Shewhart (1939), the founder of statistical quality control, said:

> It is important to realize... that there are two aspects of an operation of measurement; one is quantitative and the other qualitative. One consists of numbers or pointer readings such as the observed lengths in n measurements of the length of a line, and the other consists of the physical manipulations of the physical things by someone in accord with instructions that we shall assume to be describable in words constituting a text (130).

Shewhart's classification emphasizes the importance of measurement method and measurement process (Murphy 1961).

6.2 Gage R&R Analysis

The measurement systems are analyzed through *Gage R&R analysis*. The purposes of Gage R&R analysis are to (1) determine how much of the total observed variability comes from the gauge, instrument, or measurement system; (2) identify and isolate the variability causes in the measurement system; and (3) evaluate if the measure system is capable, adequate, accurate, and appropriate (Montgomery 2013). The idea behind Gage R&R is to differentiate measurement system variability from process variability. In Gage R&R analysis, the total variance of observed data is formulated as given in Eqs. 6.1 and 6.2 (Montgomery 2013; Al-Refaie and Bata 2010).

$$y = x + \varepsilon \tag{6.1}$$

$$\sigma^2_{\text{Total}} = \sigma^2_{\text{P}} + \sigma^2_{\text{Gauge}} \tag{6.2}$$

where

- y = the total observation value
- x = the true value of the observation
  ε = the measurement error
- σ^2_{Total} = the total observed measurement variability
- σ^2_{P} = the observed variability due to variability in the product itself
- σ^2_{Gauge} = the measurement error or gauge variability, which is called *Gage Repeatability and Reproducibility* (Gage R&R).

It is assumed that x and ε are normally distributed and independent. If the difference between x and ε is zero, total observation value (y) will equal the true value of the observation, which means that the measurement system variability is zero. The measurement error refers to the difference between true value and measured value. The estimated gauge variance is calculated as given in Eqs. 6.3 and 6.4 (Montgomery 2013; Al-Refaie and Bata 2010).

$$\hat{\sigma}_{\text{Gauge}} = \frac{\overline{R}}{d_2} \tag{6.3}$$

$$\hat{\sigma}^2_{\text{Gauge}} = \hat{\sigma}^2_{\text{repeatability}} + \hat{\sigma}^2_{\text{reproducibility}} \tag{6.4}$$

where $\hat{\sigma}^2_{\text{repeatability}}$ = the estimated repeatability variation due to the gauge operation and $\hat{\sigma}^2_{\text{reproducibility}}$ = the estimated variation caused by different operations using the same gauge to measure identical quality characteristic of the same unit.

6

Repeatability focuses on the differences when the same units are measured by the same measurement system several times. It is the variation among repeated measurements performed by one operator on a single piece and with the same measurement instrument. The lower the variation, the better the repeatability. In other words, repeatability analyzes the variability between independent measurements taken from a single quantity of homogenous material in a single measurement system by a single operator.

Reproducibility measures the differences when the same units are measured using different measurement systems. It is the variation among the average measurements performed by several operators using the same pieces and the same measurement instrument. The lower the variation, the better the repeatability. Measurement systems are expected to differentiate conformance units from nonconformance units, in other words, bad and good units. Reproducibility measures the variability between single test results randomly taken by multiple operators from a single quantity of homogeneous material in different measurement processes.

One of the two measures used for assessing the measurement system is the precision-to-tolerance ratio (*P*/*T*), as given in Eq. 6.5 (Montgomery 2013; Al-Refaie and Bata 2010).

$$P/T = \frac{k\hat{\sigma}_{\text{Gauge}}}{\text{USL} - \text{LSL}} * 100 \tag{6.5}$$

where

- $\hat{\sigma}_{\text{Gauge}}$ = the estimated measurement error or gauge variability
- k = A constant value that corresponds to number of standard deviations between tolerance limits
- USL = the upper specification limit
- LSL = the lower specification limit.

A measurement system is judged to be *capable* if the *P*/*T* is less or equal to 10%, whereas a gauge is considered *incapable* if the *P*/*T* is greater than 30%. The second measure, *R&R*% ($\hat{\rho}_{\text{M}}$), is the ratio of the measurement error to total variability as given in Eq. 6.6. The $\hat{\rho}_{\text{M}}$ in the total variation is an indicator to evaluate whether a measurement system is adequate. The capability of the measurement system is considered desirable if $\hat{\rho}_{\text{M}}$ is less than 10%. However, it is not adequate if $\hat{\rho}_{\text{M}}$ is greater than 30% (Montgomery 2013). The potential ranges and status for Gage R&R analysis results are as follows:

- Desirable: P/T and $\hat{\rho}_{\text{M}}$ (%) are between 0 and 10.
- Acceptable: P/T and $\hat{\rho}_{\text{M}}$ (%) are between 10 and 20.
- Marginally acceptable: P/T and $\hat{\rho}_{\text{M}}$ (%) are between 20 and 30.
- Unacceptable: P/T and $\hat{\rho}_{\text{M}}$ (%) are between 30 and 100.

$$\hat{\rho}_{\text{M}} = \frac{\hat{\sigma}^2_{\text{Gauge}}}{\hat{\sigma}^2_{\text{Total}}} * 100 \tag{6.6}$$

If measurement system variability ($\hat{\sigma}^2_{\text{Gauge}}$) is greater than process variability ($\hat{\sigma}^2_{\text{p}}$), it means that the data collected through that system are not suitable (appropriate) for use in further decision-making processes. Without making sure that measurement system is capable of collecting reliable and valid data, further decisions would be questionable. In this case, the measurement system should be analyzed considering such things as sample size, measurement methods, measurement requirements, environmental conditions, and operator qualifications. The root causes of high variation should be determined and eliminated to decrease the variability as much as possible, especially to the level where measurement system is considered acceptable.

The flow of Gage R&R analysis using software is as follows:

- Step 1: Identify components of measurement system.
- Step 2: Collect data.
- Step 3: Enter data in software.
- Step 4: Draw and evaluate run chart.
- Step 5: Draw and evaluate Gage R&R ANOVA.
- Step 6: Run Gage R&R ANOVA test and evaluate the results.
- Step 7: Calculate *R&R*% and *P*/*T* values.
- Step 8: Make the final decision on measurement system.

► **Example 1**

In a dishwasher plant, one of the CTQ characteristics of component A is the distance between two points on an orthogonal line, which is called K11. The measurement system consists of two operators in a well-defined measurement instructions environment. Operators take the measurements three times for 15 parts of component A selected randomly from the production line in 2 days. The operators do not know which part is measured in a sequence. Upper specification limit (USL) is 300.7 mm, lower specification limit (LSL) is 299.1 mm, and target value is 299.9 mm. The data are shown in ◘ Table 6.1. Analyze the measurement system using Gage R&R analysis, and decide if the measurement system is acceptable. ◄

▪ Solution

We will show how to go through the steps of Gage R&R analysis using Minitab 19.0 based on the steps of the flow given above.

- Step 1: Identify components of measurement system.
  First of all, let's begin with itemizing components of measurement system presented in this question. The components of the measurement system are as follows:
  - *Product*: Dishwasher
  - *Characteristic*: K11 point in internal back side of dishwasher
  - *Specifications*: 299.9 mm ± 0.8 mm
  - *Measurement system*: Two operators working with digital gauge
  - *Software*: Minitab 19.0
  - *Sample size* (*n*): Randomly selected 15 pieces
  - *Repetition*: Three.

- Step 2: Collect data.
  The data were already collected in this example as presented in ◘ Table 6.1. It is critical to collect measurements based on (1) part number, (2) operator, (3) number of repetition, and observation. In this example, we have (1) 15 pieces, (2) operators Murat and Emre, and (3) 3 repetitions for each part.
- Step 3: Enter data in software.
  The measurements are entered in a spreadsheet or software. In our example, we will be using Minitab 19.0 statistical software.
- Step 4: Draw and evaluate run chart.
  There are several components of Gage R&R analysis based on the data provided in the question. First, let's analyze the current measurement system through Run Chart that shows us the differences between and among the operators. ◘ Image 6.2 tells us that the parts were measured equally by the measurement system, except parts 5, 11, 12, and 14. There are 15 pieces measured, and each piece is measured by two employees three times, demonstrated by red and black measurements indicating which operator measured the pieces. For example, in part 3, there are six measurements that are very close to each other. However, in parts 5, 11, 12, and 14, there are differences that are visually easy to see among six measurements of each piece. This shows that run chart-based visual evaluation may not be enough to decide if the measurement system is acceptable. Further analysis seems necessary to make better decisions.

To draw run chart of Gage R&R analysis in Minitab, after transferring data set (◘ Table 6.1) in a Minitab worksheet, click on Stat→Quality Tools→Gage Study→Gage Run Chart. In the following screen, enter part numbers in "part numbers," operator in "operators," and data K11 (mm) in measurement data. Click on OK. The run chart of Gage R&R analysis will appear on Minitab output as shown in ◘ Image 6.2.

- Step 5: Draw and evaluate Gage R&R ANOVA diagram.
  The second visual analysis can be done through Gage R&R ANOVA output. To draw Gage R&R ANOVA diagram in Minitab, click on Stat→Quality Tools→Gage Study→Gage R&R (crossed). In the following screen, enter part numbers in "part numbers," operator in "operators," and data K11 (mm) in measurement data. In method of analysis, select ANOVA. Click on "Options" and enter 299.1 in

6

Table 6.1 The data set for Gage R&R analysis

Part #	Operator	K11 (mm)	Part #	Operator	K11 (mm)	Part #	Operator	K11 (mm)
1	Emre	300.2	6	Emre	300.36	11	Emre	299.17
1	Emre	300.18	6	Emre	300.34	11	Emre	299.15
1	Emre	300.18	6	Emre	300.34	11	Emre	299.34
1	Murat	300.13	6	Murat	300.28	11	Murat	299.12
1	Murat	300.15	6	Murat	300.3	11	Murat	299.13
1	Murat	300.1	6	Murat	300.35	11	Murat	299.16
2	Emre	300.1	7	Emre	300.09	12	Emre	300.55
2	Emre	300.09	7	Emre	300.11	12	Emre	300.53
2	Emre	300.08	7	Emre	300.12	12	Emre	300.43
2	Murat	300.1	7	Murat	300.16	12	Murat	300.43
2	Murat	300.07	7	Murat	300.16	12	Murat	300.41
2	Murat	300.13	7	Murat	300.1	12	Murat	300.47
3	Emre	300.64	8	Emre	300.72	13	Emre	299.59
3	Emre	300.65	8	Emre	300.7	13	Emre	299.60
3	Emre	300.64	8	Emre	300.72	13	Emre	299.57
3	Murat	300.62	8	Murat	300.70	13	Murat	299.62
3	Murat	300.63	8	Murat	300.67	13	Murat	299.61
3	Murat	300.62	8	Murat	300.70	13	Murat	299.60
4	Emre	299.81	9	Emre	299.76	14	Emre	299.86
4	Emre	299.8	9	Emre	299.75	14	Emre	299.85
4	Emre	299.79	9	Emre	299.73	14	Emre	299.68
4	Murat	299.74	9	Murat	299.71	14	Murat	299.71
4	Murat	299.75	9	Murat	299.74	14	Murat	299.70
4	Murat	299.78	9	Murat	299.76	14	Murat	299.80
5	Emre	299.08	10	Emre	299.36	15	Emre	299.60
5	Emre	299.1	10	Emre	299.33	15	Emre	299.61
5	Emre	299.09	10	Emre	299.30	15	Emre	299.58
5	Murat	299.06	10	Murat	299.29	15	Murat	299.60
5	Murat	299.1	10	Murat	299.25	15	Murat	299.62
5	Murat	299.2	10	Murat	299.31	15	Murat	299.60

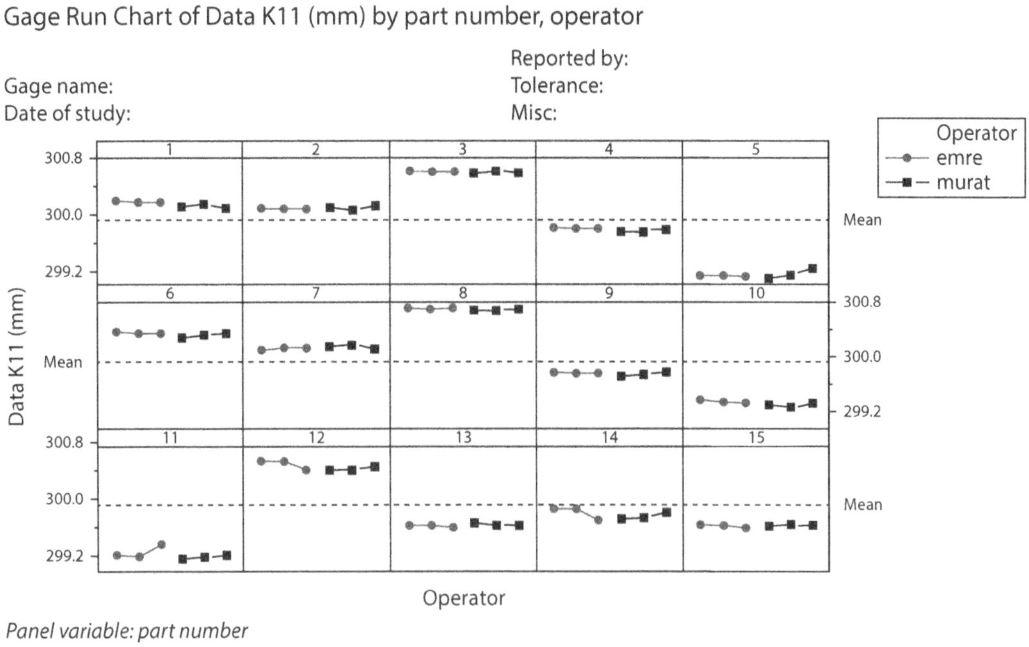

Image 6.2 Run chart of Gage R&R analysis. (Source: Author's creation based on Minitab)

"Lower spec" and 300.7 in "Upper spec" since the process specification limits are given as 299.9 mm ± 0.8 mm. For "study variation," which refers to *k* variable in calculations, we can use 6 or 5.15, depending on the decision-makers' approaches. Click on OK→OK. The Gage R&R ANOVA diagram will appear on Minitab output as shown in Image 6.3.

According to the first diagram titled "Components of variation" in the image, the majority of the variance come from *part-to-part variance*, which means that the variance caused by production lines are larger than the variance of the measurement system. Ideally, the variation caused by measurement system should be less than part-to-part variation. The second diagram named "Data K11 (mm) by part no" tells us that the average of operators by parts are close to each other, except parts 5, 11, and 14. Therefore, the measurement system is acceptable. The diagram titled "R chart by Operator" shows that the first operator's measurements for parts 11 and 14 are out of *UCL*, whereas the second operators' measurements for part 5 are above *UCL*. These are the measurements that show a decrease in the performance of the measurement system. Ideally, each point in the *R* chart should be located in between *UCL* and *LCL*, which shows the consistency of the measurements of the operators.

The "Data K11 (mm) by operator" graph represents the consistency between the averages of the two operators. The line that links two operators' averages should be flat if there are no differences between the averages of the operators. If the line has a slope, it shows that there is a difference between the averages of the operators' measurements. In the "Xbar chart by operator" graph, all points should be out of control for an acceptable measurement system meaning that the variation caused by

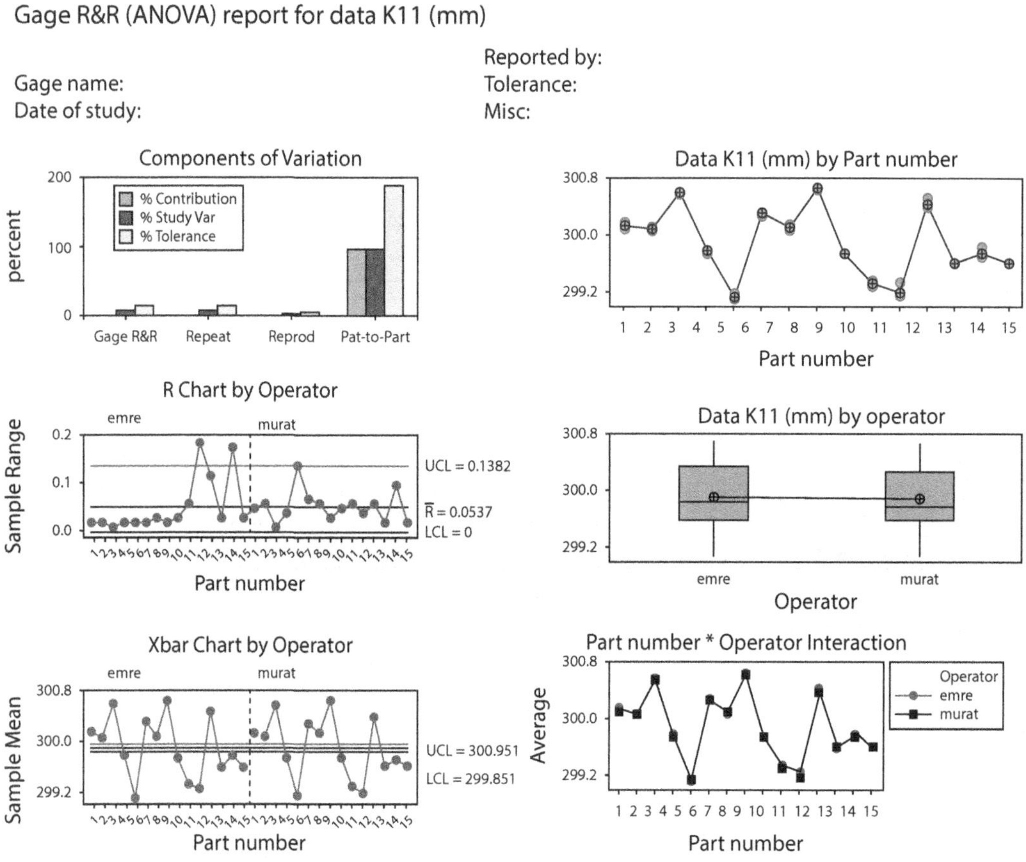

Image 6.3 Gage R&R (ANOVA) analysis visual results. (Source: Author's creation based on Minitab)

parts is higher than the variation caused by the measurement system. Lastly, the graph named "Part no*operator interaction" demonstrates the interaction causing operators to measure different parts in different ways. The flat lines are seen where there is no interaction.

- Step 6: Run Gage R&R ANOVA test and evaluate the results.

  Gage R&R ANOVA results are presented in Table 6.2. Total Gage R&R accounts for 0.71% of the total variation, whereas part-to-part variation accounts for 99.29%. That result shows that the measurement system is acceptable, since the majority of the variation is caused by the manufacturing processes and only 0.71% of the total variation is due to measurement system.

- Steps 7 and 8: Calculate *R&R*% and *P*/*T* values and make the final decision.

  R&R% is 8.41 and *P*/*T*% is 16.00, which demonstrates that the measurement system is acceptable. Recall the ranges and status for the values of *R&R*% and *P*/*T*%. Since these two values are less than 30% in this example, the measurement system is considered reliable, valid, and accurate.

If there is no software available, we can analyze measurement system manually using the following steps:

Table 6.2 Gage R&R (ANOVA) results

Gage R&R				
Variance components				
Source	*VarComp*	*% Contribution (of VarComp)*		
Total Gage R&R	0.001821	0.71		
Repeatability	0.001594	0.62		
Reproducibility	0.000227	0.09		
Operator	0.000227	0.09		
Part-to-part	0.255780	99.29		
Total variation	0.257601	1,000.00		
Process tolerance = 1.6				
Gage evaluation				
Source	*StdDev (SD)*	*Study Var (6 × SD) (%SV)*	*% Study Var (SV/Toler)*	*% Tolerance*
Total Gage R&R	0.042669	0.25601	8.41	16.00
Repeatability	0.039926	0.23956	7.87	14.97
Reproducibility	0.015051	0.09030	2.97	5.64
Operator	0.015051	0.09030	2.97	5.64
Part-to-part	0.505747	3.03448	99.65	189.66
Total variation	0.507544	3.04526	100.00	190.33
Number of distinct categories = 16				

Source: Author's creation based on Minitab

- Step 1: Identify components of measurement system.
- Step 2: Collect data.
- Step 3: Calculate range of measures (R_i) and average of ranges $\left(\overline{R}\right)$.
- Step 4: Calculate total standard deviation of data (s).
- Step 5: Estimate the standard deviation of measurement error ($\hat{\sigma}_{\text{Gauge}}$).
- Step 6: Estimate the variance of measurement error $\left(\hat{\sigma}^2_{\text{Gauge}}\right)$.
- Step 7: Estimate the variance of total variability $\left(\hat{\sigma}^2_{\text{Total}}\right)$.
- Step 8: Calculate and interpret P/T and $R\&R$ ratios.
- Step 9: Make the final decision on measurement system.

► Example 2 shows how to analyze a measurement system manually and interpret findings.

Table 6.3 Data collected from the measurement system

Part number	Measurements		Range (R_i)
	Measurement 1	Measurement 2	
1	21	24	3
2	24	23	1
3	20	21	1
4	27	27	0
5	19	22	3
6	23	25	2
7	22	21	1
8	19	19	0
9	24	23	1
10	24	23	1

Source: Author's creation

► Example 2

Assume that the Six Sigma team wants to evaluate a measurement system used in manufacturing department at a company that produces a great variety of metal wires. One of the CTQ characteristics inspected in this measurement system is length of the metal wires. The data set includes ten parts, one operator, and two measurements per part, as shown in Table 6.3 Use the following variables USL = 30 mm, LSL = 15 mm, and k = 6. Decide if this measurement system is "desirable" in Gage R&R analysis. ◄

Solution

- Step 1: Identify components of measurement system.
  Let's start by identifying the components and values of measurement system as follows:
  - *Product*: Metal wires
  - *Characteristic*: Length
  - *Specifications*: USL = 30 mm and LSL = 15 mm
  - *Measurement system*: One operator working with digital gauge
  - *Software*: Minitab 19.0
  - *Sample size* (n): Randomly selected ten pieces
  - *Repetition*: Two.
- Step 2: Collect data.
  In our example, the data are already collected as presented in Table 6.3. There are ten pieces measured twice, and a total of 20 observations are contained in data set.
- Step 3: Calculate range of measures (R_i) and average of ranges ($\overline{R}$).
  We can continue with calculating range (R_i) of each part as shown in Table 6.3. Going forward, we can calculate average of ranges ($\overline{R}$) and total standard devia-

tion of the data (s). The average of the ranges ($\bar{R}$) is

$$\bar{R} = \frac{\sum_{i=1}^{n} R}{n} = \frac{13}{10} = 1.3$$

- Step 4: Calculate total standard deviation of data (s).
  The total standard deviation of data (s) is

$$s = \sqrt{\frac{\sum_{i=1}^{n} (X_i - \bar{X})^2}{n-1}} = 2.3725$$

- Step 5: Estimate the standard deviation of measurement error ($\hat{\sigma}_{\text{Gauge}}$).
  In the next steps of the calculations, to see the components of the total variance, we use total observed measurement variability $\sigma_{\text{Total}}^2 = \sigma_{\text{P}}^2 + \sigma_{\text{Gauge}}^2$.

  The standard deviation of measurement error σ_{Gauge} can be estimated:

$$\hat{\sigma}_{\text{Gauge}} = \frac{\bar{R}}{d_2} = \frac{1.3}{1.128} = 1.1524$$

The coefficient of d_2 is found from Table A.11 as a function of sample size. In our example, sample size (n) is two, and d_2 is found to be 1.128 in Table A.11.

- Step 6: Estimate the variance of measurement error $\left(\hat{\sigma}^2_{\text{Gauge}}\right)$.
  The variance of the measurement error is estimated based on $\hat{\sigma}_{\text{Gauge}}$.

$$\hat{\sigma}^2_{\text{Gauge}} = 1.1524 = (1.1524)^2 = 1.328$$

- Step 7: Estimate the variance of total variability $\left(\hat{\sigma}^2_{\text{Total}} = s^2\right)$.
  The variance of the total variability can be calculated as the square of total standard deviation.

$$s = 2.3725$$

$$\hat{\sigma}^2_{\text{Total}} = s^2 = (2.3725)^2 = 5.63$$

Using $\hat{\sigma}^2_{\text{Gauge}}$ and $\hat{\sigma}^2_{\text{Total}}$ variables, the estimated variability of the process can be calculated.

$$\hat{\sigma}^2_{\text{Total}} = \hat{\sigma}^2_{\text{P}} + \hat{\sigma}^2_{\text{Gauge}}$$

$$5.63 = \hat{\sigma}^2_{\text{P}} + 1.328$$

$$\hat{\sigma}^2_{\text{P}} = 5.63 - 1.328 = 4.302$$

Based on the calculations above, the estimated variability of the process ($\hat{\sigma}^2_{\text{P}} = 4.302$) is greater than the estimated variability of the measurement system $\hat{\sigma}^2_{\text{Gauge}} = 1.328$, $\sigma^2_{\text{P}} > \sigma^2_{\text{Gauge}}$. For a reliable measurement system, the process variability should be greater than measurement system variability.

- Step 8: Calculate and interpret P/T and R&R ratios.
  In the precision-to-tolerance ratio (P/T), constant k has two alternative values: k = 5.15 and k = 6. The value k = 5.15 refers to the number of standard deviations between bounds of a 95% tolerance interval that contains 99% of the data, while k = 6 refers to the number of standard deviations between upper natural tolerance limit (UNTL) and lower natural tolerance limit (LNTL) of the data (Montgomery 2013). In the data set, USL = 30 mm, LSL = 15 mm, and k = 6. The P/T ratio is

$$P/T = \frac{k\hat{\sigma}_{\text{Gauge}}}{\text{USL} - \text{LSL}} = \frac{6*(1.1524)}{30-15}$$

$$= 0.46096 = 46.096\%$$

The $R\&R$ ratio is

$$\hat{\rho}_{\text{M}} = \frac{\hat{\sigma}^2_{\text{Gauge}}}{\hat{\sigma}^2_{\text{Total}}} * 100 = \frac{1.328}{5.63} * 100 = 23.58\%$$

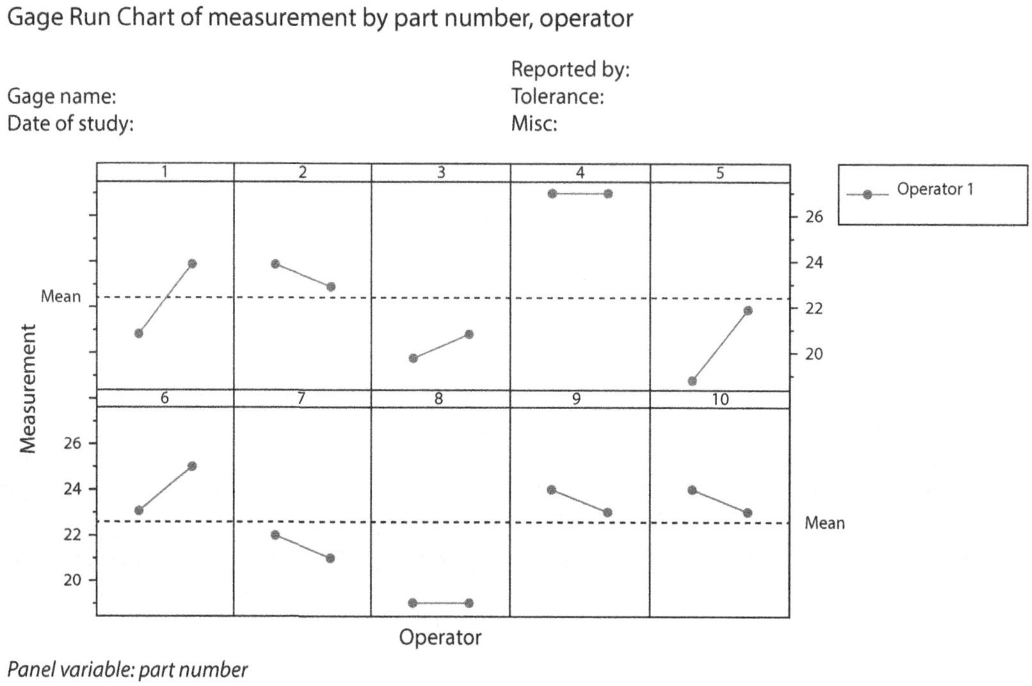

Image 6.4 Gage R&R analysis run chart output drawn in Minitab. (Source: Author's creation based Minitab)

Therefore, the variance of the measurement system accounts for 23.58% of the total estimated variance of the measurements.

- Step 9: Make the final decision on measurement system.
  As a result, *P/T,* and *R&R* variables show that the measurement system cannot be considered desirable, since *P/T* is greater than 30%. The next statistical analyses and interpretations should not rely on the data collected through this measurement system.

As an additional analysis, we can also run Gage R&R analysis in Minitab for this question. The Gage R&R analysis results are presented in Images 6.4 and 6.5 and Table 6.4. Since there is only one operator taking two measurements for each piece in this system, there will be only two points for each piece as presented in Image 6.4. There are only two pieces, #4 and #8, for which the measurements are found to be the same. For the other parts, although there is one operator and each piece is measured twice, except #4 and #8, all measurements are different, which shows the inconsistency of the measures and low reliability of the measurement system.

As seen in Image 6.5, in "Components of variation" section, part-to-part variation is higher than the measurement system variation which is presented in "Gage R&R" column. "Data by part" section shows what measurements were taken different from the first measurements of each part. It shows that there are significant differences between the first and second measurements of all pieces, except #4 and #8. "XBar Chart" shows that there are only two parts of which the averages of measurements are located in between *UCL* and *LCL*, whereas all points are supposed to be located out of *UCL* and *LCL*. Table 6.4 shows that *R&R*% is 6.9714% whereas *P/T*% is 47.97%. Since *P/T*% is greater than 30%, the measurement system should not be considered valid, reliable, and accurate.

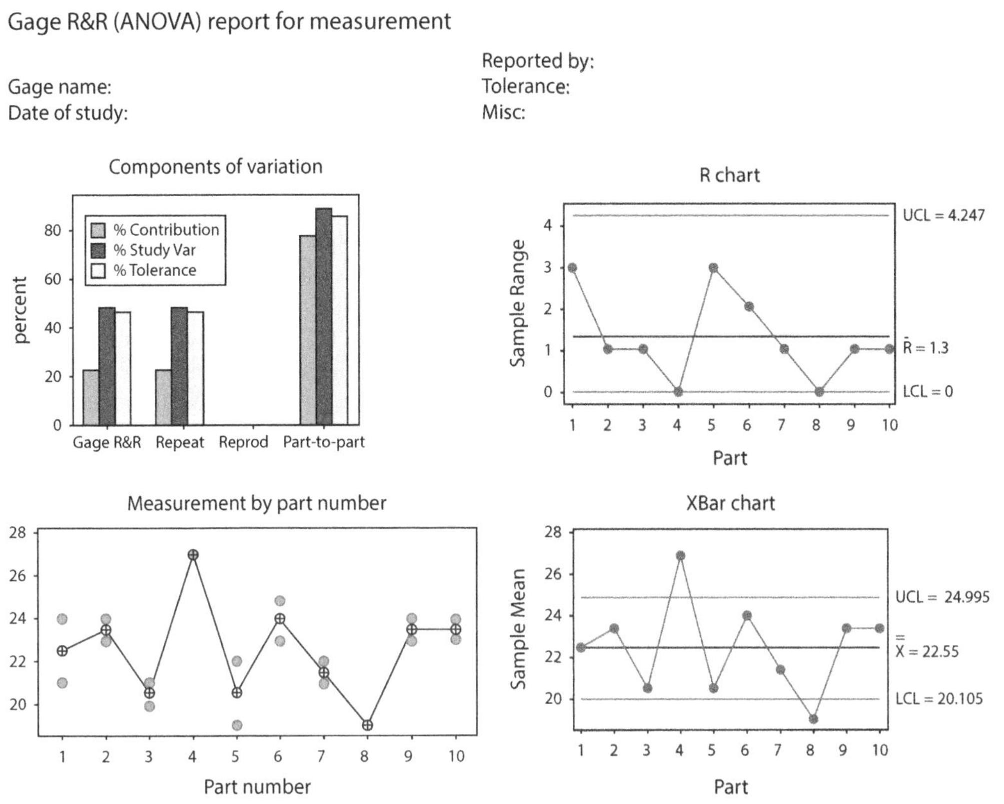

Image 6.5 Gage R&R analysis ANOVA output drawn in Minitab. (Source: Author's creation based on Minitab)

Table 6.4 Gage R&R ANOVA results in Minitab

Gage R&R variance components

Source	*VarComp*	*% Contribution (of VarComp)*
Total Gage R&R	1.35000	23.01
Repeatability	1.35000	23.01
Part-to-part	4.51667	76.99
Total variation	5.86667	100.00

Process tolerance = 15

Gage evaluation

Source	*StdDev (SD)*	*Study Var (6 × SD)*	*% Study Var (%SV)*	*% Tolerance (SV/Toler)*
Total Gage R&R	1.16190	6.9714	47.97	46.48
Repeatability	1.16190	6.9714	47.97	46.48
Part-to-part	2.12525	12.7515	87.74	85.01
Total variation	2.42212	14.5327	100.00	96.88

Number of distinct categories = 2

Source: Author's creation based on Minitab

6

Key Concepts

DMAIC, measure phase, measurement system components, Architect Amenemipt's ruler, Royal cubit, ANSI (American National Standards Institute), NIST (National Institute of Standards and Technology), ASTM (American Society for Testing and Materials), ISO (International Organization for Standardization), Gage R&R Analysis, repeatability, reproducibility, measurement system variance, total variance, process variance, measurement error, accuracy, precision, part-to-part variance, *Gauge Repeatability and Reproducibility* (Gage R&R), R&R%, P/T% (precision-to-tolerance ratio), ANOVA, run chart, upper specification limit, and lower specification limit.

Summary

Before collecting data in Six Sigma project, Gage R&R analysis aims for differentiating measurement system variability from process variability to make sure that the data are collected through valid and reliable measurement systems. This chapter presents how to conduct Gage R&R analysis in Six Sigma projects.

Practice and Discussion Questions

1. Define and discuss the goals of measurement system analysis.
2. Which components of measurement systems most affect decisions made using those measurements?
3. Define units of measure and discuss why the choice of units might affect decision-making.
4. What are the roles of ANSI and NIST in terms of reference standards?
5. What is the difference between accuracy and precision in Gage R&R Analysis? How might having accuracy without precision affect decision-making?
6. Define repeatability and discuss how variances may affect decision-making.
7. If a measurement cannot be consistently repeated, how might that affect the outcomes of production?
8. How might each measure in measurement system analysis affect decisions made using those measurements?
9. What are the acceptable ranges for *P/T%* and *R&R%*?
10. If the *P/T%* is 35% and the *R&R%* is 25%, is the measurement system acceptable?
11. In a production line at an electric wire producer, one of the CTQ characteristics monitored is the length of steel cords. The measurement system consists of two operators, a measurement instruction, a digital gauge, and parts randomly drawn from the production line. Operators take the measurements two times for randomly selected ten steel cords in a day. Upper specification limit (USL) is 300.2 mm, lower specification limit (LSL) is 299.6 mm, and target value is 299.9 mm. The data are shown in ◘ Table 6.5. Analyze the measurement system using Gage R&R analysis in Minitab, and decide if the measurement system is acceptable.
12. In the same production line given in the previous question, one of the CTQ characteristics monitored is the diameter of the steel cords. The measurement system consists of one operator, a measurement instruction, a digital gauge, and parts randomly drawn

Table 6.5 Data collected from the measurement system

Part number	Operator	Data K11 (mm)	Part number	Operator	Data K11 (mm)
1	1st	299.95	6	1st	299.96
1	1st	299.86	6	1st	299.99
1	2nd	300.12	6	2nd	299.96
1	2nd	300.11	6	2nd	299.94
2	1st	299.89	7	1st	300.23
2	1st	299.77	7	1st	300.21
2	2nd	299.95	7	2nd	300.06
2	2nd	300.02	7	2nd	300.08
3	1st	299.98	8	1st	299.71
3	1st	300.1	8	1st	299.69
3	2nd	299.86	8	2nd	299.95
3	2nd	299.95	8	2nd	299.73
4	1st	299.94	9	1st	300.05
4	1st	300.12	9	1st	300.26
4	2nd	299.75	9	2nd	300.06
4	2nd	300.09	9	2nd	299.97
5	1st	299.94	10	1st	299.76
5	1st	300.1	10	1st	299.97
5	2nd	300.12	10	2nd	299.98
5	2nd	299.76	10	2nd	300.05

Source: Author's creation

from the production line. Operator takes the measurements two times for randomly selected 12 steel cords in a day as shown in Table 6.6. The upper specification limit (USL) is 35 mm, and lower specification limit (LSL) is 39 mm. Analyze the measurement system using Gage R&R analysis in Minitab, and decide if the measurement system is acceptable.

Table 6.6 Data collected from the measurement system

	Measurements	
Part number	**Measurement 1 (mm)**	**Measurement 2 (mm)**
1	30	30
2	31	30
3	32	32
4	30	33
5	32	32
6	29	33
7	31	31
8	30	31
9	29	29
10	33	33
11	34	34
12	32	34

Source: Author's creation

6

References

Al-Refaie, A., & Bata, N. (2010). Evaluating measurement and process capabilities by GR&R with four quality measures. *Measurement, 43*, 842–851.

Juran, J. M. (1992). *Juran on quality by design: The new steps for planning quality into goods and services.* Simon and Schuster. New York: The Free Press.

Juran, J. M., & Gryna, F. M. (1980). *Quality planning and analysis*. New York: McGraw-Hill.

Montgomery, D. C. (2013). *Introduction to statistical quality*. New York: Wiley.

Murphy, R. B. (1961). On the meaning of precision and accuracy. Materials Research and Standards, ASTM, April. 1961. (Reprinted in Ku, 1969).

Shewhart, W. (1939). *The statistical method from the viewpoint of quality control.* New York: Dover Publications.

Analyze Phase: A Is for Analyze

Contents

F. Pakdil, *Six Sigma for Students*, https://doi.org/10.1007/978-3-030-40709-4_7

Learning Objectives

After careful study of this chapter, you should be able to:

- Explain Analyze phase of DMAIC
- Analyze data in Six Sigma projects
- Compute and interpret measures of central tendency
- Compute and interpret measures of variability
- Understand the basic properties of a probability distribution
- Identify type of the distribution
- Compute the mean and standard deviation of a probability distribution
- Understand common and assignable causes of variation
- Understand the theory of sampling distributions
- Learn the basics and types of inferential statistics
- Apply point and interval estimation procedures
- Construct and interpret confidence intervals, tolerance intervals, and prediction intervals
- Understand the basics of hypothesis testing
- Test hypotheses on a single variance and on the difference in two variances
- Test hypotheses on a single mean and on the difference in two means
- Test hypotheses on a single proportion and on the difference in two proportions
- Run correlation, regression, and ANOVA analyses and interpret findings
- Compute and interpret process capability indices
- Implement Taguchi's loss function and calculate and interpret the quality losses.

7.1 Introduction

In the Analyze phase, the objective is to use the data collected in the Measure phase of the DMAIC process to begin to identify the cause and effect relationships in the process and to understand the different sources of variability (Montgomery 2005). Discovering the causes of variability is crucial to identify root causes and take preventive or corrective actions in the phases of Six Sigma projects. The Analyze phase focuses on analyzing data collected in the previous phase, making data-driven decisions for the next phase of Six Sigma, discovering quality problems and issues, and identifying root causes of defects and the points where the wastes occur in the process. Pyzdek and Keller (2018: 145) states that "data-driven management provides a means of achieving organizational objectives by quantifying needs or wants of stakeholder groups relative to current baselines and acting upon data to reduce those critical gaps in performance."

Montgomery (2005: 56) identifies several issues and questions to be considered by the team members in the Analyze phase:

1. What opportunities will be targeted for investigation in the Improve (next) phase?
2. What data and analysis support the targeted opportunities?
3. How will improving/eliminating them will have the desired outcome on customer Critical-to-Quality (CTQ) characteristics that were the original focus of the project?
4. Are there other opportunities that will not be further evaluated? If so, why?
5. Is the project still on track with respect to time and anticipated outcomes? Are any additional resources required at this time?

The Analyze phase in the DMAIC process focuses on statistical analysis, which can provide precise results that can be used to make better decisions. Bruhl (2018: 6) describes statistical analysis as a collection of various mathematical procedures for generating statistics based on observed values. Statistical analysis on a data set starts with sorting and counting the observations to create a list including observed values and their frequencies. Then, it continues to analyze the data sets using a great variety of statistical analysis tools.

From a quality improvement point of view, Deming (1982) states that statistics is all about studying and understanding *variation* and interactions among the variables in processes and populations, operational definitions, and taking action to reduce variation in a process or population. Statistics are analyzed within two categories: *descriptive statistics* and *inferential statistics*. Descriptive statistics allow the analysis of various basic parameters of the observations and measurements, such as (1) measures

7

of central tendency, (2) measures of variability, (3) shape of distribution, and (4) charting the data. Inferential statistics allow managers to make statistical decisions about the population from a subset of data, called a sample or individual observations. As detailed in ▸ Chap. 5, the population or universe represents the entire group of individual observations during a certain period of time. The sample is a limited part of the population that has the high representation ability of the population. The sample taken from a population should represent the structure of the population, since further interferences will be made using the sample.

There are many useful tools and techniques that can be categorized, such as qualitative and quantitative or basic and advanced tools mentioned in the discussion of the Analyze phase. For example, measures of central tendency and variability, graphs, seven old and new tools, run charts, statistical process control (SPC), basic and advanced process flow charts, confidence intervals, hypothesis testing, regression and correlation analyses, ANOVA, FMEA, and Design of Experiment are only some of the many quantitative tools and techniques available. The following sections will discuss the use of descriptive statistics and measures; shape of the distribution; types of variation; statistical distributions; inferential statistics, including estimation of confidence intervals, tolerance intervals, and prediction intervals; hypothesis testing; correlation analysis; regression analysis; ANOVA; process capability analysis; and Taguchi's loss function.

7.2 Descriptive Statistics

Descriptive statistics are considered when decision-makers need to see what the data look like and its common trend and features. Descriptive statistics, in other words summary statistics, can better explain the nature of the data set. Bruhl (2018) states that descriptive statistics are used when the researchers would like to learn the extent to which a set of phenomena differ based on a variable of interest. Descriptive statistics summarize data based on three components:

1. Measures of central tendency
2. Measures of dispersion (variability)
3. Type of distribution.

These statistics allow decision-makers to summarize detailed data to interpret its structure. If you compute descriptive statistics from a sample, these are called *statistics*. If you calculate descriptive statistics from a population, these are called *parameters*. Knapp (2016) finds that continuous variables are generally analyzed using nine descriptive statistics: percentage, number, mean, median, mode, standard deviation, minimum, maximum, and range. Categorical variables are summarized using two descriptive statistics: number and percentage. The measures of central tendency and variability of the population are not known in reality but are estimated from a sample taken from the population.

7.2.1 Measures of Central Tendency

The central tendency demonstrates the spatial representation of a set of observations. There are three main measures of central tendency in descriptive statistics: mode, median, and mean. If the data set is normally distributed, the mode, median, and mean should be the same. "These three points are used to represent a "typical" observation for the distribution (Bruhl 2018: 94)." Because of the sensitivity, the mean is the most common used measure of central tendency. The mean considers all individual observations. The only disadvantage of mean is that it gets distorted when there are outliers. Therefore, if the data have extreme outliers, the median is used, instead of the mean, as it will be a more powerful measure of central tendency.

7.2.1.1 Mean

The arithmetic mean, usually called the average, is a ratio of summation of the numerical observations to the number of observations. As the individual observations get closer, the mean is centered in the middle of the observations. The average is calculated as shown in Eq. 7.1, where x_i is the measurement of individual observation; n is the number of observations $x_1, x_2, \ldots, x_n$; and $\overline{X}$ is the sample average. Statistically speaking, the average is referred to as the *mean* (Eq. 7.1).

$$\bar{X} = \frac{\sum_{i=1}^{n} x_i}{n} = \frac{\text{Sum of observations}}{\text{Number of observations}}$$

$$= \frac{x_1 + x_2 + \ldots + x_n}{n} \quad \text{for } i = 1, 2, \ldots, n \qquad (7.1)$$

▶ Example 1

The customer ordering system in an online shopping web site is analyzed. The decision-makers would like to know the average ordering time in the last 2 weeks. The ordering time covers the time period between when the customers start and finish ordering online. The data are presented in ascending order in ◘ Table 7.1. Calculate the average customer ordering time (seconds). ◀

▪ **Solution**

Using Eq. 7.1, the average customer ordering time of 30 observations is 62 seconds in this example (Eq. 7.2).

$$\bar{X} = \frac{\sum_{i=1}^{n} x_i}{n} = \frac{\sum_{i=1}^{30} x_i}{30} = \frac{\text{Sum of observations}}{\text{Number of observations}}$$

$$= \frac{75 + 29 + \ldots + 70}{30} = 62 \qquad (7.2)$$

7.2.1.2 Mode

The mode refers to the most frequently occurring observations in a data set, that is, the most common value or values in the data set. Extreme values do not affect the mode. The mode can be used for all types of data. However, the mode makes more sense on categorical data sets since it shows the most frequent observation. In cases where there are multiple modes, the variable is considered *bimodal*, meaning two modes. To find out the mode, we can follow the process below.

- Step 1: Collect data.
- Step 2: Sort data in an ascending order.
- Step 3: Find the individual observation/s that repeated mostly in the data set. This observation will be the mode of the data.

◘ **Table 7.1** Data set sorted in an ascending order

The number of observation	Ordering time (seconds)
1	15
2	29
3	30
4	30
5	33
6	36
7	38
8	40
9	45
10	46
11	47
12	55
13	59
14	65
15	69
16	**70**
17	**70**
18	**70**
19	73
20	**75**
21	**75**
22	**75**
23	76
24	77
25	85
26	87
27	88
28	95
29	102
30	105

Source: Author's creation

▶ Example 2

For the data set used in ▶ Example 1, find the mode of customer ordering time. ◀

■ **Solution**

To find the mode of the data set, let's follow that three-step process.

- Step 1: Collect data.

In this example, the data are given to us. If there is no available data set or historical data, the Six Sigma team needs to collect data related to CTQ characteristic analyzed by the team in the project.

- Step 2: Sort data in an ascending order.

The data set given in ▶ Example 1 is sorted in an ascending order that starts from the lowest observation as presented in ◘ Table 7.1.

- Step 3: Find the individual observation/s that repeated mostly in the data set.

After sorting the data from shortest to the longest ordering time in an ascending order as given in ◘ Table 7.1, one can see that there are two most frequent ordering times in data set: 70 seconds and 75 seconds, with three observations for each of them. It is technically possible to have more than one mode. As a reminder, this data set should be considered bimodal since it has two modes.

7.2.1.3 Median

The median represents the physical center, in other words, the middle point of the data. The median also splits a data set into two equal pieces, one half locating above the median and the other half locating below the median. It also helps identify any outliers within the data set. To calculate the median, we can follow the next steps.

- Step 1: Collect data.

First, the data are collected and recorded.

- Step 2: Sort data in an ascending order.

Then, the observations are sorted from smallest to the highest one.

- Step 3: Find the individual observation/s located in the middle in the ordered data.

Mathematically, the median ($\tilde{X}$) can be calculated as follows (Eq. 7.3):

$$\tilde{X} = X_{\left(n+1/2\right)} \text{ or}$$

$$\text{Median} = \frac{n+1}{2}\text{th ranked value} \tag{7.3}$$

where n is the number of observation and X is the individual observation. If the number of observations is an odd number, such as 11, the median is found as the observation that falls into the middle of the observations, which is the sixth one in the data set that is in an ascending order, as shown in Eq. 7.4.

$$\tilde{X} = X_{\left(n+1/2\right)} = X_{\left(11+1/2\right)} = X_6 \tag{7.4}$$

If the number of observations is an even number, the median is the average of the two observations that fall in the middle of the data set. The median is not affected by the outliers or extreme values in data set. For example, if the sample size is 30, then, the median is located in the middle of 15th and 16th observations as shown in Eq. 7.5.

$$\tilde{X} = X_{\left(n+1/2\right)} = X_{\left(30+1/2\right)} = X_{15.5} \tag{7.5}$$

Additionally, Bruhl (2018) notes that the median can be identified from the relative frequency distribution by focusing on the observation located in 50% of the relative cumulative frequency.

▶ Example 3

For the customer ordering time given in the same data set in ▶ Example 1, calculate the median. ◀

■ **Solution**

To calculate the median, let's follow three-step process one by one as follows.

- Step 1: Collect data.

First, we need to collect data and make sure that the data are reliable and valid for the relevant CTQ characteristics. In our example, the data set is ready and given in the question.

- Step 2: Sort data in an ascending order.

The data are sorted in an ascending order as presented in ◘ Table 7.1. The smallest observation is 15 seconds and the longest observation is 105 seconds.

- Step 3: Find the individual observation/s located in the middle in the ordered data.

To find out what observation represents the median, let's find the rank of the value as given in Eq. 7.6.

$$\tilde{X} = X_{(n+1/2)} = X_{(30+1/2)} = X_{15.5} \quad (7.6)$$

There are 30 observations in data set and the median is average of 15th and 16th observation. As we can see from the ranked data in ◘ Table 7.1, the 15th observation is 69 seconds, and 16th observation is 70 seconds. The median of the data set is 69.5 seconds, which is the average of 69 and 70. Because the rank ($X_{15.5}$) is not an integer number, we need to calculate the average of 15th and 16th observations (Eq. 7.7).

$$X_{15} = 69$$

$$X_{16} = 70$$

$$\tilde{X} = \frac{X_{15} + X_{16}}{2} = 69.5 \quad (7.7)$$

Let's compare mean, mode, and median of the same data set used in the examples. Measures of central tendency show a big range between 62 seconds and 75 seconds in this data set. To calculate measures of central tendency on Minitab, after transferring data set to Minitab, from menu bar click on Stat→Basic Statistics→Display Descriptive Statistics→OK. Transfer relevant data set from left box to variables box and click on Select. Click on Statistics. Then, mark Mean, Median, and, Mode. Click on OK→OK. Minitab output for measures of central tendency shows mean (62 seconds), median (69.5 seconds), and mode (70 and 75 seconds), respectively.

7.2.2 Measures of Variability (Dispersion)

As stated by Montgomery (2005: 68), "no two units of product produced by a process are identical. Some variation is inevitable." Variation may occur in any kind of process and is the amount of deviation or dispersion in a data set. Multiple measures of the same parameter often produce slightly different results, even if nothing is changed in the process (Benneyan et al. 2003). The most frequently used measures of variability are *range*, *variance*, and *standard deviation*. All three measures will always be greater than or equal to zero. The more homogenous the data set is, the smaller the measures of variability will be. Ideally, all three measures of variability are centered at zero if there are no variations in data set. Increasing measures indicate more disperse data sets. In terms of meeting specification limits, increasing variability always identifies the decreasing probability of meeting specification limits, in other words, decreasing quality level.

7.2.2.1 Range

The range (R) is known as the simplest measure of variation. Basically, the range is the difference between the maximum (highest) and minimum (lowest) observations in the data set. It gives a general idea how widely spread the observations are in a data set. Bruhl (2018: 112) states that "the range provides the most global view of the variability found among observations." The smaller the range is, the lower the variation and total spread are, which means higher quality. Continuous quality improvement and Six Sigma projects seek to minimize range. The definition of range is mathematically given in Eq. 7.8.

$$R = X_{max} - X_{min} \quad (7.8)$$

where R is the range, X_{max} is the maximum observation, and X_{min} is the minimum observation in a data set. As a measure of total spread, the range does not consider how the values distribute around the mean. Variance and standard deviation represent how far all each observation is from the mean. The range is higher in data sets that have higher outlier observations. Juran and Gryna (1980) state that range is not useful when the number of observations is large.

▶ Example 4

Calculate the range of the data given in ▶ Example 1.◀

7

▪ Solution

The range of the data set is 90 seconds as the mathematical difference between the largest and smallest observation (Eq. 7.9).

$$R = X_{max} - X_{min} = 105 - 15 = 90 \text{ seconds} \quad (7.9)$$

7.2.2.2 Standard Deviation

Standard deviation considers how each observation fluctuates around the mean of the data. The lower the standard deviation, the higher the quality level in manufacturing and service settings. The more widespread the data, the greater the standard deviation. In other words, if the observations of a variable are similar and closer to one another, the variable contains lower standard deviation, which is an advantage for obtaining higher quality. Bruhl (2018: 128) states that "it is common practice to use the standard deviation rather than the variance in describing the variability in a set of observations." The standard deviation is computed as shown in Eq. 7.10. The equation first calculates summation of the square of the difference between each individual observation and the mean. Then, the summation is divided by $n - 1$ and the equation takes the square root of the result.

$$s = \sqrt{\frac{\sum_{i=1}^{n}(X_i - \bar{X})^2}{n-1}} \quad (7.10)$$

where

- s = the sample standard deviation
- n = sample size
- X_i = ith value of observation X
- $\bar{X}$ = the mean of the data set.

To compute standard deviation, the seven-step process below can be followed:

- Step 1: Collect data and list each observation.
- Step 2: Calculate mean of the data set.
- Step 3: Compute the difference between each value and the mean.
- Step 4: Square each difference.
- Step 5: Calculate the sum of squared differences.
- Step 6: Divide the total by $n - 1$.
- Step 7: Take the square root of the product of the previous step.

The standard deviation demonstrates how a data set clusters or distributes around the mean. It is also known that all individual observations lie within a ±3 standard deviation interval.

▶ Example 5

Calculate the standard deviation of the data given in ▶ Example 1.◀

▪ Solution

The standard deviation can be computed in MS Excel, statistical software, or even manually. Let's show how to compute standard deviation using sorted data set as below by following our seven-step process.

- Step 1: Collect data and list each observation.

The data were already collected in the previous steps as presented in ◘ Table 7.1.

- Step 2: Calculate mean of the data set.

The mean of the data was found to be 62, as previously presented in ▶ Example 1. We will use $\bar{X} = 62$.

- Step 3: Compute the difference between each value and the *mean*.

The difference between each value and the mean is shown in Eq. 7.11 below. For the first observation, it is $X_1 - \bar{X} = 15 - 62$.

- Step 4: Square each difference.

This step is shown in Eq. 7.11 below. For the first observation, it is $(X_1 - \bar{X})^2 = (15 - 62)^2$.

- Step 5: Calculate the sum of squared differences.

This step is shown in Eq. 7.11 below.

- Step 6: Divide the total by n − 1.

This step is shown in Eq. 7.11 below.

- Step 7: Take the square root of the product of the previous step.

This step is shown in Eq. 7.11 below. At the end of the calculation, $s = 23.794$.

$$s = \sqrt{\frac{\sum_{i=1}^{n}(X_i - \overline{X})^2}{n-1}} = \sqrt{\frac{(15-62)^2 + (29-62)^2 + \ldots + (105-62)^2}{30-1}} = 23.794 \quad (7.11)$$

Let's compare range and standard deviation of the same data set used in the examples. The range was found to be 90 seconds, whereas standard deviation is 23.794 seconds. It is easily seen that there is a big difference between these two measures of variability.

7.2.2.3 Variance

Variance, as the other measure of variability, is the square of standard deviation (s^2). A lower variance shows that the difference between individual observation and mean is low. When the variance increases, that means the difference between these two expands. Statistically speaking in Six Sigma, increasing variance indicates potential problematic areas. To increase quality, it is necessary to decrease the variance of the relevant CTQ characteristic. The variance is computed in Eq. 7.12.

$$s^2 = \frac{\sum_{i=1}^{n}(X_i - \overline{X})^2}{n-1} \quad (7.12)$$

where

- s^2 = the sample variation
- n = sample size
- X_i = ith value of observation X
- $\overline{X}$ = the mean of the data set

To compute the variation, the next process can be followed:

- Step 1: Collect data and list each observation.
- Step 2: Calculate mean of the data set.
- Step 3: Compute the difference between each value and the mean.
- Step 4: Square each difference.
- Step 5: Calculate the sum of squared differences.
- Step 6: Divide the total by $n - 1$.

▶ Example 6

Calculate the variance of the data given in ▶ Example 1.◀

▪ Solution

The variance can be calculated in MS Excel, statistical software, or even manually. Similar to the standard deviation calculation, we can compute variance using steps given above.

- Step 1: Collect data and list each observation.

The data are given in the question. Therefore, we don't need to collect data in this example.

- Step 2: Calculate mean of the data set.

To recall, we can calculate the mean of the data as follows (Eq. 7.2):

$$\overline{X} = \frac{\sum_{i=1}^{n} x_i}{n} = \frac{\sum_{i=1}^{30} x_i}{30} = \frac{\text{Sum of observations}}{\text{Number of observations}} = \frac{75 + 29 + \ldots + 70}{30} = 62$$

The mean of the data was found to be 62, as previously presented in ▶ Example 1. We will use $\overline{X} = 62$.

- Step 3: Compute the difference between each value and the mean.

This step is presented in Eq. 7.13.

- Step 4: Square each difference.

This step is presented in Eq. 7.13.

- Step 5: Calculate the sum of squared differences.

This step is presented in Eq. 7.13.

- Step 6: Divide the total by $n-1$.

$$s = \frac{\sum_{i=1}^{n}(X_i - \overline{X})^2}{n-1} = \frac{(15-62)^2+(29-62)^2+\ldots+(105-62)^2}{30-1} = 566.14 \quad (7.13)$$

Let's compare range, standard deviation, and variance of the same data set used in the examples.

As seen from range (90 seconds), standard deviation (23.794 seconds), and variance (566.14 seconds), the variance shows the highest dispersion of the data. Standard deviation and variance present a smaller variability. It would be more meaningful to compare two different data sets' measures of variability to identify which data set has a larger or smaller dispersion.

To calculate measures of variability on Minitab, after transferring data set to Minitab, from menu bar click on Stat→Basic Statistics→Display Descriptive Statistics→OK. Transfer relevant data set from left box to variables box and click on Select. Click on Statistics. Then, mark Range, Standard Deviation, and, Variance. Click on OK→OK. Minitab output shows the measures of variability.

7.3 Other Descriptive Measures

7.3.1 Quartiles

When using quartiles, a data set is split into four equal parts. Quartile 1 ($Q1$) refers to the first 25% of the data set. Quartile 2 ($Q2$) represents the median with 50% of the observations. Quartile 3 ($Q3$) depicts 75% of the data. If the product of the number of observations (n) and percentile (p) is an even number, $Q1$, $Q2$, and $Q3$ are computed as given in Eqs. 7.14, 7.15, and 7.16, respectively. After calculating quartiles, it will be possible to draw box-and-whisker plots as well. The Y-axis on a box plot shows the output that is measured in a data set, whereas the X-axis displays the frequencies of the observations (◘ Image 7.1). A box plot includes several important variables such as *minimum* and *maximum value* and *first*, *second*, and *third quartile*. First quartile demonstrates the individual value where the 25% of the data are located in the data set, while third quartile focuses on the value where the 75% of the data are shown. Second quartile also refers to the *median* of the data set. Asterisks on box plots show the outlier on two points of the data set. In a box plot, the horizontal line in the middle of the box signifies the median. The upper and lower horizontal lines demonstrate the first and third quartiles, and maximum and minimum observations are given at the two outliers.

$$Q1 = n.p = r = n.(0.25) \Rightarrow \text{integer number}\, Q1 = \frac{x_r + x_{r+1}}{2} \quad (7.14)$$

$$Q2 = n.p = r = n.(0.50) \Rightarrow \text{integer number}\, Q2 = \frac{x_r + x_{r+1}}{2} \quad (7.15)$$

$$Q3 = n.p = r = n.(0.75) \Rightarrow \text{integer number}\, Q3 = \frac{x_r + x_{r+1}}{2} \quad (7.16)$$

where

- $Q1$ = the first quartile
- $Q2$ = the second quartile
- $Q3$ = the third quartile
- n = the number of the observations
- p = the percentage of the quartile
- x_r = the rth observation of the sorted data
- x_{r+1} = the $r+1$th observation in the sorted data.

If the product of the number of observations (n) and percentile (p) is an odd number, quartiles are computed as given in Eq. 7.17. In case the product is an odd number, quartiles are located in between *r*th and *r+1*th observations in the data set.

$$Q = n.p = r$$
$$\Rightarrow \text{non-integer number } x_r < Q < x_{r+1} \quad (7.17)$$

To calculate the quartiles, the next process can be followed:

- Step 1: Collect data and list each observation.
- Step 2: Sort data in an ascending order.
- Step 3: Determine sample size.
- Step 4: Calculate $Q1$, $Q2$, and $Q3$.
- Step 5: Identify the lowest and highest measurements in the data set as outliers.
- Step 6: Draw box-and-whisker plot and interpret the findings.

▶ Example 7

Calculate $Q1$, $Q2$, and $Q3$ for the data set given in ▶ Example 1. To remember the data, it is presented in ◘ Table 7.1. ◀

▪ Solution

Let's go through four-step process given below:

- Step 1: Collect data and list each observation.

The data are given in ◘ Table 7.1. Each observation is presented in the data set.

- Step 2: Sort data in an ascending order.

The data set is sorted in an ascending order in ◘ Table 7.1.

- Step 3: Determine sample size.

The sample size (n) is 30 in the data set.

- Step 4: Calculate $Q1$, $Q2$, and $Q3$.

Now, we can calculate $Q1$, $Q2$, and $Q3$ using Eqs. 7.14, 7.15, 7.16, and 7.17.

There are 30 observations in the data set ($n = 30$). Three quartiles can be calculated as follows:

$$Q1 = n.p = r = 30.(0.25)$$
$$Q1 = 7.5(\text{non-integer}) \Rightarrow Q1 = x_r < Q_1 < x_{r+1}$$

$$Q1 = x_7 < Q1 < x_{7+1} = 38 < Q1 < 40$$

$$Q1 = \frac{38+40}{2} = 39$$

$$Q2 = n.p = r = 30.(0.50) = 15(\text{integer})$$
$$\Rightarrow Q2 = \frac{x_r + x_{r+1}}{2} = \frac{X_{15} + X_{16}}{2}$$
$$Q2 = \frac{69+70}{2} = 69.5$$

$$Q3 = n.p = r = 30.(0.75)$$
$$Q3 = 22.5(\text{non-integer})$$
$$\Rightarrow Q3 = x_{22} < Q3 < x_{22+1}$$

$$Q3 = x_{22} < Q3 < x_{23} = 75 < Q1 < 76$$

$$Q3 = \frac{75+76}{2} = 75.5$$

As a result, it is seen that the quartiles are 39, 69.5, and 75.5, respectively.

- Step 5: Identify the lowest and highest measurements in the data set as the outliers.

In the example, the lowest measurement is 15 seconds and the highest is 105 seconds. These two outliers are added on the box plot.

- Step 6: Draw the box-and-whisker plot and interpret the findings.

To manually draw a box plot, first, minimum and maximum observations are put at two outlier points. Then, $Q1$, $Q2$, and $Q3$ are located on the *X-Y* plane. When the data sets get bigger, statistical software is used in data analysis. Minitab, JMP, and Stata are some of the most frequently used statistical software. To draw quartiles in the box plot in Minitab, the data set is entered in one column in Minitab. Click on Graph→Box Plot and select Simple→OK. Transfer "ordering time (seconds)" variable from left box to Graph variable section→OK. The box plot drawn in Minitab is presented in ◘ Image 7.1. As seen in box plot, the first quartile ($Q1$) is 39, the second quartile ($Q2$) is 69.5, and the third quartile ($Q3$) is 75.5.

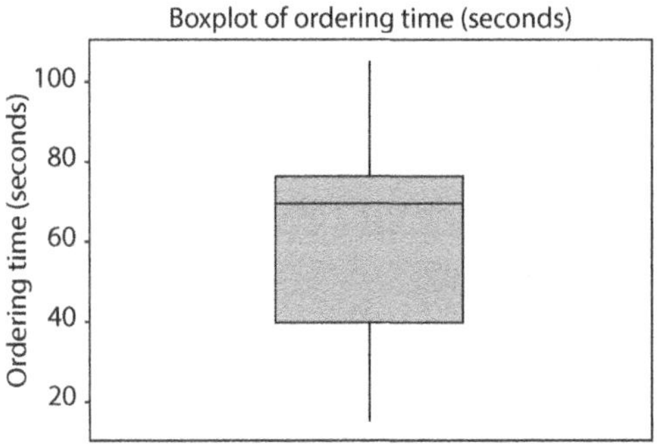

Image 7.1 Box plot showing three quartiles. (Source: Author's creation based on Minitab)

7

7.3.2 The Five-Measure Summary

This measure provides a detailed information about the shape of the distribution. A five-measure summary contains (1) smallest value, (2) $Q1$, (3) median, (4) $Q3$, and (5) largest value.

If the difference between the smallest and largest values is relatively small, which also refers to range (R), it means that the data set has a low dispersion. Additionally, the difference between each measure gives a message about the shape of the distribution. The lower differences indicate smaller variation and steeper distribution. Based on the magnitude of the differences, the skewness and symmetry of the data set are analyzed.

► Example 8

Find out five measures of data set given in ► Example 1. ◄

▪ Solution

After looking into sorted data presented in the previous questions' solutions and quartiles, we can easily find the five measures as follows: $X_{min} = 15$, $Q1 = 39$, $Q2(\text{Median}) = 69.5$, $Q3 = 75.5$, and $X_{max} = 105$.

Calculating the five-measure summary on Minitab is similar to that in previous sections. After transferring the data set to Minitab, click on Stat→Basic Statistics→Display Descriptive Statistics in menu bar. Transfer relevant data set from left box to variables box and click on Select. Click on Statistics. Then, mark first quartile, median, third quartile, minimum, and maximum. Click on OK→OK. Minitab output shows the five-measure summary.

The five-measure summary shows the general structure of dispersion of the data. It is a good way for Six Sigma practitioners to see the large or small variability of the data to understand the root causes of the problems.

7.4 The Shape of Distribution

The shape of the distribution of the data set is the third important feature of data. A histogram of the data set helps Six Sigma practitioners understand the distribution of the data. If the histogram is skewed, the distribution is shifted to the right or left side. The shape of data is analyzed based on symmetrical and skewness. To analyze the shape of the data, the mean and median are compared (Levine 2006). If these two measures are equal, the variable is considered to be symmetrical or zero-skewed ($\bar{X} = \tilde{X} \Rightarrow \textit{zero skewness}$) (Image 7.2). If the mean is smaller than the median, the variable is called left-skewed. In this case, the distribution of the variable has a long tail on the left side caused by smaller observations that take the mean to a lower value ($\bar{X} < \tilde{X} \Rightarrow \textit{left skewness}$) (Image 7.3). In case the median is smaller than mean, the variable is right-skewed. In this case, the distribution of the variable has a long tail on the right side caused by larger observations

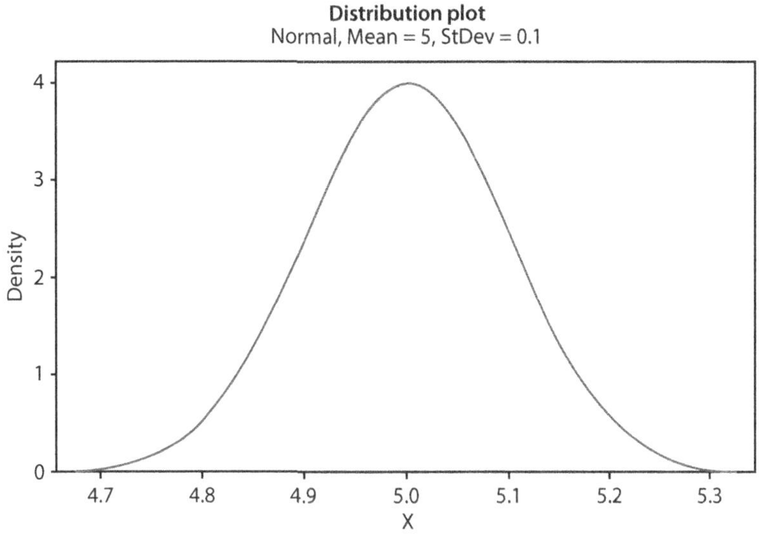

Image 7.2 An example of symmetric normal distribution. (Source: Author's creation based on Minitab)

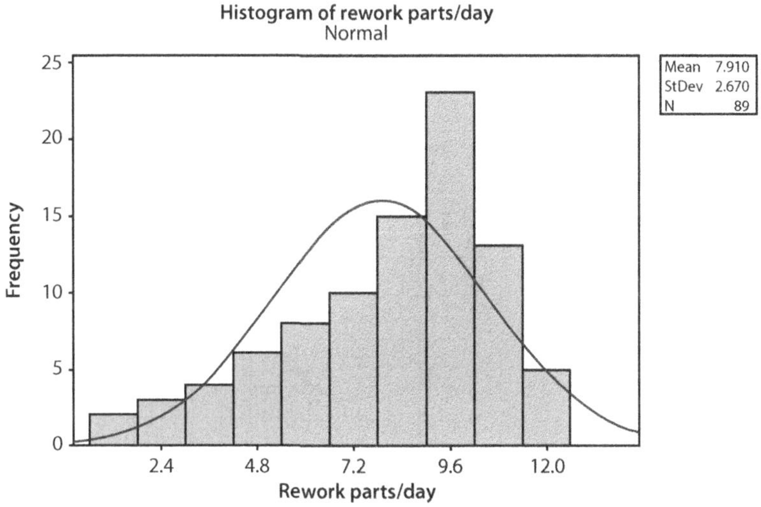

Image 7.3 An example of left-skewed normal distribution. (Source: Author's creation based on Minitab)

that increase the mean of variable ($\bar{X} > \tilde{X} \Rightarrow$ *right skewness*) (Image 7.4).

While analyzing the shape of the data, the other important factor to be considered is if the data set has a bimodal distribution (Image 7.5). If the shape of the distribution represents a bimodal structure, that means the data set has two different distributions. In the case of bimodal shape, the two data sets should be differentiated first, and then, the further analyses can be performed. Since the types of the distributions are analyzed in the "Statistical Distributions" section in this chapter, we will not detail the shape of the distribution in this section.

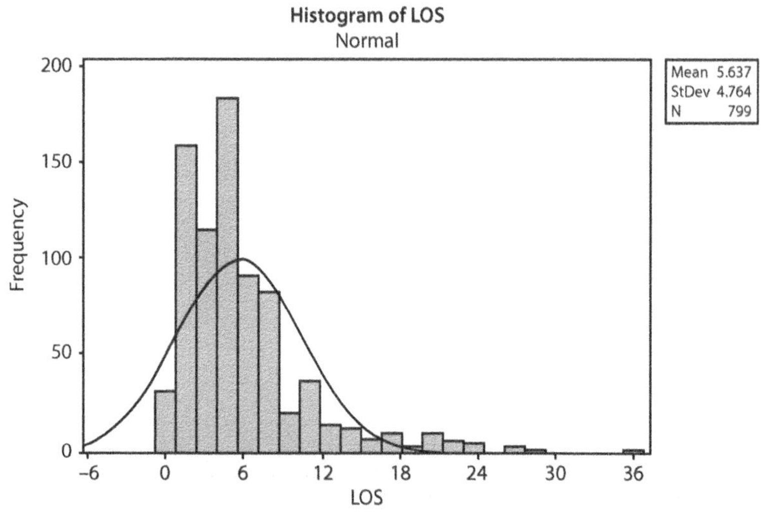

Image 7.4 An example of right-skewed normal distribution. (Source: Author's creation based on Minitab)

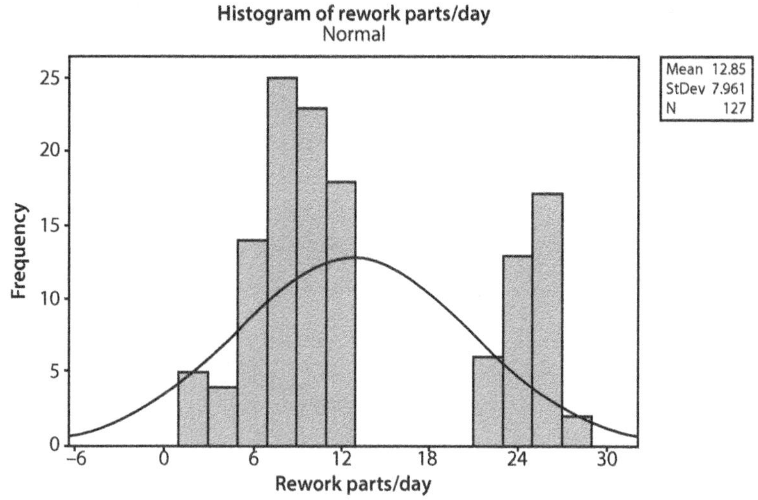

Image 7.5 Bimodal histogram. (Source: Author's creation based on Minitab)

7.5 Types of Variation

Repeated measures of the same parameter often are expected to yield slightly different results even if there is no fundamental change (Benneyan et al. 2003). However, variation may occur in any kind of process in manufacturing or service delivery processes. The fact that the product lots or batches contain multiple pieces from different production lines bring piece-to-piece and within-piece variations (Juran and Gryna 1980: 295).

Since variation is expected to incur, processes should be organized and prepared with multiple ways for decreasing variation. Preventive and corrective actions are useful for this. According to Deming (quoted in Boardman and Boardeman 1990), "Action taken on a stable system in response to variation within the control limits, in an effort to compensate for this variation, is tampering, the results of which will inevitably increase the variation and increase costs ...." Finison et al. (1993: 65) state that a "faulty item is not a signal of a special cause." Tampering, therefore, is treating common cause of variation as if it were due to special cause.

Since the components of the process such as work environment, operators, procedures, methods, equipment, technology, and raw material contribute to the variation, naturally, the outcomes of the process may vary. The types of variation are categorized into two groups: (1) common (chance) causes of variation and (2) assignable causes of variation. A certain amount of common causes of variation stems from the process itself. This variation is called "natural variability," "stable system of chance causes," or "background noise" (Montgomery 2005). All unavoidable, predictable, and natural events contribute to common causes of variation in the process. These natural variations can never be eliminated economically from the process. Hart and Hart (2002: 2) state that "the reality is that no matter how alike the inputs to the process are, the outputs will vary." If a process is run with only common causes of variation, it is considered statistically in-control, which means stability and predictability of a process. In a controlled process, the variation that does exist is not due to assignable or uncontrolled causes (Hart and Hart 2002). Finison et al. (1993: 10) state that:

> Common cause variation is the summation of all the small causes that combine on a chance basis every day to produce variation day to day (or hour to hour). Common cause of variation is variation that is random in nature and whose causes can be discovered only through systematic study of the process and removed only by changing the system.

Common variation may be seen even in standardized processes. According to Benneyan et al. (2003), a common cause of variation is expected based on the underlying statistical distribution, if its parameters remain constant over time. Common causes of variation are naturally inherent in the process. A few examples of common causes are human variation in setting control dials, slight vibration in machines, a faulty setup, and slight variation in raw material (Juran and Gryna 1980).

The second type of variation is known as assignable causes of variation. Montgomery (2005) states that this variability usually is generated through three sources: improperly adjusted or controlled machines, operator errors, and defective raw material. Assignable causes of variation also result in relatively low performance in the process. In cases where this variation exists, the process is considered statistically out-of-control, which indicates that the process shifts from the regular performance. A process is considered stable when no assignable cause of variation appears. Assignable causes refer to statistically significant differences and unnatural variation (Benneyan et al. 2003). A process with assignable variation is unpredictable (Juran and Gryna 1980: 294). The root causes of unnatural variation are not related to the process itself. Six Sigma practitioners are expected to differentiate patterns and trends indicating either common or assignable causes. Processes that appear with assignable causes of variation are not stable, and all assignable causes should be eliminated to turn the process into in-control status. Unnatural variation can be reduced by identifying the non-systemic causes of the process and standardizing the process flow (Benneyan 1998). "A process that is operating without assignable causes of variation is said to be in a state of statistical control which is usually abbreviated to in-control" (Juran and Gryna 1980: 289).

Common and assignable causes of variation are treated in different ways in order to minimize the total variation and abnormalities in the processes. If processes present an assignable cause, they should revert to in-control status by eliminating the causes pushing the process out-of-control status. If the processes contain only common causes, it is expected that these processes produce within the tolerance limits identified in the design step.

If a process shows stable performance, the variation of the process can be estimated and described using a statistical distribution (Benneyan et al. 2003). Since common variation is a predictable and natural part of the process, it is expected to have a variation based on common causes in the processes. *UCL* and *LCL* in control charts also take common causes of variation into account. The magnitude of common cause variation creates the *UCL* and *LCL* in control charts. Very tight control limits (minimal common cause variation) allow special cause of variation to be detected more quickly (Duncan et al. 2011). However, assignable variation is not expected to exist, and it pushes the process into a statistically out-of-control status, which is an indication of low performance and improvement needs in the process. When the process is in-control, in other words, when it is stable, statistics fall into *UCL* and *LCL*, whereas at least one statistic is located beyond either *UCL* or *LCL* in cases where the process is out-of-control. Ideally, both common and assignable causes should be eliminated, and overall variation should be minimized in the process.

7.6 Statistical Distributions

By Dr. Nasibeh Azadeh-Fard
Assistant Professor
Rochester Institute of Technology

In managing business processes, there are few situations where the outcomes of changes in the process within the system under study can be predicted completely. The Six Sigma practitioners analyze a system that is probabilistic rather than deterministic. There are many causes of uncertainty and variation in the system, and most of these variations occur by chance and cannot be predicted. Moreover, the Six Sigma approach is fundamentally data-driven where probability and statistics play a crucial role in analyzing a system and improving a process. Developing an appropriate statistical model by sampling, selecting a known distribution, and then making an estimate of parameters of the distribution can help the Six Sigma practitioners manage the variability in the system.

This section contains a review of probability terminology and concepts. Then, we discuss a number of discrete distributions followed by continuous distributions. The selected distributions are those that are used widely by Six Sigma practitioners and describe a wide variety of probabilistic events. Additional discussions about empirical distributions are also provided in this section.

7.6.1 Random Variables

A random variable X is defined as a variable with numerical values that are outcomes of a random event or a random process. Rolling a die and tossing a coin are two examples of random events where you can define a random variable for the outcomes. For example, random variable X could be defined to be 1 if you get head or 0 if you get tail when flipping a coin. There are two types of random variables, discrete and continuous, which we will discuss in the following sections.

7.6.1.1 Discrete Random Variables

A discrete random variable X has a finite or countable infinite number of possible values. The possible values of X are denoted by x_1, x_2, …in the range space of R_x, where $R_x = \{0, 1, 2, \ldots\}$. The probability that a discrete random variable X equals to the value of x_i is given by $p(x_i) = P(X = x_i)$ where $p(x_i)$ is called the probability mass function (PMF) of X. The following conditions hold for the PMF:

(a) $p(x_i) \geq 0$, for all i

(b) $\sum_{i=1}^{\infty} p(x_i) = 1.$

▶ Example 9

Consider the number of cars produced in a car manufacturing company. Define X as the number of cars passed the quality department per hour with $R_x = \{1,2,3,4\}$. Assume that the discrete probability distribution for this random experiment is given by:

x_i	1	2	3	4
$p(x_i)$	4/10	3/10	2/10	1/10

Are the conditions for PMF distribution being satisfied? ◀

■ Solution

Since probability of each car produced in the company is greater than 0, i.e., $p(x_i) \geq 0$, for $i = 1, 2, \ldots, 4$ and all of these probabilities add up to 1, i.e., $\sum_{i=1}^{\infty} p(x_i) = \frac{4}{10} + \frac{3}{10} + \frac{2}{10} + \frac{1}{10} = 1,$ both conditions for PMF distribution are satisfied.

7.6.1.2 Continuous Random Variables

A continuous random variable X is an interval or a collection of intervals in the range space R_x, where the probability that X is in the interval $[a, b]$ is given by

$$P(a \leq X \leq b) = \int_a^b f(x)\,dx \tag{7.18}$$

The function $f(x)$ is called the probability density function (PDF) of the random variable X. This function is used to specify the probability of the random variable X is falling within a particular range of values in R_x. The following conditions hold for the PDF:

(a) $f(x) \geq 0$ for all x in R_x.

(b) $\int_{R_x} f(x)\,dx = 1.$

(c) $f(x) = 0$ if x is not in R_x.

▶ Example 10

The life of an imaging system (its length of use) in a hospital is given by a continuous random variable X, where all values are in the range $x \geq 0$. The PDF of this system's lifetime in years is as the following:

$$f(x) = \begin{cases} \frac{1}{3} e^{-x/3}, & x \geq 0 \\ 0 & \text{otherwise} \end{cases}$$

where the random variable X has an exponential distribution with mean 3 years. What is the probability that the life of an imaging system in a hospital is between 3 and 4 years? ◀

■ Solution

The probability that the life of the imaging system is between 3 and 4 years is calculated as

$$\begin{aligned} P(3 \leq X \leq 4) &= \frac{1}{3}\int_3^4 e^{-x/3}\,dx \\ &= -e^{-4/3} + e^{-1} \\ &= -0.264 + 0.368 \\ &= 0.104 \end{aligned}$$

In other words, there is 10.4% chance that the life of an imaging system in the hospital is between 3 and 4 years.

7.6.2 Cumulative Distribution Function (CDF)

The cumulative distribution function (CDF), denoted as $F(x)$, measures the probability that the random variable X has a value less than x, i.e., $F(x) = P(X \leq x)$.

If X is a discrete random variable, then we calculate CDF as

$$F(x) = \sum_{x_i \le x} \text{all } p(x_i) \quad (7.19)$$

If X is a continuous random variable, then we calculate CDF as

$$F(x) = \int_{-\infty}^{x} f(t)dt \quad (7.20)$$

Some important properties of the CDF are:
(a) F is a non-decreasing function. If $a < b$, then $F(a) \le F(b)$.
(b) $\text{Lim}_{x \to \infty} F(x) = 1$
(c) $\text{Lim}_{x \to -\infty} F(x) = 0$
(d) $P(a \le X \le b) = F(b) - F(a)$ for all $a < b$.

7

▶ Example 11

The CDF of the system described in ▶ Example 10 is calculated as the following:

$$F(x) = \frac{1}{3}\int_0^x e^{-t/3} dt = 1 - e^{-x/3}$$

which explains the probability that the imaging system will die before a specific time, x. Calculate the probability that this system will die before the third year. ◀

■ **Solution**

The probability that the imaging system will die before the third year is given by

$$\begin{aligned} P(0 \le X \le 3) &= F(3) - F(0) \\ &= 1 - e^{-\frac{3}{3}} - \left(1 - e^{-\frac{0}{3}}\right) \\ &= 1 - e^{-1} - 1 + 1 = 1 - e^{-1} \\ &= 0.632 \end{aligned}$$

The probability that the life of the system is between 3 and 4 years is calculated as the following:

$$\begin{aligned} P(3 \le X \le 4) &= F(4) - F(3) \\ &= \left(1 - e^{-4/3}\right) - \left(1 - e^{-1}\right) \\ &= -e^{-4/3} + e^{-1} \\ &= -0.264 + 0.368 = 0.104 \end{aligned}$$

7.6.3 Discrete Distributions

A discrete probability distribution represents the properties of discrete random variables. Discrete probability distributions are built from outcomes that have finite measurable values. A well-known example is rolling a dice with six numbered sides where there can only be six possible outcomes. Therefore, the finite outcome value is six. Another example is flipping a coin which can only result in two outcomes, so the finite value is two. In this section we discuss the most common discrete distributions that are used when analyzing discrete data.

7.6.3.1 Bernoulli Distribution

A random variable X has a Bernoulli distribution if $X = 1$ occurs with probability p and $X = 0$ occurs with probability $1 - p$. The PMF for the Bernoulli distribution is as follows:

$$f(x) = \begin{cases} p, & x = 1 \\ 1 - p, & x = 0 \\ 0, & \text{otherwise} \end{cases} \quad (7.21)$$

where $0 \le p \le 1$. The mean of a Bernoulli distribution is

$$E(X) = 1(p) + 0(1-p) = p \quad (7.22)$$

and the variance of a Bernoulli distribution is

$$\begin{aligned} \text{Var}(X) &= E(X^2) - [E(x)]^2 \\ &= 1^2 p + 0^2(1-p) - p^2 \\ &= p(1-p) \end{aligned} \quad (7.23)$$

For example, in manufacturing, the Bernoulli distribution is used to model a defective manufactured products, parts, or components, i.e., if a single product is defective with probability p versus non-defective with probability $(1-p)$. Some other examples where Bernoulli distribution is useful for Six Sigma practitioners in manufacturing organizations are to model the number of customer complaints, number of returned products, number of rejected batches to suppliers, and number of reworks.

7.6.3.2 Binomial Distribution

The binomial distribution represents the number of successes and failures in n independent Bernoulli trials. Suppose that X_1, X_2, ..., X_n are n independent and identically distributed (iid) Bernoulli random variables and $X = X_1 + X_2 + \ldots + X_n$. Then, X has a binomial distribution with parameters n and p (i.e., probability of success). This distribution is associated with data that can have one of two values, e.g., pass or fail, head or tail, etc. The PMF for binomial distributions is as follows:

$$f(x) = \binom{n}{i} p^i (1-p)^{n-i} = \frac{n!}{i!(n-i)!} p^i (1-p)^{n-i}$$

$$\text{for } X_i = 0,1,2,\ldots,n, \text{ and } 0 \le p \le 1. \quad (7.24)$$

where i is the number of successes and $n - i$ is the number of failures.

Note that $X = X_1 + X_2 + \ldots + X_n$, then by additive properties of mean and variance, we have

$$\begin{aligned} E(X) &= E(X_1) + E(X_2) + \ldots + E(X_n) \\ &= np \end{aligned} \quad (7.25)$$

and variance of a binomial distribution is

$$\begin{aligned} Var(X) &= Var(X_1) + Var(X_2) + \ldots + Var(X_n) \\ &= np(1-p) \end{aligned} \quad (7.26)$$

The binomial distribution is very useful to characterize defective and nonconforming data (e.g., defective manufactured item or nonconforming service), and it helps Six Sigma practitioners to know how often a specific event would occur within a process, rather than an effect of that event. Some applications of binomial distribution include estimating the number of items demanded from an inventory, number of defective items in a batch, and sampling for attributes (acceptance sampling).

▶ Example 12

Suppose that we toss three coins. If we assume that the outcomes are independent, what is the probability that two heads and one tail are obtained?◀

- **Solution**

We define X to be the number of heads that appear and then X is a binomial random variable with parameters ($n = 3, p = \frac{1}{2}$).

Therefore, according to Eq. 7.24

$$\begin{aligned} P\{X = 2\} &= \binom{3}{2}\left(\frac{1}{2}\right)^2\left(1-\frac{1}{2}\right)^{3-2} \\ &= \frac{3!}{2!(3-2)!}\left(\frac{1}{2}\right)^2\left(\frac{1}{2}\right)^1 = \frac{3}{8} \end{aligned}$$

▶ Example 13

Suppose that products that are manufactured by a certain machine will be defective with probability 0.2, independently of any other product. What is the probability that in a sample of four products, at most one will be defective?◀

- **Solution**

Assume that X is a random variable that represents the number of defective products in the sample and then X is a binomial random variable with parameters ($n = 4, p = 0.2$). Thus

$$\begin{aligned} &P\{X = 0\} + P\{X = 1\} \\ &= \binom{4}{0}(0.2)^0(1-0.2)^4 + \binom{4}{1}(0.2)^1(1-0.2)^3 \\ &= \frac{4!}{0!(4-0)!}(0.2)^0(0.8)^4 \\ &\quad + \frac{4!}{1!(4-1)!}(0.2)^1(0.8)^3 \\ &= \frac{4!}{1(4!)}1(0.41) + \frac{4!}{1(3)!}0.2(0.51) \\ &= 0.41 + 0.41 = 0.82 \end{aligned}$$

In cases where we are interested to model the number of trials required, n, to achieve k successes before a specified number of failures (r) occur, we can use negative binomial distribution. The PMF of a random variable X that follows a negative binomial distribution is

$$f(x) = \binom{k+r-1}{k} p^k (1-p)^r$$

$$\text{for } 0 \le p \le 1 \quad (7.27)$$

where p is the probability of success and $r = n - k$.

▶ Example 14

Suppose that for a startup company, there is a 10% chance of receiving two funds from venture capitalists before four of their funding requests get rejected in a year. What is the probability that the company submits six funding requests to receive two funds in a year?◀

Solution

p = 0.1 (10% chance of receiving two funds in a year), n = 6 trials, k = 2 successes, r = 4 rejections

7

$$P\{X=6\} = \binom{2+4-1}{2}(0.1)^2(0.9)^4$$

$$= \frac{5!}{2!(5-2)!}(0.1)^2(0.9)^4$$

$$= (10)(0.01)(0.66) = 0.07$$

7.6.3.3 Hypergeometric Distribution

The hypergeometric distribution is a generalized form of binomial distribution where the trials are not independent and describes the probability of k successes in a sample of n draws from a finite population of size N, *without* replacement. Therefore, the hypergeometrically distributed random variable, X, is characterized with parameters N, K, and n. For example, X is a hypergeometric random variable if we have to draw pencils of two colors (red and blue) from a set of colored pencils without replacement. Assume that we define a red pencil as a success and a blue pencil as a failure. If N describes the number of all pencils in the box and K describes the number of red pencils in the box, then $N - K$ represents the number of blue pencils. On the other hand, assume a situation where the pencils are replaced in the box and then the number of one color has a binomial distribution. Therefore, the binomial distribution is used to describe the probability of successes in draws *with* replacement. The PMF for hypergeometric distribution is given by

$$f(x) = P\{X=k\} = \frac{\binom{K}{k}\binom{N-K}{n-k}}{\binom{N}{n}} \quad (7.28)$$

where N is the population size, K is the number of success states in the population, n is the number of draws (quantity drawn in each trial), and k is the number of observed successes.

The mean of the hypergeometric distribution is equal to np where $p = K/N$, and the variance of this distribution can be calculated by $np(1-p)\frac{(N-n)}{(N-1)}$.

▶ Example 15

A company produces ten products where six of them are product type a and four of them are product type b. Three products are randomly selected to be sent for inspection at the quality control section of the company. What is the probability that two of the selected products are type a?◀

Solution

$K = 6$, $N = 10$, $n = 3$, $k = 2$

$$P\{X=2\} = \frac{\binom{6}{2}\binom{10-6}{3-2}}{\binom{10}{3}} = \frac{\frac{6!}{2!4!}\frac{4!}{1!3!}}{\frac{10!}{3!7!}}$$

$$= \frac{\frac{30}{2}*4}{\frac{720}{6}} = 0.5$$

There is 50% chance that two of the selected products are type a.

7.6.3.4 Geometric Distribution

Suppose that independent trials, each can occur with probability p, are performed until a success (or a specific event) happens. We define the random variable, X, with parameter p to follow a geometric distribution if X represents the number of trials required until the first success. The PMF for geometric distribution is

$$(x) = P\{X = n\} = (1-p)^{n-1}\,p,$$

$$\text{for } x = 1,2,\ldots \text{and } 0 \le p \le 1. \quad (7.29)$$

where n is the number of independent trials and p is the probability of success. The mean and variance of geometric distribution are calculated by $\frac{1}{p}$ and $\frac{(1-p)}{p^2}$, respectively.

Generally, geometric distribution is used when we are interested (1) in the number of failures before the first success in a sequence of trials with probability of success p for each trial and (2) the number of items inspected before finding the first defective item.

▶ Example 16

Suppose that, in a manufacturing company, the probability that any particular product will be defective is 0.1. What is the probability that the first product found to be defective is the fifth product that was inspected? ◀

- **Solution**

$p = 0.1,\ n = 5$

$$P\{X = 5\} = (1-0.1)^{5-1}(0.1) = 0.0656$$

The probability that the first product found to be defective is the fifth product that was inspected is 6.56%.

7.6.3.5 Poisson Distribution

Poisson distribution is another discrete distribution that is very useful in quality control. It is used when we would like to estimate the probabilities of random events, e.g., number of arrivals, defects, failures, calls, etc., per some unit of measure. Unlike the binomial distribution that has only two possible outcomes (e.g., pass/fail), this distribution could have number of discrete occurrences over a defined time interval. The PMF for Poisson distribution is

$$f(x) = P\{X = i\} = \frac{e^{-\lambda}\lambda^i}{i!}, i = 0,1,2,\ldots \quad (7.30)$$

where λ is the Poisson parameter, also called rate parameter, and defined as the average number of events per interval, i is the number of times an event occurs in a specific time interval, and e is the constant number 2.718.

The expected value and variance of X, respectively, are

$$E(X) = \lambda x \quad (7.31)$$

$$Var(X) = \lambda x \quad (7.32)$$

▶ Example 17

Suppose that the number of defective products that are manufactured in a company each day is a Poisson random variable with parameter $\lambda = 2$, i.e., the average number of defective products manufactured per day. Calculate the probability that at least one defective product will be manufactured today. ◀

- **Solution**

$$P\{X \ge 1\} = 1 - P\{X = 0\} = 1 - \frac{e^{-\lambda}\lambda^i}{i!}$$
$$= 1 - \frac{e^{-2}2^0}{0!} \approx 0.86$$

Some example applications of the Poisson distribution are when studying the number of customers per minute in a grocery store, number of transactions per hour at a bank, number of sick days per month, number of defective parts in a box, etc.

7.6.4 Continuous Distributions

In this section, we discuss some of the important continuous probability distributions which are used to describe the properties of continuous random variables. As mentioned in ▶ Sect. 7.6.1.2, a continuous random variable is defined over one or more intervals of real numbers with an infinite number of possible outcomes. The probability distribution that is used to characterize the outcomes of continuous random variables is called probability density function (PDF). One important difference between continuous and discrete distributions is that since there are an infinite number of values in an interval, it is not feasible to define a probability for a specific value x. Therefore, probabilities could be calculated between two numbers (e.g., a and b). In the following, we discuss about most common continuous probability distributions that are useful for Six Sigma practitioners.

7.6.4.1 Uniform Distribution

A random variable X follows a uniform distribution over the interval $(0,1)$ if its PDF is characterized by

$$f(x)=\begin{cases}1, & 0<x<1\\ 0, & \text{otherwise}\end{cases} \tag{7.33}$$

In general, X is a uniform random variable on the interval (a, b) if its PDF is given by

$$f(x)=\begin{cases}\dfrac{1}{b-a}, & a<x<b\\ 0, & \text{otherwise}\end{cases} \tag{7.34}$$

Since $F(x)=\int_{-\infty}^{x} f(t)dt$, the CDF is given by

$$(x)=\begin{cases}0, & x\le a\\ \dfrac{x-a}{b-a}, & a<x<b\\ 1, & x\ge b\end{cases} \tag{7.35}$$

The mean and variance of uniform distribution are $\dfrac{(b+a)}{2}$ and $\dfrac{(b-a)^2}{12}$, respectively.

The uniform distribution is one of the simplest yet useful continuous distributions that could be used to model different types of CTQ variables over a specified range. For example, suppose that the number of time spent to answer the calls received in a customer service of a company is unknown, but we know the estimation of minimum and maximum of call durations (e.g., between 5 and 30 minutes); then we can use continuous uniform distribution to model the time required to answer the calls.

▶ Example 18

Suppose that the arrival of a bus that travels between cities in a specific bus station is uniformly distributed over (1, 6) in hours; calculate the probability that (a) the bus arrives in less than 3 hours, (b) it takes more than 4 hours for the bus to arrive in the bus station, (c) the bus arrives between 2 and 5 hours. ◀

- **Solution**

(a) $P\{X<3\}=\dfrac{\int_1^3 dx}{6-1}=\dfrac{2}{5}=0.4$

This implies that there is 40% chance that the bus will arrive in the bus station in less than 3 hours.

(b) $P\{X>4\}=\dfrac{\int_4^6 dx}{6-1}=\dfrac{2}{5}=0.4$

This implies that there is 40% chance that the bus will arrive in the bus station in more than 3 hours.

(c) $P\{2<X<5\}=\dfrac{\int_2^5 dx}{6-1}=\dfrac{3}{5}=0.6$

This implies that there is 60% chance that the bus will arrive in the bus station within 2 and 5 hours.

7.6.4.2 Exponential Distribution

A random variable X is said to follow an exponential distribution with parameter $\lambda > 0$, if its PDF is given by

$$f(x)=\begin{cases}\lambda e^{-\lambda x}, & x\ge 0\\ 0, & x<0\end{cases} \tag{7.36}$$

where λ is called the rate parameter and $\dfrac{1}{\lambda}$ is the mean of the exponential distribution (note that λ is the mean of the Poisson distribution), x is the time interval, and e is the number 2.718. The CDF of exponential distribution is

$$F(x)=\int_0^x \lambda e^{-\lambda t}=1-e^{-\lambda x},\ x\ge 0 \tag{7.37}$$

When calculating the CDF, note that $F(\infty)=\int_0^\infty \lambda e^{-\lambda t}=1$.

The expected value and variance of X, respectively, are

$$E(X)=\frac{1}{\lambda} \tag{7.38}$$

$$Var(X)=\frac{1}{\lambda^2} \tag{7.39}$$

Both the exponential and Poisson distributions are frequently used when studying the queues and process quality. Both of these distributions are used to model a process with rate λ; however, the important difference is that exponential distribution is used when the data is continuous. Specifically, the Poisson distribution is characterized by the number of occurrences (i.e., a discrete random variable) in a fixed period of time, where the exponential distribution deals with the time (i.e., a continuous variable) between occurrences of successive events. Some examples of exponentially distributed random variables are the time between successive incoming calls to a call center, waiting times, and the time between arrival of customers to a grocery store. According to Eq. 7.38, the mean inter-arrival time (time between successive arrivals) is $\frac{1}{\lambda}$, i.e., λ is the mean arrival *rate* per unit time.

An important characteristic of exponential distribution is the "lack of memory" or "memoryless" property. This means that the probability of an event that may happen in the future is independent of previous events (i.e., there is no relation to whether or not it has happened in the past). For example, in a grocery store, the probability of a customer arriving to the store in the next 5 minutes is the same as the probability of arriving 2 minutes ago. Therefore, the probability that a customer arrives in the future does not increase (or decrease) based on the waiting time. Mathematically, if X is exponentially distributed, then the "lack of memory" property is characterized by

$$\{X > x+s | X > s\} = P\{X > x\} \tag{7.40}$$

▶ Example 19

Suppose that customers arrive at a bank with an average arrival rate of 10 customers per hour. (a) What is probability that the time between customer arrivals will be less than 12 minutes? (b) If the bank opens at 8:00 am and it is now 8:30 am, what is the probability that a customer will arrive after 8:40 am?◀

■ Solution

(a) $\lambda = 10 \frac{\text{customers}}{\text{hour}}$

$$P\{X < 12 \text{ minutes}\}$$
$$= P\left\{X < \frac{1}{5}\text{hour}\right\} = 1 - e^{-\lambda x}$$
$$= 1 - e^{-10\left(\frac{1}{5}\right)} = 1 - (2.718)^{-2}$$
$$= 1 - 0.14 = 0.86$$

(b) According to the lack of memory property

$$P\{X > 40 \text{ minutes} | X > 30 \text{ minutes}\}$$
$$= P\left\{X > \frac{2}{3}\text{hour} \middle| X > \frac{1}{2}\text{hour}\right\}$$
$$= P\left\{X > \frac{1}{6}\text{hour}\right\}$$
$$= 1 - P\left\{X \le \frac{1}{6}\right\} = 1 - \left[1 - e^{-10\left(\frac{1}{6}\right)}\right]$$
$$= e^{-\frac{10}{6}} = 0.19$$

Note: $x + s = \frac{2}{3}$; $s = \frac{1}{2}$; therefore, $x = \frac{1}{6}$.

This implies that there is 19% chance that the customer will arrive after 8:40 am given that the current time is 8:30 am and the bank opened at 8:00 am.

7.6.4.3 Triangular Distribution

The triangular distribution is a continuous probability distribution used to model a process where only the minimum, most likely (mode), and maximum values of the distribution are known. Its PDF is shaped like a triangle where the peak value of the triangle is the mode. This distribution is commonly used by Six Sigma practitioners when the mean and standard deviation are not known. Moreover, it is very useful when the data are scare or when data collection is difficult or expensive. The PDF for triangular distribution is given by

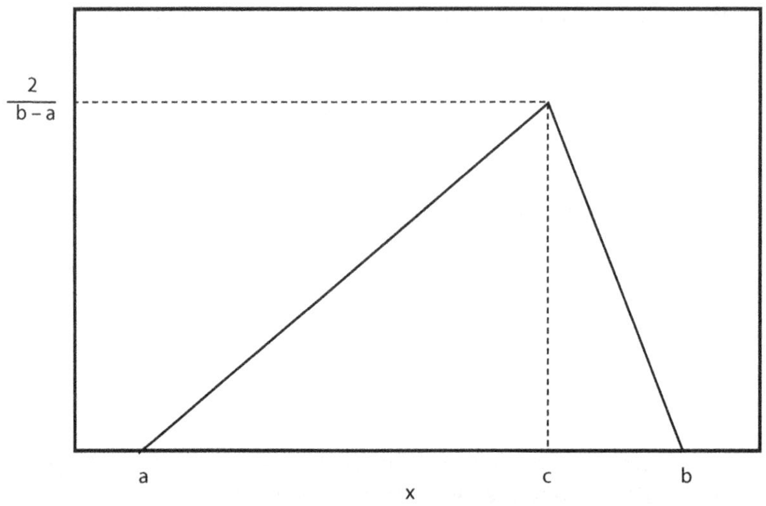

Fig. 7.1 Triangular distribution probability density function. (Source: Author's creation)

7

$$f(x)=\begin{cases}0, & x<a\\ \dfrac{2(x-a)}{(b-a)(c-a)}, & a\le x\le c\\ \dfrac{2(b-x)}{(b-a)(b-c)}, & c\le x\le b\\ 0, & x>b\end{cases} \quad (7.41)$$

where a is the minimum of lower limit, b is the maximum or upper limit, and c is the mode or the peak value (see Fig. 7.1).

The mean of triangular distribution can be calculated using $\frac{(a+b+c)}{3}$, and its variance is $\frac{(a^2+b^2+c^2-ab-ac-bc)}{18}$.

▶ Example 20

Suppose that a company is planning to open a new sales office in another state. The company wants to model the future weekly sales with a minimum of \$1,000 and maximum of \$5,000 and a peak value of \$4,000. Propose the probability density function that the company can use to model its weekly sales in the new office. ◀

Solution

A triangular distribution with parameters a = \$1,000, b = \$5,000, and c = \$4,000 could be used to model the weekly sales in the new office, and the PDF is

$$f(x)=\begin{cases}0, & x<1{,}000\\ \dfrac{2(x-1{,}000)}{(5{,}000-1{,}000)(4{,}000-1{,}000)}, & 1{,}000\le x\le 4{,}000\\ \dfrac{2(5{,}000-x)}{(5{,}000-1{,}000)(5{,}000-4{,}000)}, & 4{,}000\le x\le 5{,}000\\ 0, & x>5{,}000\end{cases}$$

7.6.4.4 Normal Distribution (Gaussian Distribution)

A random variable X is said to be normally distributed with parameters μ and σ^2 if the PDF of X is given by

$$f(x) = \frac{1}{\sqrt{2\pi}\sigma} e^{-(x-\mu)^2/2\sigma^2}, \ -\infty < x < \infty \quad (7.42)$$

where μ is the mean or expected value of the distribution, σ is the standard deviation, σ^2 is the variance, e is 2.718, and π is 3.141. A normal distribution with a mean of 0 and a standard deviation of 1 is called a standard normal distribution, and the letter z is usually used to represent this type of random variable. The PDF for the standard normal distribution is

$$f(z) = \frac{1}{\sqrt{2\pi}} e^{-z^2/2} \quad (7.43)$$

As represented in Fig. 7.2, the PDF of normal distribution is a bell-shaped curve that is symmetric around μ. Another important fact about normal distribution is that the mode and median values are also the same as the mean (i.e., μ).

When data are grouped around the mean and there is an equal probability that a data point is above or below the average, we can use normal distribution to model the data. In Six Sigma, the normal distribution is used to model a process that can be thought of as the sum of a number of component processes. For example, a time to assemble a product, that is, the sum of times required for each assembly operation, could be modeled with a normal distribution. Moreover, it is used when the goal is to conduct a probabilistic assessment of distribution of time between independent events that occur at a constant rate. Finally, normal distribution is used when comparing two process means.

The CDF for normal distribution is denoted with Φ and is given by

$$\Phi(x) = \frac{1}{\sqrt{2\pi}} \int_{-\infty}^{x} e^{-t^2/2} dt \quad (7.44)$$

Special tables of areas under the standard normal distribution curve could be used to calculate the cumulative probabilities. These tables are provided in Table A.2. If x is any value from a normal distribution with mean μ and standard deviation σ, we can convert it to an equivalent value from a standard normal distribution using the following formula:

$$z = \frac{x-\mu}{\sigma} \quad (7.45)$$

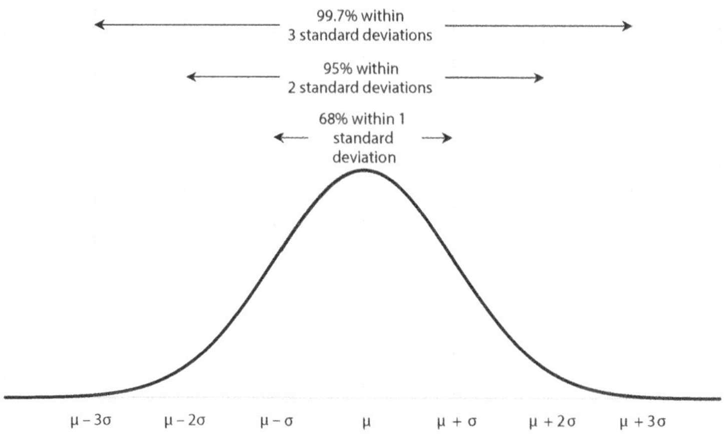

Fig. 7.2 Normal distribution probability density function. (Source: Author's creation)

▶ Example 21

A manufacturer of CT scan devices reports that the average number of days between device malfunctions is 550 days, with a standard deviation of 30 days. Assuming a normal distribution, (a) what is the probability that the number of days between adjustments will be less than 600 days? (b) What is the probability that the number of days between adjustments will be more than 450 days? (c) Calculate the number of days for which the probability that the CT scan device would not malfunction is 0.90.◀

7

Solution

(a) Let's articulate the problem in a statistical notation as follows:

$P(X \leq 600\,\text{days}) = ?$

By using the table in Table A.2, we convert the value of x to a z value. For $x = 600$ days, we have

$$z = \frac{x-\mu}{\sigma} = \frac{600-550}{30} = 1.67$$

Therefore, using Table A.2, $P(X \leq 600\text{ days}) = P(z \leq 1.67) = 0.9525$. The probability that the number of days between adjustments will be less than 600 days is 95.25%.

(b) Let's articulate the problem in a statistical notation as follows:

$P(X > 450\,\text{days}) = ?$

Note that first we need to calculate the probability that the number of days between adjustments will be less than 450 days: $P(X \leq 450\text{ days})$.

$$z = \frac{x-\mu}{\sigma} = \frac{450-550}{30} = -3.33 \rightarrow P(X \leq 450) = P(z \leq -3.33) = 0.0004$$

Therefore, $P(X > 450) = 1 - P(X \leq 450) = 1 - 0.0004 = 0.9996$. The probability that the number of days between adjustments will be more than 450 days is 99.96%.

(c) We know that $P(X \leq x) = 0.90$ which is equivalent to $P(Z \leq z) = 0.90$, where $z = \frac{x-550}{30}$. In Table A.2, we can find that corresponding z for 0.90 probability is approximately equal to 1.28. Therefore, we solve

$$1.28 = \frac{x-550}{30} \rightarrow x = 588.4$$

With 90% probability, the CT scan device would not malfunction in the next 588.4 days.

7.6.4.5 Weibull Distribution

The Weibull distribution is very useful in Six Sigma because it can model the time to failure for components or machines. For example, the time to failure of a machine in a manufacturing company could be characterized with a Weibull distribution. The Weibull distribution can be customized for the product characteristics specifically in the wear-out phases and mortality of the life cycle. If a random variable X has a Weibull distribution, its PDF is given by

$$f(x) = \begin{cases} \frac{k}{\lambda}\left(\frac{x}{\lambda}\right)^{k-1} e^{-\left(\frac{x}{\lambda}\right)^k}, & x \geq 0 \\ 0, & x < 0 \end{cases} \tag{7.46}$$

where $k > 0$ is called the shape parameter and $\lambda > 0$ is the scale parameter for the distribution. The CDF of Weibull distribution is

$$F(x) = 1 - e^{-\left(\frac{x}{\lambda}\right)^k} \tag{7.47}$$

Exponential distribution is a special case of Weibull distribution where the shape parameter, k, is equal to 1. If we assume that X is a Weibull random variable that models the time to failure, then $k > 1$ indicates that the failure rate increases with time. In other words, the probability that a machine will fail increases as the time flows. Some applications of Weibull distribution include modeling lifetime of a product, reliability engineering, failure probabilities that vary over time, etc.

▶ Example 22

The time to failure of an MRI device (in years) in a hospital is modeled as a Weibull distribution with $k = 0.90$ and $\lambda = 0.5$. What is the probability that this equipment will fail within 6 months?◀

- **Solution**

$$P(x < 1/2) = F(1/2) = 1 - e^{-\left(\frac{0.5}{0.5}\right)^{0.9}} = 0.59$$

This implies that there is 59% chance that the MRI device will fail within 6 months.

7.7 Inferential Statistics: Fundamentals of Inferential Statistics

By Dr. Aysun Kapucugil Ikiz
Associate Professor
Dokuz Eylul University

In any business, decisions are made based on incomplete information and uncertainty, and decision-makers cannot be certain of the future behavior of some factors. To measure performance of an ongoing service or production process, evaluate conformance to standards, or assist in formulating alternative courses of actions in decision-making processes, Six Sigma teams should move from only describing the nature of the data (i.e., descriptive statistics) to the ability infer meaning from data as to what will happen in the future (i.e., inferential statistics) .

In Six Sigma methodology, the tools implemented in the Define phase are used to identify all possible input variables (*X*s) that may affect the primary output variable (*Y*) of the problem under investigation. In the Measure phase, all these *X*s are refined to identify vital few variables, and the current performance of related process and the magnitude of the problem are determined. The Analyze phase presents how to determine the variables that significantly impact the primary variable (*Y*) and then identify root causes of *X* variables using inferential statistical analysis. In other words, hypotheses are developed as to why problems occur and then they are proved or disproved for verification. Inference is defined in the dictionary as:

- The act of reasoning from factual knowledge or evidence
- The act or process of deriving logical conclusions from premises known or assumed to be true
- The act of passing from one proposition, statement, or judgment considered as true to another whose truth is believed to follow logically from that of the former (Merriam-Webster Dictionary, n.d.).

Inferential statistics consists of those methods used to draw inferences about the population (or process) being studied by modeling patterns of data in a way that accounts for randomness and uncertainty in the observations. It is divided into two major areas: *estimation* and *hypothesis testing*. *Estimation* involves assessing the value of an unknown population parameter – such as the mean time that a customer spends waiting at a tourism information desk, the variance of the thickness of nickel plating on the armatures, and the proportion of defectives in a lot purchased from a vendor – by using sample data. These estimations are very useful if Six Sigma teams want only the approximate values of selected parameters for process problems or pain points. However, in some cases, knowing whether a parameter meets a certain standard would be more important than estimating its value. For example, based on observed daily pollution measurements, an environmental officer may want to know whether the mean pollutant level emitted by a plant of a chemical company exceeds the maximum allowable guidelines. *Hypothesis testing* is a framework for solving this kind of problems. It is used for making decisions about specific values of the population parameters.

This section discusses the core fundamentals of the Analyze phase: sampling distribution, properties of sampling distribution, and point estimation.

7.7.1 Sampling Distribution

Six Sigma projects usually focus on changing the output metrics of a process to reduce cycle time, lead time, the error rate, costs, investment or improve service level, throughput, and productivity. In statistical terms, this translates into shifting the process mean and/or reducing the process standard deviation. These decisions about how to adjust key process input variables are made based on sample data, not population data. Thus, they involve an element of uncertainty and bring some risks.

Here a technique for measuring the uncertainty associated with making inferences is discussed. Whenever an investigator collects data, a sample is taken from some generally unknown population. The distribution of all values in this population is represented by a random variable. It would be too ambitious to describe the entire population distribution using information in a small random sample of observations. However, quite firm inferences about the important characteristics of this population are made. In many situations, the goal is to estimate a numerical characteristic of the population, called a parameter,[1] such as the population mean (μ) and variance (σ^2).

Assume that a manufacturer of home garden equipment collects a variety of data for inspecting quality of the end product. Some in-process measurements are taken to ensure that manufacturing processes remain in control and can comply with design specifications. Also assume that the mean weight of mower blades in the population is required to be 5.01 pounds. For example, given a random sample of 15 observations of blade weights taken from the manufacturing process that produces mower blades, the sample mean is used to make inferential statements about the population mean of blade weights. Suppose the mean weight of these mower blades ($\overline{x}$) is computed as 5.07 pounds. This quantity computed from the observations in the sample is called a statistic – namely, the sample mean $\overline{x} = 5.07$ pounds. Important statistical questions here are:

1. How good is the estimation obtained from the sample?
2. Is this value ($\overline{x} = 5.07$) likely to occur, even if the true population mean is 5.01?
3. Can it be used as a tool in making an inference about the corresponding population parameter?

1 The parameters are denoted by using a symbol and usually Greek letters are used to represent them. For example, μ is used for population mean and σ for population standard deviation.

To give answers for these questions, the population attributes and the random sample attributes should be distinguished.

The value of a population parameter (e.g., the mean, μ) is constant (but usually unknown); its value does not vary from sample to sample. However, the value of a sample statistic (e.g., the sample mean, $\overline{x}$) is highly dependent on the particular sample taken. For each different sample of 15 measurements, there is a different sample mean ($\overline{x}$) that varies both above and below the population's true mean. Thus, any statistic, such as the sample mean, can be regarded as a random variable with a probability distribution. The probability distribution of a statistic is called a *sampling distribution*. Fortunately, the uncertainty of a statistic generally has characteristic properties that are known and reflected in its sampling distribution. Knowledge of the sampling distribution of a particular statistic provides Six Sigma teams with information about its performance over the long run.

7.7.2 Properties of Sampling Distributions

Let's perform a sampling experiment by considering the weight of a mower blade as a random variable which has a triangular distribution for the population. Among others, some of the characteristics of this random variable are computed as follows: the population mean is 5.015 pounds and the standard deviation is 0.052 pounds.

Suppose that a random sample of two pieces is generated from the population of mower blades and the sample mean is computed. When the average of these two values is computed, it would be expected the sample mean to be close to 5.015, but probably not exactly equal to 5.015 because of the randomness in the sample. If this experiment is repeated, say 100 times, a set of 100 sample means is obtained, where these means vary between 4.943 and 5.085.

Suppose that this experiment is now repeated by taking larger samples (say n is increased from 2 to 5, 15, 30, 50, and 100) from the population of the blade weights.

If the mean and standard deviations of 100 sample means are compared to the population parameters, different results are found for each of the sampling distributions (◻ Table 7.2). In each case, notice that as the sample size n gets larger, the average of the 100 sample means seems to be getting closer to the expected value of 5.015 (the center remains the same). Additionally, the standard deviation of the 100 sample means becomes smaller (the variation decreases), meaning that the values of the sample means are less spread out and clustered closer together around the true expected value. ◻ Image 7.6 shows histograms of the sample means for each of these cases. These illustrate also how the shape of sampling distributions changes and seems to be rather normally distributed.

◻ Table 7.2 Experimental results for sampling error and comparison of estimated standard errors to the theoretical values

Sample size (n)	Average of 100 sample means	Standard deviation of 100 sample means	Standard error of the mean (theoretical)
2	5.01616	0.0344471	0.03677
5	5.01622	0.0222285	0.02326
15	5.01690	0.0124147	0.01343
30	5.01587	0.0086345	0.00949
50	5.01569	0.0071434	0.00735
100	5.01566	0.0051892	0.00520

Source: Author's creation

In the long run, if all values of the sample mean with the same size, n, would be generated, the distributions would have been better defined. The means of all possible samples of a fixed size n from the same population form a distribution called the sampling distribution of the sample mean ($\bar{x}$). Thus, the *sampling distribution* of a sample statistic (based on n observations) is the relative frequency distribution of the values of the statistic theoretically generated by taking repeated random

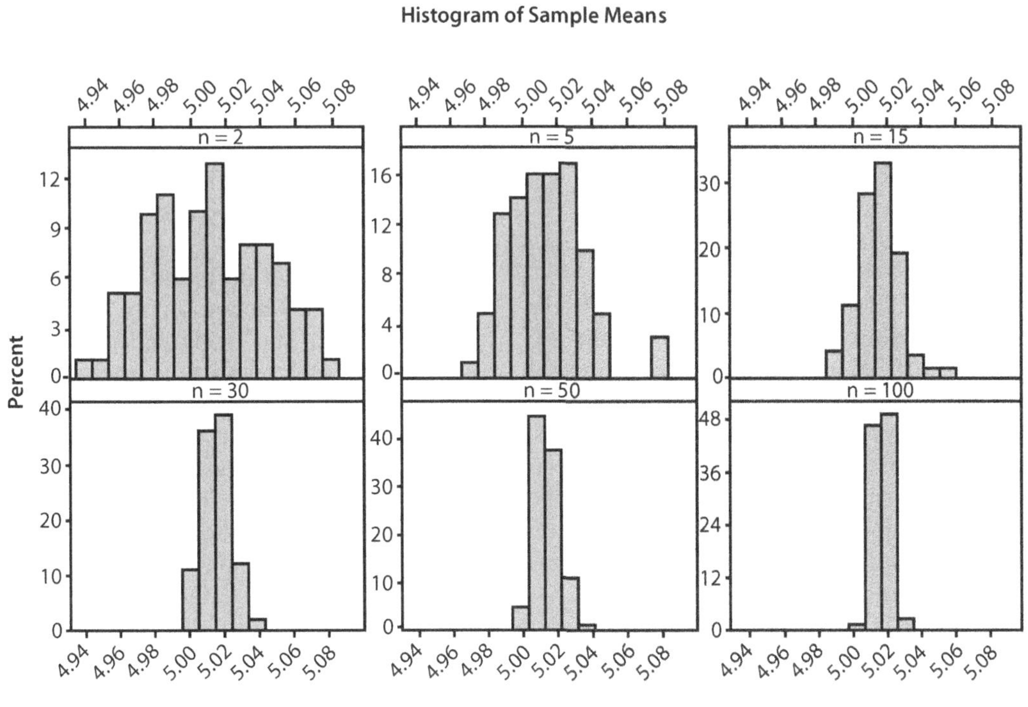

◻ Image 7.6 Histograms of 100 sample means for different sample sizes. (Source: Author's creation based on Minitab)

samples of size n and computing the value of the statistic for each sample (Sincich 1996: 312).

The sampling distribution of the sample mean has two key properties:

7.7.2.1 First Property: The Standard Error of the Mean

The standard deviation of the distribution of means is called the *standard error* of the mean ($\sigma_{\bar{x}}$) and is defined as

$$\sigma_{\bar{x}} = \frac{\sigma}{\sqrt{n}} \quad (7.48)$$

7

where σ is the standard deviation of the population from which the individual observations are drawn and n is the sample size. As seen from Eq. 7.48, the standard error decreases as n increases, and the mean is more stable than a single observation by a factor of the square root of the sample size.

For the sampling experiment above, the standard error of the mean for each of the sample size is computed theoretically by using Eq. 7.48. For example, when sample size is 2, the standard deviation of 100 sample means generated in the experiment is found as 0.0344471 pounds. The population standard deviation of the mower blade weights was 0.052 pounds. Then, the theoretical value of standard error of the mean $\sigma_{\bar{x}}$ is found as follows:

$$\sigma_{\bar{x}} = \frac{\sigma}{\sqrt{n}} = \frac{0.052}{\sqrt{2}} = 0.03677$$

◘ Table 7.2 shows the estimates of the standard error of the mean based on the 100 samples. If these estimates are compared to the theoretical values, it is seen that they are getting closer as sample size n increases. This suggests that the estimates of the mean obtained from larger sample sizes provide greater accuracy in estimating the true mean. In other words, larger sample sizes have less sampling error.

7.7.2.2 Second Property: The Central Limit Theorem

The *central limit theorem*, the fundamental importance in statistics, provides information about the actual sampling distribution of $\bar{x}$.

This theorem states that the means of all possible random samples, each of size n drawn from any distribution (an unknown probability distribution) with mean μ and variance σ^2, will have an approximately normal distribution with a mean equal to μ and a variance equal to σ^2/n (Montgomery and Runger 2014: 243). In other words, if the sample size is large enough, the sampling distribution of the mean is approximately normally distributed, regardless of the distribution of the population. This is exactly what was observed in the experiment above. The distribution of the population was triangular, yet the sampling distribution of the mean converges to a normal distribution as the sample size increases.

Although the central limit theorem works well for small samples (n = 4 or 5) in most cases, particularly where the population is continuous, unimodal, and symmetric, larger samples will be required in other situations, depending on the shape of the population. In many cases of practical interest, if $n \geq 30$, the normal approximation is satisfactory regardless of the shape of the population. If $n < 30$, the central limit theorem will work if the distribution of the population is not severely non-normal (Montgomery and Runger 2014: 246).

To sum up, the central limit theorem furthers knowledge of how to apply inferential statistics to data and helps understand the risk that Six Sigma teams take when sample data are used rather than population data. This theorem allows the use of confidence intervals, hypothesis testing, Design of Experiment, regression analysis, and other parametric techniques on data. The following example demonstrates how to use the central limit theorem.

▶ **Example 23**

Assume that a Six Sigma project focuses on reducing the setup time for a large casting and the setup time is a random variable X that has a continuous uniform distribution with

$$f(x)=\begin{cases}\frac{1}{50-40}=0.1, & 40\le x\le 50\\ 0, & \text{otherwise}\end{cases}$$

(a) What is the probability that the next setup time will last no more than 44 minutes?
(b) What is the probability that the *mean* setup time for a random sample of 35 observations does not exceed 44 minutes? ◀

■ Solution

(a) The probability that the next setup time will last no more than 44 minutes is found by calculating the expression $P(X \le 44)$ and using the CDF of the uniform distribution (i.e., $F(x) = P(X \le x_0) = P(X \le 44)$). For the uniform setup times defined in the range 40–50 minutes, the CDF in that range is as follows[2]:

$$F(x)=\begin{cases}0, & x<40\\ 0.1x, & 40\le x\le 50\\ 0, & x>50\end{cases}$$

The probability that the next setup time will last no more than 44 minutes ($P(X \le 44)$ refers to the probability that a setup time occurs between 40 minutes and 44 minutes ($P(40 \le X \le 44)$). Therefore, the area under the probability density function from 40 to 44 is calculated as

$$\begin{aligned}P(40\le X\le 44)&=F(44)-F(40)\\&=(0.1*44)-(0.1*40)\\&=0.40\end{aligned}$$

The probability that the next setup time will last no more than 44 minutes is 40%.

To calculate the CDF for this uniform distribution in Minitab, click on Calc→Probability Distributions→Uniform. Then, choose "Cumulative Probability" from box; enter 40 in the "Lower endpoint" box and 50 in the "Upper endpoint" box; enter 44 in the "Input constant" box; now click on OK. The result is shown in Session Window. The requested probability is $P(X \le 44) = 0.40$.

(b) The probability that the mean setup time does not exceed 44 minutes is symbolically expressed as $P(\bar{x} \le 44)$. To find this probability for 35 setups, the distribution of the sample mean of the setup times must be used. Remember that, if the sample size is large enough, i.e., $n \ge 30$, the sampling distribution of the mean is approximately normally distributed, regardless of the distribution of the population. Based on the central limit theorem, the distribution of the sample mean $\bar{x}$ of these uniform setup times is approximately normal with mean $\mu_{\bar{x}} = \mu$ and variance $\sigma_{\bar{x}}^2 = \frac{\sigma^2}{n}$, or shortly $\bar{x} \sim N(\mu_{\bar{x}};\sigma_{\bar{x}}^2)$. Remember that from ▶ Sect. 7.6.4.1, for a uniform distribution defined over the range from a to b, the mean and the variance are formulized as $\mu = \frac{a+b}{2}$ and $\sigma^2 = \frac{(b-a)^2}{12}$, respectively. Thus, the mean and variance of setup times are $\mu = \frac{(40+50)}{2} = 45$ and $\sigma^2 = \frac{(50-40)^2}{12} = 8.33$; and the sampling distribution of the mean setup time $\bar{x}$ is normal with mean $\mu_{\bar{x}} = 45$ and variance $\sigma_{\bar{x}}^2 = \frac{\sigma^2}{n} = \frac{8.33}{35} = 0.238$ or standard deviation $\sigma_{\bar{x}} = \sqrt{\sigma_{\bar{x}}^2} = \sqrt{0.238} = 0.488$.

To compute the probability that the mean setup time is lower than 44 minutes,

2 See uniform distribution in Continuous Distributions section in this chapter for this transition from a uniform probability density function to its cumulative distribution function.

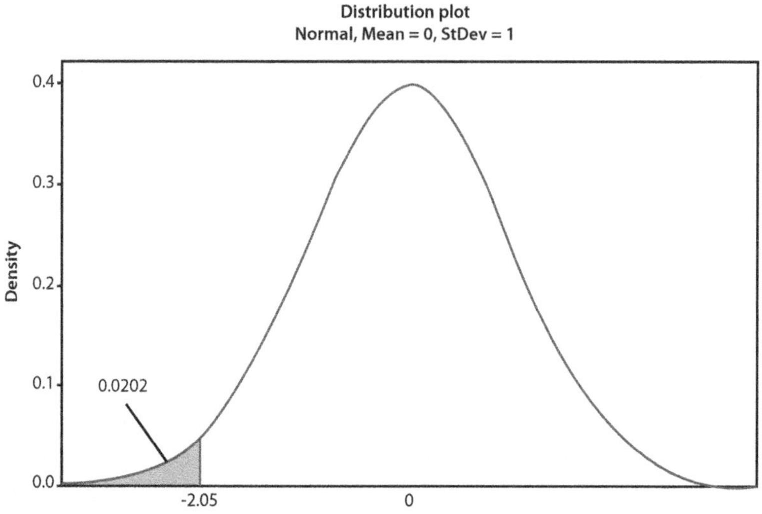

Image 7.7 Standard normal distribution with probability $P(Z \leq -2.05) = 0.0202$. (Source: Author's creation based on Minitab)

$P(\overline{x} \leq 44)$, first the random variable $\overline{x}$ is converted to the standard normally distributed random variable z by the transformation[3] $z = \dfrac{\overline{x} - \mu_{\overline{x}}}{\sigma_{\overline{x}}}$ as follows:

$$P(\overline{x} \leq 44) = P\left(\frac{\overline{x} - \mu_{\overline{x}}}{\sigma_{\overline{x}}} \leq \frac{44-45}{0.488}\right)$$
$$= P(z \leq -2.05)$$

Then the table of standard normal distribution, called z-table (Table A.2), is used to obtain the value of $P(z \leq -2.05)$. The requested probability is shown in Image 7.7 as the shaded area which lies in the interval from $-\infty$ to the upper limit of -2.05. As normal curves are symmetrical, it is an identical area to that of $P(z \geq 2.05)$. The z-table provides areas (probabilities) for only intervals starting from[4] $-\infty$ and ending at a positive value of z. Thus

$$P(z \leq -2.05) = P(z \geq 2.05)$$
$$= 1 - P(z \leq 2.05)$$
$$= 1 - 0.9798$$
$$= 0.0202$$

where the value $P(z \leq 2.05) = 0.9798$ corresponds to that of row 2.0 and column 0.05 in z-distribution (Table A.2). Therefore, the probability that the mean setup time is lower than 44 minutes, $P(\overline{x} \leq 44)$, is 0.0202. In other words, the chance that the average of 35 setup times is lower than 44 minutes is very unlikely (2.02%) for this distribution.

For this calculation in Minitab, click on Calc→Probability Distributions→Normal.... On the box, choose "Cumulative Probability," and enter 45 in "Mean," 0.488 in "Standard deviation," and 44 in "Input constant" box; now click on OK. The result is shown in Session Window. The requested probability is $P(X \leq 44) = 0.02022$.

3 Any normal distribution can be transformed into a standard normal distribution with $\mu = 0$ and $\sigma^2 = 1$ by subtracting the mean from every value of a random variable and dividing it by the standard deviation, i.e., $z = \frac{x-\mu}{\sigma} \sim N(0;1)$. This procedure is known as the z-transformation which creates standard values for making the probability calculations associated with normal distributions easier.

4 The mathematical normal curve extends indefinitely to the left and right, indicated by the infinity symbols $-\infty$ and $+\infty$, respectively.

The key to applying sampling distribution in a correct way is to understand if the probability to be computed relates to an individual observation or to the mean of a sample. For this example, while the setup time of 44 minutes or less has the 40% probability of being observed, the probability that the mean setup time is lower than 44 minutes is very unlikely (2.02%) if the true mean is 45 minutes. Assume that after applying some improvements, a sample of setup times is taken and its average is calculated as 44 minutes. The project team can get excited because this average value seems impossible if the true mean is still 45. Thus, this would be a real change in this large casting process and resulted in a 1-minute reduction for the setup time.

7.7.3 Estimation

Populations generally correspond to some aspects of processes investigated in Six Sigma projects, and they are characterized by parameters. In many processes, as it is very costly and inefficient to measure every unit of product, service, or information provided, inferences about unknown parameter values are computed on statistics such as the mean, standard deviation, and proportion computed from the information in a sample selected randomly from those processes. As shown with sampling distributions in previous sections, when a known population is sampled many times, the calculated statistics can unfortunately be different simply due to the nature of random sampling. The sampling distribution of any statistic (e.g., the sample mean or proportion) indicates how far this statistic could be from a known population parameter.

In this section, the main concern is to estimate true parameters of the process improvement opportunities and assess its reliability based on knowledge of the sampling distributions of the statistics being used. The statistical problem is to determine how far an unknown population parameter could be from the computed statistic of a simple random sample selected from that population.

In discussing the estimation of an unknown parameter, two possibilities must be considered: the first is to compute a single value, called *a point estimate*, from the sample as the best representative of the unknown population parameter. The point estimate is unlikely equal the true value of the parameter due to sampling error so that it alone is not reliable. The second to get truly meaningful information is to determine a range of values, called *confidence interval*, which most likely contains the value of the true population parameter. This is accomplished by using the characteristics of the sampling distribution of the statistic that was used to obtain the point estimate.

7.7.3.1 Point Estimates

An *estimator*, generically denoted by $\hat{\theta}$, is basically a descriptive statistic that is used to estimate an unknown parameter θ and is a random variable depending on the sample information. A point estimate is a particular value of an estimator. In estimation problems, the capability of random sampling from the physical process being studied is required. If a random sample is available, a point estimator may be constructed.

For analyzing the weights of mover blades mentioned before, remember that a random sample of 15 observations was taken to make inferential statements about the population mean of blade weights. ◘ Table 7.3 shows the sample dataset and the descriptive statistics of blade weights such as the sample mean (5.0127), the sample median (5.0200), and the average of the smallest and largest values in the sample (5.0300). These three statistics might be considered some guesses obtained from different choices for the point estimator of the true population mean (e.g., say, in this case, $\mu = 5.01$). But, which one gives the best guess or reasonable estimate? Unfortunately, no single mechanism exists for the determination of a uniquely "best" point estimator in all circumstances. What is available, instead, is a set of criteria under which particular estimators can be compared.

A point estimator possesses a variety of properties. Among others, two important

Table 7.3 Blade weights sample data and descriptive statistics

Weights of mover blades n = 15 (in pounds)			Descriptive statistics	
5.21	5.02	4.90	Sample mean Sample median	5.017 5.0200
5.00	5.16	5.03	Minimum	4.8500
4.96	5.04	4.98	Maximum Average of min	5.2100
5.07	5.02	5.08	and max value	5.03000
4.85	4.90	4.97	Sample standard deviation	0.0954

Source: Author's creation

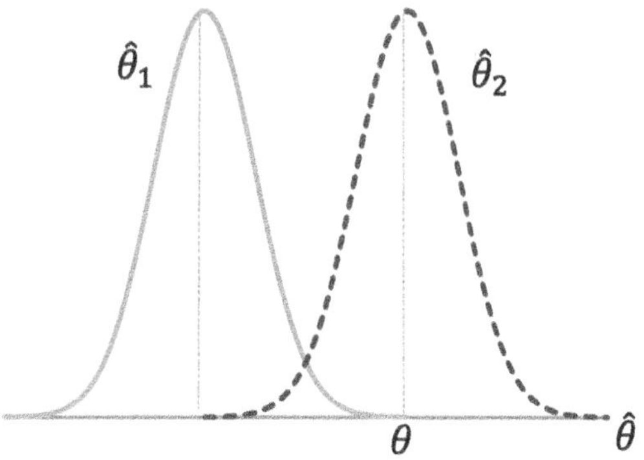

Fig. 7.3 Unbiasedness of an estimator. (Source: Author's creation)

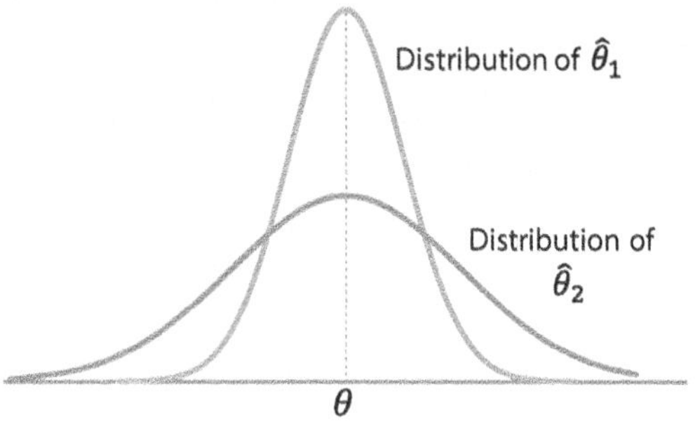

Fig. 7.4 Efficiency of an estimator. (Source: Author's creation)

properties are unbiasedness and efficiency. If possible, a point estimator which is both accurate and efficient would be preferred. A point estimator is accurate if its expected value is equal to the parameter being estimated; i.e., $E(\hat{\theta}) = \theta$. If an estimator $\hat{\theta}$ possesses this property, it is an unbiased estimator of θ, and its value is called an unbiased point estimate (Barnes 1994: 156).

Notice that unbiasedness does not mean that a particular value of θ must be exactly the same value of θ. Rather, an unbiased estimator has the capability of estimating the population parameter correctly on average. Sometimes $\hat{\theta}$ overestimates and other times underestimates the parameter, but it follows from notion of expectation that, if the sampling procedure is repeated many times, then, on the average, the value obtained for an unbiased estimator equals the population parameter. Therefore, an unbiased estimator is correct on the average (Newbold et al. 2013: 287). Figures 7.3 and 7.4 illustrate the probability density functions for two estimators, $\hat{\theta}_1$ and $\hat{\theta}_2$, of the parameter θ. It should be obvious that $\hat{\theta}_1$ is an unbiased estimator of θ and $\hat{\theta}_2$ is not an unbiased estimator of θ (Fig. 7.3).

In many practical problems, different unbiased estimators can be obtained. However, knowing that an estimator is unbiased is often not sufficient when we are searching for the best estimator to use. In this situation, it is natural to prefer the estimator whose distribution is most closely concentrated about the population parameter being estimated. Values of such an estimator are less likely to differ, by any fixed amount, from the parameter being estimated than are those of its competitors (Newbold et al. 2013: 287). Using variance as a measure of concentration, the most efficient unbiased estimator is the one whose distribution has the smallest variance, that is, $\hat{\theta}_1$ is said to be more efficient than $\hat{\theta}_2$ if $Var(\hat{\theta}_1) < Var(\hat{\theta}_2)$ where $\hat{\theta}_1$ and $\hat{\theta}_2$ are two unbiased estimators of θ, based on the same number of sample observations (Fig. 7.4).

One of the important assumptions in inferential statistics is the normality of the population under investigation. For example, if the population is not normally distributed, the sample mean may not be the most efficient estimator of the population mean. In particular, if outliers heavily affect the population distribution, the sample mean is less efficient than other estimators such as the median (Newbold et al. 2013: 288). The procedures for constructing interval estimates are illustrated in the following sections.

7.8 Inferential Statistics: Interval Estimation for a Single Population

By Dr. Aysun Kapucugil Ikiz
Associate Professor
Dokuz Eylul University

One of the major areas of inferential statistics is the estimation, which involves assessing the value of an unknown population parameter. As mentioned in the previous section, the point estimate is unlikely to equal the true value of the parameter due to sampling error. To get more meaningful information, interval estimates are constructed using the characteristics of the sampling distribution of the statistic used to obtain the point estimate. Based on these foundations of inferential statistics in the previous section, interval estimation methods for a single population are presented here.

7.8.1 Interval Estimates

A point estimate $\hat{\theta}$ varies from sample to sample because it depends on the items selected in the sample, and this variation must be taken into consideration when providing an estimate of the population characteristic. An *interval estimate* provides a range of plausible values for the population parameter θ and, thus, more information than a point estimate $\hat{\theta}$.

Let's consider the mower blades case again. Assume that estimating the proportion p of the defective mower blades is now the main CTQ in the manufacturing process, and based on a random sample, it is reported as $\hat{p} = 0.10$ with a margin of error of ±0.01. This means that the true rate of the defective mower blades, p, is *very likely* between 0.09 and 0.11. Therefore, as a small range is found, there is a confidence in predicting what the defective rate would be. If this proportion is 0.10 with a margin of error of ±0.02, the true rate is *likely* to be somewhere between 0.08 and 0.12. In this situation, the uncertainty about this proportion will increase, and the confidence in its estimation will not be as high as the one in the previous interval. These interval estimates are also described as being "very likely" or "likely" to include the true, but unknown, value of the population proportion of the defective mower blades. To increase precision, these estimates must be phrased in terms of probability statements.

There are three types of interval estimates (Montgomery and Runger 2014: 273). One of them is *confidence interval*, which allows organizations to make estimates about population or distribution parameters with a known degree of certainty. The other one is *tolerance interval*, which bounds a selected proportion of a distribution. The third one is *prediction interval*, which provides a range for future (or new) observations from the population or distribution.

The following sections introduce the concept of confidence intervals and present how to construct a confidence interval for different population parameters. The other types of interval estimates are explained at the end of this section.

7.8.2 Confidence Interval Estimation

In general, a $100(1-\alpha)\%$ probability interval is any interval $[A, B]$, such that the probability of falling between A and B is $(1-\alpha)$ (Evans 2012: 137). For example, based on empirical rules, for many large bell-shaped populations, the mean ±1 standard deviation (i.e., $\mu \pm 1\sigma$) describes an approximate 68% probability interval around the mean in a data

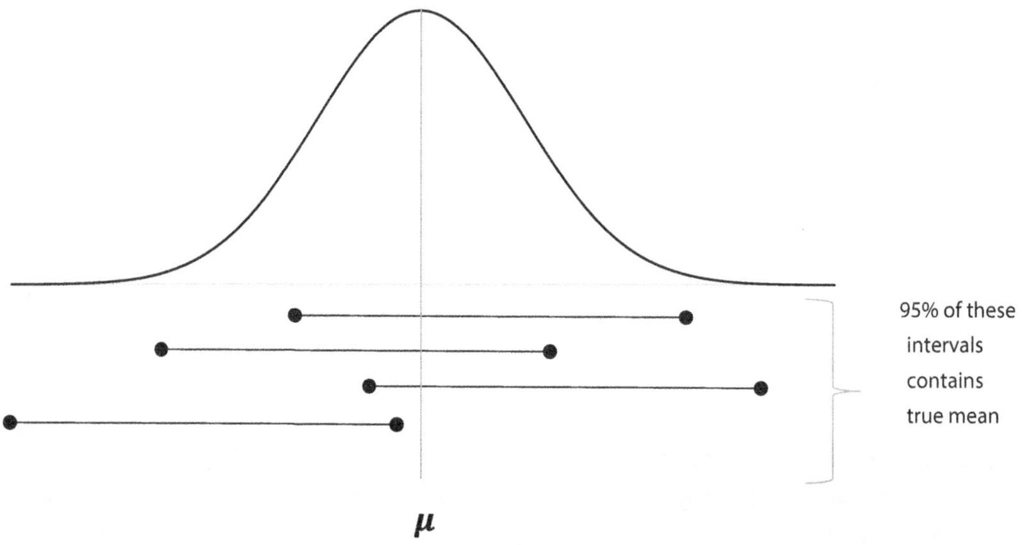

7

■ **Fig. 7.5** Schematic description of 95% confidence intervals. (Source: Author's creation)

set. Similarly, the interval $\mu \pm 2\sigma$ describes an approximate 95% probability interval (Newbold et al. 2013: 76). A *confidence interval* is an interval estimate of the likelihood that the interval contains the true population parameter. This probability is called the level of confidence, denoted by $(1 - \alpha)$, where α is a number between 0 and 1. The level of confidence is usually expressed as a percentage; common values are 90%, 95%, or 99% (Evans 2012: 137). If the level of confidence is 90%, then there is always a 10% risk that the interval does not contain the true population parameter, α.

An interpretation of these interval areas follows. Suppose that random samples are repeatedly taken from the population and intervals are calculated. In each interval constructed, the point estimate will differ, and no two intervals will be the same because confidence interval is also a random interval. Therefore, the interpretation depends on the relative frequency view of probability. It can be stated that, in the long run, $1 - \alpha$ percentage (say 95%) of these intervals include the true value of the unknown parameter and only α percent (say 5%) of them do not. (See ■ Fig. 7.5).

It is important to understand that this rationale fully embraces the possibility of the so-called bad sample, that is, that the low-probability sample is made up primarily of very low (or very high) values, all from the same low-probability tail of the population's distribution. If such an unlikely sample is obtained by random sampling procedure, the true nature of the population will not be reflected. Therefore, the interval based on this anomalous sample will be unlikely to contain the true mean. Bad samples, in the long run, are expected to occur α proportion of the time (Barnes 1994: 160).

As a general format, confidence interval estimates are centered on a point estimate $\hat{\theta}$ and constructed by adding and subtracting a margin of error (ME) as shown in Eq. 7.49.

$$\hat{\theta} \mp ME \tag{7.49}$$

The questions that arise are (1) what causes the margin of error and (2) how large this error might be. The margin of error could be a result of random sampling variation or a true difference in performance if some changes are implemented in the process. It is also called the sampling error and affected by the population standard deviation, the confidence level, and the sample size (Newbold et al. 2013: 295). The length of a confidence interval is equal to twice the margin of error, and it is a measure of the precision of estimation (Montgomery and Runger 2014: 276). The smaller margin of error results in a more precise (narrower) confidence interval for the population parameter. Keeping all the other factors constant, the more the population standard deviation reduces, the smaller the margin of error. Six Sigma efforts strive to reduce variability in the measurements of CTQs and take direct actions on the physical process being analyzed by removing assignable or common causes of variability. When possible, this should be the first step to decrease the margin of error. When this is not possible, the sample size must be increased. Larger sample sizes create tighter confidence intervals, as expected from the central limit theorem. However, increasing sample size causes additional costs. Therefore, if the confidence level decreases, then the margin of error will get smaller. Nevertheless, lowering the confidence level increases α risk. In this situation, there is less assurance that the confidence interval contains the true population parameter.

Many different types of confidence intervals may be developed. Depending on the population parameter of interest, these intervals are constructed by taking into account the known variability in the sampling distribution of the relevant sample statistics (i.e., the estimator). The characteristics or assumptions about the population will also be used in the calculation of these intervals. The most common types of confidence intervals for a single population are discussed in this section.

7.8.2.1 Confidence Interval Estimation for the Mean

The cases that must be considered in constructing confidence intervals on means are schematically detailed in ◘ Fig. 7.6. Selection of the correct distribution is determined by whether the value of population standard deviation σ is known, the sample size n, and whether the sampled population is normally

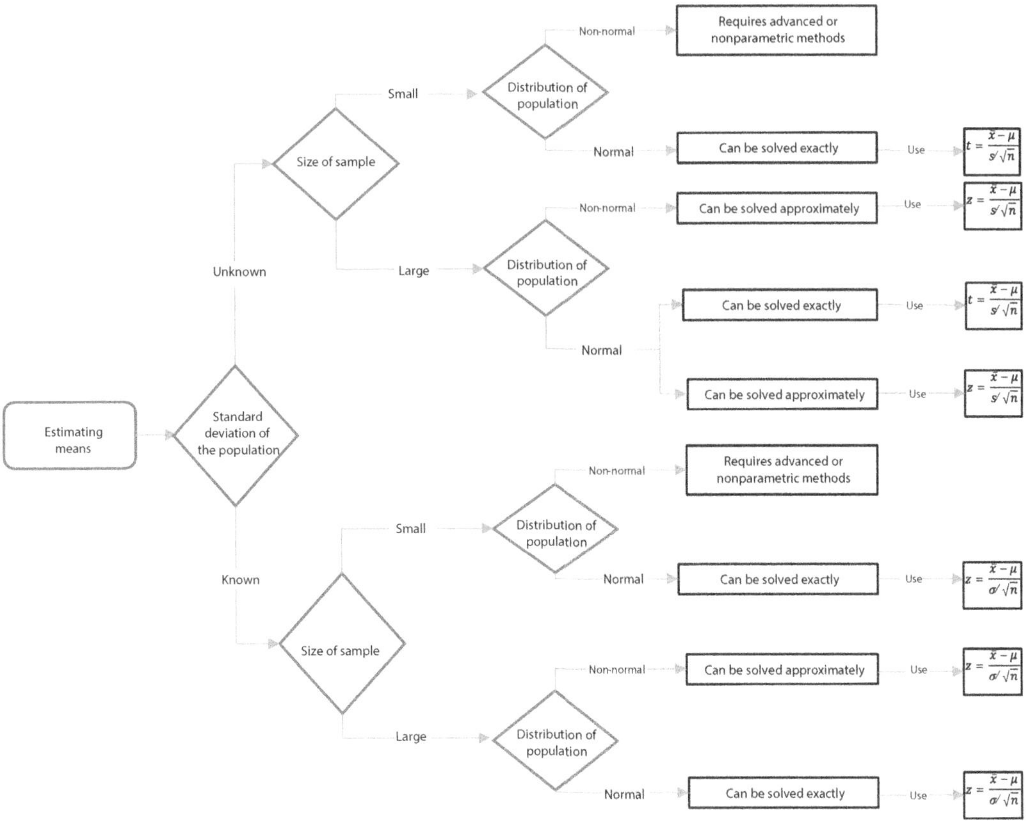

◘ **Fig. 7.6** The eight situations associated with estimating means of normally distributed random variable. (Source: Adapted from Barnes (1994))

distributed. The term large sample refers to the sample being of a sufficient size to allow the central limit theorem to be applied to identify the form of the sampling distribution of the sample mean $\bar{x}$.

Confidence Interval for the Mean (σ Is Known)

The simplest type of confidence interval is for the mean of a population where the standard deviation is assumed to be known. However, the population standard deviation will not be known in most practical sampling applications. Nevertheless, in some applications, such as measurements of parts taken from an automated machine, a process might have a very stable variance that has been established over a long history, and it can reasonably be assumed that the standard deviation is known.

7

When σ is known and the population is normally distributed, it does not matter whether the sample size is large or small. Regardless of the sample size, if $\bar{x}$ is governed by a normal distribution with the parameters μ and $\sigma/\sqrt{n}$ (i.e. $\bar{x} \sim N\left(\mu, \sigma/\sqrt{n}\right)$), the random variable

$$Z = \frac{\bar{x} - \mu}{\sigma/\sqrt{n}} \tag{7.50}$$

is distributed according to an $N(0, 1)$ distribution called standard normal distribution. A $100(1-\alpha)\%$ confidence interval for the population mean μ is given by

$$\bar{x} \mp Z_{\alpha/2}\left(\sigma/\sqrt{n}\right) \tag{7.51}$$

or

$$\bar{x} - Z_{\alpha/2}\left(\sigma/\sqrt{n}\right) < \mu < \bar{x} + Z_{\alpha/2}\left(\sigma/\sqrt{n}\right) \tag{7.52}$$

where $Z_{\alpha/2}$ is the value from the standard normal distribution, $N(0, 1)$, such that the upper tail probability is $\alpha/2$ (Barnes 1994: 158). In Eqs. 7.51 and 7.52, ME, the margin of error (also known as the sampling error), is given by $\text{ME} = Z_{\alpha/2}\left(\sigma/\sqrt{n}\right)$. The alternate way of presenting this confidence interval is as follows:

$$P\begin{pmatrix} \bar{x} - Z_{\alpha/2}\left(\sigma/\sqrt{n}\right) < \mu < \bar{x} \\ +Z_{\alpha/2}\left(\sigma/\sqrt{n}\right) \end{pmatrix} = 1 - \alpha \tag{7.53}$$

where the lower confidence limit is calculated by $\bar{x} - Z_{\alpha/2}\left(\sigma/\sqrt{n}\right)$ and the upper confidence limit is $\bar{x} + Z_{\alpha/2}\left(\sigma/\sqrt{n}\right)$.

To construct a confidence interval, the following steps are followed:
- Step 1: Determine the critical value, $Z_{\alpha/2}$.
- Step 2: Compute the standard error ($\sigma_{\bar{x}}$) and the margin of error.
- Step 3: Estimate the confidence interval and interpret the results.

Let's consider the following example for constructing a confidence interval for a mean under the assumptions that σ is known and the population is normally distributed.

▶ Example 24

A manufacturer of home garden equipment collects a variety of data for controlling quality. Some in-process measurements (CTQs) are taken to ensure that manufacturing processes remain in control and can produce goods according to design specifications. One of the CTQs is the weights of mower blades produced in the manufacturing process. The mean weight of mower blades in the population is required to be 5.01 pounds. At periodic intervals, samples are selected to determine whether the mean of blade weights is still equal to 5.01 pounds or whether something has gone wrong in the manufacturing process to change the weights. If such a situation has occurred, corrective action is needed. Suppose a random sample of 20 blade weights, shown in Table 7.4 below, is taken from the manufacturing process. It is known, from long experience in working with similar mower blades, that the standard deviation is 0.1164 pound and the blade weights are normally distributed. Construct a 95% confidence interval for the population mean of blade weights. ◀

Table 7.4 Weights of mower blades (in pounds)

5.21	5.02	4.90	5.00	5.16	5.03	4.96	5.04	4.98	5.07
5.02	5.08	4.85	4.90	4.97	5.09	4.89	4.87	5.01	4.97

Source: Author's creation

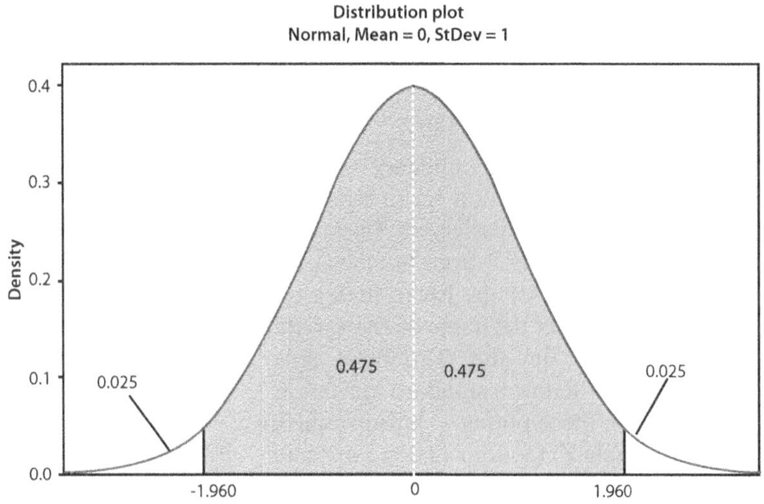

Image 7.8 Standard normal distribution with probability $P(-1.96<Z<1.96) = 0.95$ and values of $Z_{\alpha/2}$ for some confidence levels. (Source: Author's creation based on Minitab)

▪ **Solution**

The sample mean $(\bar{x})$ of these 20 weights was found to be 5.0010 pounds. The information given is $\sigma = 0.1164$ and $n = 20$ from a normal distribution.

- Step 1: Determine the critical value, $Z_{\alpha/2}$.

The value of $Z_{\alpha/2}$ selected for constructing a confidence interval is called the *critical value* for the distribution. There is a different critical value for each confidence level. The value of $Z_{\alpha/2}$ should be determined for the 95% confidence level. This value may be found from the standard normal distribution table or computed in Minitab. As the level of confidence is 95%, then $\alpha = 1 - 0.95 = 0.05$ and $\alpha/2 = 0.025$. As Image 7.8 shows, the value of $Z_{0.025}$ corresponding to an area of $0.95/2 = 0.4750$ from the center of the standardized normal distribution is 1.96 because there is 0.025 in the upper tail of the distribution, and the cumulative area less than $+Z_{0.025}$ is 0.975 (Image 7.8).

To calculate the value of $Z_{0.025}$ for standard normal distribution in Minitab, click on the following menus: Calc→Probability Distributions→Normal…. Then, in the next screen, select "Inverse Cumulative Probability"; enter 0 for "Mean," 1 for "Standard deviation," and 0.975 for "Input constant"; now click on OK. The result is shown in Session Window. The requested value is 1.95996, which means that $P(X \leq 1.96) = 0.975$. For the 95% confidence interval, Eq. 7.51 is

$$\bar{x} \mp 1.96\left(\sigma / \sqrt{n}\right)$$

- Step 2: Compute the standard error $(\sigma_{\bar{x}})$ and the margin of error.

$$\text{Standard error} = \sigma_{\bar{x}} = \frac{\sigma}{\sqrt{n}} = \frac{0.1164}{\sqrt{20}} = 0.0260$$

$$\text{Margin of error} = \text{ME} = Z_{\alpha/2}\left(\sigma/\sqrt{n}\right) = 1.96(0.026) = 0.0510$$

- Step 3: Estimate the confidence interval and interpret the results.

The 95% confidence interval for the mean blade weights is calculated using Eq. 7.52 as follows:

$$\bar{x} - Z_{\alpha/2}\left(\sigma/\sqrt{n}\right) < \mu < \bar{x} + Z_{\alpha/2}\left(\sigma/\sqrt{n}\right)$$

$$5.0010 - 0.0510 < \mu < 5.0010 + 0.0510$$

$$4.950 < \mu < 5.052$$

Using Minitab, this confidence interval estimate for the population mean when σ is known is obtained with selecting Stat→Basic Statistics→1-Sample Z... from the menu bar. The One-Sample Z for the Mean dialog box appears. Then, select the name of the variable "blade weights" in the Variables; enter 0.1164 in the "Known standard deviation" box; click on the Options... button and in the One-Sample Z: Options dialog box; enter 95.0 in the "Confidence Level" box. Click on OK to return to the previous dialog box. Click the OK. The results will appear on Session Window.

Thus, with 95% confidence, the population mean of blade weights is estimated to be between 4.95 and 5.05 pounds. Because 5.01 pounds, the value which indicates that the manufacturing process is working properly, falls within this interval, there is no reason to believe that anything is wrong with the manufacturing process.

One-Sided Confidence Interval for the Mean (σ Is Known)

A confidence interval gives likely minimum and maximum values for a population parameter. In some cases, only a maximum or minimum value would be considered by Six Sigma teams (e.g., the highest value of expenditures in a credit card or the lowest limit for the content of authentic drug in a crude product).

A one-side confidence interval for the population mean μ is 100(1 − α)% upper confidence limit, which has the form

$$\mu \le \text{Upper Confidence Limit} = \bar{x} + Z_{\alpha}\left(\sigma/\sqrt{n}\right) \quad (7.54)$$

or a 100(1 − α)% lower confidence limit, which has the form

$$\text{Lower Confidence Limit} = \bar{x} - Z_{\alpha}\left(\sigma/\sqrt{n}\right) \le \mu \quad (7.55)$$

▶ Example 25

The breaking strength of strand used in manufacturing cotton fabric is required to be at least 6.9 bars. From past experience, breaking strength is known to be normally distributed, and its standard deviation is 0.14 bars. A random sample of 10 specimens is tested, and the average breaking strength is found to be 6.76 bars. Construct a 95% lower confidence limit on the true mean breaking strength. ◀

▪ **Solution**

The purpose of this analysis is to generate a lower limit for the true mean breaking strength. The information given is $\bar{x} = 6.76$ bars, σ = 0.14, and n = 10 from a normal distribution. Using Eq. 7.55, a 95% lower confidence limit is obtained, when α = 1 − 0.95 = 0.05 and $Z_{0.05}$ = 1.64, as follows:

$$\bar{x} - Z_{\alpha}\left(\sigma/\sqrt{n}\right) \le \mu$$

$$6.76 - 1.64\left(0.14/\sqrt{10}\right) \le \mu$$

$$6.687 \le \mu$$

A 95% one-sided confidence interval is obtained by following the menus in Minitab: Stat→Basic Statistics→1-Sample Z.... Once the One-Sample Z for the Mean dialog box appears, set the upper box on the right-hand side to "Summarized data," and simply enter 10 in "Sample size," 6.76 in "Sample mean," and 0.14 in the "Known standard deviation" box. Then, click on Options... and in the One-Sample Z: Options dialog box; enter 95.0 in the "Confidence Level" box. In the "Alternative hypothesis" box, the default would be a two-sided interval, but select the relevant option for the one-sided version. As the lower confidence limit is required, select here mean → hypothesized mean option for this example. "Greater than" is used because the true population mean is assumed to be greater than the figure generated. Click on OK to return to the previous dialog box. Click on OK. The procedure then produces just one limit, as presented as "95% Lower Bound" in Minitab. The results appear on Session Window.

Thus, with 95% confidence, the population mean of breaking strength is unlikely to be smaller than 6.6872 bar, and no maximum is reported. This shows that the average breaking strength is expected to be at least 6.6872 bar and that the strand used in manufacturing fulfills the requirement of having at least 6.9 bar as strength.

Confidence Interval for the Mean (σ Is Unknown, Large Sample)

In reality, there would be no historical information about the population standard deviation, but the population mean is still required to be estimated. This may be the case if the mower blade is a new kind of blade and little experience has been gained in its production or if the average productivity (in units per hour) for operators using a particular manufacturing process is to be estimated. Thus, a random sample is now taken from a population with unknown mean μ and standard deviation σ. To develop a confidence interval for this situation, one of the two paths on ◘ Fig. 7.6 can be followed based on the sample size.

When the sample size n is large enough, the procedures developed for the case with population standard deviation σ known can be used, even if that population standard deviation has to be estimated from the sample. This case can also be described as a large-sample confidence interval. If the sample size n is large, the central limit theorem implies that $\bar{x}$ has approximately a normal distribution with mean μ and variance σ^2/n. Therefore $Z = (\bar{x} - \mu)/\left(\sigma/\sqrt{n}\right)$ has approximately a standard normal distribution. This ratio could be manipulated by replacing σ by the sample standard deviation s to produce an approximate confidence interval for μ. Thus, regardless of the shape of the population distribution, with a confidence level of $100(1-\alpha)\%$, a large-sample confidence interval for μ can be obtained approximately by

$$\bar{x} \pm z_{\alpha/2}\left(s/\sqrt{n}\right) \quad (7.56)$$

Generally, n should be at least 40 to use this result reliably. The central limit theorem generally holds for $n \ge 30$, but the larger sample size is recommended here because replacing σ by s in Z results in additional variability (Montgomery and Runger 2014: 279).

▶ Example 26

In a Six Sigma project, the procurement manager considers changing suppliers of a particular part that has a CTQ, hardness, and a proposed new supplier claims that they can maintain the hardness of 5 units on their product. A random sample of 50 of their product was selected from the lot they provided, and the hardness was measured. The hardness measurements resulted in a sample mean ($\bar{x}$) of 4.85 and a sample standard deviation (s) of 0.25. Set up a 99% confidence interval for the required (population) mean of hardness. Determine if the new supplier's claim is accurate. ◀

▪ **Solution**

For this example, the population standard deviation σ is unknown; however, the sample size is 50 and large enough, i.e., $n \ge 30$, to use the sample standard deviation s instead σ to compute the interval. The required confidence coefficient is 0.99. Then, $\alpha = 1 - 0.99 = 0.01$, $\alpha/2 = 0.005$ and $Z_{0.005} = 2.58$ from the standard normal distribution table (Table A.2). The general form of a large-sample 99% confidence interval for the mean of hardness is given as follows:

$$\bar{x} \pm z_{\alpha/2}\left(s/\sqrt{n}\right) = \bar{x} \pm 2.58\left(s/\sqrt{n}\right)$$

The values of the sample mean and sample standard deviation are $\bar{x} = 4.85$ and $s = 0.25$, respectively. Substituting these values into the formula, the approximate 99% confidence interval is

7

$$4.85 \pm 2.58\left(0.25/\sqrt{50}\right)$$

$$\left(4.759;4.941\right)$$

Using Minitab, the 99% confidence interval for the mean of hardness can be constructed by following this path Stat→Basic Statistics→1-Sample Z… from the menu bar, and after setting the upper box on the right hand side to "Summarized data," enter 50 in "Sample size," 4.85 in "Sample mean," and 0.25 in "Known standard deviation"; then click on "Options" to determine confidence level and alternative hypothesis. To form a 99% confidence interval with two tails, enter 99.0 in the "Confidence level" box, and select "mean≠hypothesized mean." Finally, click on OK on the Options box, and again Click on OK on Dialog Box to obtain results. The results appear on Session Window.

With 99% confidence, the interval (4.759, 4.941) encloses the true mean of hardness. As this interval does not include the target mean of 5 for the hardness, the new supplier's claim is incorrect. The procurement manager should not consider this candidate for a supplier of this particular part.

Confidence Interval for the Mean (σ Is Unknown, Small Sample)

The more practical situation with population standard deviation unknown will be introduced now based on a family of probability distributions called Student's t. In cases where σ is unknown, the random variable $Z = \left(\bar{x} - \mu\right)/\left(\sigma/\sqrt{n}\right)$ cannot be used directly to develop a confidence interval estimate for the mean. If the population has normal distribution and σ is replaced by the sample standard deviation s, a new random variable is given by

$$t = \frac{\bar{x} - \mu}{s/\sqrt{n}} \tag{7.57}$$

It is obtained as a result of the extra uncertainty caused by the replacement. The distribution of this random variable does not follow a standardized normal distribution, although its appearance is similarly a bell-shaped and symmetrical. A Student's t distribution is characterized by a quantity called $v = n - 1$ degrees of freedom. The number of degrees of freedom is the amount of information available for estimating, in addition to μ, the unknown quantity σ. As the number of degrees of freedom v increases, the t distribution gradually approaches the standard normal distribution until the two distributions are virtually identical.

To generate inferences based on the t distribution, the critical values should be determined for the appropriate degrees of freedom. As similar to $Z_{\alpha/2}$ value, $t_{(v=n-1),\alpha/2}$ is the value from the Student's t distribution for $n - 1$ degrees of freedom such that the upper tail area is $\alpha/2$. For example, a t value of 2.78 with 4 degrees of freedom means that the probability that t exceeds +2.78 is 0.025 or 2.5%. The t-value for a given tail area ($\alpha/2$) is greater than the z-value for the same area ($\alpha/2$) (◘ Image 7.9).

The 100(1 − α)% confidence interval for the mean with σ is unknown is given by

$$\bar{x} - t_{v,\alpha/2}\left(\frac{s}{\sqrt{n}}\right) \le \mu \le \bar{x} + t_{v,\alpha/2}\left(\frac{s}{\sqrt{n}}\right) \tag{7.58}$$

where n is the sample size, $\bar{x}$ is the sample mean, s is the sample standard deviation, and $t_{v,\alpha/2}$ is the critical value of the t distribution

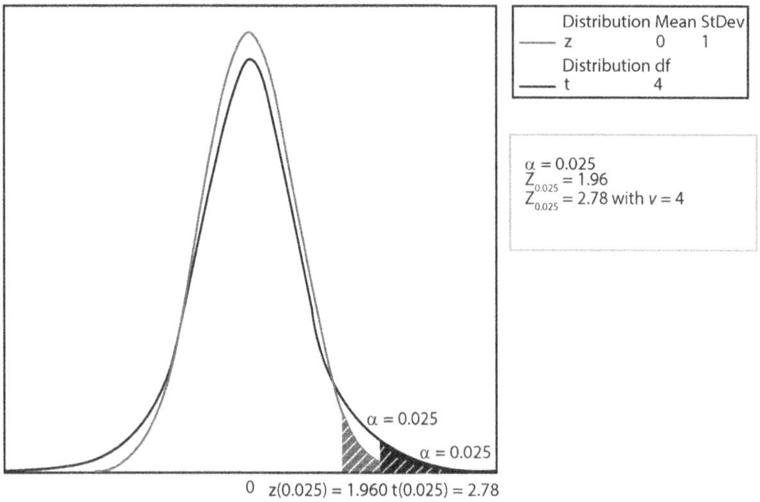

Image 7.9 Critical values of z and t ($df = 4$) for the same tail area $\alpha/2 = 0.025$. (Source: Author's creation based on Minitab)

with $v = n - 1$ degrees of freedom for an area of $\alpha/2$ in the upper tail.

To construct a confidence interval for the mean, the following steps are followed:
- Step 1: Check the assumptions.
- Step 2: Determine the critical value, $t_{\alpha/2}$.
- Step 3: Compute the standard error $(s_{\bar{X}})$ and the margin of error.
- Step 4: Estimate the confidence interval and interpret the results.

▶ Example 27

The environmental protection officer of a large industrial plant sought to determine the mean daily amount of sulfur oxides (a pollutant) emitted by the plant. Because measurement costs are high, only a random sample of 10 days' measurements was obtained. These emissions were 8, 7, 10, 15, 11, 6, 8, 5, 13, and 12 in tons per day. Estimate the mean amount of sulfur oxides emitted per day, using a confidence interval with a confidence coefficient of 0.95. ◀

Solution

The sample mean $(\bar{X})$ and sample standard deviation (s) of these ten sulfur oxide measurements are calculated as 9.5 and 3.24, respectively.

- Step 1: Check the assumptions.

The t distribution assumes that the population being studied is normally distributed. In practice, however, as long as the sample size is large enough and the population is not very skewed, the t distribution can be used to estimate the population mean when σ is unknown. When dealing with a small sample size and a skewed population distribution, the validity of the confidence interval should be considered primarily (Levine et al. 2015: 248). However, the assumption of normality in the population is assessed by evaluating the shape of the sample data through a histogram, stem-and-leaf diagram, box-and-whisker plot, or normal probability plot.

For this example, the normality of sample data is assessed using normal probability plot. To draw probability plot in Minitab, after entering data in worksheet, select Graph→ Probability Plot..., and choose "Single" and click on OK. Then, in the next screen, select variable from left box to "Graph variable"; click on OK. The probability plot will be shown in Session Window (Image 7.10). If the plotted points in a normal distribution plot lie either on or close to an imaginary straight line rising from the lower-left corner of the graph to the upper-right corner, then

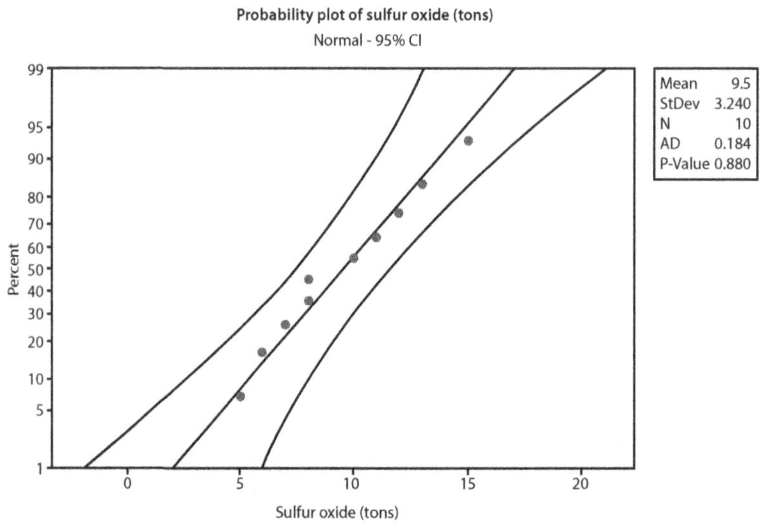

7

Image 7.10 Minitab "Probability Plot of Sulfur Oxide" for assessing normality. (Source: Author's creation based on Minitab)

the data are (at least approximately) normally distributed. On the other hand, if the plotted points deviate from this imaginary straight line in some patterned fashion, then the data are not normally distributed. As seen from Image 7.10, most of the data are in the straight line which indicates that the data are approximately normally distributed.

On this plot, there are some useful statistics to confirm this conclusion. Anderson Darling (AD) is a measure (test statistic) for assessing the fitness of a data set to a normal distribution. This normality test is conducted based on hypothesis testing procedures which will be discussed in the next section. The value of AD is used to draw a conclusion on whether the data come from a normal distribution. The AD test statistic (0.184) and the probability value (p-value) of this statistic are given on the output (Image 7.10). Simply, small p-values (less than $\alpha = 0.05$) mean that the data do not come from the normal distribution. The larger the p-value, the closer the data to the normal distribution. For this example, AD test statistic (AD = 0.184) confirms that the sulfur oxide amounts are normally distributed with a pretty large p-value = 0.88 > 0.05.

- Step 2: Determine the critical value, $t_{v,\alpha/2}$. The value of $t_{v,\alpha/2}$ is determined for the 95% confidence level and $v = n - 1 = 10 - 1 = 9$ degrees of freedom. This value is found from the t-table (Table A.4) or is computed in Minitab. As the level of confidence is 95%, then $\alpha = 0.05$ and $\alpha/2 = 0.025$. By looking up the column corresponding to 0.025 and row 9 from the Table A.4, the value of $t_{9;\,0.025}$ is found to be 2.262. This means that the probability that a Student's t random variable with 9 degrees of freedom exceeds 2.262 is 0.025.

To calculate the value of $t_{9;\,0.025}$ for Student t distribution in Minitab, click on the following menus: Calc→Probability Distributions→t.... In the dialog box, select "Inverse cumulative probability"; enter 0.0 in the "Noncentrality parameter[5]" box; enter 9 in the "Degrees

5 When your sampling distribution is derived from a normal distribution with a non-zero mean, you have a non-central distribution. Then non-centrality parameter helps define your non-central distribution and represents the degree to which the mean of the sampling distribution of the test statistic departs from its mean when the null is true. In this example, as the center of a t distribution is located on zero, this parameter is entered as "0.0."

of freedom" box; enter 0.975 in the "Input constant," and click on OK. The result will be shown in Session Window. The requested value is 2.26216, which means that $P(X \leq 2.26) = 0.975$.

- Step 3: Compute the standard error ($s_{\bar{x}}$) and the margin of error.

Standard error $= s_{\bar{x}} = \frac{s}{\sqrt{n}} = \frac{3.24}{\sqrt{10}} = 1.02$ and

$$\text{Margin of error} = \text{ME} = t_{v,\frac{\alpha}{2}}\left(\frac{s}{\sqrt{n}}\right)$$
$$= t_{9;0.025}\left(s/\sqrt{n}\right)$$
$$= 2.262(1.02) = 2.318$$

- Step 4: Estimate the confidence interval and interpret the results.

Based on the Eq. 7.58, the 95% confidence interval for the mean sulfur oxides are calculated as follows:

$$\bar{x} - t_{v,\alpha/2}\left(\frac{s}{\sqrt{n}}\right) \leq \mu \leq \bar{x} + t_{v,\alpha/2}\left(\frac{s}{\sqrt{n}}\right)$$

$$9.5 - 2.318 < \mu < 9.5 + 2.318$$

$$7.18 < \mu < 11.82$$

To obtain the confidence interval estimate of the population mean with unknown standard deviation in Minitab, follow the path Stat→Basic Statistics→1-Sample t... from the menu bar. Once the "One-Sample t for the Mean" dialog box appears, choose column "sulfur oxides" as sample data set, and click Options. Enter 95 in the Confidence Level box, and select mean≠hypothesized mean for Alternative Hypothesis. Click on OK on Options Box and Dialog Box to see the outcome on the Session Window.

Hence, with the 95% confidence, the mean amount of sulfur oxides emitted per day by the plant falls between 7.18 and 11.82 tons.

7.8.2.2 Confidence Interval Estimation for the Variance and Standard Deviation

In Six Sigma projects, understanding variability is pivotal to ensure that a process consistently produces output that meets desired specifications. If the variance of the processes, components, and products fluctuates significantly, then it is quite possible that undesirable defects will be detected throughout the system. Thus, obtaining point and interval estimates for the variance or standard deviation is critical.

Sample variance s^2 is the unbiased and most efficient estimator of population variance σ^2 (Barnes 1994: 163). The sampling distribution of s^2 is not normal but is related to a skewed distribution called the chi-square (χ^2) distribution. Suppose that a random sample of n observations from a normally distributed population with variance σ^2 is taken and the sample variance s^2 is calculated. Then the random variable

$$\chi^2_{n-1} = \frac{(n-1)s^2}{\sigma^2} \tag{7.59}$$

follows a chi-square distribution with (n – 1) degrees of freedom. This result forms the basis for the derivation of confidence intervals for the population variance when sampling from a normal distribution (Newbold et al. 2013: 332; Montgomery and Runger 2014: 283).

The 100(1 – α)% confidence interval for population variance σ^2 is given by

$$\frac{(n-1)s^2}{\chi^2_{n-1,\left(\frac{\alpha}{2}\right)}} < \sigma^2 < \frac{(n-1)s^2}{\chi^2_{n-1,\left(1-\frac{\alpha}{2}\right)}} \tag{7.60}$$

where s^2 is the sample variance from a random sample of n observations from a normal distribution; $\chi^2_{n-1,\left(\frac{\alpha}{2}\right)}$ and $\chi^2_{n-1,\left(1-\frac{\alpha}{2}\right)}$ are the upper and lower 100(α/2) percentage points

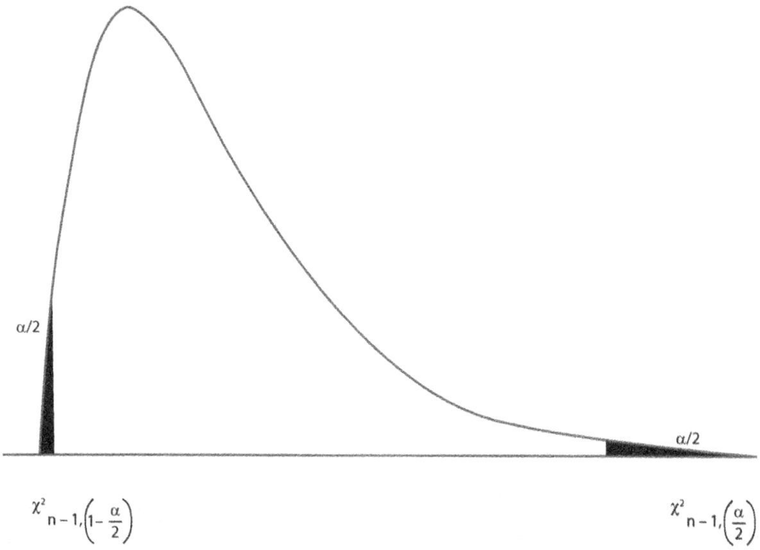

Image 7.11 Chi-square distribution for 100(1 − α)% confidence level and n − 1 degrees of freedom. (Source: Author's creation based on Minitab)

(critical values) of the chi-square distribution with (n − 1) degrees of freedom, respectively. These values are obtained from the chi-square table (Table A.3). The confidence interval is neither symmetric nor the shortest possible interval for this level of confidence (Image 7.11). In other words, this interval is not simply a point estimate plus or minus some number of standard errors. Because the χ^2 distribution is skewed to the right, the lower confidence limit is necessarily closer to the point estimate than the upper confidence limit.

If a confidence interval is desired for population standard deviation σ, simply taking the square root of the limits of (Eq. 7.60) will generate this interval.

To construct a confidence interval for the variance and standard deviation, the following steps are followed:

- Step 1: Check the assumptions.
- Step 2: Determine the critical values, $\chi^2_{n-1,\left(\frac{\alpha}{2}\right)}$ and $\chi^2_{n-1,\left(1-\frac{\alpha}{2}\right)}$.
- Step 3: Estimate the confidence interval and interpret the results.

▶ **Example 28**

The quality control manager of a manufacturing company randomly sampled 75 injection syringes from their third shift. When the manufacturing process is working well, the standard deviation in the syringe lengths should be no greater than 0.03 cm. The sample standard deviation is found to be 0.0212 cm. Find a 95% confidence interval for the population standard deviation in the lengths of injection syringes and decide whether the process produces the syringes consistently in the third shift. Assume that the syringe length is approximately normally distributed. ◀

▪ **Solution**

For this example, sample size n is 75 and sample standard deviation s is 0.0212.

- Step 1: Check the assumptions.

The validity of the interval estimate depends more critically on the assumption of normality. The following procedure is too risky to apply when the population distribution is not normal. Because the normality is assumed in this example, there is no need to check this assumption. Otherwise, before obtaining the interval estimate, a normality test should be performed on data as demonstrated in ► Example 27.

- Step 2: Determine the critical values, $\chi^2_{n-1,\left(\frac{\alpha}{2}\right)}$ and $\chi^2_{n-1,\left(1-\frac{\alpha}{2}\right)}$.

The values of $\chi^2_{n-1,(\alpha/2)}$ and $\chi^2_{n-1,(1-\alpha/2)}$ should be determined for the 95% confidence level and $v = n - 1 = 75 - 1 = 74$ degrees of freedom. These values may be found from the chi-square table or computed in Minitab.

As the level of confidence is 95%, then $\alpha = 0.05$ and $\alpha/2 = 0.025$. This is the probability in the right (upper) tail of the chi-square distribution. By looking up Table A.3, the column corresponding to 0.025 and row 74 gives the critical value of $\chi^2_{74,0.025}$ as 99.678. The probability is 0.025 that a chi-square random variable with 74 degrees of freedom is greater than 99.678. To find the value on the left tail, we subtract $\alpha/2$ from 1, i.e., $1 - \alpha/2 = 1 - 0.025 = 0.975$. The value of $\chi^2_{74,0.975}$ is 52.103 which corresponds to the column 0.975 and the row 74 in Table A.3.

To calculate the value of $\chi^2_{74,0.025}$ in Minitab, click on the following menus: Calc→Probability Distributions→Chi-Square.... In the next box, select "Inverse Cumulative Probability"; enter 0.0 in the "Noncentrality parameter" box; enter 74 in the "Degrees of freedom" box; enter 0.975 in the "Input constant" box because 0.025 is the area in the upper tail of the distribution, and then click on OK. The requested value is 99.678, which means that the cumulative area less than $\chi^2_{74,0.025}$=99.678 is 0.975.

- Step 3: Estimate the confidence interval and interpret the results.

Based on Eq. 7.60, the 95% confidence interval for the population variance in the lengths of injection syringes is calculated by substituting the relevant values as follows:

$$\frac{(n-1)s^2}{\chi^2_{n-1,\left(\frac{\alpha}{2}\right)}} < \sigma^2 < \frac{(n-1)s^2}{\chi^2_{n-1,\left(1-\frac{\alpha}{2}\right)}}$$

$$\frac{(75-1)(0.0212)^2}{99.678} < \sigma^2 < \frac{(75-1)(0.0212)^2}{52.103}$$

$$0.0003337 < \sigma^2 < 0.0006383$$

In this example, a 95% confidence interval is desired for the population standard deviation. Simply taking the square root of the limits of the interval obtained for the variance above yields the following confidence interval estimate for the population standard deviation in the lengths of injection syringes.

$$0.0183 < \sigma < 0.0253$$

To obtain 95% confidence interval for the standard deviation in the lengths of injection syringes in Minitab, select Stat→Basic Statistics→1 Variance... from the menu bar; a dialog box will appear. In the upper box on the right-hand side, Minitab provides two alternatives in the drop-down menu for the estimation which are formed based on either sample standard deviation or sample variance. Set this box to "Sample standard deviation" and enter the given figures into appropriate boxes. Then, click on Options, enter 95 for confidence level, and select "standard deviation≠hypothesized standard deviation" for Alternative hypothesis. Click on OK for Options Box and for "One-Sample Variance" Dialog Box to see the results. The Session Window will show the confidence interval estimation for the standard deviation.

The 95% confidence interval for the population standard deviation in the lengths of injection syringes is estimated between 0.0183 and 0.0253 cm. A 95% confidence interval limit for the variance (0.0003; 0.0006) is the square of those for the standard deviation. With 95% confidence, the process produces the syringes consistently in the third shift. It is because there is a very low probability (less than 0.025) that the population standard deviation will exceed 0.03 cm.

7.8.2.3 Confidence Interval Estimation for the Proportion (Large Sample)

The concept of the confidence interval is extended to categorical variables having only two possible outcomes, such as good or bad, yes or no, defective or non-defective, and go or no-go. In such situations, the main concern is to estimate the proportion of observations in a population that has a certain characteristic, which is also called a proportion of successes, p, for a binomial population. Examples include determining the frequency of occurrence of various types of errors that violate a policy or determining the proportion of products that have a nonconforming attribute.

The best point estimator for the population proportion is the statistic $\hat{p} = x/n$, where x is the number of desired characteristic in a sample of size n. The binomial distribution or hypergeometric distribution is the correct probability distribution to use in constructing confidence interval estimates of proportion; however, the calculations are complicated and time-consuming. Instead, for large n and p that are not near 0 or 1, the random variable

$$Z = \frac{\hat{p} - p}{\sqrt{\hat{p}(1-\hat{p})/n}} \qquad (7.61)$$

has, to a close approximation, a standard normal distribution, $N(0,1)$. If n is large enough, a $100(1 - \alpha)\%$ confidence interval for the population proportion p is given by

$$\hat{p} - Z_{\alpha/2}\sqrt{\hat{p}(1-\hat{p})/n} < p < \hat{p} + Z_{\alpha/2}\sqrt{\hat{p}(1-\hat{p})/n}. \qquad (7.62)$$

Therefore, if the number of successes (p) and failures ($1 - p$) is sufficiently large, the normal distribution provides an excellent approximation for the binomial distribution. The rule of thumb suggests that both $n\hat{p}$ and $n(1-\hat{p})$ should be greater than 10, i.e., $n\hat{p} \geq 10$ and $n(1-\hat{p}) \geq 10$. However, if the sample size is not large or the percentage of successes is either very low or very high, then the exact solution utilizes the binomial distribution rather than the normal distribution. Fisher and Yates (1957) tabulated the exact confidence intervals for various sample sizes and proportions of successes (Levine et al. 2015: 257) which are not in the scope of this chapter. Further, Minitab allows an exact or approximate solution for the analysis.

To construct a confidence interval for the proportion, the following steps are followed:

- Step 1: Check the assumptions.
- Step 2: Determine the critical value, $Z_{\alpha/2}$.
- Step 3: Compute the standard error ($s_{\hat{p}}$) and the margin of error.
- Step 4: Estimate the confidence interval and interpret the results.

▶ Example 29

When choosing a product to purchase, what do you consider most: price or quality? Assume that a supplier of industrial equipment conducted a customer survey for one of its products in its principal marketing regions. In the survey, 130 out of 300 customers claim that they mainly base their buying decisions on price. Construct a 99% confidence interval for the true percentage of customers who base their buying decisions more on price than quality. ◀

▪ Solution

It is known that, of 300 customers, 130 of them consider price when making buying decisions. The sample size is 300 and therefore, the sample proportion is $\hat{p} = \frac{x}{n} = \frac{130}{300} = 0.43$.

- Step 1: Check the assumptions.

To use the normal approximation to solve this question, the only criterion is a large enough sample. The sample size criterion, $n\hat{p} \geq 10$ and $n(1-\hat{p}) \geq 10$, must be checked before applying the procedure to form the confidence interval. The sample proportion $\hat{p}$ is 0.43 and the sample size n is 300; then

$$n\hat{p} = 300 * 0.43 = 129 \geq 10$$

and

$$n(1-\hat{p}) = 300 * 0.57 = 171 \geq 10$$

which satisfy of being n large enough. Instead of the binomial distribution, the normal distribution provides an approximate interval estimate for the true proportion of customers buying based on price.

- Step 2: Determine the critical value, $Z_{\alpha/2}$.

As the level of confidence is 99%, then $\alpha = 0.01$ and $\alpha/2 = 0.005$. The value of $Z_{0.005}$ is 2.58 from Table A.2.

- Step 3: Compute the standard error ($s_{\hat{p}}$) and the margin of error.

$$\text{Standard error} = s_{\hat{p}} = \sqrt{\hat{p}(1-\hat{p})/n}$$
$$= \sqrt{0.43(1-0.43)/300}$$
$$= 0.029$$

$$\text{Margin of error} = \text{ME} = Z_{\alpha/2}\sqrt{\hat{p}(1-\hat{p})/n}$$
$$= Z_{0.005}\sqrt{\hat{p}(1-\hat{p})/n}$$
$$= 2.58(0.029) = 0.074$$

- Step 4: Estimate the confidence interval and interpret the results.

Based on Eq. 7.62, the 99% confidence interval for the true percentage of customers who base their buying decisions more on price than quality is calculated as follows:

$$\hat{p} - Z_{\alpha/2}\sqrt{\hat{p}(1-\hat{p})/n} < p < \hat{p} + Z_{\alpha/2}\sqrt{\hat{p}(1-\hat{p})/n}$$

$$0.43 - 0.074 < p < 0.43 + 0.074$$

$$0.35 < p < 0.50$$

To calculate the confidence interval in Minitab, select Stat→ Basic Statistics→ 1 Proportion.... Then, select "Summarized Data" and input the sample information. It is known that 130 customers out of 300 consider price when making buying decisions. Therefore, enter 130 in the "Number of events" box and 300 in the "Number of trials" box. Click on Options to choose the solution method. As the sample size is large enough for this example, an approximate solution is obtained by setting 99% in the "Confidence level" box, Proportion≠hypothesized proportion as "Alternative Hypothesis," and Normal approximation in the "Method" box. Click on OK for Options and then for Dialog Box. The result is shown in Session Window.

The 99% confidence interval for the true proportion of customers is ranging from 0.36 to 0.51 approximately. Strictly speaking, in the long run, approximately 43% (with a 7% margin of error at the 99% confidence level) of the population of all customers of this supplier make their buying decisions based on price than quality.

7.8.3 Tolerance Interval Estimation

Six Sigma practitioners generally require that the output from a process has measurements of CTQs (e.g., tensile strength, viscosity, length, etc.) that fall within the specification limits. Specification limits, in other words, engineering specifications, are not determined by sampling the process. As previously discussed, instead, acceptable values of the CTQs are set by customers, standard setting institutions (e.g., ISO and NIST), and/or product/service designers and engineers. Even if the process is deemed to be in control, however, a single manufactured item may not always meet the specifications (Sincich 1996: 912). Therefore, in addition to process control, it is often important to predict a range of likely individual measurements of CTQs with a high degree of confidence. Such an interval that includes a certain percentage of measurements with a known probability is called tolerance interval. The endpoints of this interval are called tolerance limits. Tolerance intervals are identical to the confidence intervals, but focus is on capturing a proportion of measurements in a population rather than a population parameter.

A confidence interval's width is affected by the sampling error and population variance. If the sample size is closer to the entire population, the confidence interval narrows, and the estimated percentiles approximate the true population percentiles (The Minitab Blog 2017). For example, to determine where

90% of the population falls, data values corresponding to the 5th and 95th percentiles are estimated. As these percentile estimates have sampling error, the team cannot be 100% confident that a tolerance interval truly contains the specified proportion γ. Therefore, tolerance intervals have a confidence level.

Suppose that a manager in a hospital pharmacy wants to establish tolerance limits for 99% of the time required to fill a certain type of prescription, using a 95% confidence interval. Here, the confidence coefficient is $1 - \alpha = 0.95$, and the proportion of the fill times that the manager wants to capture is $\gamma = 0.99$. The confidence coefficient, 0.95, means that approximately 95 out of every 100 similarly constructed tolerance intervals will contain 99% of the fill times in the prescription process.

7

When the population of measurements that characterize the product is normally distributed with known mean μ and standard deviation σ, tolerance limits are easily constructed. In fact, such an interval is a 100% tolerance interval, i.e., the confidence coefficient is 1.0. Suppose that the measurements taken from a process have a normal distribution with $\mu = 5$ and $\sigma = 1$. From the standard normal distribution, 99% of these measurements will fall within $z = 2.58$ standard deviations of the mean. In other words, as the level of confidence is 99%, then $\alpha = 0.01$ and $\alpha/2 = 0.005$. The value of $z_{0.005}$ is 2.58 from the z-table in Table A.2. Therefore, a 100% tolerance interval for 99% of the measurements is $\mu \pm 2.58\sigma = 5 \pm 2.58(1) = (2.42 \text{ and } 7.58)$ which means that, with 100% confidence, 99% of the measurements will fall between 2.42 and 7.58.

In practice, Six Sigma practitioners will rarely know the true values of μ and σ, but they can obtain corresponding sample estimates, i.e., $\bar{x}$ and s, respectively. In this case, a tolerance interval for capturing at least γ% of the values in a *normal distribution* with confidence level $100(1 - \alpha)$% is given by

$$\left(\bar{x} - ks, \bar{x} + ks\right) \tag{7.63}$$

where k is a tolerance interval factor and is in Tables A.12 and A.13 (Table K) based on the values of γ, the confidence coefficient $(1 - \alpha)$, and the sample size n.

If the normality assumption is not met, a "distribution-free" or non-parametric method is used for setting a tolerance interval. A non-parametric tolerance interval uses only the smallest and largest measurements in the sample data, and this tolerance interval is applied to any set of CTQs, regardless of the shape of the probability distribution. Thus, in a sample of size n from *any distribution*, the smallest (x_{min}) and largest (x_{max}) observations form a $100(1 - \alpha)$% tolerance interval for at least $100(\gamma)$% of the population. The values of n are determined by the values of γ and the confidence coefficient $(1 - \alpha)$. Therefore, if γ is the minimum proportion of the population contained between the smallest and largest sample observations with confidence $(1 - \alpha)$, n is approximately determined by (Montgomery and Runger 2014: 304)

$$n = \frac{1}{2} + \left(\frac{1+\gamma}{1-\gamma}\right)\left(\frac{\chi^2_{\alpha,4}}{4}\right). \tag{7.64}$$

▶ **Example 30**

To be 95% confident that at least 90% of the population will be included between the extreme values of the sample, what sample size will be required? ◀

- **Solution**

For the values of $(1 - \alpha) = 0.95$ and $(\gamma) = 0.90$, Eq. 7.64 yields

$$n = \frac{1}{2} + \left(\frac{1+\gamma}{1-\gamma}\right)\left(\frac{\chi^2_{\alpha,4}}{4}\right)$$

$$n = \frac{1}{2} + \left(\frac{1+0.90}{1-0.90}\right)\left(\frac{\chi^2_{0.05,4}}{4}\right)$$

$$n = 45.5775$$

where the value of $\chi^2_{0.05,4}$ is 9.49 from χ^2 distribution table in Table A.3. Then, the required sample size is $n = 46$. It means that, if a random sample of size of 46 is taken, the smallest and largest measurements in this sample will represent the lower and upper tolerance limits, respectively, for at least 90% of the measurements with 95% confidence.

To construct a tolerance interval, the following steps are followed:

- Step 1: Check the assumptions.
- Step 2: Determine the tolerance interval factor k.
- Step 3: Estimate the tolerance interval and interpret the results.

▶ Example 31

Let's reconsider the hardness CTQ and measurements in ▶ Example 26. A random sample of 50 of the supplier product was selected from the lot they provided, and the hardness was measured as shown in ◻ Table 7.5. Set up a 99% tolerance interval for the measurements in hardness process that has confidence level 99%. ◀

Solution

The true values of μ and σ for the hardness measurements are unknown, but corresponding sample estimates resulted in a sample mean $\bar{x}$ of 4.853 and a sample standard deviation s of 0.248, respectively.

- Step 1: Check the assumptions.

For identifying the appropriate procedure to obtain a tolerance interval, first the normality assumption should be assessed. Tolerance interval is very sensitive to the normality assumption. Normal probability plot shows that the sample data for hardness is approximately normal because the values are almost lying on the straight line. There is no evidence for the violation of normality. The Anderson Darling (AD) normality test also confirms this conclusion with AD = 0.418 (p-value = 0.318 > 0.05).

- Step 2: Determine the tolerance interval factor k.

As the normality assumption has been satisfied, the procedure with Eq. 7.63 is followed to form the required tolerance interval. The tolerance factor, k, is determined for $n = 50$, $\gamma = 0.99$, and 99% confidence. Looking up these parameters in Table A.13, the tolerance factor is $k = 3.385$.

- Step 3: Estimate the tolerance interval and interpret the results.

Based on Eq. 7.63, the tolerance interval for the measurements in hardness process that includes 99% of the values in the population with 99% confidence is formed as follows:

$$\left(\bar{x} - ks,\ \bar{x} + ks\right)$$

$$\left(4.853 - 3.385(0.248),\ 4.853 + 3.385(0.248)\right)$$

or

$$\left(4.01,\ 5.69\right).$$

To obtain the Tolerance Intervals in Minitab, select following steps on menu bar: Stat→Quality Tools→ Tolerance Intervals (Normal Distribution).... On the opening dialog box, select the sample "hardness" from the Variables box, and set 99% in the "Minimum percentage of population in interval:" box. Click on Options..., and set 99% in the "Confidence level" box and Two-Sided in "Tolerance Interval" box. From Graphs..., mark "the Tolerance interval plots." Click on OK. The resulting graphical summary is shown in Session Window. The graph presents tolerance interval estimation for both a normal and any shape of distribution, as well as tools for assessing the normality for hard-

◻ **Table 7.5** Hardness measurements of supplied parts

4.60871	5.10742	4.54217	5.28888	4.58794	5.10228	5.21343	5.05593	5.06924	4.99318
4.64855	4.36941	5.14516	4.71636	5.23959	4.74204	4.64849	4.53822	5.15963	4.78726
5.05996	4.79383	4.79702	5.24392	5.24695	4.52660	4.43782	4.70624	4.91439	4.89560
4.38128	4.89481	4.97903	4.93652	4.54319	4.81917	5.09362	4.63277	5.13136	4.67362
4.63944	4.96979	5.01294	4.71409	4.90951	4.56838	4.73537	4.88745	4.95919	4.96226

Source: Author's creation

ness measurements. There is no evidence for the violation of normality as mentioned above ($AD = 0.418, p = 0.318 > 0.05$). Consequently, the first tolerance interval formed based on the normal distribution [4.012, 5.693] is considered. It can be said with 99% confidence that at least 99% of the hardness measurements lie between 4.01 and 5.69, the lower and upper tolerance limits, respectively.

The information provided by tolerance intervals is often used to determine whether product specifications (or customer requirements) are being satisfied. To determine whether the specifications are realistic, the specification limits are compared to the "natural" tolerance limits of the process, *UNTLs* and *LNTLs*, i.e., the tolerance limits obtained from sampling. If the tolerance limits do not fall within the specification limits, a review of the production process is strongly recommended. An investigation may reveal that the specifications are tighter than necessary for the functioning of the production and, consequently, should be widened. Or, if the specifications cannot be changed, a fundamental change in the production process may be necessary to reduce product variability (Sincich 1996: 915).

Assume that the specification limits were set as (4.78, 5.32) for the hardness characteristic by the customers. When the tolerance limits [4.01, 5.69] are compared to these specification limits, it is clearly seen too much product variation for the new supplier as having tolerance limits wider than the customer specifications. Therefore, the new supplier is not a promising one and needs to review its processes to lower this excessive variability.

7.8.4 Prediction Interval Estimation

Another useful interval estimate is the prediction interval which provides a range for bounding the value of a single future observation (X_{n+1}) from the same population. This is different from a confidence interval, which estimates a population parameter. A confidence interval is associated with the sampling distribution of a statistic, whereas a prediction interval is associated with the distribution of the random variable itself. Prediction interval is also different from a tolerance interval bounding unknown elements of a distribution. In cases with normally distributed data and a large sample size, the prediction interval tends to the tolerance interval constructed with a known mean μ and standard deviation σ. However, for more modest sample sizes, the prediction and tolerance intervals are different.

A key aspect of the prediction interval is that it is calculated before a replication study is conducted creating an a priori statement of expectations (from Immanuel Kant 1781). An a priori statement of expected results enables planned replications and replication proposals to specify, in advance, the criteria that will be used to evaluate the replication, independent of the results of the replication (Spence and Stanley 2016: 3).

When the population standard deviation is unknown, a $100(1-\alpha)\%$ prediction interval for a single future observation (X_{n+1}) from a normal distribution is given by (Montgomery and Runger 2014: 329)

$$\bar{x} - t_{\frac{\alpha}{2},n-1}\, s\sqrt{1+\frac{1}{n}} \le X_{n+1} \le \bar{x} + t_{\frac{\alpha}{2},n-1}\, s\sqrt{1+\frac{1}{n}} \tag{7.65}$$

The prediction interval is wider than the confidence interval given in Eq. 7.58 by virtue of the additional value of 1 under the square root. This is because, in addition to estimating the population mean, the variability of the new observation around the mean must also be taken into account. Although the t distribution-based confidence interval for μ is robust to the normality assumption when n is small, on the other hand, the prediction intervals are very sensitive to the normality assumption, and Eq. 7.65 should not be used unless being very comfortable with the normality assumption.

To construct a prediction interval, the following steps are followed:

- Step 1: Check the assumptions.
- Step 2: Find the critical value of $t_{\alpha/2, n-1}$.
- Step 3: Estimate the prediction interval and interpret the results.

▶ Example 32

Consider again the hardness measurements in ▶ Examples 26 and 31. A random sample of 50 of the supplier product was selected from the lot they provided, and the hardness was measured. The hardness measurements resulted in a sample mean of 4.85 and a sample standard deviation of 0.25. The 99% confidence interval on the population mean of hardness was [4.7089, 4.8911] and doesn't satisfy this key characteristic requirement (i.e., 5) of the part. Now, the procurement manager plans to test one more part from the lot that new supplier provided. Set up a 99% prediction interval on the hardness measurement of this new observation. ◀

▪ Solution

In this question, the procurement manager plans to conduct a replication study with one more part from the lot that new supplier provided and wants to know what range of hardness values she can reasonably expect. This question is answered by computing a prediction interval. Using the sample size $n = 50$ and the sample mean $\bar{x} = 4.85$, 99% prediction interval for the new hardness data will be estimated based on Eq. 7.65. As the population standard deviation is unknown, its estimate is obtained as $s = 0.25$. The required steps are given below.

- Step 1: Check the assumptions.

Prediction interval requires the normally distributed data. This assumption was checked in ▶ Example 31 for the hardness data set and found that there is no evidence for the violation of normality $(\text{AD} = 0.418, p = 0.318 > 0.05)$.

- Step 2: Find the critical value of $t_{\alpha/2,\, n-1}$.

We need two-tail t-value for $\alpha = 0.01$ significance level and $df = 49$ degrees of freedom, that is, $t_{\frac{\alpha}{2},n-1} = t_{0.005,49} = 2.68$.

- Step 3: Estimate the prediction interval and interpret the results.

Based on Eq. 7.65, the prediction interval for the measurements in hardness process with 99% confidence and substituting relevant values is formed as follows:

$$\bar{x} - t_{\frac{\alpha}{2},n-1}\, s\sqrt{1+\frac{1}{n}} \le X_{n+1} \le \bar{x} + t_{\frac{\alpha}{2},n-1}\, s\sqrt{1+\frac{1}{n}}$$

$$4.85 - 2.68 * 0.25 * \sqrt{1+\frac{1}{50}} \le X_{51} \le 4.85 + 2.68 * 0.25 * \sqrt{1+\frac{1}{50}}$$

$$4.173 \le X_{51} \le 5.527$$

In this question, hardness is measured with 50 sample values. Given the whole hardness values that were already observed, the hardness of the 51st *new sample* will be between 4.173 and 5.527 with a confidence of 99%. In general, if we would repeat our sampling process infinitely, 99% of such constructed prediction intervals would contain the new hardness measurement. Thus, the prediction interval constructed here indicates that, due to chance (i.e., sampling error), the procurement manager can expect to see a hardness measurement as low as 4.173 or as high as 5.527 in the replication study.

7.9 Inferential Statistics: Hypothesis Testing for a Single Population

By Dr. Aysun Kapucugil Ikiz
Associate Professor
Dokuz Eylul University
Dr. Fatma Pakdil
Eastern Connecticut State University

As stated in previous sections, there are two general methods for making inferences about population parameters: a Six Sigma team can *estimate* their values using confidence intervals or can *make decisions* about them. Making decisions about specific values of the population parameters – testing hypotheses about these values – is the topic of this section.

A statistical hypothesis is a claim or assertion about one or more parameters or the form of the population distribution. Because it is too expensive or too impractical and not to mention theoretically impossible to collect population data, the Six Sigma team will make decisions based on sample data. When the team is dealing with sample data, there will be some uncertainty about the population's true parameters. *Hypothesis testing* helps us make fact-based decisions about whether there are different population parameters or if the differences are simply due to expected sample variation. In a Six Sigma project, this turns into the question: How will the Six Sigma team know that their key process output variable really changed and that it is not just an unrepresentative sample?

7.9.1 Concepts and Terminology of Hypothesis Testing

Hypothesis testing starts with a claim or question about a particular CTQ characteristic analyzed by the Six Sigma team. The parameters of the CTQ characteristics help the Six Sigma team test these claims or questions developed by the team. Hypothesis testing helps Six Sigma teams in many types of data analysis procedures in the Analyze phase of DMAIC.

Hypotheses should be about parameters, not statistics. The hypothesis testing concept is implemented in statistical analyses to see whether these claims or questions are accepted or rejected, using statistics. In some cases, hypothesis testing is used to compare populations using parameters. Hypothesis testing procedures are constructed upon several components:

1. Assumptions and conditions
2. Formulation of null and alternative hypotheses
3. Decisions and errors in a hypothesis test
4. Test statistics and rejection regions
5. Reporting test results: *p*-values.

7.9.1.1 Assumptions and Conditions

There are several assumptions and conditions that need to be checked and met prior to starting hypothesis testing.

- The *independence assumption* requires that the sampled values are independent of each other.
- The *sample size assumption* means that the sample size is large enough ($n \geq 30$). This assumption is related to the *normal population assumption* or *nearly normal condition*. The normality assumption is important when the sample size is small ($n \leq 30$). When the sample size is greater than 30, the central limit theorem approximates the data to normality. Six Sigma teams need to make sure that the data are normally distributed. Any violation of the normality assumption should be checked before proceeding in hypothesis testing. Skewness or outliers may contribute to violating this assumption.
- The *randomization condition* requires that the sample is randomly selected.
- The *10% condition* says that if samples are not selected with replacement, the sample size should be no larger than 10% of the population.
- The success/*failure condition* in proportion tests expects that the sample size must be big enough so that numbers of "successes" (np) and "failures" ($np-(1\ p)$) are at least 10.

7.9.1.2 Formulation of Null and Alternative Hypotheses

After checking assumptions and conditions, a set of two hypotheses is articulated by the team in hypothesis testing: the null hypothesis and alternative hypothesis. The *null hypothesis* articulates the assertion that is initially assumed to be true. It generally states that there are no differences between the parameters of the populations being compared. The null hypothesis may also claim that the parameter equals a numerical value. The null hypothesis is denoted with H_0, specifies a population model parameter, and proposes a value for that parameter.

After articulating the null hypothesis, the alternative hypothesis is developed. An alternative hypothesis is the claim that is contradictory to the null hypothesis. The *alternative hypothesis* is denoted with H_1 and is accepted if the null hypothesis is rejected.

A critical part of hypothesis testing is to correctly identify the parameters and their values in the null and alternative hypotheses. The target values of the CTQs in design specifications or previous performance of the parameters may be helpful for identifying the values of the parameters. The null hypothesis statement always includes an "equal sign" for the value of the parameters, while the alternative hypothesis does not. For example, the parameter of the CTQ characteristic may be assumed to equal a certain numerical value, such as 20-minute processing time (μ = 20 minutes), in the null hypothesis. The alternative hypothesis would be the opposite: processing time is not 20 minutes.

The alternative hypotheses can be two-tailed or one-tailed. In a *two-tailed alternative hypothesis*, we are equally interested in deviations on either side of the null hypothesis value. An alternative hypothesis that focuses on deviations from the null hypothesis value in only one direction is called *one-tailed alternative hypothesis*.

▶ Example 33

In a manufacturing firm, one of the CTQ characteristics of a round metal part is the diameter. The target value of the CTQ has been determined as 0.5 cm. The Six Sigma team is concerned that the mean of the CTQ of the last 200 batches that include 1200 parts/batch did not meet the target value over the last 2 weeks. State the null and alternative hypotheses. ◀

▪ Solution

In the null hypothesis, we expect to find that the mean of the population equals 0.5 cm. Therefore, we can articulate null hypothesis as follows:

$$H_0 : \mu = 0.5 \text{ cm}$$

For the alternative hypothesis, we have three ways to articulate it. For example, although it is not given in the question, the mean of the population may be "not equal," "less," or "greater" than the target value. Therefore, we can develop three different alternative hypotheses as follows:

$$H_1 : \mu \neq 0.5 \text{ cm}$$

$$H_1 : \mu > 0.5 \text{ cm}$$

$$H_1 : \mu < 0.5 \text{ cm}$$

The first alternative hypothesis is two-tailed, whereas the last two alternative hypotheses are one-tailed ones. After articulating null and alternative hypotheses, we can construct three sets of hypotheses for testing as shown below:

Two-tailed hypothesis testing

$$H_0 : \mu = 0.5 \text{ cm}$$
$$H_1 : \mu \neq 0.5 \text{ cm}$$

One-tailed hypothesis testing

$$H_0 : \mu = 0.5 \text{ cm}$$
$$H_1 : \mu > 0.5 \text{ cm}$$

or

$$H_0 : \mu = 0.5 \text{ cm}$$
$$H_1 : \mu < 0.5 \text{ cm}$$

▶ Example 34

Let's recall the Six Sigma team's concern given in ▶ Example 33. The CTQ characteristic of a round metal part is the diameter. The target value of the CTQ has been determined as 0.5 cm. The team is concerned that the mean of the CTQ of the last 200 batches that include 1200 parts/batch is *less* than target value over the last 2 weeks. State the null and alternative hypotheses. ◀

▪ Solution

In the null hypothesis, we claim that the mean of the population equals 0.5 cm. Therefore, we can articulate the null hypothesis as follows:

$$H_0 : \mu = 0.5 \text{ cm}$$

In the alternative hypothesis, we want to test whether the mean of the diameters is *less* than 0.5 cm as follows:

$$H_1 : \mu < 0.5 \text{ cm}$$

▶ **Example 35**

Let's recall the Six Sigma team's concern given in ▶ Example 33. The CTQ characteristic of a round metal part is the diameter. The target value of the CTQ has been determined as 0.5 cm. The team is concerned that the mean of the CTQ of the last 200 batches that include 1200 parts/batch is *greater* than target value over the last 2 weeks. State the null and alternative hypotheses. ◀

■ **Solution**

In the null hypothesis, we claim that the mean of the population equals 0.5 cm. Again, we can articulate null hypothesis as follows:

$$H_0 : \mu = 0.5\ \text{cm}$$

In the alternative hypothesis, we want to test whether the mean of the diameters is *greater* than 0.5 cm as follows:

$$H_1 : \mu > 0.5\ \text{cm}$$

7.9.1.3 Decisions and Errors in a Hypothesis Test

The hypothesis testing may conclude in two ways: (1) fail to reject the null hypothesis and (2) reject the null hypothesis. First, the searchers may fail to reject the null hypothesis. If the sample does not strongly contradict the null hypothesis, it is believed that the null hypothesis is true. Second, the null hypothesis is rejected if the sample evidence does support that the null hypothesis is false. In other words, rejecting the null hypothesis means that we have statistical evidence indicating that the alternative hypothesis is true. If the null hypothesis is not rejected, it shows that the alternative hypothesis has not been proved.

Testing a hypothesis using statistical methods is equivalent to making an educated guess based on the probabilities associated with being correct. When a Six Sigma team makes a decision based on a statistical test of a hypothesis, the team can never know for sure whether the decision is right or wrong, because of sampling variation. In hypothesis testing, two types of error occur: Type I error and Type II error. These two errors can be analyzed using level of significance and the power of the test. Six Sigma teams aim to minimize the chance of committing either type of error. The probabilities of these two errors are

$$\begin{aligned}\alpha &= P\{type\ I\ error\} \\ &= P\{reject\ H_o | H_o\ is\ true\}\end{aligned}$$

$$\begin{aligned}\beta &= P\{type\ II\ error\} \\ &= P\{fail\ to\ reject\ H_o | H_o\ is\ false\}\end{aligned}$$

Type I error occurs when the null hypothesis is true, but it is rejected in hypothesis testing. Type I error is known as α and equals the *level of significance* of the test. The level of significance may be 0.01, 0.05, and 0.10, depending on how much risk the Six Sigma team may be willing to accept that the team is wrong when the null hypothesis is rejected. Prior to running hypothesis testing, the significance level is determined by the team. Type I error, α, is also known as *producer's risk*, since type I error refers to the probability that a batch is rejected when it is acceptable within the supply chain. Rejecting that batch is a cost to the producer of the batch, since the producer will run additional quality control tests and inspections after the batch is rejected and returned by the customer.

Type II error occurs when the null hypothesis is false, but the hypothesis testing fails to reject it. The power of a statistical test determines the probability of Type II error, which is known as β. When the sample size of the test is large enough, Type II error decreases. Type II error is also known as *consumer's risk* since type II error refers to the probability that a batch is not rejected because of poor quality when it should be rejected and returned to the producer. Failing to reject that batch will be a cost to the consumer of the batch, since the consumer takes a significant risk for its own processes, operations, and ultimately finished goods. The probability of rejecting the null hypothesis when it is false equals $1 - \beta$, which is known as the power of a statistical test. The power of a statistical test can be denoted as follows:

$$\text{Power} = 1 - \beta = P\{reject\ H_o | H_o\ is\ false\}$$

▶ Example 36

Let's recall ▶ Example 33. The CTQ characteristic of a round metal part is the diameter. The target value of the CTQ has been determined as 0.5 cm. The Six Sigma team is concerned that the mean of the CTQ of the last 200 batches that include 1200 parts/batch did not meet the target value over the last 2 weeks. Identify Type I and Type II errors in this example. ◀

▪ Solution

Let's recall the null and alternative hypotheses articulated in ▶ Example 33.

$$H_0 : \mu = 0.5 \text{ cm}$$

$$H_a : \mu \neq 0.5 \text{ cm}$$

A Type I error occurs when a Six Sigma team rejects the null hypothesis and decides that the mean of the diameters is not equal to 0.5 cm while the mean equals 0.5 cm. Type I error refers to a wrong decision made by the team. A Type II error occurs when a Six Sigma team fails to reject the null hypothesis when it should be rejected. In other words, the Six Sigma team will falsely accept that the mean of the diameters equal 0.5 cm when it is not.

7.9.1.4 Test Statistics and Rejection Regions

In hypothesis testing, a *test statistic* is a function of the data computed and used to decide which hypothesis will be true. The hypotheses are tested using randomly selected samples from the population. After selecting the sample and collecting data, the sampling distribution of the parameter of interest is identified and the appropriate test statistic is computed based on the sample. For example, the sampling distribution of the test statistic for testing the population mean is generally expected to have a normal distribution or t distribution. When the sample size is large

$$Z = \frac{\bar{X} - \mu}{\sigma / \sqrt{n}} \sim N(0,1) \tag{7.66}$$

where Z is the test statistic normally distributed with zero mean and 1 standard deviation, $\bar{X}$ is the mean of the sample, μ is the mean of the population, σ is the standard deviation of the population, and n is the sample size. The type of test statistic varies based on the distribution type of the data and hypotheses tested as shown in ◘ Fig. 7.7. For example, an appropriate test statistic for testing $H_0 : \mu = \mu_0$ is

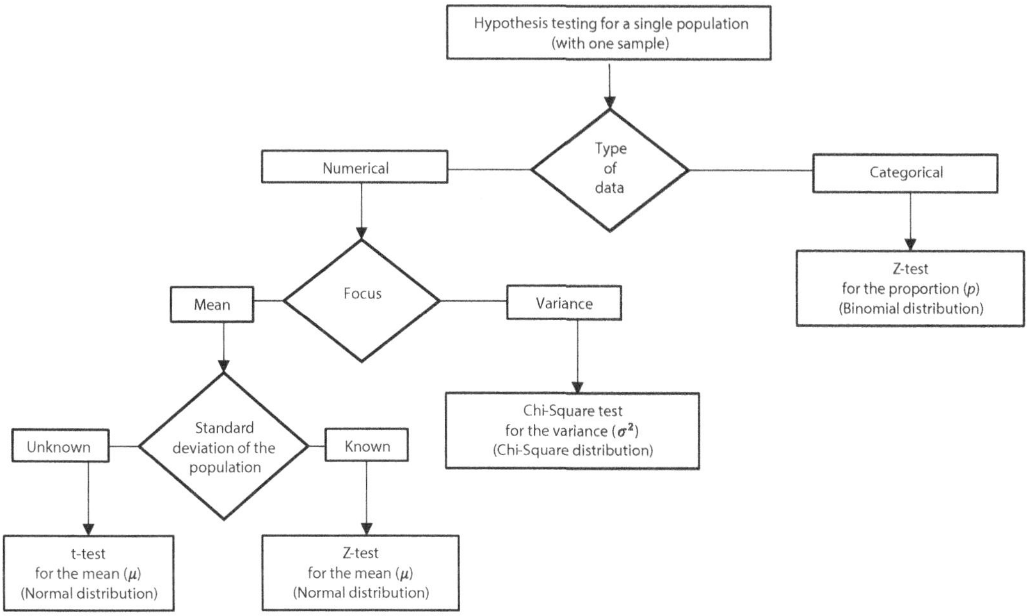

◘ **Fig. 7.7** Flow chart to select a hypothesis test for a single population. (Source: Author's creation)

$$Z = \frac{\bar{X} - \mu_0}{s/\sqrt{n}} \quad (7.67)$$

where μ_0 is the hypothesized value, $\bar{X}$ is the estimated value for μ, s is the standard deviation of the sampling distribution, and n is the sample size.

A *critical value of a test statistic* is a point that shows where the null hypothesis is rejected based on the distribution of the test statistic. The critical value of the test statistic divides the distribution into two regions: rejection region and non-rejection region. One-sided tests have only one critical value, while two-sided tests have two critical values on the distribution of the test statistic.

7

The probability that the test statistic falls into the *rejection region* when the null hypothesis is true should equal the level of the significance of the α test identified in the previous steps. ◘ Images 7.12 and 7.13 show that the results of a one-tailed test will be significant if the test statistic is less than −1.282 in ◘ Image 7.12 and if it equals or greater than the critical value that is presented at 1.645 in ◘ Image 7.13, respectively. The dark-colored areas demonstrate the rejection areas at the levels of significance $\alpha = 0.10$ and $\alpha = 0.05$, respectively. It also refers to Type I error of the test.

◘ Image 7.14 shows that the results of a two-tailed test will be significant with $\alpha = 0.10$ if the absolute test statistic equals or greater than the critical value that is presented at 1.645. The dark-colored areas demonstrate the rejection regions and the half of the significance level $\alpha/2 = 0.05$ on each side.

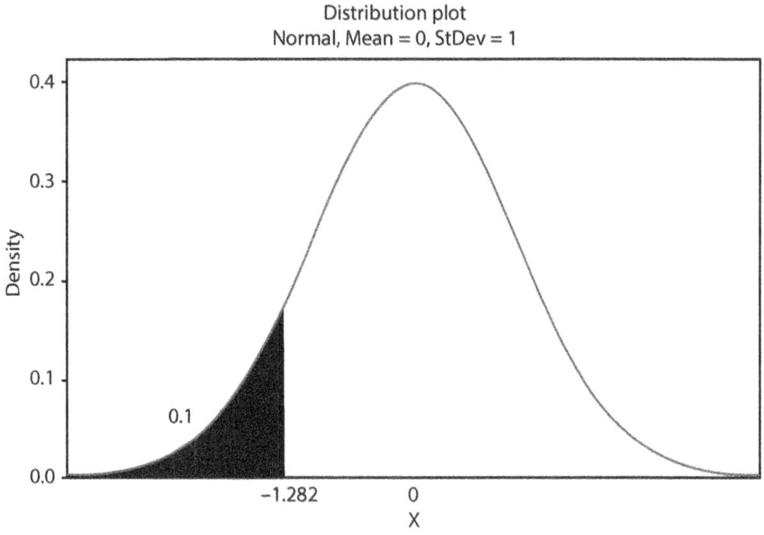

◘ Image 7.12 The lower-tailed test of a population mean with $\alpha = 0.10$, i.e., $P(Z \leq -1.282) = 0.10$. (Source: Author's creation based on Minitab)

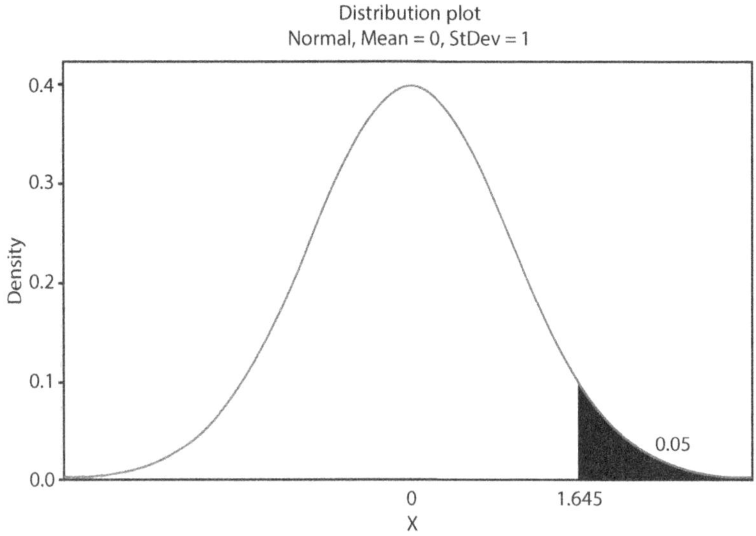

Image 7.13 The upper-tail test of a population mean with $\alpha = 0.05$, i.e., $P(Z \geq 1.645) = 0.05$. (Source: Author's creation based on Minitab)

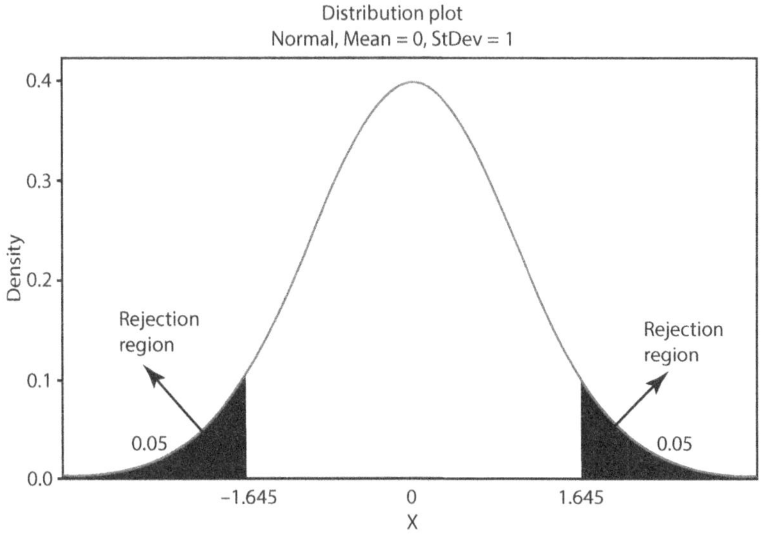

Image 7.14 Location of rejection regions for two-tailed test with $\alpha = 0.10$, i.e., $P(-1.645 \leq Z \leq 1.645) = 0.90$. (Source: Author's creation based on Minitab)

7.9.1.5 Reporting Test Results: *p*-Values

Hypothesis testing is conducted based on a certain level of significance, which is also known as the α (alpha) value. It is relatively easier for the Six Sigma team to control Type I error by determining the risk level of α that will be tolerated while rejecting the null hypothesis when it is true. The team can directly control the risk of Type I error by identifying the α value before starting hypothesis testing. The significance level of a test is often examined by *the p-value of the test*. After identifying the α value, the rejection region is automatically described by the team. The *p-value* is also known as observed significance level. It is the probability of obtaining a test statistic at least as contradictory to the null hypothesis as the value that actually resulted, assuming that the null hypothesis is true. "The p-value refers to the smallest level of significance that would lead to rejection of the null hypothesis. The p-value is the probability that the test statistic will take on a value that is at least as extreme as the observed value of the statistic when the null hypothesis is true" (Montgomery 2013: 121–122). The significance level can be 0.10, 0.05, or 0.01 depending on the error risk that the Six Sigma team is willing to accept.

For hypothesis testing that includes normally distributed data, p-values are calculated as follows, where P is the p-value and $Z_{computed}$ is the computed value of the test statistic Z:

- Two-tailed test: $P = 2P(Z > |Z_{computed}|)$
- One-sided test: $P = P(Z > Z_{computed})$ or $P = P(Z < Z_{computed})$

Depending on the type of the distribution, different distributions can be used to calculate the p-value in hypothesis testing. For example, if the data have a t distribution, t distribution table is used for calculating the p-value. Calculating the p-value manually may not always be easy in hypothesis testing. Statistical software packages help researchers calculate p-value.

As a general decision rule, when the p-value is less than the α value ($\alpha > p$), the null hypothesis is rejected. When the p-value is greater than the α value ($\alpha < p$), the hypothesis testing fails to reject the null hypothesis. The calculated test statistic is compared with the critical value of the test statistic. If the test statistic falls into the rejection region, the null hypothesis is rejected. If the test statistic is in the non-rejection region, then the null hypothesis is not rejected.

The steps that need to be followed in hypothesis testing can be listed as follows:

- Step 1: Check the assumptions and conditions.
- Step 2: State the null hypothesis H_0 and the alternative hypothesis H_1.
- Step 3: Choose the level of significance, α.
- Step 4: Choose the sample size, n.
- Step 5: Collect the data.
- Step 6: Compute the appropriate test statistic.
- Step 7: Calculate the p-value and identify the critical values of test statistic.
- Step 8: Apply the decision rule, and express the statistical finding in the scope of the question.

7.9.2 Hypothesis Tests for a Single Population

There are different types of hypothesis tests depending on the population parameter of interest. The key to correctly diagnosing a hypothesis test is then to determine the target parameter based on the phenomenon of interest and the variable measured. For numerical data, the target parameter would be either a population mean or variance; for categorical or attribute data, the parameter will be a population proportion. To determine the target parameter, the Six Sigma team can look for some key words or phrases in the statement of the problem when conducting a hypothesis test as follows (Sincich 1996):

- Mean; average: μ
- Difference in means or averages; mean difference; comparison of means or averages: $\mu_1 - \mu_2$
- Proportion; percentage; fraction; rate: p
- Difference in proportions, percentages, fractions, or rates; comparison of proportions, percentages, fractions, or rates: $p_1 - p_2$
- Variance; variation; spread; precision: σ^2
- Ratio of variances; difference in variation; comparison of variances: σ_1^2 / σ_2^2.

The testing procedure and related test statistic to be used in a hypothesis test of a population parameter also change based on the distributional characteristics of the data. When the underlying distribution of the data is known or can be assumed, parametric tests which subscribe the data to that distribution are used. For example, the data used for t-testing should subscribe to the normal distribution. If there is no assumption of a specific underlying distribution of the data, then non-parametric (or distribution free) tests are used. The following includes only parametric tests as non-parametric tests are out of scope of this section. ◘ Figure 7.7 shows a flow chart as a guide for selecting the appropriate type of parametric hypothesis test for a single parameter of interest. The next section presents testing procedures for two populations.

7.9.3 Testing of the Population Mean

Several hypothesis tests for the population mean are developed based on the characteristics or assumptions about the population. Selection of the appropriate test procedure for a single mean is determined by whether the value of population standard deviation σ is known, the sample size n, and the sampled population is normally distributed.

7.9.3.1 Tests of the Mean of a Normal Distribution (Population Standard Deviation Known)

The simplest type of a hypothesis test is for the mean of a population where the standard deviation is assumed to be known. Indeed, this is rare case in practice, but possible, for example, if the Six Sigma team takes measurements of parts from an automated machine or from a very stable process. If the underlying distribution of the data subscribes to the normal distribution, the test procedure will use z-statistic (see ◘ Fig. 7.7). Formally, if the population standard deviation is known, testing of the population mean of a normal distribution will be as shown in the following box.

Testing Procedure for the Mean of a Normal Distribution (σ is known)

A random sample of n observations was obtained from a normally distributed population with mean μ and known standard deviation σ. If the observed sample mean is $\bar{x}$, the population mean is tested, with the significance level α, by using the procedure summarized in ◘ Table 7.6, where μ_0 is the hypothesized value (i.e., particular numerical value specified for μ in H_0), z_α is the z-value such that $P(z > z_\alpha) = \alpha$, and $z_{\alpha/2}$ is the z-value such that $P\left(z > z_{\frac{\alpha}{2}}\right) = \alpha/2$.

Assumptions: randomly selected sample, normally distributed population, known population standard deviation.

◘ **Table 7.6** Hypothesis testing for the population mean with known σ

One-tailed test	*Two-tailed test*
$H_0 : \mu = \mu_0$ $H_1 : \mu > \mu_0$ (*or* $H_1 : \mu < \mu_0$)	$H_0 : \mu = \mu_0$ $H_1 : \mu \neq \mu_0$
Test statistic: $Z = \dfrac{\bar{x} - \mu_0}{\sigma / \sqrt{n}}$ (7.68)	
Rejection region Reject H_0 if $z > z_\alpha$ (*or* $z < z_\alpha$)	*Rejection region* $\lvert z \rvert > z_{\alpha/2}$

Source: Author's creation

7

▶ Example 37

According to the policy of a certain branch of the H&A bank, customer complaints must be resolved in a courteous and timely manner. One of the frequent complaints is that the customers cannot withdraw enough cash from the ATMs over the weekend. From previous experience, the amount of money withdrawn from ATMs per customer transaction over the weekend period is known as normally distributed with a mean \$180 and a standard deviation \$20. To solve this complaint, this H&A branch has started a project to analyze the withdrawals made on weekends. A random sample of 25 transactions made by customers during weekends is selected as given in ◘ Table 7.7. At the 0.05 level of significance, is there sufficient evidence to conclude that the true mean withdrawal at this branch is greater than \$180? ◀

Solution

The parameter of interest is μ, the mean amount of money withdrawn from the ATMs. The purpose is to obtain strong evidence that the true mean is greater than \$180. To solve this example, the steps will be followed as given below.

- Step 1: Check the assumptions and conditions.

In testing of the population mean when the population standard deviation is known, the main assumptions include randomization and normality. This example gives a historical experience about the amount of money withdrawn from ATMs per customer transaction over the weekend period and states the value of standard deviation as \$20. The 25 transactions are drawn randomly from all transactions made by customers; thus, this ensures that the condition of *randomization* is in place. The normal probability plot shows that the sample data for money withdrawals is approximately normal because the values are almost lying on the straight line. There is no evidence for the violation of normality. The Anderson Darling (AD) normality test also confirms this conclusion with AD = 0.533 (p-value = 0.156 > 0.05).

- Step 2: State the null hypothesis H_0 and the alternative hypothesis H_1.

$$H_0 : \mu = 180$$

$$H_1 : \mu > 180$$

- Step 3: Choose the level of significance, α.

The testing rule is designed with $\alpha = 0.05$; therefore, we know that rejecting the null hypothesis provides strong evidence that the mean withdrawal is greater than \$180, because the probability of error is a small value, i.e., only 5%.

- Step 4: Choose the sample size, n.

The sample size is 25 transactions.

- Step 5: Collect the data.

We assume that the data are collected from the transactions made by customers during weekends in the context of a Six Sigma project which aims at resolving related customer complaints in a timely manner (See ◘ Table 7.7).

- Step 6: Compute the appropriate test statistic.

The sample mean ($\overline{x}$) of these 25 money withdrawals is \$191.2. The population standard deviation is \$20. Substituting these values, the z-statistic can be calculated as follows:

$$Z = \frac{\overline{x} - \mu_0}{\sigma / \sqrt{n}} = \frac{191.2 - 180}{20 / \sqrt{25}} = \frac{10}{4} = 2.8$$

- Step 7: Calculate the p-value and identify the critical values of test statistic.

We know that the significance level is $\alpha = 0.05$. Because we're looking for the withdrawal amount, which is greater than \$180, we're interested in the upper tail. Therefore, the critical value is $Z_{0.05}$ = 1.645, as shown in ◘ Image 7.13. By using the critical value approach, the null hypothesis H_0 will be rejected if $Z > 1.645$.

◘ Table 7.7 Money withdrawn (\$) from the ATMs of H&A bank

185	190	220	195	180
215	195	200	155	195
200	190	205	170	190
200	205	160	225	180
195	155	200	200	175

Source: Author's creation

For this upper-tailed test, the computed value of the test statistic is $z_{computed} = 2.8$, and the associated p-value is $P = P(Z > Z_{\text{computed}}) = P(Z > 2.8) = 0.0025$, found from the standard normal distribution table (Table A.2). This means that the probability that a Z value exceeds 2.8 is 0.0025 (◘ Image 7.15). Based on the p-value approach, the null hypothesis H_0 will be rejected if $\alpha = 0.05 > p$.

To perform a hypothesis test for a single population mean in Minitab, follow the path: Stat→Basic Statistics→1-Sample Z…. Once "One-Sample Z for the Mean" dialog box appears, choose column "money withdrawn ($)" as the sample data set, enter Known standard deviation (20), click on "Perform hypothesis test.," and enter Hypothesized mean (180). Click on "Options…." Enter 95 on Confidence Level box, and select "mean>hypothesized mean" for Alternative Hypothesis. Click on OK on Options Box and then click "Graphs…" to select "Histogram." Click OK on Graphs… box and OK on Dialog Box to see the results and requested Histogram on the Session Window.

- Step 8: Apply the decision rule, and express the statistical finding in the scope of the question.

Our decision rule based on the critical value approach rejects H_0 if $Z > 1.645$, and we found that $Z = 2.8 > Z_{0.05} = 1.645$. Therefore, we reject the null hypothesis $H_0 : \mu = 180$ at the 0.05 significance level. We thus conclude that the true mean withdrawal at this H&A branch is greater than $180. The probability of Type I error (rejecting the null hypothesis when, in fact, it is true) is 0.05. Besides, the p-value was found as 0.0025, which also gives the minimum α value that leads to a rejection of the null hypothesis. Consequently, we will reject H_0 for any α value exceeding 0.0025. Therefore, the Six Sigma team of this H&A branch who desires a Type I error rate less than 0.05 has very strong evidence to say that the true mean withdrawal of customers exceeds $180 since $\alpha = 0.05 > p = 0.0025$. As seen from Minitab output, the hypothesized value $180 is outside the population one-sided lower confidence limit, which tells us

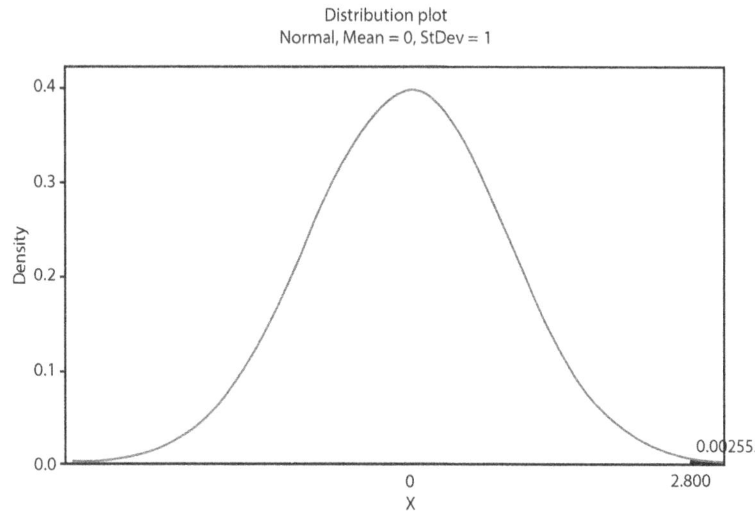

◘ **Image 7.15** Observed significance level (p-value) of $z_{\text{computed}} = 2.8$ for the upper-tailed test. (Source: Author's creation based on Minitab)

there is a significant difference between the true central point of the amounts of money withdrawals and the hypothesized value.

Thus, the H&A bank should take alternative courses of actions for resolving the customer complaint about the cash stocks in their ATMs over the weekend period.

7.9.3.2 Tests of the Mean of a Normal Distribution (Population Standard Deviation Unknown)

In testing the population mean, the population standard deviation will not be known in most practical sampling applications. This may be the case when a new part/service is offered with little previous use or experience or when a performance parameter is set under new operating conditions. Thus, a random sample is taken from a population with unknown mean μ and unknown standard deviation σ. If the underlying distribution of the data subscribes to the normal distribution, the test procedure will use the t-statistic (see ◘ Fig. 7.7).

As we established in ▶ Sect. 7.8.2.1, when n is sufficiently large, the sample mean can be assumed as approximately normally distributed based on the central limit theorem, and the sample standard deviation s, the large sample estimate of σ, is sufficiently accurate for applying testing procedures.

When the sample size n is not large and there is no prior knowledge of the value of population standard deviation, σ, the population under study must be assumed as having normal distribution. If this assumption cannot be made, the procedures presented in this chapter cannot be employed. In such cases special techniques called non-parametric techniques must be employed. If the population is normal, the t distribution may be used in the same way that the standardized normal distribution has been used. All we need to do is to replace $t_{n-1,\alpha}$ wherever z_α was used before, and to substitute the sample estimate s wherever σ was used.

In the case where the population standard deviation is unknown, testing of the population mean of a normal distribution will be as shown below in the box.

Testing Procedure for the Mean of a Normal Distribution (σ is unknown)

A random sample of n observations was obtained from a normally distributed population with mean μ. Using the observed sample mean $\bar{x}$ and sample standard deviation s, the procedure in ◘ Table 7.8 can be used for testing the population mean with the significance level α, where μ_0 is the hypothesized value (i.e., particular numerical value specified for μ in H_0), t_α is the t-value such that $P(t > t_\alpha) = \alpha$, and $t_{\alpha/2}$ is the t-value such that

$$P\left(t > t_{\frac{\alpha}{2}}\right) = \alpha/2.$$

Assumptions: randomly selected sample, approximately normally distributed population, unknown population standard deviation.

◘ Table 7.8 Hypothesis testing for the population mean with unknown σ

One-tailed test	*Two-tailed test*
$H_0: \mu = \mu_0$ $H_1: \mu > \mu_0$ $(or\ H_1: \mu < \mu_0)$	$H_0: \mu = \mu_0$ $H_1: \mu \neq \mu_0$
Test statistic: $t = \dfrac{\bar{x} - \mu_0}{s/\sqrt{n}}$	(7.69)
Rejection region Reject H_0 if $t > t_\alpha$ $(or\ t < t_\alpha)$	*Rejection region* $\lvert t \rvert > t_{\alpha/2}$

Source: Author's creation

▶ Example 38

Let's reconsider the situation related to money withdrawals from ATMs of H&A bank over an entire weekend in ▶ Example 37. Suppose that H&A bank has launched a new branch and wants to guarantee that ATMs will have enough cash stock on every weekend to keep its customers satisfied. Thirty-two randomly selected transactions (◘ Table 7.9) are examined to test whether the mean amount of money withdrawn by customers over the weekend equals the expected (population) mean of $180 as experienced from other similar branches. Is there any evidence to believe that the mean amount of money withdrawn from ATMs of this new branch is not different from other branches with 0.05 significance level?◀

Solution

The parameter of interest is μ, the mean amount of money withdrawn from the new ATMs. The purpose is to obtain strong evidence that the true mean amount of money withdrawn by customers of the new branch over the weekend period is not different from that of other branches and equals $180.

- Step 1: Check the assumptions and conditions.

In testing of the population mean when the population standard deviation is unknown, the main assumptions include randomization and normality. As the 32 transactions are drawn randomly from all transactions made by customers, this satisfies the *randomization* condition. The sample size is greater than 30, and based on the central limit theorem, the normality assumption is also plausible for this example. As seen from the normal probability plot, the values of money withdrawals are lying on an almost straight line, and there is no evidence for the violation of normality. The Anderson Darling (AD) normality test also confirms this conclusion with AD = 0.461 (p-value = 0.244 > 0.05).

- Step 2: State the null hypothesis H_0 and the alternative hypothesis H_1.

$$H_0 : \mu = 180$$

$$H_1 : \mu \neq 180$$

- Step 3: Choose the level of significance, α.

The testing rule is designed with $\alpha = 0.05$. Therefore, we know that rejecting the null hypothesis provides strong evidence that the mean withdrawal is different from $180 because the probability of error is a small value, i.e., only %5.

- Step 4: Choose the sample size, n.

The sample size is 32 transactions.

- Step 5: Collect the data.

The data are collected from the transactions made by customers during weekends and presented in ◘ Table 7.9.

- Step 6: Compute the appropriate test statistic.

The sample mean ($\bar{x}$) and the sample standard deviation of these 32 money withdrawals are found as $183.44 and 16.48, respectively. Using these values, the t statistic can be calculated as follows:

$$t = \frac{\bar{x} - \mu_0}{s/\sqrt{n}} = \frac{183.44 - 180}{16.48/\sqrt{32}} = \frac{3.44}{2.91} = 1.18$$

- Step 7: Calculate the p-value and identify the critical values of test statistic.

We're looking for a withdrawal amount which is different from $180; in this case, the mean amount of money withdrawn could be either too large or too small. Therefore, there are two tails of the distribution to be tested.

◘ **Table 7.9** Money withdrawn ($) from the ATMs of new branch of H&A bank

160	145	180	190	180	205	145	180
185	205	160	180	195	190	170	175
185	200	190	185	175	205	200	180
200	190	180	215	190	170	170	190

Source: Author's creation

The critical value of $t_{v,\alpha/2}$ should be determined for the significance level $\alpha = 0.05$ and $v = n - 1 = 32 - 1 = 31$ degrees of freedom. This value may be found from the *t*-table (Table A.4) or may be computed in Minitab. As the significance level is 0.05, then $\alpha/2 = 0.025$. By looking in the column corresponding to 0.025 and row 31 from the *t*-table, the value of $t_{31, 0.025}$ is found to be 2.04 (◘ Image 7.16). This means that the probability that a Student's *t* random variable with 31 degrees of freedom exceeds 2.04 is 0.025. Using the critical value approach, the null hypothesis H_0 will be rejected if $|t| > 2.04$.

For the two-tailed test, the computed value of the test statistic is $t_{computed} = 1.18$ and the associated *p*-value is

$P = 2P(t > |t_{computed}|) = 2P(t > 1.18) = 2 * (0.1235) = 0.247$. Based on the *p*-value approach, the null hypothesis H_0 will be rejected if $\alpha = 0.05 > p$.

In Minitab, to perform a hypothesis test for a single population mean with unknown standard deviation, follow the path: Stat→Basic Statistics→1-Sample t.... Once "One-Sample t for the Mean" dialog box appears, choose column "money withdrawn ($)" as the sample data set, click on "Perform hypothesis test.," and enter Hypothesized mean (180). Click on "Options...." Enter 95 on Confidence Level box, and select "mean ≠ hypothesized mean" for Alternative Hypothesis. Click on OK on Options Box and then click "Graphs..." to select "Histogram." Click OK on Graphs... box and OK on Dialog Box to see the results and requested Histogram on the Session Window.

- Step 8: Apply the decision rule and express the statistical finding in the scope of the question.

Based on the critical value approach, H_0 fails to reject the null hypothesis $H_0 : \mu = 180$, since $t = 1.18 < t_{31, 0.025} = 2.04$. We can

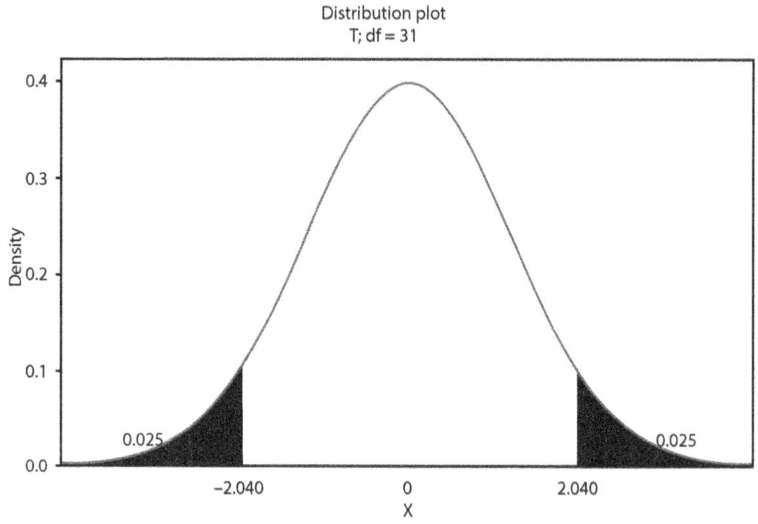

◘ **Image 7.16** Location of rejection regions for the two-tailed t-test with $\alpha = 0.05$. (Source: Author's creation based on Minitab)

conclude that the true mean withdrawal at this new H&A branch is not different from \$180. The p-value was found to be 0.247. Consequently, we can reject H_0 for the α values exceeding 0.247. Therefore, the Six Sigma team of this H&A branch who desires a Type I error rate less than 0.05 has no strong enough evidence to say that the true mean withdrawal of customers is different from \$180 since $\alpha = 0.05 < p = 0.247$. The hypothesized value \$180 is within the population confidence limits, which tell us there is no significant difference between the true mean and the hypothesized value of the amounts of money withdrawals.

7.9.4 Testing the Population Variance of a Normal Distribution

In Six Sigma projects, understanding variability is particularly important because a process that, for example, has a largely fluctuating variance can produce many undesirable defects. In such cases, the team may want to determine whether the variance of the processes, components, or products is a particular value or set of values.

When the focused parameter is the population variance, then testing procedure for the population variance of a normal population is summarized in the following box.

Testing Procedure for the Variance of a Normal Distribution

A random sample of n observations was obtained from a normally distributed population with variance σ^2. If the observed sample variance is s^2, the population variance is tested, with the significance level α, by using the procedure summarized in Table 7.10, where σ_0^2 is the hypothesized value (i.e., particular numerical value specified for σ^2 in H_0) and χ_α^2 and $\chi_{1-\alpha}^2$ are values of χ^2 that locate an area of α to the right and α to the left, respectively, of a chi-square distribution based on $(n - 1)$ degrees of freedom.

Assumptions: randomly selected sample, normally distributed population.

Table 7.10 Hypothesis testing for the population variance

One-tailed test	*Two-tailed test*
$H_0 : \sigma^2 = \sigma_0^2$	$H_0 : \sigma^2 = \sigma_0^2$
$H_1 : \sigma^2 > \sigma_0^2$	$H_1 : \sigma^2 \neq \sigma_0^2$
(or $H_1 : \sigma^2 < \sigma_0^2$)	
Test statistic: $\chi^2 = \dfrac{(n-1)s^2}{\sigma_0^2}$	(7.70)
Rejection region Reject H_0 if $\chi^2 > \chi_\alpha^2$ (or $\chi^2 < \chi_{1-\alpha}^2$)	*Rejection region* $\chi^2 < \chi_{1-\frac{\alpha}{2}}^2$ or $\chi^2 > \chi_{\alpha/2}^2$

Source: Author's creation

▶ Example 39

The quality control manager of a manufacturing company randomly sampled 75 injection syringes from their third shift. The collected data are given in Table 7.11. When the manufacturing process is working well, the standard deviation in the syringe lengths should be no greater than 0.03 cm. Test, at the 10% significance level, whether the population standard deviation in the lengths of injection syringes is at most 0.03 cm and decide whether the process produces the syringes consistently in the third shift. ◀

Table 7.11 The lengths of injection syringes from the third shift production process

12.573	12.535	12.611	12.624	12.581	12.586	12.591	12.606	12.598	12.601
12.604	12.573	12.624	12.609	12.619	12.593	12.573	12.543	12.598	12.588
12.583	12.576	12.604	12.586	12.611	12.619	12.583	12.629	12.619	12.606
12.555	12.624	12.619	12.614	12.616	12.598	12.616	12.639	12.624	12.616
12.606	12.626	12.624	12.634	12.601	12.639	12.637	12.560	12.588	12.588
12.593	12.598	12.604	12.634	12.588	12.619	12.606	12.596	12.616	12.576
12.573	12.591	12.598	12.619	12.604	12.598	12.609	12.616	12.619	12.601
12.611	12.614	12.601	12.591	12.596					

Source: Author's creation

7

Solution

The parameter of interest is σ, the standard deviation in the lengths of injection syringes produced in a manufacturing company. The purpose is to determine whether the true standard deviation of these lengths is less than 0.03 cm. To solve this example, the steps will be followed as given below.

- Step 1: Check the assumptions and conditions.

The most critical assumption in testing the population variance is normality. This assumption of a normal population is required regardless of whether the sample size n is large or small. Even moderate departures from normality can result in the χ^2 test statistic having a distribution that is very different from chi-square. The normal probability plot shows that the sample data for syringe lengths is approximately normal because the values are lying in almost a straight line. There is no evidence for the violation of normality. The Anderson Darling (AD) normality test also confirms this conclusion with AD = 0.611 (p-value = 0.108 > 0.05).

- Step 2: State the null hypothesis H_0 and the alternative hypothesis H_1.

We wish to test the hypothesis that the standard deviation of a normal population of the syringe lengths equals $\sigma_0 = 0.03$ cm or, equivalently, that the population variance of the syringe lengths is equal to $\sigma_0^2 = 0.0009\ \text{cm}^2$. Since the null and alternative hypotheses must be stated in terms of σ^2 (rather than σ), the hypotheses are as follows for this example:

$$H_0 : \sigma_0^2 = 0.0009\ \text{cm}^2$$

$$H_1 : \sigma_0^2 < 0.0009\ \text{cm}^2$$

- Step 3: Choose the level of significance, α.

The testing rule is designed with $\alpha = 0.10$. Therefore, we know that rejecting the null hypothesis provides evidence that the true variance of syringe lengths is less than $0.0009\ \text{cm}^2$, or equivalently the true standard deviation is less than 0.03 cm, because the probability of error is relatively a small value, i.e., 10%.

- Step 4: Choose the sample size, n.

The quality control manager of this company randomly sampled 75 injection syringes from their third shift. The sample size is 75.

- Step 5: Collect the data.

The collected data was shown in Table 7.11.

- Step 6: Compute the appropriate test statistic.

The sample variance (s^2) of these 75 measurements is found as $0.000448\ \text{cm}^2$. The hypothesized value is $\sigma_0^2 = 0.0009$ cm^2. Substituting these values in Eq. 7.70, the test statistic can be calculated as follows:

$$\chi^2 = \frac{(n-1)s^2}{\sigma_0^2} = \frac{(75-1)(0.000448)}{0.0009} = 36.87$$

- Step 7: Calculate the p-value and identify the critical values of test statistic.

Because we're looking for the variance of syringe lengths which is at most 0.0009, we're interested in the lower tail. The smaller the value of s^2 we observe, the stronger the evidence in favor of H_1. Thus, H_0 will be rejected for small values of the test statistic, i.e., reject H_0 if $\chi^2 < \chi^2_{1-\alpha}$. With the significance level $\alpha = 0.10$ and $(n-1) = (75-1) = 74$ degrees of freedom, the critical value $\chi^2_{1-0.10}$ is the value of χ^2 that locates an area of 0.10 to the left of a chi-square distribution based on 74 degrees of freedom. This value can be obtained from the chi-square table (Table A.3), and the value of $\chi^2_{0.90}$ is 58.90, which corresponds to the column 0.90 and the row 74 in the chi-square table. Using the critical value approach, we will reject H_0 if $\chi^2 < 58.90$ (◘ Image 7.17).

For this lower-tailed test, the computed value of the test statistic is $\chi^2_{\text{computed}} = 36.87$ as given in Step 6, and the associated p-value is $P = P(\chi^2 < \chi^2_{\text{computed}}) = P(\chi^2 < 36.87) = 0.0000919$ that may be found from chi-square table or chi-square cumulative distribution function in Minitab. This means that the probability that a χ^2 value is less than 36.87 is 0.0000919. Based on the p-value approach, the null hypothesis H_0 will be rejected if $\alpha = 0.05 > p$.

To perform a hypothesis test for a single population standard deviation in Minitab, follow the path: Stat→Basic Statistics→1 Variance.... On the opened window, check "Perform Hypothesis Test," and enter the value of hypothesized standard deviation (0.03). Click "Options." Set 90.0 as Confidence Level, and because we're looking for the standard deviation in syringe length which is at most 0.03, select "standard deviation < hypothesized standard deviation" for Alternative Hypothesis. Click on OK on Options Box and OK on Dialog Box to see the results (◘ Table 7.12) on the Session Window.

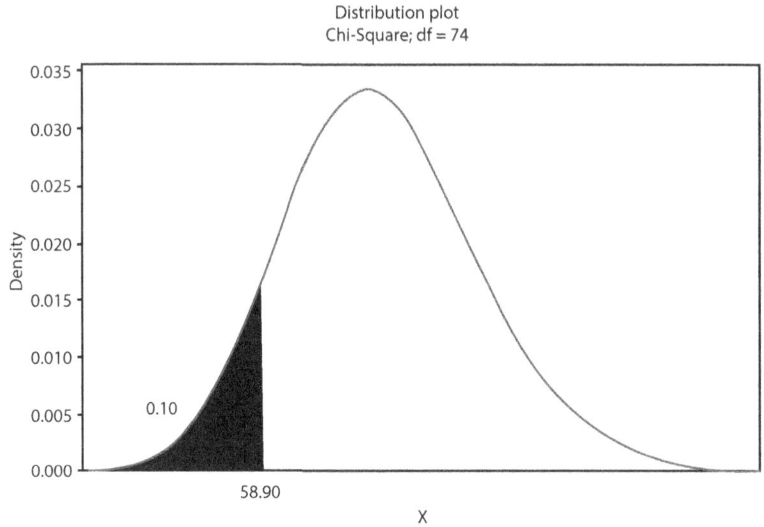

◘ **Image 7.17** Critical value of χ^2 for the lower-tail area $\alpha = 0.10$. (Source: Author's creation based on Minitab)

Table 7.12 Minitab output of "One-Sample Variance" for the syringe lengths (cm)

Test and CI for one variance: syringe lengths (cm)				
Method				
σ: Standard deviation of syringe lengths (cm)				
The Bonett method is valid for any continuous distribution.				
The chi-square method is valid only for the normal distribution.				
Descriptive statistics				
N	**StDev**	**Variance**	**90% upper bound for σ using Bonett**	**90% upper bound for σ using chi-square**
75	0.0212	0.000448	0.0243	0.0237
Test				
Null hypothesis			H_0: $\sigma = 0.03$	
Alternative hypothesis			H_1: $\sigma < 0.03$	
Method	**Test statistic**	**DF**	***P*-value**	
Bonett	–	–	0.001	
Chi-square	36.87	74	0.000	

Source: Author's creation based on Minitab

- Step 8: Apply the decision rule, and express the statistical finding in the scope of the question.

Since the test statistic, $\chi^2_{\text{computed}} = 36.87$, is less than the value of $\chi^2_{0.90} = 58.90$, the quality control manager of this manufacturing company can conclude that the standard deviation of the population of all syringe lengths is less than 0.03 cm (or the variance is less than 0.0009) with 90% confidence. In other words, if this χ^2 test procedure is repeatedly used, it will incorrectly reject H_0 only 10% of the time. The *p*-value was found to be 0.0000919, and it is too small when compared to the significance level $\alpha = 0.10$. Based on the *p*-value approach, the null hypothesis H_0 is rejected. Thus, the quality control manager is confident in the decision that the manufacturing process is operating consistently within the desired limits of variability.

7.9.5 Testing the Population Proportion (Large Samples)

If the Six Sigma team is working with categorical variables having only two possible outcomes, such as good or bad, yes or no, defective or non-defective, go or no-go, and so on, the target parameter will be the proportion of observations in a population that has a certain characteristic. This parameter is also called a proportion of successes, *p*, for a binomial population. For testing the population proportion, the test procedure will use *z*-statistic (Fig. 7.7). The procedure described below is used to test a population proportion *p* based on a large sample from the target population. If the sample size is not large or the percentage of successes is either very low or very high, then the following procedure can be employed. Instead, the exact solution must be obtained with binomial distribution rather than the normal distribution, which are not in the scope of this section. Further, Minitab allows finding an exact or approximate solution for the analysis.

Testing Procedure for the Proportion (Large Samples)

A random sample of n observations was obtained from a population that has a proportion p whose members possess a particular attribute. If the sample size is large and the observed sample proportion is $\hat{p}$, with the significance level α, the test procedure will be as follows (◘ Table 7.13), where p_0 is the hypothesized value (i.e., particular numerical value specified for p in H_0), z_α is the z-value such that $P(z > z_\alpha) = \alpha$, and $z_{\alpha/2}$ is the z-value such that $P\left(z > z_{\frac{\alpha}{2}}\right) = \alpha/2$.

Assumptions: the large sample size, randomly selected sample.

◘ **Table 7.13** Hypothesis testing for the population proportion

One-tailed test	*Two-tailed test*
$H_0: p = p_0$ $H_1: p > p_0$ (*or* $H_1: p < p_0$)	$H_0: p = p_0$ $H_1: p \neq p_0$
Test statistic: $z = \dfrac{(\hat{p} - p_0)}{\sqrt{\dfrac{p_0(1-p_0)}{n}}}$	(7.71)
Rejection region Reject H_0 if $z > z_\alpha$ (*or* $z < -z_\alpha$)	*Rejection region* $\lvert z \rvert > z_{\alpha/2}$

Source: Author's creation

▶ **Example 40**

In a Six Sigma project, the team has monitored the customer returns and found that 1120 of 37,300 customers made returns during the last 2 months. The true proportion of customer returns over the last 5 years was estimated to be 3% ($p_0 = 0.03$). The team is interested in identifying whether the proportion of customer returns over the last 2 months is greater than the true proportion of customer returns. It is assumed that the customer returns are normally distributed. Test the hypothesis using one-proportion z test.◀

- Step 1: Check the assumptions and conditions.

For proportions, the model for the sampling distribution of the statistic is expected to be normally distributed. Since all models require some assumptions and conditions, we need to articulate and check these assumptions and conditions in this hypothesis testing. In a test of a proportion, we will work on the independence assumption, sample size assumption, randomization condition, 10% condition, and success/failure condition, respectively. In this example, we assume that the customers randomly return the products and are independent of each other. The sample size with 1120 customers is large enough ($n \geq 30$). We also assume that the sample size meets the 10% condition, due to the assumption that the company has large enough number of customers and that 1120 customers are less than 10% of all customers (1120/37,300 < 0.10). For success/failure condition, number of failures ($np(1 - p)$) for 1120 customers returning the products is $np(1 - p) = 1120(0.03)(1 - 0.03) = 32.59 > 10$. This condition is met since the number of failures is greater than ten customers. Since the conditions and assumptions are satisfied, we can assume that the sampling distribution of the proportion is normally distributed.

- Step 2: State the null hypothesis H_0 and the alternative hypothesis H_1.

$$H_0 : p = 0.03$$

$$H_1 : p > 0.03$$

- Step 3: Choose the level of significance, α.

Let's use $\alpha = 0.05$ in this hypothesis testing.

- Step 4: Choose the sample size, n.

The sample size is 37,300 customers.

- Step 5: Collect the data.

We assume that the data are collected in this Six Sigma project.

- Step 6: Compute the appropriate test statistic.

Since we have one proportion and all assumptions and conditions are met in this example, the test called one-proportion z test statistic will be used as follows:

$$z = \frac{(\hat{p} - p_0)}{\sqrt{\frac{p_0(1-p_0)}{n}}} = \frac{0.03002 - 0.03}{\sqrt{\frac{0.03(1-0.03)}{37,300}}} = \frac{0.00002681}{0.000883} = 0.030$$

where

- $\hat{p}$ = the proportion of customer returns in the sample
- p_0 = the hypothesized value of the proportion of customer returns
- n = sample size.

- Step 7: Calculate the p-value and identify the critical values of test statistic.

For this upper-tailed test, the computed value of the test statistic is $z_{\text{computed}} = 0.0303$, and the associated p-value is $P = P(Z > Z_{\text{computed}}) = P(Z > 0.0303) = 0.4880$ found from the standard normal distribution table (Table A.2). As seen in ◘ Image 7.18, this means that the probability that a Z value exceeds 0.0303 is 0.4880. Based on the p-value approach, the null hypothesis H_0 will be rejected if $\alpha = 0.05 > p$.

To run one-proportion z test in Minitab, click on Stat→Basic Statistics→1 Proportion. In the next input screen, select "summarized data"; and enter the variables number of event (1120), number of trials (37,300), and hypothesized proportion (0.03); and click on "perform hypothesis test." Click on "Options," and select alternative hypothesis as "Proportion>hypothesized proportion"; select "normal approximation" in "Method." Click on OK→OK. The results are shown in Session Window.

- Step 8: Apply the decision rule, and express the statistical finding in the scope of the question.

Applying the decision rule shows that we cannot reject the null hypothesis since α (0.05) is less than the p-value ($p = 0.488$). We can conclude that there is insufficient evidence to reject the null hypothesis. Let's remember the rule: if the level of significance was greater than p-value ($\alpha > p$), the null hypothesis would be rejected. In other words, we have compelling evidence in favor of the null hypothesis. As a result, we can say that if the true proportion of customer returns equals 3%, we are 95% confident that the proportion of customer returns over the last 2 months equals 3% (p-value = 0.4880).

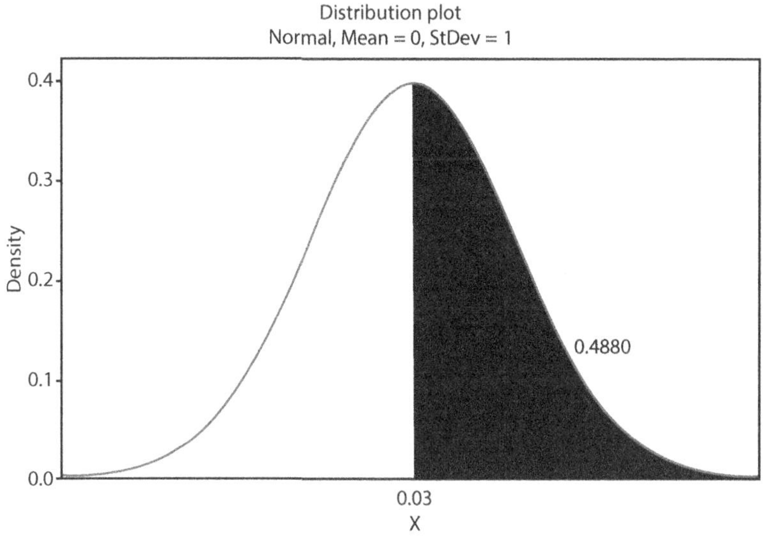

Image 7.18 Observed significance level (p-value) of $z_{\text{computed}} = 0.0303$ for the upper-tailed test. (Source: Author's creation based on Minitab)

7.10 Inferential Statistics: Comparing Two Populations

By Dr. Aysun Kapucugil Ikiz
Associate Professor
Dokuz Eylul University

Hypothesis testing is used in comparing two populations for differences in means, proportions, or variances. The hypothesis testing procedures are similar to those discussed in the previous section. The formulas for the test statistics are more complicated than for single population tests.

In Six Sigma, hypothesis testing is used to help determine whether the variation between groups of data is due to true differences between the groups or is the result of common cause of variation, in other words, due to chance, the natural variation in a process (GOAL/QPC 2002: 142). This tool is most commonly used in the Analyze step of the DMAIC to determine whether different levels of a discrete process setting (x) result in significant differences in the output (y). In other words, it helps determine whether the difference observed between groups is larger than expected from common cause of variation alone.

For example, assume that the quality control manager of a syringe manufacturing company wants to see whether the manufacturing process works well both in the first and second shifts. If the CTQ is the rate of defective syringes (y), the main question is to identify whether the shifts (x) have an effect on the observed rate of defective syringes. If the observed difference is large because of common cause of variation, then it can be said that this difference is not statistically significant. A two-sample test of proportions will help to answer this question.

In another example, the regional sales manager of a product is interested in comparing the sales volume of the product when it is displayed in the knee-level shelf as compared to an eye-level shelf. The question becomes: Does the location of shelf used (x) in a store affect the sales of products (y)? In this situation, the sales of the product placed on knee-level shelf represent one population, and the ones placed in the eye-level shelf the other. To investigate the question, we select a random sample from each population and compute the mean of the two samples. If the two population means are the same, we would expect the difference between the two sample means to be zero. But what if our sample results yield a difference other than zero? Is that difference due to chance or is it because there is a real difference in sales? A two-sample test of means will help to answer this question.

Hypothesis testing is also used to compare two dependent or paired groups of data. In this case, some of the characteristics of the pairs are similar, and thus, that portion of the variability is removed from total variability of the differences between means (Newbold et al. 2013: 387). For example, the dimensions of the parts produced on the same specific machine will be closer than the dimensions of the parts produced on two different, independently selected machines. By using dependent samples, we are able to reduce the variation in the sampling distribution. Thus, its standard error is always smaller. That, in turn, leads to a larger test statistic and a greater chance of rejecting the null hypothesis (Lind et al. 2012: 396). Therefore, whenever possible we prefer to use paired data to compare measurements from two populations.

7

The statistical theory of comparing two population means requires determining the difference between the sample means and studying the distribution of differences in the sample means. As discussed in the section on Fundamentals of Inferential Statistics, a distribution of sample means approximates the normal distribution. We assume that a distribution of sample means will follow the normal distribution. It can be shown mathematically that the distribution of the differences between sample means for two normal populations is also normal. If we find the mean of the distribution of differences is zero, this implies that there is no difference in the two populations. On the other hand, if the mean of the distribution of differences is equal to some value other than zero, either positive or negative, then we conclude that the two populations do not have the same mean.

Another point that should be emphasized is that we need to know something about the variability of the distribution of differences. To put it another way, what is the standard deviation of the distribution of differences? Statistical theory shows that, when we have independent populations, the distribution of the differences has a variance that equals to the sum of the two individual variances (Eq. 7.72).

$$\sigma^2_{\bar{x}_1-\bar{x}_2} = \frac{\sigma_1^2}{n_1} + \frac{\sigma_2^2}{n_2} \tag{7.72}$$

The term $\sigma^2_{\bar{x}_1-\bar{x}_2}$ looks complex, but it is not difficult to interpret. The σ^2 portion indicates that it is a variance, and the subscript $\bar{x}_1 - \bar{x}_2$ identifies it as a distribution of differences in the sample means (Lind et al. 2012: 374). We can put this equation in a more usable form by taking the square root, so that we have the standard deviation of the distribution or standard error of the differences. ◘ Figure 7.8 shows a flow chart as a guide for selecting the appropriate type of parametric hypothesis test for comparing two parameters of interest.

7.10.1 Connection Between Hypothesis Test and Confidence Interval Estimation

Hypothesis testing and confidence interval (CI) estimation are two general methods for making inferences for population parameters. The CI provides a range of likely values for a population parameter at a stated confidence level, whereas hypothesis testing is an easy framework for displaying the risk levels such as the p-value associated with a specific decision. Although each provides somewhat different insights, these methods are related, and both can be used to make decisions about parameters.

Formally, there is a close relationship between the test of a hypothesis about any parameter, say θ, and the CI for θ. If lower confidence limit (L) or upper confidence limit (U) is a $100(1 - \alpha)\%$ CI for the parameter θ, the test of size α of the hypothesis

$$H_0 : \theta = \theta_0$$

$$H_1 : \theta \neq \theta_0$$

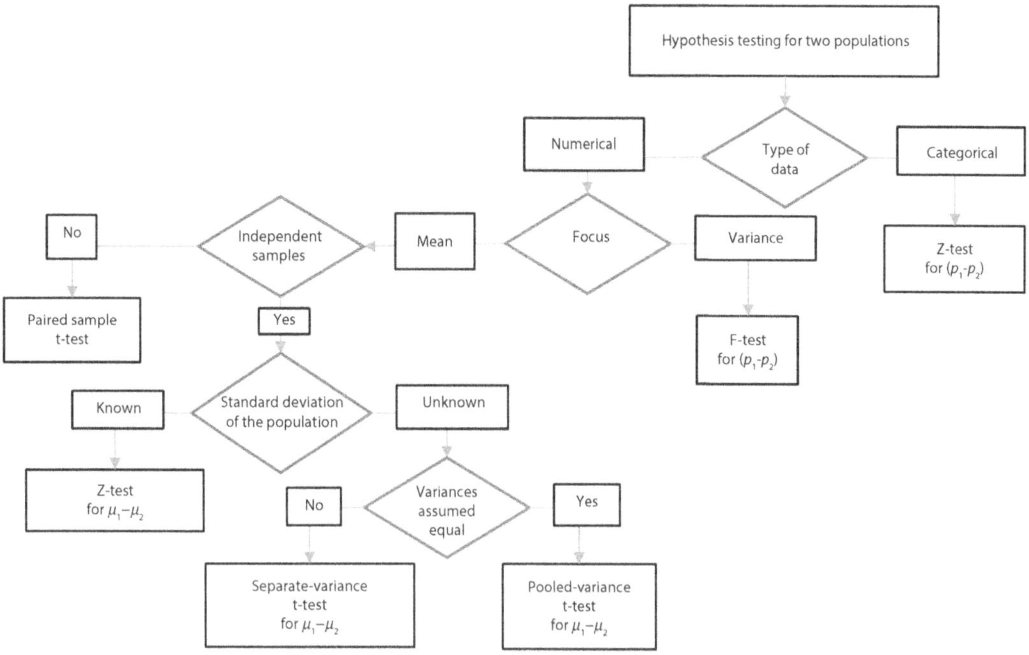

Fig. 7.8 Flow chart to select a hypothesis test for comparing two populations. (Source: Author's creation)

will lead to rejection of H_0 if and only if θ_0 is not in the $100(1-\alpha)\%$ CI [L, U] (Montgomery and Runger 2014: 293).

For example, a company purchases plastic pipes in lots of 10,000, and a plastic pipe manufacturer competes for being a supplier of the company. The company manager wants assurance that no more than 1% of the pipes in any given lot are defective. Since the company cannot test each of the 10,000 pipes in a lot, the manager must decide whether to accept or reject a lot based on an examination of a sample of pipes selected from the lot. If the number x of defective pipes in a sample of, say $n = 100$, is large, the manager will reject the lot and send it back to the manufacturer. Thus, the manager wants to decide whether the proportion p of defectives in the lot exceeds 0.01, based on the information contained in a sample, i.e., $H_0: p = 0.01$ and $H_1: p > 0.01$. If a CI for p falls below 0.01 (or sample proportion $\hat{p}$ is not included in the CI), then the company will accept the lot and can be confident with a specific level that the proportion of defectives is less than 1%; otherwise, he will reject it. This example shows a one-tailed hypothesis test.

Other cases would use a two-tailed hypothesis test. Recall from the previous section that, in finding the value of z (or t) used in a $(1-\alpha)100\%$ CI, the value of α is divided in half and $\alpha/2$ is placed in both the upper and lower tails of the z (or t) distribution. Consequently, CIs are designed to be two-directional. Using a two-directional technique when a one-directional method is utilized leads the Six Sigma team to understate the level of confidence associated with the method (Sincich 1996). However, hypothesis tests are appropriate for either one- or two-directional decisions about a population parameter.

7.10.2 Comparing Two Population Means: Independent Samples

In ► Sect. 7.8.2.1, we stated that the standard deviation of the population under study is not known in every case. Likewise, when we take a random sample from each of two independent populations, we do not know the standard deviation of either population. However, we need to know whether we can assume that variances in the two populations are equal, because the method used to compare the means of each population depends on whether we can assume that the variances of the two populations are equal as shown above in ◘ Fig. 7.8.

7.10.2.1 Population Variances Unknown and Assumed to Be Equal

If we assume that the random samples are independently selected from two populations and that the populations are normally distributed and have equal variances, a *pooled-variance t-test* is used to determine whether there is a significant difference between the means of the two populations (See the path in ◘ Fig. 7.8). If the populations are not normally distributed, the pooled-variance *t*-test can still be used if the sample sizes are large enough (typically more than 30 for each sample). The following box describes the procedure.

Testing Procedure for Comparing Two Independent Population Means

Assume that two independent random samples of size n_1 and n_2 observations were obtained from normally distributed populations with means μ_1 and μ_2 and a common variance. If the observed sample variances are s_1^2 and s_2^2 and the observed sample means are $\bar{x}_1$ and $\bar{x}_2$, the procedure summarized in ◘ Table 7.14 can be used for testing the difference between two population means with the significance level α, where $\mu_1 - \mu_2$ is the hypothesized difference between the means or 0 (i.e., particular numerical value specified for $\mu_1 - \mu_2$ in H_0), t_α is the t-value such that $P(t > t_\alpha) = \alpha$, $t_{\alpha/2}$ is the t-value such that $P\left(t > t_{\frac{\alpha}{2}}\right) = \alpha/2$, and s_p^2 is a commonly pooled estimator of the equal population variance and is computed as the weighted average of the two sample variances s_1^2 and s_2^2 as follows:

$$s_p^2 = \frac{(n_1 - 1)s_1^2 + (n_2 - 1)s_2^2}{(n_1 + n_2 - 2)} \quad (7.74)$$

◘ **Table 7.14** Hypothesis testing for two population means (*pooled-variance t-test*)

One-tailed test	*Two-tailed test*
$H_0 : \mu_1 = \mu_2$ or $\mu_1 - \mu_2 = 0$ $H_1 : \mu_1 - \mu_2 > 0$ (*or* $H_1 : \mu_1 - \mu_2 < 0$)	$H_0 : \mu_1 = \mu_2$ or $\mu_1 - \mu_2 = 0$ $H_1 : \mu_1 \neq \mu_2$ or $\mu_1 - \mu_2 \neq 0$
Test statistic: $t = \dfrac{(\bar{x}_1 - \bar{x}_2) - (\mu_1 - \mu_2)}{\sqrt{s_p^2\left(\dfrac{1}{n_1} + \dfrac{1}{n_2}\right)}}$ (7.73)	
Rejection region Reject H_0 if $t > t_\alpha$ (*or* $t < t_\alpha$)	*Rejection region* $\lvert t \rvert > t_{\alpha/2}$

Source: Author's creation

Assumptions: randomly selected sample, approximately normally distributed population, unknown population variance, variances assumed equal.

The following example shows how to use the pooled-variance t-test.

▶ Example 41

A Six Sigma team investigates customer complaints about the round metal parts produced in the plant over the last month. The team wants to determine whether the average diameter of the round metal parts produced by the day shift is 3 cm larger than the average of the diameter of the parts produced by the evening shift. The quality inspectors randomly select ten pieces from each shift. The mean diameter was 32.5 cm in the first shift and 37.3 cm in the second shift. Standard deviations were 5.2 cm and 9.6 cm in the two shifts, respectively. Normality was plausible for the means of the diameter in both shifts. It is assumed that samples are independent and have common variance. Perform a hypothesis test to answer the Six Sigma team's question at $\alpha = 0.05$ significance level. ◀

■ **Solution**

Let's follow the steps of the hypothesis testing to answer this question.

- Step 1: Check the assumptions and conditions.

For the independence assumption, we assume that samples taken in each shift are independent of each other. The sample size assumption is that the sample size is large enough ($n \geq 30$) and normality is plausible. Because the sample size is 10, we will assume that the sampling distribution of the statistic has t distribution. For the randomization condition, we know that samples have been randomly selected. For the 10% condition, we will assume that the 10 samples taken in each shift are not larger than 10% of the population.

- Step 2: State the null hypothesis H_0 and the alternative hypothesis H_1.

Let μ_1 represent the mean of the diameters produced in day shift and μ_2 represent the mean of the diameters produced in the evening shift. We can state the null hypothesis and alternative hypothesis as follows:

$$H_0: \mu_1 - \mu_2 = 3\,\text{cm}$$

$$H_0: \mu_1 - \mu_2 > 3\,\text{cm}$$

As presented in the hypotheses, this is a one-tailed test. The null hypothesis claims that the difference between the means of diameters produced in day and evening shifts is 3 cm, while the alternative hypothesis claims that the difference is greater than 3 cm.

- Step 3: Choose the level of significance, α.

Generally, the default level of significance is 0.05 in statistical software packages, and in this example, it is given as 0.05 ($\alpha = 0.05$).

- Step 4: Choose the sample size, n.

The sample size is already given in our example. The ten randomly selected parts are measured for quality inspection.

- Step 5: Collect the data.

We assume that the data were already collected since the question gives us the means and standard deviations of the two shifts.

- Step 6: Compute the appropriate test statistic.

Since we have two samples with less than 30 observations in each shift and having a common variance, we use pooled-variance t-test statistic for two samples.

$$t = \frac{(\bar{x}_1 - \bar{x}_2) - (\mu_1 - \mu_2)}{\sqrt{s_p^2\left(\frac{1}{n_1} + \frac{1}{n_2}\right)}} = \frac{(32.5 - 37.3) - 3}{\sqrt{59.6\left(\frac{1}{10} + \frac{1}{10}\right)}}$$

$$= \frac{-7.8}{3.4525} = -2.259 \cong -2.26$$

where

$$s_p^2 = \frac{(n_1 - 1)s_1^2 + (n_2 - 1)s_2^2}{(n_1 + n_2 - 2)}$$

$$= \frac{(10-1)5.2^2 + (10-1)9.6^2}{(10+10-2)} = 59.6$$

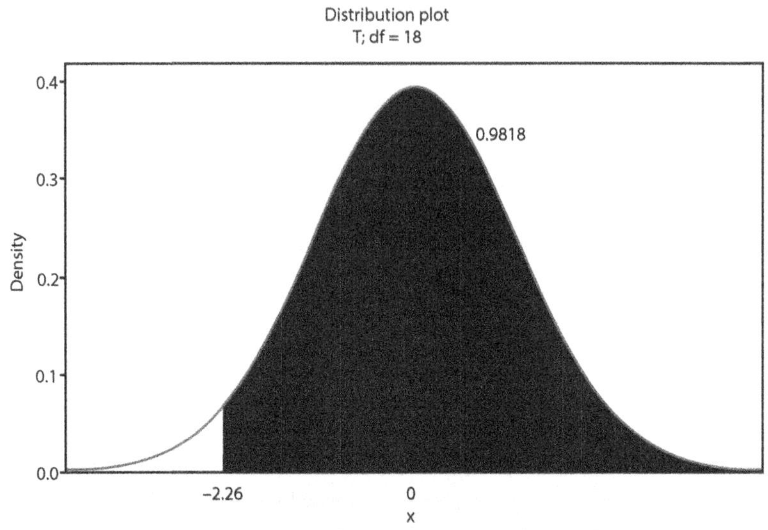

Image 7.19 Observed significance level of $t_{\text{computed}} = -2.26$ for the upper-tailed test. (Source: Author's creation based on Minitab)

- Step 7: Calculate the *p*-value and identify the critical values of test statistic.

Before calculating the *p*-value, let's compute the degrees of freedom (*df*) for the *t*-test for two samples. The *df* is $n + n - 2 = 18$. Since the hypothesis testing is one-tailed, we will calculate *p*-value as follows:

$$P = P\left(t > \left|t_{\text{computed}}\right|\right) = P\left(t > -2.26\right) = 0.982$$

In the percentage points of the *t* distribution table (Table A.4), the probability of $t = -2.26$ with $df=18$ is approximately 0.982. Therefore, the *p*-value of the test statistic will be 0.982.

To run two-sample *t*-test in Minitab, click on Stat→Basic Statistics→2-Sample t. In the next input screen, select "summarized data," and enter the variables sample size, sample mean, standard deviation. Click on "Options," and select alternative hypothesis as "Difference > hypothesized difference"; select "assume equal variances." Click on OK→OK. The results are shown in Session Window. The probability distribution of the *t*-test is presented in Image 7.19.

- Step 8: Apply decision rule, and express the statistical finding in the scope of the question.

As the main decision rule, let's compare α (0.05) with the *p*-value. If $\alpha >$ *p*-value, the null hypothesis is rejected. Since the *p*-value (0.982) is greater than the α value (0.05), the hypothesis testing fails to reject the null hypothesis. In other words, there is not enough evidence to reject the null hypothesis. As a result, we are 95% confident that the difference between the averages of the diameter of the round metal parts produced by the day shift and the night shift equals 3 cm (*p*-value = 0.982).

7.10.2.2 Population Variances Unknown and Assumed to Be Unequal

If the assumption that the two independent populations have equal variances cannot be made, a commonly pooled estimator of the two sample variances cannot be calculated. Instead, the *separate-variance t-test* is used. The following box describes the procedure.

Testing Procedure for Comparing Two Independent Population Means

Assume that two independent random samples of size n_1 and n_2 observations were obtained from normally distributed populations with means μ_1 and μ_2 and unequal variances. If the observed sample variances are s_1^2 and s_2^2 and the observed sample means are $\bar{x}_1$ and $\bar{x}_2$, the procedure summarized in Table 7.15 can be used for testing the difference between two population means with the significance level α, where $\mu_1 - \mu_2$ is the hypothesized difference between the means or 0 (i.e., particular numerical value specified for $\mu_1 - \mu_2$ in H_0), t_α is the t-value such that $P(t > t_{v,\alpha}) = \alpha$, and $t_{\alpha/2}$ is the t-value such that $P\left(t > t_{v,\frac{\alpha}{2}}\right) = \alpha / 2$.

The degrees of freedom v for the Student's t statistic is given by the following:

$$v = \frac{\left(\frac{s_1^2}{n_1} + \frac{s_2^2}{n_2}\right)^2}{\left(\frac{s_1^2}{n_1}\right)^2 / (n_1 - 1) + \left(\frac{s_2^2}{n_2}\right)^2 / (n_2 - 1)} \quad (7.76)$$

Table 7.15 Hypothesis testing for two population means (*separate-variance t-test*)

One-tailed test	*Two-tailed test*
$H_0 : \mu_1 = \mu_2$ or $\mu_1 - \mu_2 = 0$ $H_1 : \mu_1 - \mu_2 > 0$ (*or* $H_1 : \mu_1 - \mu_2 < 0$)	$H_0 : \mu_1 = \mu_2$ or $\mu_1 - \mu_2 = 0$ $H_1 : \mu_1 \neq \mu_2$ or $\mu_1 - \mu_2 \neq 0$
Test statistic: $t = \frac{(\bar{x}_1 - \bar{x}_2) - (\mu_1 - \mu_2)}{\sqrt{\frac{s_1^2}{n_1} + \frac{s_2^2}{n_2}}}$ (7.75)	
Rejection region Reject H_0 if $t > t_\alpha$ (*or* $t < t_\alpha$)	*Rejection region* $\lvert t \rvert > t_{\alpha/2}$

Source: Author's creation

Assumptions: randomly selected sample, approximately normally distributed population, unknown population standard deviation, assumed unequal variances.

► Example 42

The manufacturing firm given in ► Example 41 wants to test whether there is a statistically significant difference between the averages of means of diameter of the round metal parts produced by the day and evening shifts. The quality inspectors randomly select six pieces from each shift. The mean of the diameter was 32.5 cm in the first shift and 37.3 cm in the second shift. Standard deviations were 5.2 cm and 9.6 cm in both shifts, respectively. Normality was plausible for the means of the diameter in both shifts. It is assumed that samples are independent and variances are unequal. Perform a hypothesis test to determine whether there is any statistically significant difference between the averages of the means of diameter of the round metal parts produced by the day and evening shifts at $\alpha = 0.05$ level of significance.◄

Solution

Let's follow the eight-step process given above to perform a hypothesis test to answer this question.

- Step 1: Check the assumptions and conditions.

For the independence assumption, we assume that samples taken in each shift are independent of each other. The sample size assumption expects that the sample size must be large enough ($n \geq 30$). In this example, since the sample size is 6, we will assume that the sampling distribution of statistic has t distribution. For the normality assumption, it is given that the normality is plausible. For the randomization condition, we know that the samples have been randomly selected. For the 10% condition, six samples taken in each shift are not larger than 10% of the population.

- Step 2: State the null hypothesis H_0 and the alternative hypothesis H_1.

Let μ_1 represent the mean of the day shift and μ_2 represent the mean of the evening shift. We can state the null hypothesis and alternative hypothesis as follows:

$$H_0 : \mu_1 = \mu_2 \text{ or } H_0 : \mu_1 - \mu_2 = 0$$

$$H_1 : \mu_1 \neq \mu_2 \text{ or } H_1 : \mu_1 - \mu_2 \neq 0$$

As presented in the hypotheses, this is a two-tailed hypothesis testing. The null hypothesis claims that there is no difference between the means of diameters produced by the day and evening shifts, while the alternative hypothesis claims the opposite.

7

- Step 3: Choose the level of significance, α.

The level of confidence is given as $\alpha = 0.05$ for our hypothesis testing.

- Step 4: Choose the sample size, *n*.

The sample size is already given in our example. The randomly selected six parts from each shift are measured during the quality inspection. Thus, the sample size from the day shift is $n_1 = 6$, and the sample size from the evening shift is $n_2 = 6$.

- Step 5: Collect the data.

We assume that the data were already collected, since the question gives us the means and standard deviations.

- Step 6: Compute the appropriate test statistic.

Since we have two independent samples that have less than 30 observations in each shift, let's use the separate-variance *t*-test statistic for two samples.

$$t = \frac{(\bar{X}_1 - \bar{X}_2) - (\mu_1 - \mu_2)}{\sqrt{\frac{s_1^2}{n_1} + \frac{s_2^2}{n_2}}} = \frac{(32.5 - 37.3) - 0}{\sqrt{\frac{5.2^2}{6} + \frac{9.6^2}{6}}}$$

$$= -1.08$$

In the *t*-test statistic, the hypothesized difference between the two means is $\mu_1 - \mu_2 = 0$. Since we expect that the hypothesized difference between the two means is zero, the difference will be zero in the *t*-test statistic calculation.

- Step 7: Calculate the *p*-value and identify the critical values of test statistic.

Before calculating the *p*-value, let's compute the degrees of freedom (*df*) for the *t*-test for two samples as follows:

$$df = n + n - 2 = 6 + 6 - 2 = 10$$

Since the hypothesis testing is two-tailed, we will calculate the *p*-value as follows:

$$P = 2P\left(t > \left|t_{computed}\right|\right)$$

$$P = 2P\left(t > \left|\ 1.08\right|\right) = 2P(t > 1.08)$$

$$P = 2(0.1539) = 0.3077$$

In the percentage points of the *t* distribution table (Table A.4), the probability of $t = 1.08$ with $df = 10$ is 0.1539. Because the hypothesis testing is two-tailed, we will multiply the probability by two, and the *p*-value of the test statistic will be 0.3077.

To run two-sample hypothesis testing in Minitab, click on Stat→Basic Statistics→2-sample t. In the next input screen, enter 6 for both sample sizes, 32.5 and 37.3 for sample means, and 5.2 and 9.6 standard deviation for sample 1 and sample 2, respectively. Enter 95 for confidence interval, and select difference ≠ hypothesized difference in alternative hypothesis. Click on OK→OK. The results are shown in Session Window. The probability distribution of the *t*-test is presented in ◘ Image 7.20.

- Step 8: Apply decision rule, and express the statistical finding in the scope of the question.

As the main decision rule, let's compare the α (0.05) with the *p*-value. If $\alpha >$ *p*-value, the null hypothesis is rejected. The manually calculated t-test statistic ($t = -1.08$), $df = 10$, and *p*-value (0.307) are presented above. Since the *p*-value (0.307) is greater than the α value (0.05), the hypothesis testing fails to reject the null hypothesis. In other words, we don't have enough evidence to reject the null hypothesis. As a result, it can be concluded that we are

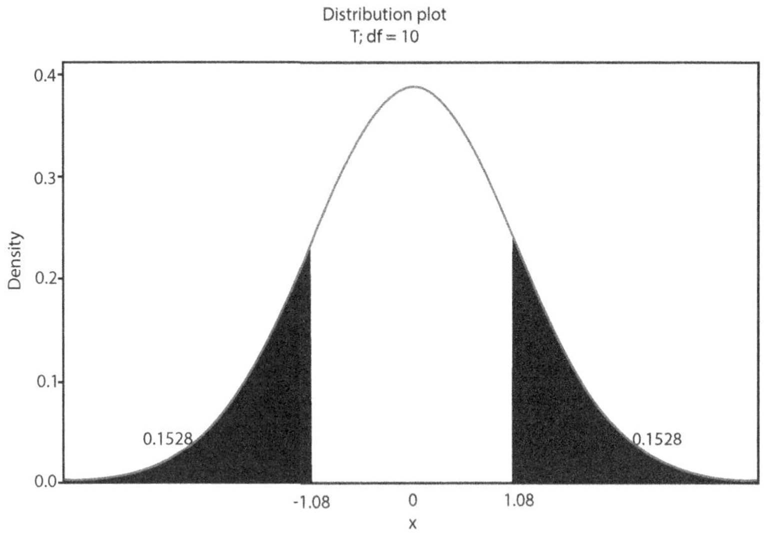

Image 7.20 Observed significance level (p-value) of $t_{computed} = 1.08$ for the two-tailed test. (Source: Author's creation based on Minitab)

95% confident that the day and evening shifts do not produce statistically significantly different diameters at $\alpha = 0.05$.

7.10.3 Comparing Two Population Means: Dependent (Paired) Samples

The hypothesis testing procedures presented in ► Sect. 7.10.2 examine differences between the means of two independent populations. If the samples are collected from related populations or, in other words, when the results of the first population are not independent of the results of the second population, then the difference between these two populations is tested by using paired t-test. There are two situations that involve related data:

1. *Repeated measurements (before and after the experiment)*: those characterized by a measurement, an intervention of some type, and then another measurement
2. *Matched samples*: those paired together according to some characteristic of interest.

When repeated measurements are taken on the same items or individuals, it is assumed that the same items or individuals will behave alike if treated alike. In this kind of experiment, each item generates two data values, one before an operation and one after an operation is performed. In this kind of experiment, we assume that the individual differences between the items are controlled and will not bias the results of the experiment. Suppose that we want to determine whether a training program will increase labor productivity. To do so, we would record before- and after-training outputs of a random sample of employees. Thus, the before and after pair of numbers for an employee are dependent and form a paired sample.

The second situation is to pair or match together according to some characteristic of interest. Suppose that a footwear provider develops a new environmentally friendly sole material for children's shoes. The wear rate is the CTQ characteristic and is evaluated by measuring the change in sole thickness with a sensitive thickness gage. The provider wants the new material to have the same capabilities as the current sole material while providing lower wearing rates than the current material. To compare these two types of material, an experiment is designed where 20 particular children try each type of sole material. To avoid any bias and to account for variation in activity of children, each child wears shoes where the new material is used in one shoe and the current material in the other shoe. The new material is randomly assigned to either the left or right shoe. This approach provides a kind of experimental control; each sample application of the new sole material is paired with a sample application with the current sole material. In this way, the difference in the usability could be studied with the knowledge that both kinds of soles would have been exposed to exactly the same ravages of weather, heat, usage, and other variables. Regardless of whether the related data are obtained as repeated measurements or matched samples, the objective is to examine the difference between two measurements by reducing the effect of the variability due to the items or individuals themselves.

7

For hypothesis testing, we are interested in the distribution of the differences in the measurements of each sample. Hence, there is only one sample. We are investigating whether the mean of the distribution of differences in the measurements (denoted as μ_d) is 0. We assume the distribution of the population of differences follows the normal distribution. The test statistic follows the t distribution. If $\bar{d}$ is the mean of the difference between the paired or related observations, s_d is the standard deviation of the differences between the paired or related observations, and n is the number of paired observations; then the value of t statistic with $n - 1$ degrees of freedom is calculated through Eqs. 7.77 and 7.78.

$$t = \frac{\bar{d} - \mu_d}{s_d / \sqrt{n}} \tag{7.77}$$

where

$$s_d = \sqrt{\frac{\Sigma\left(d - \bar{d}\right)^2}{n-1}} \tag{7.78}$$

Formally, the paired t-test is defined in the following box.

Testing Procedure for Comparing Two Dependent Population Means

A random sample of n matched pairs of observations was obtained from distributions with means μ_1 and μ_2. Let $\bar{d}$ and s_d denote the observed sample mean and standard deviation for the n differences, and μ_d refers to the population mean of the distribution of differences, i.e., $\mu_d = \mu_1 - \mu_2$. If the population distribution of the differences is normal distribution, the following procedure can be used for testing the equality of means with the significance level α (Table 7.16), where μ_d is the hypothesized population mean of the distribution of differences or 0 (i.e., particular numerical value specified for μ_d in H_0), t_α is the t-value such that $P(t > t_\alpha) = \alpha$, and $t_{\alpha/2}$ is the t-value such that

$$P\left(t > t_{\frac{\alpha}{2}}\right) = \alpha / 2.$$

Table 7.16 Hypothesis testing for the paired population means

One-tailed test	*Two-tailed test*
$H_0 : \mu_d = 0$ $H_1 : \mu_d > 0$ $(or\ H_1 : \mu_d < 0)$	$H_0 : \mu_d = 0$ $H_1 : \mu_d \neq 0$
Test statistic: $t = \frac{\bar{d} - \mu_d}{s_d / \sqrt{n}}$ (7.77)	
Rejection region Reject H_0 if $t > t_\alpha$ $(or\ t < t_\alpha)$	*Rejection region* $\|t\| > t_{\alpha/2}$

Source: Author's creation

Assumptions: randomly selected sample, approximately normally distributed population of differences, unknown population standard deviation.

It is inappropriate to apply the paired t-test when the sample size is small, and the population of differences is decidedly non-normal. In this case, an alternative non-parametric procedure is followed. The non-parametric techniques are not in the scope of this chapter.

▶ Example 43

A group of chefs want to compare the height of the cakes baked with two eggs or three eggs. Keeping other conditions and the ingredients constant, they made 16 consecutive experiments and measured the heights of 16 cakes. The measurement shows that the average of differences between the heights of the cakes in two groups of cakes baked with two eggs and three eggs was 6.75 mm and the standard deviation was 8.234 mm. Normality was plausible for the differences between the heights of the cakes. Perform an appropriate hypothesis testing, and show whether the height of cakes baked with two eggs differs from cakes baked with three eggs.◀

■ Solution

Because there are two groups of cakes compared in this example, we can use paired t-test to answer this question.

- Step 1: Check the assumptions and conditions.

For the independence assumption, we assume that each experiment is independent of each other. The sample size assumption expects that the sample size must be large enough ($n \geq 30$). Since sample size is 16 in this example, we will assume that the sampling distribution of statistic has t distribution. For normality assumption, normality was plausible for the differences between the heights of the cakes. For randomization condition, we know that each experiment has been randomly run. For a 10% condition, 16 experiments are not larger than 10% of the population.

- Step 2: State the null hypothesis H_0 and the alternative hypothesis H_1.

$$H_0 : \mu_d = 0$$

$$H_1 : \mu_d \neq 0$$

As presented in the hypotheses, this is a two-tailed test, since the chefs are investigating whether the averages of differences of the heights of the cakes are different if two or three eggs are used. The null hypothesis claims that there is no difference between the means of differences of the cake heights, and the alternative hypothesis claims the opposite. In our hypotheses above, the μ_d variable refers to the population mean of the distribution of differences. We could state the relationship between the variables as follows: $\mu_d = \mu_1 - \mu_2$, where μ_1 is the average of the heights of the cakes baked with two eggs and μ_2 is the average of the heights of the cakes baked with three eggs. To simplify the notations, let's use μ_d variable in our calculations.

- Step 3: Choose the level of significance, α.

Let's use $\alpha = 0.05$ in this hypothesis testing.

- Step 4: Choose the sample size, n.

The question already states that they baked 16 cakes in the experiments ($n = 16$).

- Step 5: Collect the data.

We have repeated measurements in the experiments by changing two eggs to three eggs in the ingredients. We were given the average $(\bar{d} = 6.75 \text{ mm})$ and standard deviation of the differences ($s_d = 8.234$ mm).

- Step 6: Compute the appropriate test statistic.

Since we have less than 30 observations, let's focus on t distribution and calculate test statistic using t distribution. μ_d is the hypothesized population mean of the distribution of differences. Since we are investigating whether the means of differences of the heights of the two groups of cakes are different, μ_d will be zero.

$$t = \frac{\bar{d} - \mu_d}{s_d / \sqrt{n}} = \frac{6.75 - 0}{8.234 / \sqrt{16}} = 3.28$$

$$df = n - 1 = 16 - 1 = 15$$

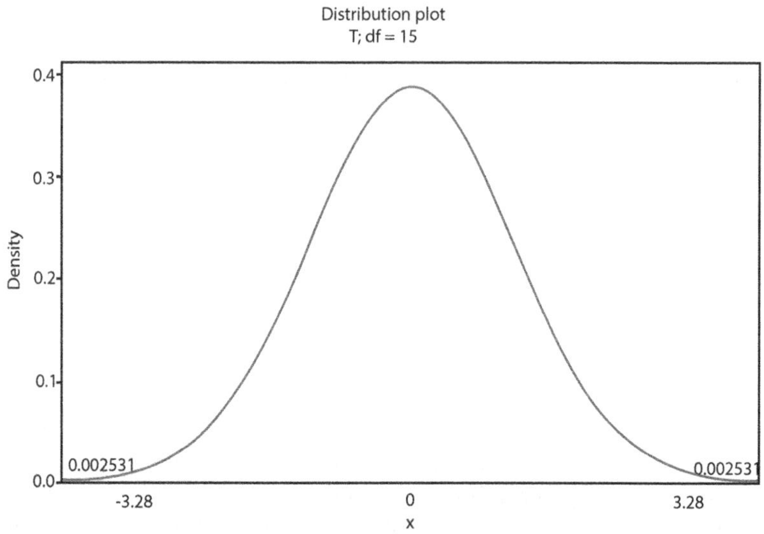

Image 7.21 Observed significance level (p-value) of $t_{\text{computed}} = 3.28$ for the two-tailed test. (Source: Author's creation based on Minitab)

- Step 7: Calculate the p-value and identify the critical values of test statistic.

$$P = 2P\left(t > \left|t_{\text{computed}}\right|\right) = 2P\left(t > 3.28\right)$$

$$P = 2\left(0.0025\right) = 0.005$$

In the percentage points of the t distribution table (Table A.4), the probability of $t = 3.28$ with $df = 15$ is 0.0025. Because the hypothesis testing is a two-tailed one, we will multiply the probability by two, and the p-value of the test statistic will be 0.005.

To run paired t-test in Minitab, click on Stat→Basic Statistics→Paired t. In the next input screen, select "summarized data (differences)," and enter the variables sample size (16), sample mean (6.75), and standard deviation (8.234); then click on "Options," and select alternative hypothesis as "Difference ≠ hypothesized difference." Click on OK→OK. The results are shown in Session Window. The probability distribution of the t-test is presented in Image 7.21.

- Step 8: Apply decision rule, and express the statistical finding in the scope of the question.

As the main decision rule, let's compare α (0.05) with the p-value. If $\alpha > p$-value, the null hypothesis is rejected. Manually calculated t-test statistic ($t = 3.28$) and p-value (0.005) are presented above. Since the p-value (0.005) is lower than the α value (0.05), the hypothesis testing rejects the null hypothesis. In other words, we have compelling evidence against the null hypothesis. As a result, the chefs can say that they are 95% confident the average of the differences of the cake heights baked with two eggs and three eggs will be statistically different (p-value = 0.005). The amount of the eggs used in cake baking process impacts the cake heights.

7.10.4 Comparing Two Normally Distributed Population Variances

There are many situations which require comparing the variances of two normally distributed populations in Six Sigma projects. For example, the Six Sigma team may be concerned with which process or production line has the smaller variance in a specific output. Although the means of the outputs produced in two processes or production lines may be satisfactory, this process has more tendency to produce outputs that do not meet specifications, if one of the processes has larger variance. In some cases, assuming equal variances is a technical requirement of conducting the Student's *t*-test for comparing the means of two small independent samples ($\mu_1 - \mu_2$), and thus it is subject to be confirmed by implementing statistical hypothesis testing procedures. If the two population variances are greatly different, any inferences derived from the *t*-test are suspect. Consequently, it is important to detect a significant difference between the two variances, if it exists, before applying the small-sample *t*-test (Sincich 1996: 562).

The probability distribution used in this analysis is *F* distribution to test whether two samples are from the populations that have the equal variances. It was named to honor Sir Ronald Fisher, one of the founders of modern-day statistics. The test statistic that follows the *F* distribution is quite sensitive to the normality assumption and requires that the data must be interval-scale.

Assume that two random samples of size n_1 and n_2 observations were obtained from two normally distributed populations. If the observed sample variance of the first sample drawn from the population with variance σ_1^2 is s_1^2 and the observed sample variance of the second sample drawn from the population with variance σ_2^2 is s_2^2, then the random variable

$$F = \frac{s_1^2 / \sigma_1^2}{s_2^2 / \sigma_2^2} \tag{7.79}$$

follows the *F* distribution with $v_1 = n_1 - 1$ and $v_2 = n_2 - 1$ degrees of freedom, which is denoted by F_{v_1,v_2}. With the significance level α, the number $F_{v_1,v_2,\alpha}$ refers to the critical value of the *F* distribution. This value is obtained from the *F* tables (Tables A.5, A.6, A.7, A.8, and A.9).

To test the hypothesis of equality of variances, the test statistic is computed as shown in Eq. 7.80.

$$F = \frac{s_1^2}{s_2^2} \tag{7.80}$$

To reduce the size of the table of critical values of *F* in practical applications, the larger sample variance is placed in the numerator; hence, the tabulated *F*-ratio is always larger than 1.00. Thus, the upper-tail critical value is the only one required. Under this condition, it is not necessary to divide the significance level in half. For conducting a two-tailed *F* test, the critical value of *F* is found by dividing the significance level in half ($\alpha/2$) and then referring to the appropriate degrees of freedom in Tables A.5, A.6, A.7, A.8, and A.9. Formally, comparing two normally distributed population variances is defined in the following box.

Example 44 will show the use of the *F* test.

7

Testing Procedure for Comparing Two Population Variances

Two random samples of n_1 and n_2 observations were obtained from two normally distributed populations. The observed sample variance of the first sample is s_1^2 and the observed sample variance of the second sample is s_2^2, two population variances can be compared, with the significance level α, by using the procedure summarized in ◘ Table 7.17, where $F_{v_1,v_2,\alpha}$ and $F_{v_1,v_2,\alpha/2}$ are values that locate an area of α *and* $\alpha/2$, respectively, in the upper tail of the F distribution with the degrees of freedom for the sample variance in the numerator, $v_1 = n_1 - 1$, and the degrees of freedom for the sample variance in the denominator, $v_2 = n_2 - 1$.

Assumptions: randomly selected samples from two populations, normally distributed two populations.

◘ **Table 7.17** Hypothesis testing for two population variances

One-tailed test	*Two-tailed test*
$H_0: \sigma_1^2 = \sigma_2^2$ $H_1: \sigma_1^2 > \sigma_2^2$	$H_0: \sigma_1^2 = \sigma_2^2$ $H_1: \sigma_1^2 \neq \sigma_2^2$
Test statistic: $F = \frac{s_1^2}{s_2^2} = \frac{\text{Larger sample variance}}{\text{Smaller sample variance}}$ (7.80)	
Rejection region Reject H_0 if $F > F_{v_1,v_2,\alpha}$	*Rejection region* $F > F_{v_1,v_2,\alpha/2}$

Source: Author's creation

▶ Example 44

The quality control manager of a syringe manufacturing company wants to see whether the manufacturing process works well both in the first and second shifts. Thirty-five injection syringes are randomly sampled from each shift, and the lengths of the syringes are measured with results shown in ◘ Table 7.18.

Test, at 5% significance level, whether the population standard deviations for the lengths of injection syringes produced were the same for the first shift as for the second shift. If the standard deviation in the syringe lengths should be no greater than 0.03 cm for the quality standards, evaluate the performance of these shifts. ◀

◘ **Table 7.18** The lengths of injection syringes

The lengths of the syringes from the first shift production process						
12.6035	12.6162	12.5538	12.5959	12.5908	12.6238	12.6162
12.5730	12.5984	12.6086	12.5984	12.5984	12.5933	12.6111
12.6035	12.5832	12.6009	12.5887	12.6187	12.5834	12.5349
12.5984	12.6162	12.6060	12.5730	12.6060	12.6338	12.6086
12.5959	12.6365	12.6340	12.6309	12.6187	12.6387	12.6162
The lengths of the syringes from the second shift production process						
12.6136	12.6009	12.6060	12.6009	12.5933	12.6086	12.5908
12.5730	12.5538	12.6035	12.5832	12.6340	12.5806	12.5832
12.5857	12.6009	12.6035	12.6086	12.5552	12.5730	12.5984
12.5882	12.5908	12.6111	12.6136	12.5908	12.6162	12.6238
12.6060	12.6162	12.5857	12.6390	12.5425	12.6387	12.5933

Source: Author's creation

- **Solution**

This example requires designing a study that compares the population standard deviations of the lengths of injection syringes produced in both the first and second shifts. For this example, the steps will be followed as given below.

- Step 1: Check the assumptions and conditions.

The most critical assumption in testing the equality of the two variances is the normality. The assumption of a normal population is required regardless of whether the sample size is large or small for each data set taken from two shifts. The normal probability plots for the data sets from the first and second shift, respectively. The sample data for syringe lengths observed in the first shift is approximately normal because the values are almost placed on a straight line, and the p-value (0.195) of Anderson Darling (AD) normality test exceeds the significance level of 0.05. Similarly, the sample data for syringe lengths observed in the second shift seem to be normally distributed, with the greater p-value of the Anderson Darling statistic than the predetermined significance level, i.e., p-value = 0.457 > α = 0.05. Thus, there is no evidence for the violation of normality for each sample data set. For the independence assumption, it is assumed that the lengths of syringes produced in two shifts are independent of each other. Thirty-five injection syringes are randomly sampled from each of these two shifts.

- Step 2: State the null hypothesis H_0 and the alternative hypothesis H_1.

The example investigates whether the population standard deviations of the lengths of injection syringes produced in the two shifts are the same, i.e., $\sigma_1 = \sigma_2$. Since the null and alternative hypotheses must be stated in terms of σ^2 (rather than σ), the equivalent hypotheses are articulated as follows:

$$H_0 : \sigma_1^2 = \sigma_2^2$$

$$H_1 : \sigma_1^2 \neq \sigma_2^2$$

- Step 3: Choose the level of significance, α.

The testing rule is designed with the significance level $\alpha = 0.05$.

- Step 4: Choose the sample sizes, n_1 and n_2.

The quality control manager randomly sampled 35 injection syringes from each shift. The sizes of these two random samples are $n_1 = 35$ and $n_2 = 35$.

- Step 5: Collect the data.

The collected data were given in ◘ Table 7.18.

- Step 6: Compute the appropriate test statistic.

A random sample of 35 syringe lengths produced in the first shift resulted in a sample variance $s_1^2 = 0.00050$, while an independent random sample of 35 syringe lengths produced in the second shift resulted in a sample variance $s_2^2 = 0.00049$. The test statistic or F-ratio is calculated as follows:

$$F = \frac{s_1^2}{s_2^2} = \frac{0.00050}{0.00049} = 1.03818$$

- Step 7: Calculate the p-value and identify the critical values of test statistic.

Because we're investigating the equality of two normally distributed population variances, we're interested in a two-tailed hypothesis test. If the variances differ significantly, we would expect the test statistic F to be much larger than 1. As the F-ratio approaches 1, it is more likely that the variances are the same and that there is stronger evidence in favor of H_0.

The significance level is $\alpha = 0.05$; then $\alpha/2 = 0.025$. With the degrees of freedom for the sample variance in the numerator, $v_1 = 35 - 1 = 34$, and the degrees of freedom for the sample variance in the denominator, $v_2 = 35 - 1 = 34$, the critical value $F_{34,34,0.025}$ is the value of F that locates an area of 0.025 in the upper tail of an F distribution. Usually the F tables are restricted to several significance levels, such as 0.05 and 0.01 for one-tailed tests and 0.10 and 0.02 for two-tailed tests (See Tables A.5, A.6, A.7, A.8, and A.9). For computing the F statistic, a more complete table is consulted, or Minitab is used.

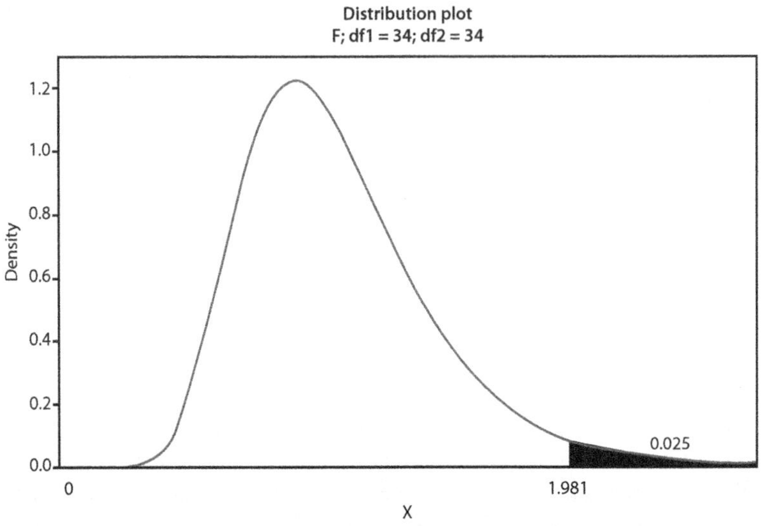

Image 7.22 Critical value of $F_{34,34,0.025}$ for the upper tail area $\alpha = 0.025$. (Source: Author's creation based on Minitab)

7

By looking up F table, the column corresponding to 34 and row 34 gives the critical value $F_{34,34,0.025}$ as 1.98112. The probability is 0.025 that an F random variable with $v_1 = 34$ and $v_2 = 34$ is greater than 1.98112.

To compute the value of $F_{34,34,0.025}$ for the F distribution in Minitab, click on Calc→Probability Distributions→ F…. In the next screen, select "Inverse Cumulative Probability"; enter 0 in "Noncentrality parameter" and 34 in the "Numerator degrees of freedom" and "Denominator degrees of freedom" boxes; and enter 0.975 for "Input constant," and click on OK. The result will be shown in Session Window. The requested value is 1.98112, which means that $P(F \leq 1.98) = 0.975$ or $P(F > 1.98) = 0.025$. Thus, using the critical value approach H_0 is rejected if $F > F_{34,34,0.025}$ (Image 7.22).

For this F test, the computed value of the test statistic is $F_{computed} = 1.03818$. For this two-tailed test, the p-value (or observed significance level) is $P = 2P(F > F_{34,34,0.025}) = 2P(F > 1.03818)$. To calculate the associated p-value, we must first calculate the cumulative distribution function (CDF), i.e., $P(F \leq 1.03818)$. To do so in Minitab, click on Calc→Probability Distributions→F. In the next screen, select "Cumulative Probability," enter 0 in "Noncentrality parameter," enter 34 in "Numerator degrees of freedom," enter 34 in "Denominator degrees of freedom"; and choose "Input constant" and enter 1.03818. Now click on OK. The result will be shown in Session Window. The requested p-value is $2[1 - P(F \leq 1.03818)]$. Therefore, from the Minitab, we get observed significance level for the upper-tail $[1 - P(F \leq 1.03818)] = [1 - 0.543176] = 0.4568$ (Image 7.23). Since the test is designed as two-tailed, this result is multiplied by 2. Thus, the p-value is $2(0.4568) = 0.9136$. Based on the p-value approach, the null hypothesis H_0 is rejected if $\alpha = 0.05 > p$.

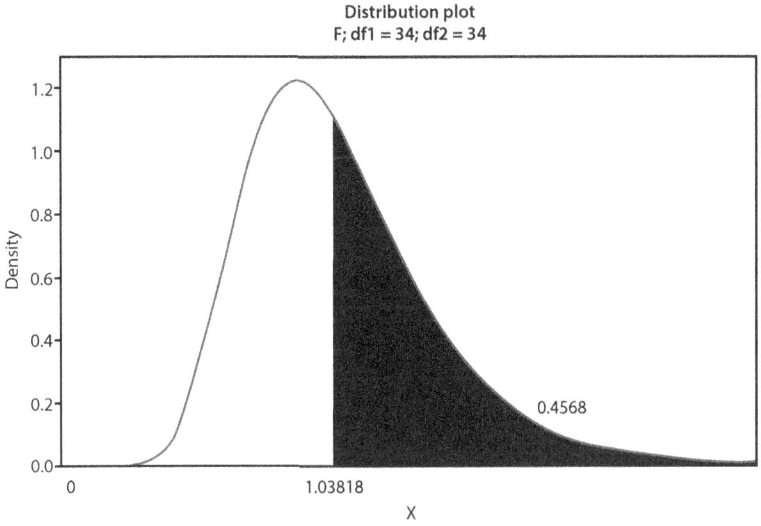

Image 7.23 Observed significance level of $F_{computed} = 1.03818$ for the upper-tailed test. (Source: Author's creation based on Minitab)

To perform a hypothesis test for comparing two population variances in Minitab, click on Stat→Basic Statistics→2 Variances.... In the next screen "Two-Sample Variance" window, select the variables for "Sample 1: (1st Shift – lengths)" and "Sample 2: (2nd Shift – lengths)," and then click on the Options. In Options tab, the upper box "Ratio" provides two alternatives in the drop-down menu for the test which can be formed based on either the ratio of two sample standard deviations or the ratio of two sample variances. Set this box to "sample 1 standard deviation/ sample 2 standard deviation," enter Confidence level (95.0), enter Hypothesized ratio (1), and select "ratio ≠ hypothesized ratio" for Alternative hypothesis; check the box "Use test and CIs based on normal distribution." Click on OK on Options Box and then click "Graphs..." to select "Summary Plot." Click OK in the Graph box and OK in the Dialog Box to see the results. The summary plot is presented in Session Window (Image 7.24).

- Step 8: Apply decision rule, and express the statistical finding in the scope of the question.

Using the critical value approach, H_0 is not rejected since the test statistic $F_{computed} = 1.03818$ is less than the critical value of $F_{34,34,0.025} = 1.98112$. The quality control manager concludes that the two standard deviations in the syringe lengths are the same in both shifts with 95% confidence or at the 5% significance level. Using *p*-value approach, *p*-value = 0.9134 indicates that the null hypothesis H_0 is not rejected as *p*-value is greater than the significance level $\alpha = 0.05$ ($p = 0.9134 > \alpha = 0.05$). Therefore, the quality control manager is confident in the decision that the manufacturing processes operate in both shifts with similar performance.

Based on the quality standards of the company, the standard deviation in the syringe lengths produced in both the first and second shifts should be no greater than 0.03 cm. To

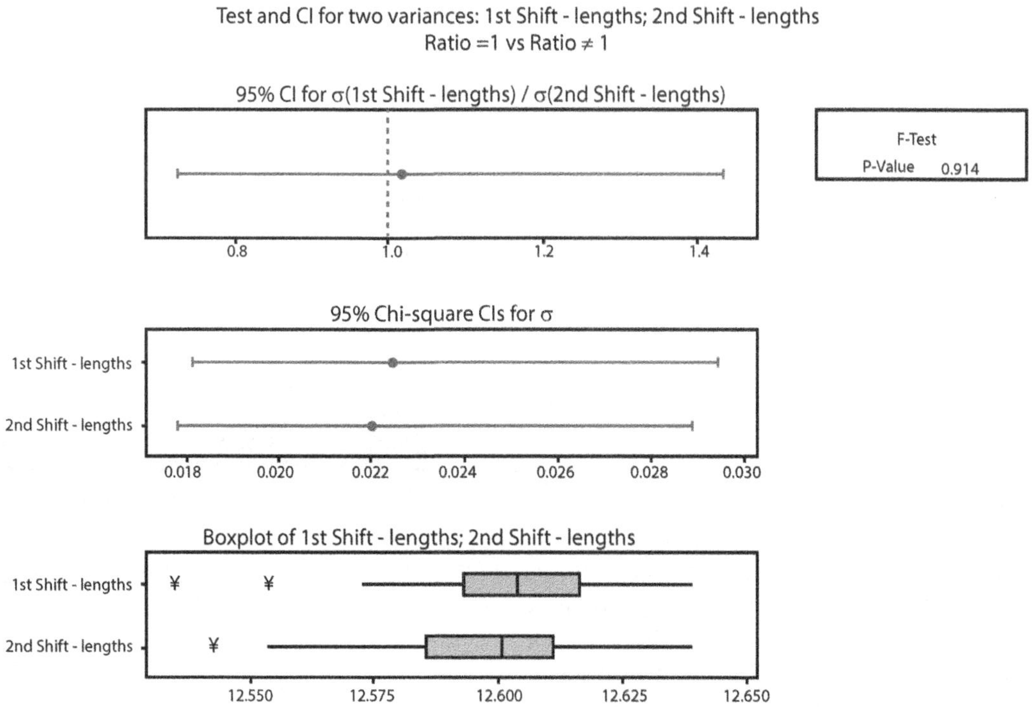

Image 7.24 Summary plot of "Two-Sample Variance" for the syringe lengths from the two shifts. (Source: Author's creation based on Minitab)

determine whether these standard deviations are consistently within the desired limits of variability, the summary plot in Image 7.24 can be examined. From the 95% CI for σ, it is determined that the true standard deviation of the 1st Shift lengths lies within the interval (0.01815; 0.02941) and the true standard deviation of the 2nd Shift lengths lies within the interval (0.01782; 0.02886). Thus, neither of the upper limits of these two intervals exceeds 0.03 cm. The summary plot also shows that the CIs are overlapped. This means that two shifts operate consistently at very similar performance levels.

7.10.5 Comparing Two Population Proportions (Large Samples)

When Six Sigma teams need to compare the number of nonconforming or defectives in two populations, hypothesis testing for the difference between two proportions is used in the analysis. If the sample size is large enough, the difference between two proportions is expected to follow a normal distribution. The statistic used in this hypothesis testing is the "proportions of the events" analyzed in the projects. Tests are summarized in the following box.

Testing Procedure for Comparing Two Proportions (Large Samples)

Assume that two independent random samples of size n_1 and n_2 with proportion of successes $\hat{p}_1$ and $\hat{p}_2$ and the population proportions p_1 and p_2 are equal. For large samples, the procedure in Table 7.19 can be used for testing the difference between two population proportions with the significance level α, where $p_1 - p_2$ is the hypothesized difference between the proportions or 0 (i.e., particular numerical value specified for $p_1 - p_2$ in H_0), z_α is the z-value such that $P(z > z_\alpha) = \alpha$, $z_{\alpha/2}$ is the z-value such that $P\left(z > z_{\frac{\alpha}{2}}\right) = \alpha/2$, and $\bar{p}$ is a pooled estimator of the equal population proportions and is computed as the weighted average of the two sample proportions $\hat{p}_1$ and $\hat{p}_2$ as follows:

$$\bar{p} = \frac{n_1\hat{p}_1 + n_2\hat{p}_2}{(n_1 + n_2)} \quad (7.82)$$

Table 7.19 Hypothesis testing for two population proportions

One-tailed test	*Two-tailed test*
$H_0: p_1 = p_2$ or $p_1 - p_2 = 0$ $H_1: p_1 - p_2 > 0$ (or $H_1: p_1 - p_2 < 0$)	$H_0: p_1 = p_2$ or $p_1 - p_2 = 0$ $H_1: p_1 \neq p_2$ or $p_1 - p_2 \neq 0$
Test statistic: $z = \dfrac{(\hat{p}_1 - \hat{p}_2) - (p_1 - p_2)}{\sqrt{\bar{p}(1-\bar{p})\left(\dfrac{1}{n_1} + \dfrac{1}{n_2}\right)}}$ (7.81)	
Rejection region Reject H_0 if $z > z_\alpha$ (or $z < -z_\alpha$)	*Rejection region* $\lvert z\rvert > z_{\alpha/2}$

Source: Author's creation

Assumptions: the large sample sizes, randomly selected samples.

The next example demonstrates this procedure.

▶ Example 45

The managers of two assembly lines in an automobile plant wonder whether the nonconforming or defectives occurring in the last station of their assembly lines are statistically different from each other. To test this hypothesis, they monitored the number of nonconforming in the last stations over the last 2 hours in each assembly line. The numbers of nonconforming and conforming in assembly line 1 are 5 and 195, respectively, while assembly line 2 generates 7 nonconforming and 189 conforming parts. The daily production rate is 2,500 in each assembly line. Using a level of significance at $\alpha = 0.05$, test whether the proportions of the nonconforming in these assembly lines are the same in this automobile plant. ◀

- Step 1: Check assumptions and conditions.

For the independence assumption, it is assumed that the nonconforming identified in the last stations in two assembly lines are independent of each other. For the sample size assumption, the total numbers of conforming and nonconforming in 2 hours are large enough ($n \geq 30$) in each assembly line. The randomization condition is met, since the nonconforming occurs randomly in each assembly line. For 10% condition, total numbers of conforming and nonconforming in assembly line 1 ((200/2,500) < 10%) and assembly line 2 ((196/2,500) < 10%) are less than 10% of the daily production rate, for each line. For the success/failure condition in both assembly lines, the number of successes (np = 396(384/396) = 384) and failures ($np(1 - p)$ = 396(12/396)(1 – 12/396) = 11.63) is at least 10. For the success/failure condition, it is also possible to say that there are 384 successes and 12 failures in both assembly lines, both of which are greater than 10.

- Step 2: State the null hypothesis H_0 and the alternative hypothesis H_1.

$$H_0 : p_1 = p_2 \text{ or } p_1 - p_2 = 0$$

$$H_1 : p_1 \neq p_2 \text{ or } p_1 - p_2 \neq 0$$

As presented in the hypotheses, this is a two-tailed test, since the managers of the two assembly lines in the automobile plant want to analyze whether the proportions nonconforming in the two lines are statistically different. The null hypothesis claims that there is no difference between the proportions nonconforming, while the alternative hypothesis claims the opposite. In our hypotheses above, p_1 refers to the proportion nonconforming of assembly line 1, while p_2 refers to the proportion of nonconforming of assembly line 2.

- Step 3: Choose the level of significance, α.

The level of significance is $\alpha = 0.05$ in this hypothesis testing.

- Step 4: Choose the sample size, n.

The sample size is 200 items (n_1) in the first assembly line and 196 (n_2) in the second.

- Step 5: Collect the data.

The collected data are presented in the question.

- Step 6: Compute the appropriate test statistic.

Since sample size in each assembly line is greater than 30, let's calculate the Z test statistic for the difference between two proportions as shown in Eq. 7.81 above.

$$z = \frac{(\hat{p}_1 - \hat{p}_2) - (p_1 - p_2)}{\sqrt{\bar{p}(1-\bar{p})\left(\frac{1}{n_1} + \frac{1}{n_2}\right)}}$$

where the pooled estimator $\bar{p}$ is computed as the weighted average of the two sample proportions $\hat{p}_1$ and $\hat{p}_2$ as shown in Eq. 7.82.

$$\bar{p} = \frac{n_1\hat{p}_1 + n_2\hat{p}_2}{(n_1 + n_2)}$$

or

$$\bar{p} = \frac{X_1 + X_2}{n_1 + n_2} \tag{7.83}$$

If X_1 and X_2 refer to the number of nonconforming in assembly line 1 and assembly line 2, respectively, the sample proportion of each assembly line is computed as shown in the following.

The proportion nonconforming in the assembly line 1 is

$$\hat{p}_1 = \frac{X_1}{n_1} = \frac{5}{200} = 0.025$$

The proportion nonconforming in the assembly line 2 is

$$\hat{p}_2 = \frac{X_2}{n_2} = \frac{7}{196} = 0.036$$

Then, the average of proportion nonconforming in both assembly lines is

$$\bar{p} = \frac{X_1 + X_2}{n_1 + n_2} = \frac{5+7}{200+196} = 0.0303$$

Therefore, the Z test statistic is

$$Z = \frac{(\hat{p}_1 - \hat{p}_2) - (p_1 - p_2)}{\sqrt{\bar{p}(1-\bar{p})\left(\frac{1}{n_1} + \frac{1}{n_2}\right)}} = \frac{0.025 - 0.036 - (0)}{\sqrt{0.0303(1-0.0303)\left(\frac{1}{200} + \frac{1}{196}\right)}}$$

$$Z = \frac{-0.011}{\sqrt{0.0303(0.9697)(0.0101)}} = \frac{-0.011}{0.000297} = -0.62$$

- Step 7: Calculate the p-value and identify the critical values of test statistic.

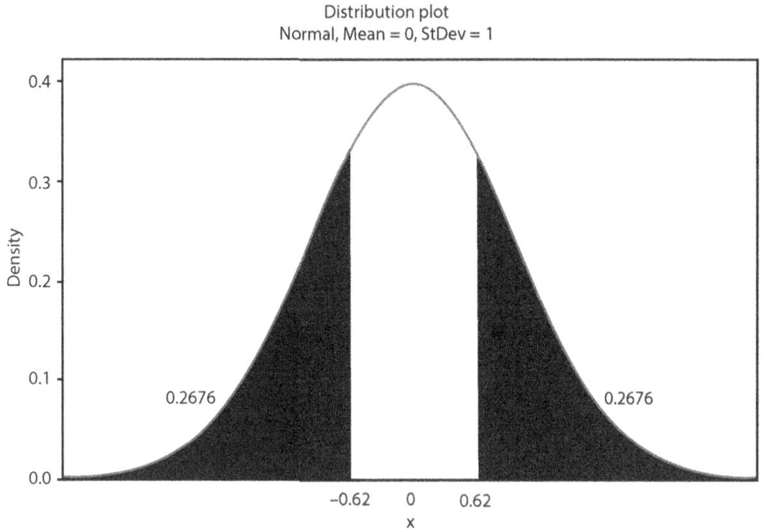

Image 7.25 Observed significance level (p-value) of $z_{\text{computed}} = 0.62$ for the two-tailed test. (Source: Author's creation based on Minitab)

$$P = 2P\left(Z > \left|Z_{\text{computed}}\right|\right)$$
$$= 2P\left(Z > \left|-0.62\right|\right) = 2P\left(Z > 0.62\right)$$
$$= 2\left(0.2676\right) = 0.535$$

In the standard normal distribution table, since the corresponding probability value for $Z = 0.62$ is 0.2676 and the hypothesis testing is two-tailed, p-value of the test is found to be 0.535.

To run two proportions z test in Minitab, click on Stat→Basic Statistics→2 Proportions. In the next input screen, select "summarized data," and enter the variables number of events and number of trials in the next screen. Click on "Options," and select alternative hypothesis as "Difference ≠ hypothesized difference"; select "estimate the proportions separately" in "Test Method." Click on OK→OK. The results are shown in Session Window. The probability distribution of the z test is presented in Image 7.25.

- Step 8: Apply decision rule, and express the statistical finding in the scope of the question.

Applying the decision rule shows that we cannot reject the null hypothesis since α (0.05) is less than the p-value ($p = 0.535$). We can conclude that there is insufficient evidence to reject the null hypothesis. Let's remember the rule: if the level of significance is greater than p-value ($\alpha > p$), the null hypothesis will be rejected. Therefore, we have compelling evidence in favor of the null hypothesis. As a result, we can say that the proportions nonconforming in two assembly lines are equal. In other words, we can conclude that there is insufficient evidence of a significant difference in the proportions of nonconforming between assembly line 1 and 2 (p-value = 0.535).

Acknowledgement

I would like to thank my graduate student Merve Gündüz who helped me in preparing Minitab outputs in my sections.

7.11 Correlation Analysis

When researchers and decision-makers are interested in finding a relationship between two continuous variables, they use correlation analysis and regression analysis. The question answered in regression and correlation analysis is: How does the value of one variable change when the value of another one changes? For example, what is the relationship between age and the frequency of doctor visits? Income and work hours? Seniority and salary?

In correlation analysis, correlation coefficient (r) is a measure of the extent to which X and Y are linearly related. The correlation coefficient is unitless and varies between −1 and +1. If the variables are X and Y, correlation coefficient is denoted as r_{xy}. *Bivariate correlation* refers to the correlation between two variables, such as the ones we just mentioned above. Correlation analysis produces information about the *direction* and *strength* of the relationship between two variables. *Direction* is demonstrated by the sign of the correlation coefficient as either negative (−) or positive (+). As two variables move toward the same direction, these variables affect each other positively, and the correlation between two variables appears as direct correlation or positive correlation. For example, it could be that as the outside temperature increases, the sale of t-shirts is more likely to increase. The correlation coefficient is expected to be positive in this example. The correlation coefficient in a positive relationship varies between 0 and +1. Negative correlation or indirect correlation can be detected in the relationship between these two variables if the correlation of two variables varies in two different directions. For example, let's say it is expected that as the outside temperature increases, the total sales of heavy coats are more likely to decrease. In this case, the correlation coefficient varies between −1 and 0. The interpretation of the correlation coefficient is detailed as follows:

- $r = -1$: All points lie on a straight line with negative slope.
- $0.0 < r < 0.2$: Very weak or no relationship.
- $0.2 < r < 0.4$: Weak relationship.
- $0.4 < r < 0.6$: Moderate relationship.
- $0.6 < r < 0.8$: Strong relationship.
- $0.8 < r < 1.0$: Very strong relationship.
- $r = 1$: All points lie on a straight line with positive slope.

Bruhl (2018) classifies correlation coefficients based on both the magnitude and the direction. According to Bruhl's classification, a correlation coefficient between −1.00 and −0.60 refers to a strong inverse relationship between the two variables. A correlation coefficient between −0.59 and −0.01 shows a weak inverse relationship between the two variables. A correlation coefficient of zero is interpreted as there is no association between the two variables, which also refers to stochastic independence of the two. If the correlation coefficient is between 0.01 and 0.59, the relationship between the two variables is considered a weak and direct relationship. A correlation coefficient between 0.60 and 1.00 refers to a strong direct association between the two variables.

The *strength* of the relationship between two variables can be measured by the magnitude of the correlation coefficient and is demonstrated by the numerical value of the correlation coefficient. As the r gets closer to zero, it shows that the correlation between two variables does not exist. When the r differs from zero and starts increasing, the strength of the relationship becomes stronger. For example, if r is 0, there is no statistical relationship between two variables. If r equals +0.80, there is a strong and positive relationship between two variables. The higher the correlation coefficient, the stronger the relationship. For example, if the correlation coefficient is 0.50, that means the relationship between variables is moderate, not strong or weak enough to state that there is no correlation between the variables. Potential relationships between two variables are presented in ◘ Table 7.20. Technically, scatter plots show how X variable behaves against Y variable. Scatter plots display the strength and direction of the correlation between the variables.

The correlation coefficient is checked for its significance by the following hypothesis testing:

- H_0: There is no correlation between two variables. $\rho_{xy} = 0$.
- H_1: There is a correlation between two variables. $\rho_{xy} \neq 0$.

Table 7.20 Types of correlation

X variable	Y variable	Type of correlation	Correlation coefficient value	Example
X variable's value increases	Y variable's value increases	Direct or positive	Positive, between 0 and +1	As the temperature goes up, the water consumption increases
X variable's value decreases	Y variable's value decreases	Direct or positive	Positive, between 0 and +1	As the temperature goes down, flip flap sales go down, too
X variable's value increases	Y variable's value decreases	Indirect or negative	Negative, between −1 and 0	As the temperature goes up, winter coat sales go down
X variable's value decreases	Y variable's value increases	Indirect or negative	Negative, between −1 and 0	As the temperature goes down, winter coat sales go up

Source: Author's creation

If the *p*-value of the hypothesis testing is lower than the α value ($p < \alpha$), then null hypothesis is rejected. That means that the alternative hypothesis is accepted and the correlation coefficient (r) between two variables is statistically significant. If the *p*-value of the hypothesis testing is greater than the α value ($p > \alpha$), then the null hypothesis is failed to be rejected. That means that the alternative hypothesis is rejected and the correlation coefficient (r) between two variables is not statistically significant.

Correlation analysis can be performed by Spearman test or Pearson test. The formula for Pearson correlation coefficient is shown in Eq. 7.84.

$$r_{xy} = \frac{n\sum XY - \sum X \sum Y}{\sqrt{\left[n\sum X^2 - \left(\sum X\right)^2\right]\left[n\sum Y^2 - \left(\sum Y\right)^2\right]}} \quad (7.84)$$

where

- r_{xy} = the correlation coefficient
- n = the sample size
- X = the individual observations on X variable
- Y = the individual observations on Y variable
- XY = the product of X and Y variables
- X^2 = the squared individual X variable
- Y^2 = the squared individual Y variable.

Correlation analysis is required to pass a set of pre-tests including *normality*, *linearity*, and *homoscedasticity*. In terms of *normality*, variables are expected to be distributed normally. Normality can be tested on histograms and probability plots. Symmetrically bell-shaped and no skewed histograms refer to normally distributed data. *Linearity* explores whether the scatter plots of the variables form a straight line. The line that goes through the dots in scatter plot should minimize the distances between the line and dots. Linearity test can be done on a scatter plot. *Homoscedasticity* refers to the density of the dots on the regression line that is seen in the scatter plot. The homoscedasticity requirement is met when the majority of dots are located in the middle of the regression line.

To compute correlation coefficient, the following steps can be used by Six Sigma teams:

- Step 1: Collect data pairs for X and Y variables.
- Step 2: Compute the sum of X values.
- Step 3: Compute the sum of Y values.
- Step 4: Compute the square of each X value.
- Step 5: Compute the square of each Y value.
- Step 6: Compute the sum of XY each product.

Table 7.21 The details of correlation coefficient

Week	Ice cream sales (X)	ENT doctor visits (Y)	X^2	Y^2	XY
1	315	127	99,225	16,129	40,005
2	340	130	115,600	16,900	44,200
3	355	135	126,025	18,225	47,925
4	367	139	134,689	19,321	51,013
5	375	142	140,625	20,164	53,250
6	378	143	142,884	20,449	54,054
7	380	143	144,400	20,449	54,340
8	390	147	152,100	21,609	57,330
9	414	155	171,396	24,025	64,170
Total	3,314	1,261	1,226,944	177,271	466,287

Source: Author's creation

7

- Step 7: Computer correlation coefficient r_{xy}.
- Step 8: Test the statistical significance of the correlation coefficient r_{xy}.
- Step 9: Interpret correlation coefficient r_{xy}.

▶ Example 46

The researchers analyze the relationship between ice cream sales and the number of ENT doctor visits for 10-week period in New England area. The first three columns of Table 7.21 represents the values for two variables. Calculate the correlation coefficient and interpret the finding. ◀

Solution

Let's follow the work steps given above to compute and interpret the correlation coefficient for these variables.

- Step 1: Collect data pairs for X and Y variables.

The independent variable (X) is ice cream sales, and dependent variable (Y) is ENT doctor visits in this question. The 9-week data are already collected and presented in Table 7.21.

- Step 2: Compute the sum of X values.

The sum of ice cream sales (X) is presented as $\sum_{i=1}^{9} X_i = 3{,}314$ as presented in Table 7.21.

- Step 3: Compute the sum of Y values.

The sum of ENT doctor visits (Y) is presented as $\sum_{i=1}^{9} Y_i = 1{,}261$ as presented in Table 7.21.

- Step 4: Compute the square of each X value.

Each X value is squared as presented in the fourth column of Table 7.21.

- Step 5: Compute the square of each Y value.

Each Y value is squared as presented in the fifth column of Table 7.21.

- Step 6: Compute the sum of XY each product.

The products of X and Y values are presented in the last column of Table 7.21. The sum of the product of X and Y is $\sum_{i=1}^{9} XY = 466{,}287$.

- Step 7: Computer correlation coefficient r_{xy}.

$$r_{xy} = \frac{n\sum XY - \sum X \sum Y}{\sqrt{\left[n\sum X^2 - \left(\sum X\right)^2\right]\left[n\sum Y^2 - \left(\sum Y\right)^2\right]}}$$

$$= \frac{9(466{,}287) - (3{,}314)(1{,}261)}{\sqrt{\left[9(1{,}226{,}944) - (3{,}314)^2\right]\left[9(177{,}271 - (1{,}261)^2\right]}}$$

$$= 0.987734$$

As well as manual computation, correlation coefficient can be calculated using MS Excel or software such as Minitab, JMP, SPSS, and Stata. We can easily use MS Excel and enter the function for correlation coefficient as "=CORREL(variable 1,variable 2)."

To calculate the correlation coefficient on Minitab, after transferring data set to Minitab, click on Stat→Basic Statistics→Correlation. Transfer variables "ice cream sales" and "doctor visits" from left box to variables box, and click on Select. Click on Method to select either Pearson correlation or Spearman rho correlation tests. Select "display *p*-values." Click on OK. Minitab output for correlation analysis is shown in Session Window ($r_{xy} = 0.988$).

- Step 8: Test the statistical significance of the correlation coefficient r_{xy}.

In addition to the correlation coefficient, the statistical significance of the coefficient should be tested. As explained above, the correlation analysis contains a hypothesis testing as follows:

- H_0: There is no correlation between two variables. $\rho_{xy} = 0$.
- H_1: There is a correlation between two variables. $\rho_{xy} \neq 0$.

Again, in cases where the *p*-value of the hypothesis testing is lower than the α value ($\alpha > p$), then the null hypothesis is rejected, meaning that alternative hypothesis is accepted and the correlation coefficient (r) between two variables is statistically significant. If the *p*-value of the hypothesis testing is greater than the α value ($p > \alpha$), then the null hypothesis is failed to be rejected, meaning that alternative hypothesis is rejected and the correlation coefficient (r) between two variables is not statistically significant. The significance of the correlation analysis ($p = 0.000$) is presented in Minitab output.

- Step 9: Interpret correlation coefficient r_{xy}.

As seen in Minitab output, the Pearson correlation coefficient was found to be 0.988 ($p = 0.000$) at 0.01 significance level. In this case, the α value is greater than the *p*-value ($\alpha = 0.01 > p = 0.000$); then null hypothesis is rejected. That means that the alternative hypothesis is accepted and the correlation coefficient (r) between two variables is statistically significant.

The correlation coefficient ($r_{xy} = 0.988$) tells us that there is a *strong* and *positive* correlation between ice cream sales and the ENT doctor visits. As the ice cream sales increase, the number of doctor visits goes up, too. The significance level tells us if this correlation is statistically significant. The *p*-value and the hypothesis test show that that correlation coefficient is statistically significant.

If there are more than two variables associated with the research interest, a correlation matrix is created to present the relationship between variable pairs. Let's assume that we have four variables involved in the statistical analysis such as age, calories consumed per day, water consumption per day, and daily sleep hours per person in ▶ Example 46. As presented in ◻ Table 7.22, correlation matrix displays correlation coefficients for each pair of variables.

◻ **Table 7.22** Correlation matrix for four variables

	Age	**Calories consumed/day**	**Water consumption/day**	**Daily sleep hours/person**
Age	1.00	0.39	−0.25	−0.60
Calories consumed/day		1.00	0.75	0.03
Water consumption/day			1.00	0.10
Daily sleep hours/person				1.00

Source: Author's creation

In Table 7.22, when the variables are correlated by themselves, the correlation coefficients are 1.00, as presented in the diagonal of the correlation matrix. These correlation coefficients are not taken into consideration. But the correlations between variable pairs will be meaningful for the Six Sigma teams. For example, the correlation coefficient between age and calories consumed/day is $r_{xy} = 0.39$, meaning that age and calories consumed/day are positively correlated. However, there is no strong correlation between these two variables. The correlation coefficient between age and water consumption/day is $r_{xy} = -0.25$, meaning that there is a weak negative relationship between two variables. In other words, as the people involved in the research get older, their water consumption/day gets lower, but the correlation is weak.

There is another way to interpret the correlation coefficients. The correlation coefficient can be used to build a regression model that describes the relationship between two variables. By squaring the correlation coefficient, we can calculate "coefficient of determination" (R^2), which tells us the percentage of variance in one variable that is shared with the variance in the other variable. The more two variables share in common, the more associated they will be. The coefficient of determination (R^2) tells us how much of the variance in one variable is accounted for by the variance in the other variable (Salkind 2016: 145). For example, if $r_{\text{age.water consumption}}$ is 0.80, R^2 is 0.64, which means that 64% of the variance in "age" can be explained by the variance in "water consumption." If the correlation coefficient is higher between two variables, the coefficient of determination of these two variables will be higher as well. The higher the correlation coefficient and coefficient of determination are, the stronger relationship between the two variables occurs.

7.12 Regression Analysis

Technically, regression analysis assesses the associations between variables. The main idea behind regression analysis is to predict the value of Y variable (dependent or response variable) depending on the value of X variables (independent variables). In other words, the regression analysis explores whether there is any statistically significant relationship between dependent and independent variables. The regression analysis is used in forecasting and prediction, identifying critical and impactful variables, and locating optimum operating conditions (Juran and Gryna 1980). If there is one dependent and one independent variable in the analysis, the regression analysis is called simple regression. If there is a linear relationship between the two variables, the regression analysis is called simple linear regression. If the relationship between the two variables is not linear, the regression analysis is called a nonlinear regression analysis. If there are multiple dependent and independent variables, the regression analysis turns into a multiple regression analysis. Linearity or nonlinearity may exist in the relationship of the variables as well.

In regression analysis, a hypothetical line is assumed to be placed among the values located in scatter diagram to represent the best fit among the values. The distance between the hypothetical line and each observation is measured and aimed to be minimized. The hypothetical line that minimizes those distances for each pair of observation is considered "the best fit" in a regression line. In a perfect fit condition, each observation is located around the regression model. In less perfect fit, the differences between individual observations and regression line are small, whereas these differences get larger in a poor fit condition (Bruhl 2018). Many applications of regression models are implemented in quality and process improvement projects (Montgomery 2013). Regression analysis focuses on the impact of the independent variable on the dependent variable. For example, a researcher may want to analyze if there are any relationships between the college GPA of students majoring in Business Administration and their first-year salary level in a full-time position after graduation. There are two questions that we can answer using regression analysis for this research question:

1. Is college GPA a good predictor of the Business Administration graduates' first-year salary?

2. Is there any statistically significant relationship between college GPA of students majoring in Business Administration and their first-year salary in a full-time position?

The model that may describe the relationship between these two variables is given in Eq. 7.85.

$$y = \beta_0 + \beta_1 x_1 + \varepsilon \quad (7.85)$$

where

- y = the graduates' first-year salary in a full-time position
- x_1 = the graduates' college GPA
- β_0 = Y-intercept of regression line
- β_1 = the regression coefficient of x_1 variable, which shows the expected change in response variable (y) per unit change in x_1.

There are several assumptions of regression analysis such as:

1. Linear relationship
2. Multivariate normality
3. No or little multicollinearity
4. No autocorrelation
5. Homoscedasticity.

First, in simple linear regression analysis, there must be a linear relationship between dependent and independent variables. This assumption can be checked through scatter plots. Second, all variables added in regression analysis should be normally distributed. Histogram, probability plots, and regular normality tests, such as chi-square, Kolmogorov-Smirnov, Anderson-Darling, Shapiro-Wilk, and goodness of fit, can be used in testing this assumption. If the data are not normally distributed, data transformation can be used as a way to have the data normally distributed. Various transformations of data such as a) logarithmic transformations either base 10 logs or natural log (base-e), (b) square root transformations, (c) arcsine transformations, and (d) other Box-Cox transformations are employed in data transformation process (McDonald 2009).

To eliminate the multicollinearity issue, the independent variables are expected to have no correlation between each other. In multiple regression, if there is a particular independent variable with a high variation inflation factor (*VIF*), this variable can be excluded from the analysis to exclude multicollinearity. Also, the data are expected to have no or little autocorrelation. Autocorrelation is detected when the residuals are not independent from each other. Scatter plots of residuals and Durbin-Watson tests are used to test autocorrelation in the data set. To detect homoscedasticity, scatter plots of residuals versus predicted variables (dependent) and scatter plot of residuals versus independent variable can be implemented in the analysis. In these scatter plots, errors that get larger in one direction by a significant amount should not occur. A certain pattern of the residuals indicates a bad fit, while residuals with no pattern shows a good fit. A log transformation of variables may be a way to eliminate the homoscedasticity issue.

If there are two variables questioned, higher absolute correlation coefficient (r) results in better prediction of the other variable. To predict the first-year salary level in our question, a simple regression analysis is performed. Using the regression analysis, the predictions of first-year salary for a number of students can be made. Let's assume that a data set including 15 students' GPA and salary levels is presented in the third columns of ◘ Table 7.23. To predict the first-year salary level for various college GPA values, first we need to check the assumptions of regression analysis given above before running the regression analysis. In a simple linear regression analysis (Eq. 7.88), (1) slope of the line (b) and (2) Y-intercept of the line (a) are computed as presented in Eqs. 7.85 and 7.86.

$$b = \frac{\sum XY - \left(\frac{\sum X \sum Y}{n}\right)}{\sum X^2 - \left[\frac{(\sum X)^2}{n}\right]} \quad (7.86)$$

$$a = \frac{\sum Y - b\sum X}{n} = \bar{Y} - b\bar{X} \quad (7.87)$$

$$Y = a + bX \quad (7.88)$$

where

- Y = the predicted value of Y variable based on X variable

7

Table 7.23 The details of the regression equation calculations

Number of student	College GPA (X_1)	First-year salary level ($) ($Y$)	Age (X_2)	X_1Y	X_1^2
1	2.5	45,000	22	112,500	6.25
2	2.6	47,000	24	122,200	6.76
3	2.7	48,000	28	129,600	7.29
4	2	45,000	22	90,000	4
5	2.1	45,000	22	94,500	4.41
6	3.5	55,000	23	192,500	12.25
7	3.9	60,000	22	234,000	15.21
8	3.7	58,000	22	214,600	13.69
9	3.6	57,000	23	205,200	12.96
10	2.7	48,500	22	130,950	7.29
11	2.7	48,000	22	129,600	7.29
12	2.9	49,000	22	142,100	8.41
13	2.3	45,500	21	104,650	5.29
14	3.8	59,000	22	224,200	14.44
15	3.5	55,500	23	194,250	12.25
TOTAL	44.5	765,500	340	2,320,850	137.79

Source: Author's creation

- a = the Y-intercept of the line
- X = the value of X variable
- $\bar{X}$ = the average of X variable
- $\bar{Y}$ = the average of Y variable
- n = the sample size.

The steps in regression analysis are as follows:
- Step 1: Define the goal of the regression analysis by focusing on potential dependent (response) and independent variables.
- Step 2: Collect data.
- Step 3: Check the assumptions.
- Step 4: Draw scatter diagram to show the potential relationships between potential variables.
- Step 5: Compute regression equation.
- Step 6: Analyze the regression equation and how it fits the data.

The steps in computing the linear regression equation are as follows:
- Step 1: Compute coefficient b.
- Step 2: Compute coefficient a.
- Step 3: Build a regression line.

▶ Example 47

Let's go back to our example given above that includes college GPA and first-year salary level variables. The data for the example are presented in Table 7.23. Let's build regression equation and estimate the associated first-year salary level for a GPA of 4.0. ◀

Solution

Let's follow our steps given above to run regression analysis and build regression equation.

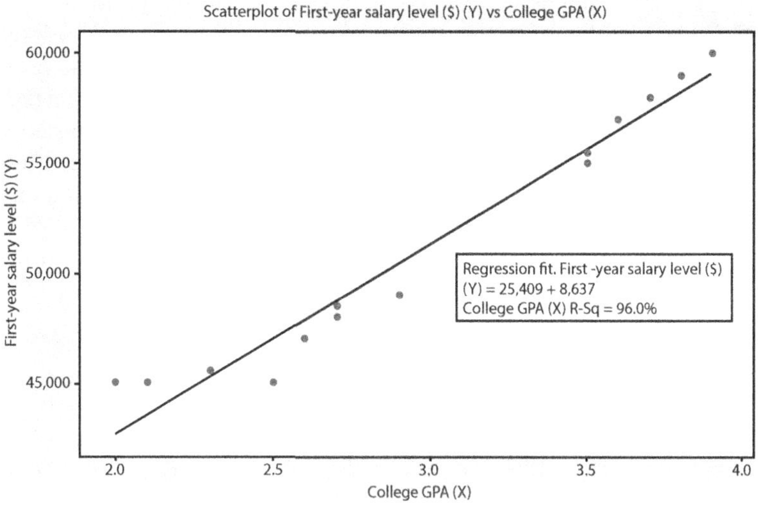

Image 7.26 Scatter plot for first-year salary and college GPA in Minitab. (Source: Author's creation based on Minitab)

- Step 1: Define the goal of the regression analysis by focusing on potential dependent (response) and independent variables.

The goal of the regression analysis is to (1) discover if there is any statistically significant relationship between college GPA (independent variable) and first-year salary level (dependent variable) and (2) find out if college GPA is a good predictor of the Business Administration graduates' first-year salary.

- Step 2: Collect data.

In this example, the data are already collected as presented in Table 7.23.

- Step 3: Check the assumptions.

In this step, Six Sigma practitioners check linear relationship, multivariate normality, multicollinearity, autocorrelation, and homoscedasticity assumptions. For example, the linear relationship assumption is checked through scatter plots as detailed in Step 4. For normality assumption, probability plots of the variables are used for normality tests as detailed in the previous sections. We assume that Six Sigma team already checked all these assumptions.

- Step 4: Draw scatter diagram to show the potential relationships between potential variables.

The scatter plot of two variables is presented in Image 7.26. To draw scatter plot in Minitab, transfer data to Minitab worksheet, and click on Graph→Scatterplot. Then, select the type of plot and click on it. To exemplify, let's go with "with regression" alternative. Click on "with regression" →OK. Then, transfer *Y* and *X* variables from left box to designated areas in the right box in the next screen. Click on OK. The scatter plot will be drawn as shown in Image 7.26.

As seen from the scatter plots, there is a relationship between the two variables. Scatter plots help decision-makers analyze the relationship between variables visually. Image 7.26 demonstrates a positive relationship between first-year salary level and college GPA variables.

- Step 5: Compute regression equation.

Table 7.23 demonstrates the details of the calculations of the variables that need to be used in computations in this step.

(a) Compute coefficient b (slope of the regression line).

$$b = \frac{\Sigma XY - \left(\frac{\Sigma X \Sigma Y}{n}\right)}{\Sigma X^2 - \left[\frac{(\Sigma X)^2}{n}\right]}$$

$$= \frac{2,320,850 - \left(\frac{44.5 * 765,500}{15}\right)}{137.79 - \left[\frac{44.5^2}{15}\right]} = 8,637.46$$

(b) Compute coefficient a (Y-intercept of the regression line).

$$a = \frac{\Sigma Y - b\Sigma X}{n}$$

$$= \frac{765,500 - (8,637.46 * 44.5)}{15} = 25,408.87$$

(c) Build regression line.

After calculating a and b coefficients as presented above, we can build regression line as follows:

$$Y = a + bX = 25,408.87 + 8,637.46X \quad (7.89)$$

The scatter plot in Image 7.26 shows regression line's equation in the box with 96% $R\text{-}Sq$ value. Regression analysis can be done using software such as Minitab, JMP, SPSS, and Stata. To perform regression analysis in Minitab, click on Stat→Regression→Regression→Fit Regression Model. Then, transfer response variable (Y variable) and continuous predictors (X variable) from left box to the designated areas on the right side in the next step. Click on OK. Minitab provides regression analysis output as shown in Table 7.24.

Table 7.24 demonstrates R^2-adjusted as 95.68%, which means that college GPA accounts for 95.68% of the variance among first-year salary levels. R^2 coefficient of determination is the proportion of variation of Y explained by having X in the linear model. A higher R^2-adjusted value indicates a good fit of the regression model. Regression equation is given as follows in Table 7.24:

$$\textit{First year salary level}(\$)(Y) = 25,409 + 8,637\,\textit{College GPA}(X)$$

- Step 6: Analyze the equation and how it fits the data.

To forecast a value for dependent variable, substitute the value questioned for the independent variable. To predict the first-year salary level based on regression line, when college GPA is 4.00, the estimated first-year salary can be calculated as $59,958.522 as presented in Eq. 7.90.

$$\begin{aligned} Y = a + bX &= 25,408.87 + 8,637.413X \\ &= 25,408.87 + 8,637.413(4.00) \\ &= \$59,958.522 \end{aligned} \quad (7.90)$$

The slope of the line (b) and Y-intercept (a) can be computed on MS Excel as well. The SLOPE function is entered as "=SLOPE (known y's, known x's)," and dependent and independent variables' observations are entered in the function, respectively. To calculate Y-intercept, the INTERCEPT function is entered as "=INTERCEPT (known y's,known x's)," and dependent and independent variables' values are entered in the function, respectively. To run a forecast on Excel, the FORECAST function is also used by entering the function as "=FORECAST(x,known y's,known x's)." In the forecast function, x refers to the value for which the team runs a forecast. For example, if the team wants to run a forecast for college GPA of 4.0, x variable in forecast function should be 4.00. After entering 4.00 in the function, dependent and independent variables are entered, respectively.

To evaluate the accuracy of forecasting (prediction) through regression analysis, several statistical variables can be utilized.

Table 7.24 Regression analysis output in Minitab

Regression analysis: first-year salary level ($) (Y) versus college GPA (X)					
Regression equation					
First-year salary level ($) (Y) =			25,409 + 8,637 college GPA (X)		
Coefficients					
Term	**Coef**	**SE coef**	**T-value**	***P*-value**	**VIF**
Constant	25,409	1,485	17.11	0.000	
College GPA (x)	8,637	490	17.63	0.000	1.00
Model summary					
S	**R-sq**	**R-sq(adj)**	**R-sq(pred)**		
1,177.16	95.99%	95.68%	94.27%		
Analysis of variance					
Source	**DF**	**ADJ SS**	**Adj MS**	**F-value**	***P*-value**
Regression	1	430,719,015	430,719,015	310.83	0.000
College GPA (X)	1	430,719,015	430,719,015	310.83	0.000
Error	13	18,014,319	1,385,717		
Lack-of-fit	10	17,722,652	1,772,265	18.23	0.018
Pure error	3	291,667	97,222		
Total	14	448,733,333			
Fits and diagnostics for unusual observations					
Obs	**First-year salary level ($) (Y)**		**Fit**	**Resid**	**Std resid**
4	45,000		42,684	2,316	2.24 R

Source: Author's creation based on Minitab

These variables are mean absolute deviation (MAD), mean square error (MSE), and tracking signal. The forecast error is the difference between estimated and actual value of the variable used in forecasting as shown in Eq. 7.91. Forecast error should be measured over time and minimized to increase forecast accuracy. The lower the forecast error detected, the better the estimations.

$$E_t = A_t - F_t \tag{7.91}$$

where

- E_t = forecast error for period t
- A_t = actual value for period t
- F_t = forecast value for period t.

MAD is the average of the sum of the absolute errors and is calculated as shown in Eq. 7.92.

$$\text{MAD} = \frac{\sum_{i=1}^{n} \left|\text{actual}_i - \text{forecast}_i\right|}{n} \tag{7.92}$$

where

- actual_i = actual value of the ith observation
- forecast_i = forecast value of the ith observation
- n = sample size.

MSE is the average of the squared errors as shown in Eq. 7.93.

$$\text{MSE} = \frac{\sum_{i=1}^{n}(\text{actual}_i - \text{forecast}_i)^2}{n} \quad (7.93)$$

where
- actual_i = actual value of the ith observation
- forecast_i = forecast value of the ith observation
- n = sample size.

Tracking signal (TS) is computed as the ratio of sum of the forecast errors divided by MAD as shown in Eq. 7.94.

7

$$\text{TS} = \frac{\sum_{i=1}^{n}(\text{actual}_i - \text{forecast}_i)}{\text{MAD}} \quad (7.94)$$

where
- actual_i = actual value of the ith observation
- forecast_i = forecast value of the ith observation
- MAD = mean absolute deviation
- n = sample size.

▶ Example 48

Astro Sun Corporation is comparing the accuracy of two linear regression models that estimate the total sales for the next 5 months. Forecasting results generated through two models are shown in ◘ Table 7.25. Which model provides a better forecast accuracy? ◀

▪ Solution

To find out which regression model provides a better forecast accuracy, let's calculate MAD, MSE, and TS for both models. MAD values reveal that model 2 results in better estimations since second model's MAD value with 2.2 is lower than the first model as shown below:

$$\text{MAD}_{\text{model1}} = \frac{\sum_{i=1}^{5}|\text{actual}_i - \text{forecast}_i|}{n} = \frac{3+3+2+2+3}{5} = 2.6$$

$$\text{MAD}_{\text{model2}} = \frac{\sum_{i=1}^{5}|\text{actual}_i - \text{forecast}_i|}{n} = \frac{5+4+2+0+0}{5} = 2.2$$

MSE values tell us that first model generates better forecast since its MSE with 7 is lower than the second model as shown below:

$$\text{MSE}_{\text{model1}} = \frac{\sum_{i=1}^{5}(\text{actual}_i - \text{forecast}_i)^2}{n} = \frac{9+9+4+4+9}{5} = 7$$

$$\text{MSE}_{\text{model2}} = \frac{\sum_{i=1}^{5}(\text{actual}_i - \text{forecast}_i)^2}{n} = \frac{25+16+4+0+0}{5} = 9$$

◘ Table 7.25 The results of the regression models

		Regression model 1			Regression model 2		
	Actual sales (A_i)	Forecast (F_i)	Absolute error $\lvert A_i - F_i \rvert$	Error² $(A_i - F_i)^2$	Forecast (F_i)	Absolute error $\lvert A_i - F_i \rvert$	Error² $(A_i - F_i)^2$
June	35	32	3	9	40	5	25
July	33	30	3	9	29	4	16
August	29	31	2	4	27	2	4
September	27	25	2	4	27	0	0
October	25	22	3	9	25	0	0

Source: Author's creation

When we analyze tracking signals for two regression models, it is seen that model 2 estimates better total sales for 5 months than the estimations of model 1.

$$TS_{model1} = \frac{\sum_{i=1}^{5} (actual_i - forecast_i)}{MAD_{model1}} = \frac{3+3+2-2+3}{2.6} = 3.46$$

$$TS_{model2} = \frac{\sum_{i=1}^{5} (actual_i - forecast_i)}{MAD_{model2}} = \frac{-5+4+2+0+0}{2.2} = 0.45$$

We have analyzed simple linear regression so far. Regression analysis can also be done when more than one dependent (response) variable and/or independent variable are included in the research question to estimate a particular dependent variable. Let's go back to our research question in this chapter. In addition to college GPAs, there may be some other variables affecting first-year salary level such as age, type of major, or marital status. When there is more than one independent variable that are potentially impactful on dependent variable, simple regression turns to multiple regression. The multiple linear regression equation is shown as presented in Eq. 7.95.

$$Y = a + b_1X_1 + b_2X_2 + \ldots b_nX_n \tag{7.95}$$

where

- Y = the dependent variable value
- a = the Y-intercept
- b = regression weight coefficient for independent variables
- X_1 = the value of the first independent variable
- X_2 = the value of the second independent variable
- X_n = the value of the nth independent variable.

▶ Example 49

Let's go back to our example given above that includes college GPA and first-year salary level. As the second independent variable, let's add "age" in the analysis. The data for the example are presented in ◘ Table 7.23. Let's build regression equation and estimate the associated first-year salary level for college GPA of 4.0 for students who are 25 years old. ◀

▪ Solution

Let's follow the steps of regression analysis given above.

- Step 1: Define the goal of the regression analysis by focusing on potential dependent (response) and independent variables.

The goal of the regression analysis is to explore if "college GPA" and "age" of the students are good predictors of the first salary of students majoring in Business Administration. It also aims to estimate the first-year salary of Business Administration graduates when GPA is 4.0 for students who are 25 years old.

- Step 2: Collect data.

The data are presented in ◘ Table 7.23.

- Step 3: Check the assumptions.

Six Sigma team checks linear relationship, multivariate normality, multicollinearity, autocorrelation, and homoscedasticity assumptions in this step. We assume that all these assumptions have been checked by the team.

- Step 4: Draw scatter diagrams to show the potential relationships between potential variables.

Using Minitab, Six Sigma team draws a scatter diagram to see the relationship between dependent and independent variables (◘ Image 7.26).

- Step 5: Compute regression equation.

Multiple linear regression analysis can be done using software such as Minitab, JMP, SPSS, and Stata. To perform regression analysis in Minitab, click on Stat→Regression→Regression→Fit Regression Model in menu bar. Then, transfer "first-year salary level" (response variable or Y variable) from left box to "responses" box and "age" and "college GPA" (continuous predictors or X variable) variables to continu-

Table 7.26 Regression analysis output in Minitab

Regression analysis: first-year salary level ($) (Y) versus age (X2), college GPA (X1)					
Regression equation					
First-year salary level ($) (Y) = 29,426–177 age (X2) + 8,638 college GPA (X1)					
Coefficients					
Term	**Coef**	**SE coef**	**T-value**	***P*-value**	**VIF**
Constant	29,426	4,639	6.34	0.000	
Age (X2)	−177	194	−0.91	0.378	1.00
College GPA (X1)	8,638	493	17.52	0.000	1.00
Model summary					
S	**R-sq**	**R-sq(adj)**	**R-sq(pred)**		
1,184.62	96.25%	95.62%	94.15%		
Analysis of variance					
Source	**DF**	**Adj SS**	**Adj MS**	**F-value**	***P*-value**
Regression	2	431,893,331	215,946,665	153.88	0.000
Age (X2)	1	1,174,316	1,174,316	0.84	0.378
College GPA (X1)	1	430,818,926	430,818,926	307.00	0.000
Error	12	16,840,002	1,403,334		
Lack-of-fit	10	16,590,002	1,659,000	13.27	0.072
Pure error	2	250,000	125,000		
Total	14	448,733,333			
Fits and diagnostics for unusual observations					
Obs	**First-year salary level ($) (Y)**	**Fit**	**Resid**	**Std resid**	
3	48,000	47,784	216	0.46	X
4	45,000	42,801	2,199	2.13	R

Source: Author's creation based on Minitab

ous predictors. Click on OK. Minitab provides regression analysis output as shown in Table 7.26.

- Step 6: Analyze the regression equation and how it fits the data.

As presented in Minitab output in Table 7.26, R^2-adjusted is 95.62% which shows that independent variables (age and college GPA) highly explain the variability of the dependent variable (first-year salary level). Regression equation is presented as follows:

$$\text{First year salary level}(\$)(Y) = 29{,}426 + 8{,}638 * \text{College GPA} - 177 * \text{Age}$$

However, the *p*-value of age variable (0.378) is greater than alpha value (0.05), which shows that age variable is not statistically significant in this regression analysis. Therefore, age variable is not included as a predictor variable in this analysis. Note that if the age variable was a statistically significant variable, we would substitute the value questioned for the independent variables to forecast a value for dependent variable. To predict the first-year salary level based on regression line, when high school GPA was 4.00 and age was 30, the estimated first-year salary would be calculated as \$58,668 as presented below:

$$\begin{aligned}&\text{First year salary level}(\$)(Y)\\&\quad = 29,426 + 8,638 * (4.00) - 177 * (30)\\&\quad = \$58,668\end{aligned}$$

Again, since the age is not a statistically significant predictor variable in this multiple regression analysis, age variable is not included in the analysis as follows:

First year salary level (\$)(Y)
= 29,426 + 8,638 * (4.00)
= 63,978

Six Sigma team either prefers to use simple linear regression with college GPA or consider using other independent variables in the multiple regression analysis.

7.13 ANOVA – Analysis of Variance

The Analysis of Variance (ANOVA) is used to test whether the means of two or more populations are equal. ANOVA is performed when the analysis includes a categorical factor or independent variable with two or more levels and a continuous response variable. ANOVA assumes that the data are approximately normally distributed with equal variances between factor levels. When the data are not normally distributed, data transformation is performed prior to conducting ANOVA. It tests two hypotheses: (1) all population means are equal in the null hypothesis versus (2) at least one population is different from the others in alternative hypothesis. The hypotheses tested in ANOVA are statistically stated as follows:

$$H_o = \mu_1 = \mu_2 = \mu_3 \ldots = \mu_k$$

$$H_1 = \textit{Means are not all equal, and at least one population mean is different.}$$

where μ is the mean of the kth population and k is the number of populations compared.

Prior to conducting ANOVA, Six Sigma teams need to make sure that several requirements are met:

1. Continuous dependent variable
2. Categorical independent variable
3. Corresponding dependent variable for each independent variable
4. Independent observations
5. Random samples from the populations
6. Normally distributed dependent variable in each population
7. Homogenous variances across the populations
8. No outliers
9. A model with a good fit to the data.

There are two types of ANOVA: one-way ANOVA and two-way ANOVA. In this section, one-way ANOVA is analyzed. Two-way ANOVA will be analyzed in the Design of Experiment section in ▶ Chap. 10.

7.13.1 One-Way ANOVA

One-way ANOVA is used when Six Sigma teams are interested in analyzing whether the population means of two or more groups significantly differ from each other, using one categorical factor (independent variable) and one response variable. In other words, one-way ANOVA tests the statistical differences among the means of two or more populations. When one-way ANOVA results show that at least one group is different, the results are analyzed in detail to determine which pairs of groups differ from the other groups. To identify which groups or populations are different, Tukey's test is conducted. ANOVA compares the variance between group means and the variance within groups to assess whether the groups come from the same or different populations.

To conduct a one-way ANOVA, the following requirements should be met prior to conducting the analysis:

1. The data have only one categorical factor or independent variable. If the data have one continuous independent variable, simple regression is recommended in the analysis. If the data have two categorical factors, then two-way ANOVA is recommended in the analysis. If there are multiple continuous independent variables, multiple regression is utilized in the analysis.
2. The response variable is continuous, not categorical.
3. Observations are independent from each other. One observation should not have any information about any of the other observations in the data set. In this case, the observations are considered independent.
4. All groups have the same standard deviation.
5. The data are normally distributed. The sample size of the groups is expected to be at least 15 in one-way ANOVA.
6. The model generated through ANOVA fits the data.

The test statistic of one-way ANOVA is denoted as *F.* The test statistic evaluates whether group means are significantly different from each other. One-way ANOVA results are represented in a framework, as shown in Table 7.29, where *SSA* = the sum of squares of factor, *SSE* = the sum of squares of error, *SST* = the total sum of squares, $r-1$= the factor degrees of freedom, $r(k-1)$= the error degrees of freedom, $rk-1$= the total degrees of freedom, *MSA* = the mean square of factor, *MSE* = the mean square of error, *F* = test statistic of one-way ANOVA, *r* = number of levels, *k* = the total number of groups, and *n* = the total number of observations.

The sum of squares of factor (*SSA*), the sum of squares of error (*SSE*), and the total sum of squares (*SST*) are calculated as shown in Eqs. 7.96, 7.97, and 7.98.

$$\mathrm{SST} = \mathrm{SSA} + \mathrm{SSE} = \sum_{i=1}^{k}\sum_{j=1}^{r}\left(y_{ij} - \bar{y}_{..}\right)^2 \quad (7.96)$$

$$\mathrm{SSA} = k\sum_{i=1}^{k}\left(\overline{y_{i.}} - \bar{y}_{..}\right)^2 \quad (7.97)$$

$$\mathrm{SSE} = \sum_{i=1}^{k}\sum_{j=1}^{r}\left(y_{ij} - \overline{y_{i.}}\right)^2 \quad (7.98)$$

where

- y_{ij} = *j*th observation of the *i*th group
- $\bar{y}_{..}$ = the mean of all observations
- $\bar{y}_{i.}$ = the average of observations in the *i*th group.

To perform one-way ANOVA, the following steps can be followed by Six Sigma teams:

- Step 1: Identify the continuous response variable.
- Step 2: Choose at least one categorical factor or independent variable.
- Step 3: Identify the levels of the factor.
- Step 4: Prepare a data collection plan.
- Step 5: Collect data.
- Step 6: Check the requirements.
- Step 7: Enter data in Minitab.
- Step 8: Run one-way ANOVA.
- Step 9: Interpret results and findings.

▶ Example 50

A Six Sigma team studies the effects of "temperature" on "tensile strength of metal cable" in a multinational metal wire manufacturing plant in New England area. It is assumed that temperature is the most significant factor of the tensile strength in the process and that the range of temperature is between 510 °F and 540 °F. The team decides to analyze the impact of four levels of temperature at 510 °F, 520 °F, 530 °F, and 540 °F in a set of experiments. The limited budget in the project allows team to make six experiments at each temperature level. The data collected are presented in Table 7.27. Using ANOVA, test the hypothesis that different levels of temperature do not affect the tensile strength of the metal cable. ◀

Table 7.27 Data collected in six experiments for tensile strength of metal cable in psi

	Temperature levels				
Measurements	**510 °F**	**520 °F**	**530 °F**	**540 °F**	
1	135	137	139	143	
2	132	138	140	144	
3	133	137	140	144	
4	135	139	140	145	
5	135	139	139	145	
6	132	139	141	145	
Total	802	829	839	866	3,336
Averages	133.7	138.2	139.8	144.3	139

Source: Author's creation

- **Solution**

- Step 1: Identify the continuous response variable.

The continuous response variable is "tensile strength of metal cable" in pounds per square inch-psi.

- Step 2: Identify at least one categorical factor or independent variable.

The categorical variable or independent variable is "temperature" in this set of experiments.

- Step 3: Identify the levels of the factor.

Since the team decided to test four levels of temperature, the level of factor is four at 510 °F, 520 °F, 530 °F, and 540 °F.

- Step 4: Prepare data collection plan.

The Six Sigma team organizes a data collection plan that identifies *when*, *where*, *how*, and *by whom* the experiments will be conducted, and the records will be kept. If there are any technical requirements of the experiments, these requirements are taken into consideration while preparing data collection plans.

- Step 5: Collect data.

The collected data are presented in Table 7.27.

- Step 6: Check the requirements.

Six Sigma team needs to make sure that all requirements of ANOVA are being met in this experiment prior to analyzing the data. The requirements previously mentioned above are also reviewed by the team before designing the experiments and collecting data.

- Step 7: Enter data in Minitab.

The data set needs to be entered in three columns, (1) runs, (2) temperature (°F), and (3) tensile strength (psi), in Minitab.

- Step 8: Run one-way ANOVA.

To run one-way ANOVA in Minitab, first transfer data to Minitab worksheet, and click on Stat→ANOVA→One-Way…. Then, in the next screen, transfer "tensile strength" to "response" box and "temperature" to "factor" box. On the same screen, click on "Comparisons," and select "Tukey"; select "tests" in "results" section at the bottom, and click on OK→OK. The results will be presented as shown in Table 7.28 and Images 7.27, 7.28, 7.29, 7.30, and 7.31.

- Step 9: Interpret results and findings.

The p-value for the temperature in "Analysis of Variance" section is less than 0.05 ($\alpha > p$) in Table 7.28. The p-value (0.000) shows that the null hypothesis is rejected, meaning that

Table 7.28 One-way ANOVA results

One-way ANOVA: tensile strength (psi) versus temperature (°F)					
Method					
Null hypothesis		All means are equal			
Alternative hypothesis		Not all means are equal			
Significance level		$\alpha = 0.05$			
Equal variances were assumed for the analysis					
Factor information					
Factor	**Levels**	**Values**			
Temperature (°F)	4	510, 520, 530, 540			
Analysis of variance					
Source	**DF**	**Adj SS**	**Adj MS**	**F-value**	***P*-value**
Temperature (°F)	3	349.67	116.556	104.38	0.000
Error	20	22.33	1.117		
Total	23	372.00			
Model summary					
S	**R-sq**	**R-sq(adj)**	**R-sq(pred)**		
1.05672	94.00%	93.10%	91.35%		
Means					
Temperature (°F)	**N**	**Mean**	**StDev**	**95% CI**	
510	6	133.667	1.506	(132.767, 134.567)	
520	6	138.167	0.983	(137.267, 139.067)	
530	6	139.833	0.753	(138.933, 140.733)	
540	6	144.333	0.816	(143.433, 145.233)	
Pooled StDev = 1.05672					
Tukey pairwise comparisons					
Grouping information using the Tukey method and 95% confidence					
Temperature (°F)	**N**	**Mean**	**Grouping**		
540	6	144.333	A		
530	6	139.833	B		
520	6	138.167	B		
510	6	133.667	C		
Means that do not share a letter are significantly different					

Table 7.28 (continued)

Tukey simultaneous tests for differences of means

Difference of levels	Difference of means	SE of difference	95% CI	T-value	Adjusted *P*-value
520–510	4.500	0.610	(2.792, 6.208)	7.38	0.000
530–510	6.167	0.610	(4.458, 7.875)	10.11	0.000
540–510	10.667	0.610	(8.958, 12.375)	17.48	0.000
530–520	1.667	0.610	(−0.042, 3.375)	2.73	0.057
540–520	6.167	0.610	(4.458, 7.875)	10.11	0.000
540–530	4.500	0.610	(2.792, 6.208)	7.38	0.000

Individual confidence level = 98.89%

Source: Author's creation based on Minitab

Table 7.29 One-way ANOVA results

Source of variation	Degree of freedom	Sum of squares	Variance (mean square)	*F* test statistic
Factor Temperature	$r - 1 = 4 - 1 = 3$	SSA = 349.67	$MSA = \frac{SSA}{r-1} = \frac{349.67}{4-1} = 116.556$	$F = \frac{MSA}{MSE} = \frac{349.67}{1.117} = 104.38$
Error	$r(k-1) = 4(6-1) = 20$	SSE = 22.33	$MSE = \frac{SSE}{r(k-1)} = \frac{22.33}{4(6-1)} = 1.117$	
Total	$rk - 1 = 4 * 6 - 1 = 23$	SST = 372.00		

Source: Author's creation

mean differences between the temperatures are statistically significant. When the p-value is less than or equal to the level of significance ($\alpha = 0.05$), that means the null hypothesis is rejected, and the team concludes that population means of four levels of temperature are not equal. To determine which pairs have statistically significant different means, the team needs to analyze Tukey comparison outcomes. The Analysis of Variance section is constructed upon the frame shown in Table 7.29.

In Table 7.29, SSA can be manually calculated as follows:

$$SSA = k\sum_{i=1}^{k}(\bar{y}_{i.} - \bar{y}_{..})^2$$

$$= 6[(133.67 - 139)^2 + (138.17 - 139)^2 + (139.83 - 139)^2 + (144.33 - 139)^2]$$

$$= 349.67$$

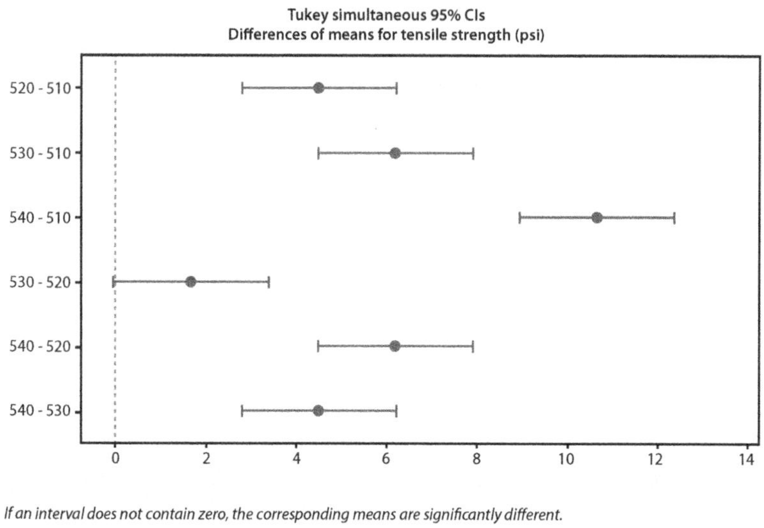

Image 7.27 The confidence intervals (CI) for the difference between the means of 520°F and 510°F. (Source: Author's creation based on Minitab)

In "Model Summary" section, "*S*" variable is used to analyze how well the model explains the variances on the response. *S* shows how far the individual observations are from the fitted model generated in the analysis. A low *S* value indicates that the model is a good fit of the responses.

R^2 represents the percentage of the variation of the response variable explained by the model generated through the data set used in the analysis. Higher R^2 values refer to a better model fit. R^2 varies between 0% and 100%. Adjusted R^2 (R-sq(adj)) demonstrates that the model generated through the data set presents a good fit to the data. The higher the adjusted R^2, the better fit of the model to the data. Predicted R^2 (R-sq(pred)) helps identify how well the model predicts the response variable when new observations are integrated into the data. A higher predicted R^2 shows that the model is more likely to correctly predict the new response variable. Predicted R^2 is preferred to adjusted R^2 in predictive studies. In our example, "temperature" explains 94% of the variation in the tensile strength. *S* variable shows that the standard deviation between the data and fitted model is 1.05672 units.

The "Grouping Information Using the Tukey…" section shows the levels of the temperature and sample size for each level. This section also demonstrates the means of each level. Where the *p*-value is statistically significant ($\alpha > p$) in one-way ANOVA, we can identify which specific pairs of groups are statistically significantly different, using the grouping information. If differences between means share a letter, that means that means are not significantly different.

In our analysis results, none of the levels share a letter, and this shows that all levels are statistically significant. The grouping information table shows that group A is only comprised of 540 °F, group B contains 530 °F and 520 °F, and group C is comprised of 510 °F. Group letters also help identify which level of the factor has a higher mean. In our case, 540 °F shows the highest mean among all these four levels of temperature, whereas 510 °F has the lowest mean in the data set.

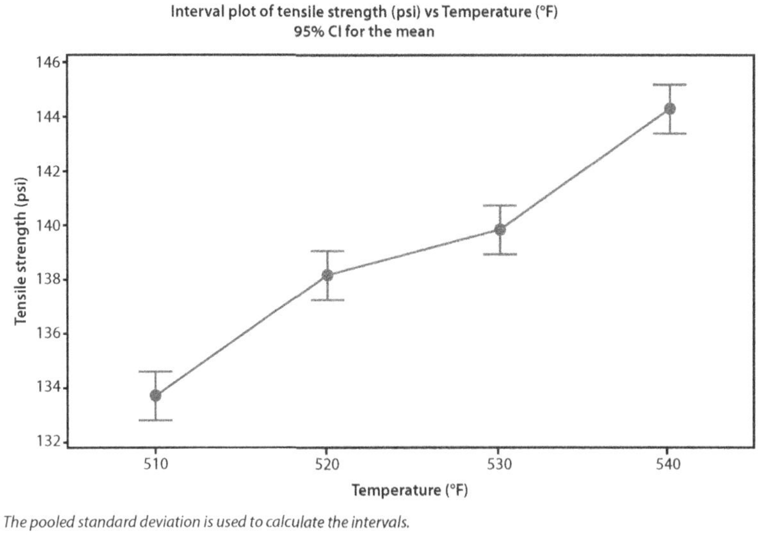

Image 7.28 Interval plot of tensile strength and temperature. (Source: Author's creation based on Minitab)

"Tukey Simultaneous Tests for Differences of Means" in Table 7.28 and "Tukey Simultaneous 95% CIs" graph in Image 7.27 show the confidence intervals (CI) for the difference between the means of 520 °F and 510 °F are 2.792 and 6.208. Since CI does not include zero, the difference between the means of 520 °F and 510 °F is considered statistically significant. This significance is also presented with the *p*-value (0.000) in the analysis results. However, the confidence intervals for the difference between the means of 530 °F and 520 °F are – 0.042 and 3.375. Since CI includes zero, the difference between the means of 530 °F and 520 °F is not considered statistically significant. As presented in the graph and results, the confidence intervals for the pairs of means are statistically significant except the pair of 530 °F and 520 °F.

Using the interval plot in Image 7.28, we can also examine the group means, central tendency, variability, and shape of the distribution and outliers of the observations. In the interval plot data, each dot demonstrates the mean of temperatures. As presented in the interval plot, 510 °F minimizes the tensile strength, whereas 540 °F maximizes it.

Normal probability plot of residuals in Image 7.29 identifies whether the residuals are normally distributed in the model. If the residuals are normally distributed, the plot shows that the placed points are located around a straight line. If the residuals do not follow the straight line, it shows that the residuals are not normally distributed. Where residuals are not normally distributed, the results of the analysis will not be considered accurate and acceptable.

The versus fits plot in Image 7.30 shows how individual residuals are placed on *Y*-axis against corresponding response variables (fitted value) on *X*-axis. The versus fits plot should show that there is no pattern or systematic behavior of the residuals and that all residuals are placed randomly on *X*-*Y*-axes. Residuals are expected to be

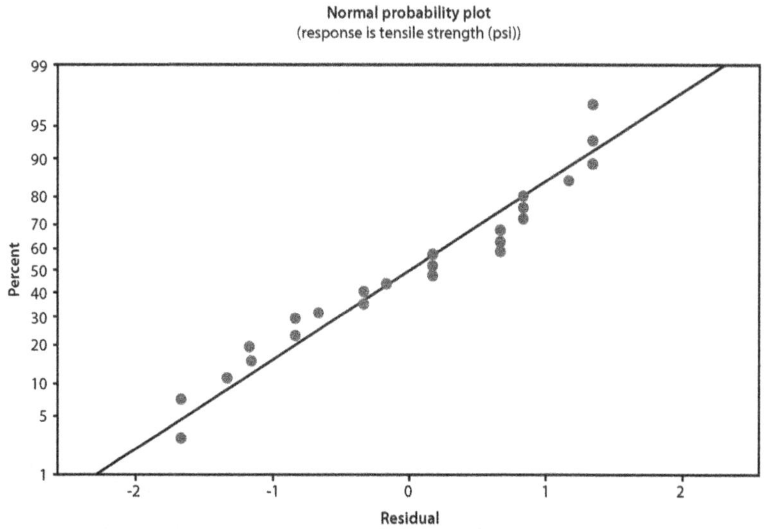

Image 7.29 Normal probability plot of residuals. (Source: Author's creation based on Minitab)

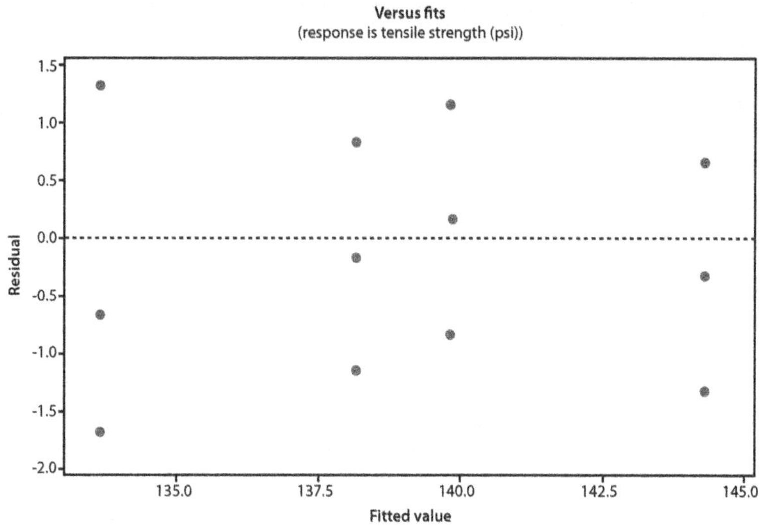

Image 7.30 The versus fits plot of residuals. (Source: Author's creation based on Minitab)

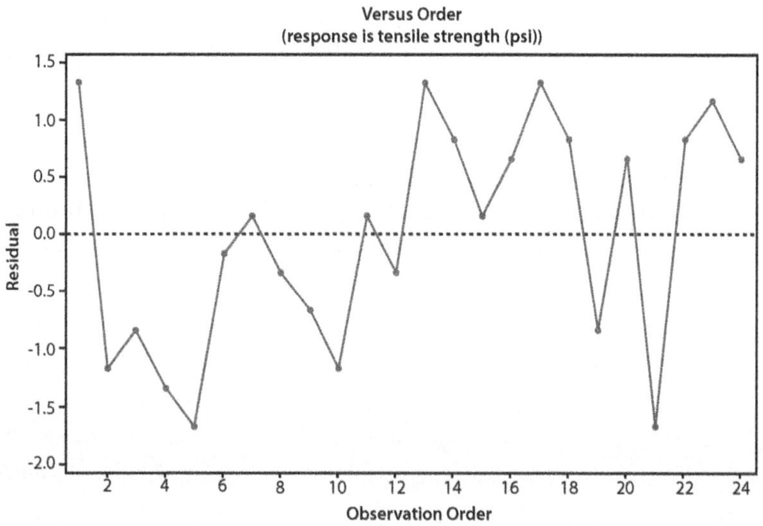

Image 7.31 The versus order plot of residuals. (Source: Author's creation based on Minitab)

randomly distributed on both sides of zero point on Y-axis. The versus fits plot in our analysis does not depict any pattern or systematic behavior. Therefore, we can say that the residuals are randomly distributed.

The versus order plot in ■ Image 7.31 shows the residuals in a time series that contain the residuals, based on the order of the data collection. The plot tests whether there are any interdependencies among the residuals. To accept that the residuals are independent, the residuals are expected to be placed randomly around the line that shows the zero point in the plot. The versus order plot shows that residuals are randomly located in the plot in our analysis.

7.14 Process Capability Analysis

Capability analysis measures the variability of CTQ characteristics of a product or service in a process. Technically speaking, the process variability is compared with the technical requirements and specifications of the product or service in process capability analysis. As a comprehensive analysis approach, capability analysis is implemented in Analyze and Improve phases of DMAIC process in Six Sigma.

After the assignable causes of variation are eliminated from a process, that process can be considered "stable." Prior to running a capability analysis, the team should make sure that the process is stable. Juran and Gryna (1980: 294) suggest that a process capability prediction should not be made until a process is in statistical control, since assignable causes may never be eliminated from the process if statistical control is not checked first. To analyze the stability of the process, the process is monitored and evaluated through control charts such as $\bar{X}-R$ charts, $\bar{X}-S$ charts, or control charts for attributes. Once the process is stable, three main questions should be answered in process control:

1. Is the machine/process capable of meeting customer specifications?
2. Are the data normally distributed?
3. Will the process produce any quality problems?

To answer these questions, capability analysis produces invaluable information for decision-makers. Normally distributed data are expected in capability analysis. When the data are not normally distributed, straightforward capability analysis cannot be implemented. First, raw data are required to be transformed into normal distribution. Then, the normally distributed (transformed) data set is used in capability analysis.

Juran (1992: 293) stated that, in some industries, process capability means "conformance to specifications" rather than meeting end customers' needs and expectations. Typical examples are seen when government agencies and regulatory bodies are clients, such as Department of Defense, Food and Drug Administration (FDA), NASA, and Department of Health and Human Services. In these cases, the customers are not those in consumer markets. Instead, they are industrial customers or governmental bodies, standard setters, or regulators that set the product or process specifications and that use the end products or services in their own manufacturing or service delivery processes.

Capability analysis predicts process adequacy, as a measure of the inherent uniformity of a manufacturing process (Juran and Gryna 1980). Capability analysis is considered a vital part of continuous quality improvement activities and categorized into two main categories: machine capability and process capability. Machine capability refers to variation in measurements caused by machine or operations, whereas process capability measures variation caused by all sources. According to Montgomery (2013: 356), process capability analyzes the uniformity of the process, where the variability of CTQ characteristics is a measure of the uniformity.

Juran and Gryna (1980: 295) emphasize that, as the processes become more complex, process capability analysis becomes more complex, too. In Six Sigma projects, it is vital to understand the current performance of the process. There are several performance indicators that can be used to measure and manage the performance of the process. The relevant literature refers practitioners to SPC methodology, and specifically capability analysis, to monitor and evaluate the production processes and the conformity of customer specifications for each QTC characteristic of the products. Variables (measurable or continuous) are used in capability analysis for each CTQ characteristic. After making sure that the data are normally distributed, the next steps of capability analysis are followed. Particularly, C_{p} and C_{pk} indices are utilized in capability analysis for continuous data. The C_{p} and C_{pk} are also utilized in various industries to measure and decrease the variance among products. Process potential capability ($\hat{C}_{\mathrm{p}}$) and actual capability indices ($\hat{C}_{pk}$) are given in Eqs. 7.99 and 7.100, respectively.

7

$$\hat{C}_{\mathrm{p}} = \frac{\mathrm{USL} - \mathrm{LSL}}{6\hat{\sigma}} \tag{7.99}$$

$$\begin{aligned}\hat{C}_{pk} &= \min\left(C_{pl}, C_{pu}\right) \\ &= \min\left\{\frac{\hat{\mu} - \mathrm{LSL}}{3\hat{\sigma}}, \frac{\mathrm{USL} - \hat{\mu}}{3\hat{\sigma}}\right\}\end{aligned} \tag{7.100}$$

where

- $\hat{\mu}$ = the estimated mean of the population
- USL = the upper specification limit
- LSL = the lower specification limit
- $\hat{\sigma}$ = the estimated standard deviation of the population.

In real cases, the standard deviation of the population is not known and estimated by unbiased estimator of $\hat{\sigma} = \frac{\overline{R}}{d_2}$ or $\hat{\sigma} = \frac{\overline{s}}{c_4}$, where $\hat{\sigma}$ is the unbiased estimator of standard deviation of the population, $\overline{s}$ is the standard deviation of the sample, $\overline{R}$ is average of the sample ranges, and d_2 and c_4 are coefficients used in control charts that vary with regard to sample size n.

The potential capability index (C_{p}) measures the six-sigma distance of the process compared to allowed variance, which is represented between USL and LSL. In other words, C_{p} is the ratio of specification spread to process spread (or natural tolerance). If the six-sigma distance is greater than the difference between USL and LSL, that means the process itself is located out of specification limits. Although C_{p} demonstrates the potential capability of the process, it does not show where the process is centered with regard to tolerance of specifications. In the actual capability index (C_{pk}), the location of the process is analyzed compared to specification limits. C_{pk} represents the distance of the mean of the process to the nearest specification limit (Wedgwood 2007). The process quality can be categorized as (1) inadequate, (2) capable, (3) satisfactory, (4) excellent, or (5) super on the basis of C_{p} and C_{pk} values. Ideally, both C_{p} and C_{pk} values are expected to be greater than 1.33 to consider a process "capable." If the indices are in between 1.00 and 1.33, the process is considered "marginally capable." If the indices are lower than 1.00, the process is considered "incapable." According to Montgomery (2013), recommended minimum values of C_{p} and C_{pk} for two-sided and one-sided specifications are as follows:

- Two-sided specifications:
  - Existing processes: 1.33
  - New processes: 1.50
  - Safety, strength, or critical parameter, existing process: 1.50
  - Safety, strength, or critical parameters and new processes: 1.67
- One-sided specifications:
  - Existing processes: 1.25
  - New processes: 1.45
  - Safety, strength, or critical parameter, existing process: 1.45
  - Safety, strength, or critical parameters and new processes: 1.60.

Based on the C_{p} value, the probability that the system produces defective products may be estimated as well. ◘ Table 7.30 represents

Table 7.30 C_p and ppm levels

C_p	Defective rates (ppm) for one-sided specifications	Defective rates (ppm) for two-sided specifications
0.25	226,628	453,255
0.50	66,807	133,614
0.60	35,931	71,861
0.70	17,865	35,729
0.80	8,198	16,395
0.90	3,467	6,934
1.00	1,350	2,700
1.10	484	967
1.20	159	318
1.30	48	96
1.40	14	27
1.50	4	7
1.60	1	2
1.70	0.17	0.34
1.80	0.03	0.06
2.00	0.0009	0.0018

Source: Adapted from Montgomery (2013)

expected defective or nonconforming units of product per million (ppm) for several C_p values – under the assumptions that the CTQ characteristic is normally distributed, the process is statistically in-control, and the process mean is located in the middle of *LSL* and *USL*. Violation of those assumptions may result in errors in decision-making processes based on C_p and ppm values (Montgomery 2013).

Capability analysis also represents short-term performance of the process. However, processes change over time. Therefore, long-term variation in the process is likely to be larger than short-term variation. As an outcome of this assumption, long-term capability indices are expected to be lower.

The process capability ratio (*PCR*) (Eq. 7.101) demonstrates the percentage of the specification band or interval used by the process. *PCR* is the ratio of process spread (6σ) to the specification interval that covers the distance between *USL* and *LSL*. The higher the percentage, the higher the variability. If the process is managed effectively and efficiently, it is expected to use relatively smaller part of the specification interval. In a general analysis, *PCR* shows the size of the variation occurring in the process. If *PCR* is greater than 100%, it means that the process is not capable of meeting specifications and produces defective items.

$$PCR = \frac{1}{\hat{C}_p} * 100 = \frac{6\hat{\sigma}}{\text{USL} - \text{LSL}} * 100 \quad (7.101)$$

▶ Example 51

The diameter of a metal cylinder part has to fit in an assembly. The specifications for the diameter are 5 ± 0.015 cm. It is estimated that the mean of the population ($\hat{\mu}$) is 4.99 cm and standard deviation of the population is ($\hat{\sigma}$) 0.004 cm. Calculate process capability indices (C_p and C_{pk}) and the percentage of the specification band or interval used by the process (*PCR*). Interpret findings and decide whether the process is capable of meeting customer requirements. ◀

- **Solution**

$$\text{USL} = 5 + 0.015 = 5.015\,\text{cm}$$

$$\text{LSL} = 5 - 0.015 = 4.985\,\text{cm}$$

$$\hat{\mu} = 4.99\text{ cm}$$

$$\hat{\sigma} = 0.004\text{ cm}$$

$$\hat{C}_p = \frac{\text{USL} - \text{LSL}}{6\hat{\sigma}} = \frac{5.015 - 4.985}{6(0.004)} = 1.25$$

C_p (Eq. 7.98) demonstrates that the process is "marginally capable" since it is less than 1.33.

$$\hat{C}_{pk} = \min\left\{C_{Pu} = \frac{\hat{\mu} - \text{LSL}}{3\hat{\sigma}}, C_{Pl} = \frac{\text{USL} - \hat{\mu}}{3\hat{\sigma}}\right\}$$
$$= \min\left\{C_{Pu} = \frac{4.99 - 4.985}{3(0.004)}, C_{Pl} = \frac{5.015 - 4.99}{3(0.004)}\right\}$$
$$= 0.4167$$

C_{pk} (Eq. 7.99) shows that the process is "not capable" of meeting the design requirements since it is less than 1.00.

7

$$PCR = \frac{1}{\hat{C}_p} * 100 = \frac{1}{1.25} = 80\%$$

According to *PCR* as presented in Eq. 7.101, the process uses 80% of the specification interval. If *PCR* increases, the likelihood of having defectives increases, since the process intends to exceed one of the specification limits. If *PCR* decreases, it means that the process uses up a smaller part of the specification, which means lower variation.

▶ Example 52

Calculate C_p and C_{pk} for the processes given in ◘ Figs. 7.9, 7.10, 7.11, 7.12, and 7.13 below and interpret the values. Decide which process shows the best performance. It is assumed that *USL* is 68, *LSL* is 44, and standard deviation of the process is 2 (USL = 68, LSL = 44, σ = 2). ◀

▪ Solution

The first process (◘ Fig. 7.9) is centered at 56 where the middle point of *USL* and *LSL* is located. As seen from the calculations below, C_p and C_{pk} equal 2, which shows that the process is capable of meeting specifications. This process shows ideal performance in terms of C_p and C_{pk}.

$$\hat{C}_p = \frac{\text{USL} - \text{LSL}}{6\hat{\sigma}} = \frac{68 - 44}{6(2)} = 2.00$$

$$\hat{C}_{pk} = \min\left\{C_{Pu} = \frac{\text{USL} - \hat{\mu}}{3\hat{\sigma}}, C_{Pl} = \frac{\hat{\mu} - \text{LSL}}{3\hat{\sigma}}\right\}$$
$$= \min\left\{C_{Pl} = \frac{68 - 56}{3(2)}, C_{Pu} = \frac{56 - 44}{3(2)}\right\} = 2$$

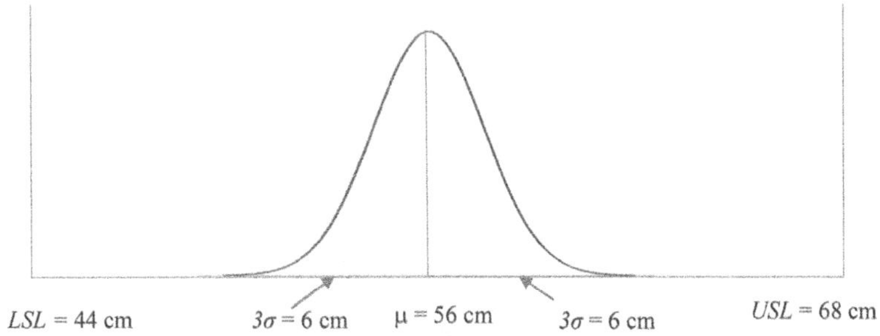

◘ **Fig. 7.9** The appearance of the first process. (Source: Author's creation)

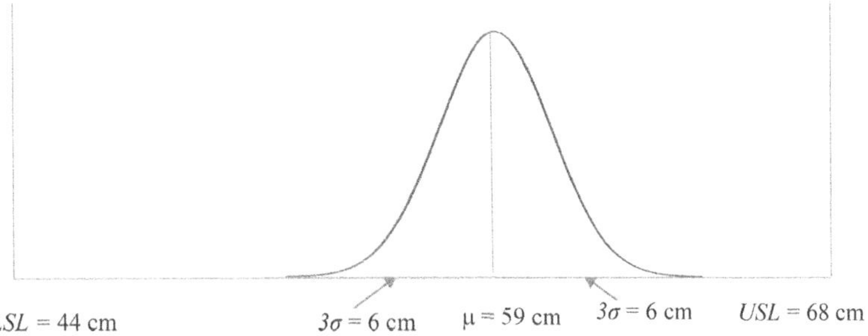

Fig. 7.10 The appearance of the second process. (Source: Author's creation)

The second process (Fig. 7.10) is centered at 59, which is closer to *USL*. C_p is still 2.00, whereas C_{pk} decreases to 1.5, as presented in the diagram. The process is still considered capable of meeting specifications.

$$\hat{C}_p = \frac{\text{USL} - \text{LSL}}{6\hat{\sigma}} = \frac{68 - 44}{6(2)} = 2.00$$

$$\begin{aligned}\hat{C}_{pk} &= \min\left\{C_{Pu} = \frac{\text{USL} - \hat{\mu}}{3\hat{\sigma}}, C_{Pl} = \frac{\hat{\mu} - \text{LSL}}{3\hat{\sigma}}\right\} \\ &= \min\left\{C_{Pl} = \frac{68 - 59}{3(2)}, C_{Pu} = \frac{59 - 44}{3(2)}\right\} = 1.5\end{aligned}$$

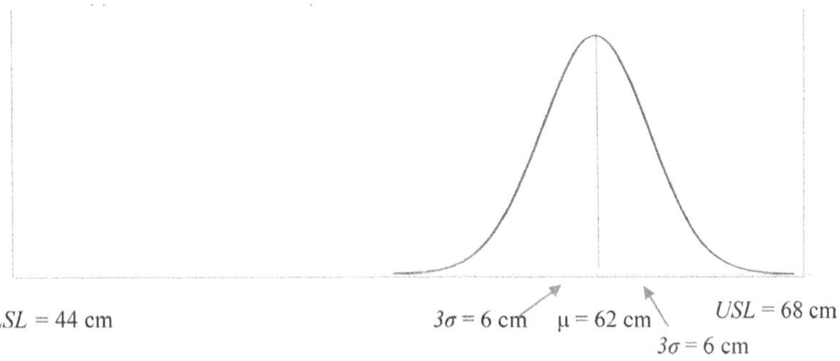

Fig. 7.11 The appearance of the third process. (Source: Author's creation)

The third process (Fig. 7.11) is centered at 62, and the distribution of data is more skewed toward *USL*. C_p is still computed as 2.00, and the potential capability index (C_p) does not show the impact of changing the center of the process on the capability index. However, C_{pk} is calculated as 1.00, which is the lowest acceptable level of capability indices. The process is still considered capable of meeting specifications.

$$\hat{C}_p = \frac{\text{USL} - \text{LSL}}{6\hat{\sigma}} = \frac{68 - 44}{6(2)} = 2.00$$

$$\begin{aligned}\hat{C}_{pk} &= \min\left\{C_{Pu} = \frac{\text{USL} - \hat{\mu}}{3\hat{\sigma}}, C_{Pl} = \frac{\hat{\mu} - \text{LSL}}{3\hat{\sigma}}\right\} \\ &= \min\left\{C_{Pl} = \frac{68 - 62}{3(2)}, C_{Pu} = \frac{62 - 44}{3(2)}\right\} = 1.0\end{aligned}$$

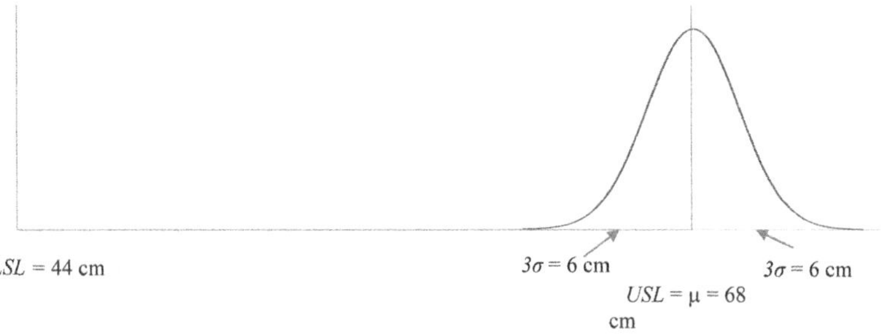

Fig. 7.12 The appearance of the fourth process. (Source: Author's creation)

The fourth process (**Fig. 7.12**) is centered at 68 where the *USL* is located. Half of the process is placed out of one of the specification limits, which means that the process does not produce the items within specification limits. Even though C_p is still 2.00, C_{pk} is zero, which shows that the process is not considered capable of meeting specifications.

$$\hat{C}_p = \frac{\text{USL} - \text{LSL}}{6\hat{\sigma}} = \frac{68-44}{6(2)} = 2.00$$

$$\hat{C}_{pk} = \min\left\{C_{Pu} = \frac{\text{USL} - \hat{\mu}}{3\hat{\sigma}}, C_{Pl} = \frac{\hat{\mu} - \text{LSL}}{3\hat{\sigma}}\right\}$$
$$= \min\left\{C_{Pl} = \frac{68-68}{3(2)}, C_{Pu} = \frac{68-44}{3(2)}\right\} = 0$$

Fig. 7.13 The appearance of the fifth process. (Source: Author's creation)

The last process shows (**Fig. 7.13**) that the majority of the items measured are located out of specification limits, specifically beyond *USL*, and the process is located at 71, out of specification limits. That indicates that the process is not capable of meeting specification limits with −0.50 C_{pk} even though C_p is 2.00.

$$\hat{C}_p = \frac{\text{USL} - \text{LSL}}{6\hat{\sigma}} = \frac{68-44}{6(2)} = 2.00$$

$$\hat{C}_{pk} = \min\left\{C_{Pu} = \frac{\text{USL} - \hat{\mu}}{3\hat{\sigma}}, C_{Pl} = \frac{\hat{\mu} - \text{LSL}}{3\hat{\sigma}}\right\}$$
$$= \min\left\{C_{Pl} = \frac{68-71}{3(2)}, C_{Pu} = \frac{71-44}{3(2)}\right\} = -0.50$$

As seen from these scenarios, several rules can be generated through C_p and C_{pk} values.

1. If the process mean is in the midpoint of the specification limits, C_p equals C_{pk}.
2. When the process mean is far away from the midpoint of the specification limits, C_{pk} gets lower than C_p ($C_{pk} < C_p$).
3. When C_{pk} is lower than 1, the majority of the process is located out of specification limits.
4. When the C_{pk} is zero, half of the process is out of the specification limits.

▶ **Example 53**

Compare C_p values and compute *PCRs* for the following four processes:
- Process A $C_p = 2.00$
- Process B $C_p = 1.33$
- Process C $C_p = 1.00$
- Process D $C_p = 0.5$. ◀

Solution

Let's calculate *PCRs* using Eq. 7.101 for each process as follows:
- Process A:
  $C_p = 2.00$, $PCR = (1/C_p)100 = (1/2)100 = 50\%$
- Process B
  $C_p = 1.33$, $PCR = (1/C_p)100 = (1/1.33)100 = 75.19\%$
- Process C
  $C_p = 1.00$, $PCR = (1/C_p)100 = (1/1.00)100 = 100\%$
- Process D
  $C_p = 0.5$, $PCR = (1/C_p)100 = (1/0.5)100 = 200\%$.

The results show that the higher C_p value results in better performance, and the process uses up a smaller proportion of the specification intervals. As the process capability gets higher, the lower percentage of the specification interval is used in the process. For example, compare C_p and *PCR* values of Process A and Process B. In Process A, *PCR* is 50%, which means the process uses half of the specification limits, and C_p is 2.00. When the C_p decreases, *PCR* increases, with 75.19% in Process B. That means that a bigger proportion of the specification limits is used by Process B, resulting in higher variation and worse quality performance. If *PCR* increases in the process, the likelihood of exceeding *USL* or *LSL* increases. The process is expected to use up a smaller proportion of the specification interval if the target is to decrease the percentage of defectives and nonconformities.

7.15 Taguchi's Loss Function

Taguchi's loss function is a powerful tool used in Measure and Analyze phases of DMAIC process because of its potential capacity to measure costs of poor quality of the process, product, or service analyzed in Six Sigma projects. Taguchi acknowledged that the traditional view of costs of conformance to specifications would not reflect the real cost of the products. He statistically proved that, as the measurements of the CTQ characteristic move away from the target value, a related cost occurs, as shown in Fig. 7.14. As long as the measurements are accumulated right on the target value, the cost related to loss function will be around zero. As the measurements start moving away from the target value, the loss of the relevant CTQ characteristic will increase. The lower the variation around the target value, the better the quality level. In other words, the greater the deviation from the target of CTQ, the greater the quality loss. As seen in Fig. 7.14, as measurements of the relevant CTQ characteristics move away from the target value, the total loss to all sides increases as a quadratic function. The concept of *loss* in Taguchi's loss function contains many relevant stakeholders, such as manufacturers, customers, and suppliers, who are negatively affected by low quality and increasing costs of the products or services.

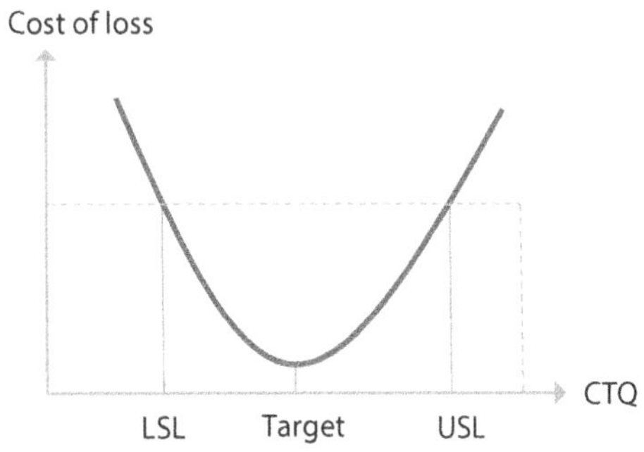

Fig. 7.14 Taguchi's loss function. (Source: Author's creation based on Taguchi (1986))

Taguchi's loss function is a way to measure the quality of products and services. It is also utilized as a method to measure and estimate the financial cost (loss) of any deviations from the target value of a CTQ characteristic. In Taguchi's loss function, the *X*-axis represents the performance of the relevant CTQ characteristic, whereas the *Y*-axis demonstrates the total monetary loss of deviation to the stakeholders. In traditional quality control activities, the performance of the CTQ characteristic is expected to remain in between lower and upper specification limits. Taguchi's loss function expects all measurements of the CTQ characteristic to be located on the target value, with no deviations to minimize the total cost of poor quality, including customer dissatisfactions, reworks, and scraps.

While only scraps and reworks are considered as a loss in the traditional approach, Taguchi's loss function takes societal loss into account as well, looking into both customers' and producers' perspective, in loss calculations. In a traditional quality approach, the financial loss that is related to low quality only occurs when the measures are out of specification limits. However, Taguchi's loss function does not accept this approach, instead stating that any deviations from the target value result in a certain amount of loss for society. Taguchi states that "as performance varies from the target, financial losses because of customer dissatisfaction increase - Loss to Society" (Wang 2005: 149). According to Taguchi (1986), "quality is the minimum loss imparted by the product to society from the time the product is shipped."

Sources of variability are called noise by Taguchi. Products and processes are subject to many sources of variation such as environmental considerations, deterioration, and manufacturing variability. Those factors result in quality loss (Wang 2005: 151). Wang (2005: 149) states that "Taguchi's quality loss function is a mathematical model of monetary loss (due to suboptimal product performance) as a function of the deviation in the quality characteristic (response) from a target value. It includes the cost of lost opportunity as a result of customer dissatisfaction and loss of resources used to rectify the failure mode. The systems with less response variability result in lower losses."

Taguchi methods were developed for quality improvement efforts and applied to the process design rather than the product design in the beginning. Taguchi methods were regarded as methods for Design of Experiments rather than a design methodology. In Taguchi methods, robustness can be measured by the magnitude of variation (Park et al. 2006: 183). The target value of the performance generally exists in the design phases. The target value can be zero, infinite, or a specific nominal value. Thus, Taguchi methods suggest a design with minimum variation around the target value (Park et al. 2006: 193). Taguchi's loss function represents robustness, as shown in Eq. 7.102. Robust design minimizes the average loss (Park et al. 2006: 184).

$$L(f) = k(f - T)^2 \tag{7.102}$$

where k is the constant to define the loss, f is the measured value, and T is the target value. The expected value of the loss function is defined as given in Eq. 7.103.

$$Q = E\left[L(f)\right] = k\left[\sigma^2 + (\mu - T)^2\right] \tag{7.103}$$

where k is the constant to define the loss, μ is the mean of variable f, σ^2 is the variance of variable f, and T is the target value. The symbol f represents the response or objective function in a general design, where the symbol y is used for f in the Taguchi method. The Taguchi methods also determine the optimum setting of control factors so that the variation of a response is minimized. Eq. 7.102 is used as an index for reaching robust design. Signal-to-noise ratio, as given in Eq. 7.104, is the ratio of the power of signal factors μ and the power of noise factors σ. Maximizing Eq. 7.104 is equivalent to minimizing loss function in Eq. 7.103. A robust design is obtained by maximizing signal-to-noise ratio given in Eq. 7.104.

$$\frac{S}{N} = 10\log_{10}\left(\frac{\mu^2}{\sigma^2}\right) \tag{7.104}$$

Taguchi's loss function categorizes CTQ characteristics into three groups: (1) nominal is the best (N type model), (2) smaller is the best (S type model), and (3) larger is the best (L type model). The next sections present the details of each group.

7.15.1 Nominal Is the Best

The response variable that has a target value T is accepted as the nominal-the-best type characteristic. The model including that type of CTQ characteristics is called "N type model" and produces a nominal target value setup prior to production. For example, the diameter of a metal cylinder item, the length of a plastic stick, the width of a wood plaque, and the top diameter of a coffee mug are N type of CTQ characteristics. Total loss function is calculated using the function in Eq. 7.105, expected value of loss is calculated using Eq. 7.106, and loss coefficient is calculated using Eq. 7.107.

$$L(y) = k(y - T)^2 \quad (7.105)$$

$$E\left[L(y)\right] = k\left[s^2 + (\bar{y} - T)^2\right] \quad (7.106)$$

$$k = \frac{A}{\Delta^2} \quad (7.107)$$

where y is the value of CTQ characteristic, $\bar{y}$ is the mean of CTQ characteristic, T is target value, k is loss coefficient, s^2 is the sampling variance, Δ is economic loss/unit, and A is one-sided tolerance interval.

7.15.2 Smaller Is the Best

The response variable targeting zero is called the smaller-the-better type and known as "S type model." The S type model is used in cases where the goal is to minimize the occurrences of undesirable product characteristics. For example, the roughness on a plastic part, scratches on a metal part, defective products, extended lead time, and high standard deviation of the measurements are targeted to be minimized. The total loss function is calculated using the function in Eq. 7.108, expected value is calculated using Eq. 7.109, and loss coefficient is calculated using Eq. 7.110.

$$L(y) = k(y)^2 \quad (7.108)$$

$$E\left[L(y)\right] = k\left[s^2 + (\bar{y})^2\right] \quad (7.109)$$

$$k = \frac{A}{\Delta^2} \quad (7.110)$$

where y is the value of CTQ characteristic, $\bar{y}$ is the mean of CTQ characteristic, k is loss coefficient, s^2 is the sampling variance, Δ is economic loss/unit, and A is one-sided tolerance interval.

7.15.3 Larger Is the Best

The response variable that proposes to increase the value up to infinite level is called the larger-the-better type and known as "L type model." L type model is used for CTQ characteristics that aim to maximize occurrences of desirable product characteristics. For example, the resistance of a concrete block, strength of a plastic part, and life cycle of product are considered L type CTQ characteristics. L type of quality loss function is shown in Eq. 7.111, expected value is shown in Eq. 7.112, and loss coefficient is calculated using Eq. 7.113, respectively.

$$L(y) = k\left[\left(\frac{1}{y}\right)^2\right] \quad (7.111)$$

$$E\left[L(y)\right] = k(1/\bar{y})^2 (1 + \left(3s^2/\bar{y}^2\right) \quad (7.112)$$

$$k = A\Delta^2 \quad (7.113)$$

where y is the value of CTQ characteristic, $\bar{y}$ is the mean of CTQ characteristic, k is loss coefficient, s^2 is the sampling variance, Δ is economic loss/unit, and A is one-sided tolerance interval.

▶ Example 54

The CTQ characteristic monitored in a manufacturing process for circle metal plaques are diameter and thickness. The inspectors measured 20 pieces for diameter and found the deviances in mm from the target value (T) as given in ◘ Table 7.31. The tolerance interval is ± 0.6 mm in diameter. Daily production rate is 80,000 units, and production cost/unit is 30 cents. Technically, it is not possible to rework defective pieces.

(a) What is the total loss when the deviance from the target value is 0.3 mm?
(b) What is the total loss when the deviance from the target value is 0.0 mm?
(c) Using the data given in ◘ Table 7.31, calculate the loss per unit in dollars.
(d) Given that the manufacturer works 250 days in a year, calculate the total annual loss. ◀

(a) Since there are a target value and tolerance limits given in this question, the CTQ characteristic is an N type one. In other words, the question is based on nominal-is-the-best model. Let's calculate k and $L(0.3)$ as follows:

$$k = \frac{A}{\Delta^2} = \frac{30 \text{ cents}}{0.6^2} = 83.34 \text{ cents}$$

$$L(y) = k(y-T)^2$$

$$L(0.3) = 83.34(0.3)^2 = 7.5 \text{ cents/unit}$$

In the total loss formula ($L(y)$), since ($y - T$) shows the difference between target value and the individual measurement, it will be 0.3 mm in part a.

(b) If the deviance from the target value (T) is 0.0 mm, the total loss is 0 cents.

$$k = \frac{A}{\Delta^2} = \frac{30}{0.6^2} = 83.34 \text{ cents}$$

$$L(y) = k(y-T)^2$$

$$L(0.3) = 83.34(0.0)^2 = 0 \text{ cents}$$

(c) The standard deviation of the measurements given in ◘ Table 7.31 is 0.180351 mm. The mean standard deviation (MSD) is 0.075 mm as follows:

$$\text{MSD} = \frac{\sum_{i=1}^{n}(y_i - T)}{n} = \frac{0.3^2 + 0.0^2 + \ldots + 0.2^2}{20} = \frac{1.5}{20} = 0.075$$

The loss per unit in dollars is

$$E[L(y)] = k\left[s^2 + (\bar{y} - T)^2\right]$$

$$E[L(y)] = 83.34\left[0.180351^2 + 0.075\right] = 8.96 \text{ cents/unit}$$

(d) Total annual loss

$$= 80{,}000 \frac{\text{units}}{\text{day}} * 8.96 \frac{\text{cents}}{\text{unit}} * 250 \frac{\text{days}}{\text{year}} = 1{,}792{,}000 \text{ dollars/year}$$

◘ **Table 7.31** Deviances (Y-T) from the target value (T) for 20 pieces (mm)

0.3	0.0	−0.1	0.3	0.2	0.1	−0.2	0.0	0.6	0.4
−0.2	0.1	0.0	−0.4	0.5	0.4	−0.2	0.0	0.0	0.2

Source: Author's creation

▶ Example 55

A desk chair manufactured by Blue Sea Manufacturing Co. has several CTQ characteristics monitored by quality inspection department. One of the CTQ characteristics is the surface smoothness of the arm holders. Because of the industry-wide standards, the CTQ characteristic of surface smoothness is expected to be no more than 12 mm. The arm holders that have more than 12 mm smoothness are classified as defective. The products that are out of tolerance cost \$80/unit. The daily production rate is 2000, and the company works for 250 days with two machines. The quality inspectors took several measurements from both machines as presented in Table 7.32.

(a) What is the total loss in dollars if the surface smoothness is 0 mm?
(b) What is the total loss in dollars if the surface smoothness is 5 mm?
(c) What is the total loss in dollars per unit occurred in both machines?
(d) Calculate the total daily loss in both machines.
(e) Calculate the total annual loss occurred in both machines. ◀

Table 7.32 Measures from desk chair production line

Machine 1									
0	5	4	2	3	1	7	6	8	4
6	0	3	10	4	5	3	2	0	7
Machine 2									
5	4	0	4	2	1	0	2	5	3
2	1	3	0	2	4	1	6	2	1

Source: Author's creation

Solution

The CTQ characteristic in this question is an S type one. Since there is a lowest expected value (T) in this question, the question has an S type model. In other words, the question is based on the smaller-the-better model.

(a) First of all, let's list the variables given in the question. Then, we can calculate k and $L(y)$ function for 0 mm surface smoothness.

- $\Delta = 12\,\text{mm}$
- $A = \dfrac{\$80}{\text{unit}}$
- Daily production rate = 2000 units/day

$$k = \frac{A}{\Delta^2} = \frac{80}{12^2} = \$0.55/\text{unit}$$

$$L(y) = k(y)^2$$

$$L(0) = k(0)^2 = 0.55(0) = 0 \text{ dollar.}$$

$L(y) = 0$ means there will be no economic loss in cases where the surface smoothness is 0 mm which is the target value of the CTQ characteristic in this manufacturing process. That scenario represents the ideal condition where total loss is expected to be zero for each relevant side.

(b) $L(y) = k(y)^2$

$$L(5) = k(5)^2 = 0.55(5)^2 = \$13.75.$$

In case that the surface smoothness is 5 mm, the total loss for this particular piece will be \$13.75.

(c) For machine 1, the total loss in dollars per unit:

$$\text{MSD} = \frac{\sum_{i=1}^{n}(y_i)^2}{n} = \frac{0^2 + 5^2 + \ldots + 7^2}{20} = \frac{468}{20} = 23.4$$

The variance of machine 1 is calculated as $s^2 = 7.79$.

$$E[L(y)] = k[s^2 + (\bar{y})^2]$$
$$= 0.55[7.79 + 23.4]$$
$$= \$17.1545/\text{unit}$$

For machine 2, the total loss in dollars per unit:

$$\text{MSD} = \frac{\sum_{i=1}^{n} (y_i)^2}{n}$$
$$= \frac{5^2 + 4^2 + \ldots + 1^2}{20}$$
$$= \frac{176}{20} = 8.8$$

7

The variance of machine 2 is calculated as $s^2 = 3.2$.

$$E[L(y)] = k[s^2 + (\bar{y})^2]$$
$$= 0.55[3.2 + 8.8] = \$6.6/\text{unit}$$

(d) For the first machine

$$\text{Total daily loss} = \$17.1545 * 2000 = \$34,309$$

For the second machine

$$\text{Total daily loss} = \$6.6 * 2000 = \$13,200$$

(e) The total annual loss occurred in the first machine:

$$\text{Total annual loss} = \$34,309 * 250 \frac{\text{days}}{\text{year}}$$
$$= \$8,577,250/\text{year}$$

The total annual loss occurred in the second machine:

$$\text{Total annual loss} = \$13,200 * 250 \frac{\text{days}}{\text{year}}$$
$$= \$3,300,000/\text{year}$$

▶ Example 56

A company manufactures two types of adhesive products. The manufacturing cost of the first product (S1) is \$50, while the second product costs \$60. The adhesion strength of the products is one of the CTQ characteristics and is expected to be more than 5 kgf. The cost of the products that are out of specifications is \$70. Annual production rate is 120,000 units. The quality inspection department takes 16 pieces from each product and tests the adhesion strength. ◘ Table 7.33 represents the results of the adhesion strength for the sample.

(a) Calculate the total loss if the adhesion strength is 10.2 kgf for the first product.
(b) Calculate the total loss if the adhesion strength is 0 kgf for the first product.
(c) Calculate the total loss if the adhesion strength is 5 kgf for the first product.
(d) Using the data given in ◘ Table 7.33, calculate the potential annual quality loss in dollars for each product. ◀

◘ **Table 7.33** The results of the adhesion strength for the sample

S1 (adhesion strength-kgf)							
10.2	5.8	4.9	16.1	15.0	9.4	4.8	10.1
14.6	19.7	5.0	4.7	16.8	4.5	4.0	16.5
S2 (adhesion strength-kgf)							
7.6	13.7	7.0	12.8	11.8	13.7	14.8	10.4
7.0	10.1	6.8	10.0	8.6	11.2	8.3	10.6

Source: Author's creation

- **Solution**

The CTQ characteristic in this question is an L type one. Since there is a highest expected value in this question, the question has an L type model. In other words, the question is based on the larger-the-better model. Let's list the variables given in the question as follows:

- $\Delta = 5$ kgf
- $A = \$70/\text{unit}$
- Annual production rate = 120,000 units.

(a) When the adhesion strength is 10.2 kgf, the total loss would be \$1393.656.

$$k = A\Delta^2 = \$70 * 5^2 = 1750$$

$$L(y) = k\left[\left(\frac{1}{y}\right)^2\right]$$

$$L(10.2) = 1750\left[\left(\frac{1}{10.2}\right)^2\right] = \$1393.656$$

(b) When the adhesion strength is 0 kgf, the total loss would not be able to be calculated as seen below:

$$L(y) = k\left[\left(\frac{1}{y}\right)^2\right]$$

$$L(0) = 1750\left[\left(\frac{1}{0}\right)^2\right] = \text{undefined}$$

(c) When the adhesion strength is 5 kgf, the total loss would be \$70.

$$L(y) = k\left[\left(\frac{1}{y}\right)^2\right]$$

$$L(5) = 1750\left[\left(\frac{1}{5}\right)^2\right] = \$70$$

(d) The potential quality loss in dollars for the first product (S1) is \$442.3763 as detailed below:

The variance of sample 1 is calculated as $s^2 = 30.4076$.

The mean of sample 1 is calculated as $\bar{y} = 10.13$.

$$\text{MSD} = \frac{\sum_{i=1}^{n}\left(\frac{1}{y}\right)^2}{n}$$
$$= \frac{\left(\frac{1}{10.2}\right)^2 + \left(\frac{1}{5.8}\right)^2 + \ldots + \left(\frac{1}{16.5}\right)^2}{20} = 0.1338$$

$$E\left[L(y)\right] = k(1/\bar{y})^2 (1 + \left(\frac{3s^2}{\bar{y}^2}\right)$$
$$= 1750(0.1338)(1 + \left(\frac{3(30.4076)}{10.13^2}\right)$$
$$= \$442.3763$$

The potential quality loss in dollars for the second product (S2) is \$215.8602 as detailed below:

The variance of sample 2 is calculated as $s^2 = 6.7407$.

The mean of sample 2 is calculated as $\bar{y} = 10.275$.

$$\text{MSD} = \frac{\sum_{i=1}^{n}\left(\frac{1}{y}\right)^2}{n} = \frac{\left(\frac{1}{7.6}\right)^2 + \left(\frac{1}{13.7}\right)^2 + \ldots + \left(\frac{1}{10.6}\right)^2}{20} = 0.10352$$

$$E[L(y)] = k(1/\bar{y})^2 (1+\left(\frac{3s^2}{\bar{y}^2}\right)$$
$$= 1750(0.10352)(1+\left(\frac{3(6.7407)}{10.275^2}\right)$$
$$= \$215.8602$$

▶ Example 57

In a production process, the target value of the CTQ characteristic is 1000 mm with ± 1 mm tolerance interval. The average of the observations is 999.63 mm, and the cost of the products that are out of specifications is $5/unit.

(a) Calculate k coefficient.
(b) If the standard deviation of the sample is 0.215 mm, calculate the total cost per unit.
(c) If the average of the sample is 999 mm and standard deviation of the sample is 0.75 mm, calculate the total cost per unit.◀

7

▪ Solution

Since there are a target value and tolerance limits given in this question, the CTQ characteristic is an N type one. In other words, the question is based on nominal-is-the-best model. Let's list the variables given in the question:

- $T = 1{,}000$ mm
- $\Delta = 1$ mm
- $\bar{X} = 999.63\,\text{mm}$
- $A = \$5$/unit.

(a) k coefficient in this question is calculated as follows:

$$k = \frac{A}{\Delta^2} = \frac{\$5}{1^2} = 5$$

(b) $s = 0.215\,\text{mm}$

$$E[L(y)] = k\left[s^2 + (\bar{y} - T)^2\right]$$
$$= 5\left[0.215^2 + (999.63 - 1{,}000)^2\right]$$
$$= \$0.9156/\text{unit}$$

When the standard deviation of the sample is 0.215 mm and the average of the sample is 999.63 mm, the total loss per unit is $0.9156.

(c) $s = 0.75$ mm and $\bar{X} = 999\,\text{mm}$

$$E[L(y)] = k\left[s^2 + (\bar{y} - T)^2\right]$$
$$= 5\left[0.75^2 + (999 - 1{,}000)^2\right]$$
$$= \$7.8125/\text{unit}$$

When the standard deviation of the sample increases three times (to 0.75 mm) and the average of the sample shifts from the target value with 999 mm, the total loss per unit increases and becomes $7.8125.

Key Concepts

Analyze phase, DMAIC, descriptive statistics, measures of central tendency, mean, median, mode, measures of variability, range, standard deviation, variance, types of variation, shape of distribution, symmetrical distribution, right-skewed distribution, left-skewed distribution, bimodel distribution, independent variable, dependent variable, statistic, parameter, quartile, the five-measure summary, common (chance) causes of variation, assignable causes of variation, parameters of a distribution, continuous and discrete random variables, probability mass function (PMF), cumulative distribution function (CDF), probability density function (PDF), discrete probability distributions, Bernoulli distribution, binomial distribution, hypergeometric distribution, geometric distribution, Poisson distribution, continuous probability distributions, uniform distribution, exponential distribution, triangular distribution, normal distribution, t distribution, F distribution, chi-square distribution, central limit theorem, Weibull distribution, statistical distributions, inferential statistics, inference, estimation, point estimation, estimator, point estimator, interval estimation, sampling distribution, standard error of the mean, confidence interval, tolerance interval, prediction interval, standard error, margin of error, upper confidence limit, lower confidence limit, natural tolerance limit, specification limit, hypothesis testing, null hypothesis, alternative hypothesis, one-tailed hypothesis, two-tailed hypothesis, Type I error, Type II error, consumer's risk, producer's risk, test statistic, rejection region, observed significance level, level of significance, critical value, *p*-value, normality, assumptions and conditions, pooled-variance t-test, degrees of freedom, separate-variance t-test, paired-t test, graphs, SPC, regression analysis, correlation analysis, multicollinearity, homoscedasticity, mean absolute deviation, mean absolute error, tracking signal, ANOVA, Tukey's test, Design of Experiment, process capability analysis, actual capability index, potential capability index, product per million (ppm), process capability ratio, Taguchi's loss function, robust design, nominal is the best, smaller is the best, larger is the best.

Summary

The Analyze phase aims to identify the cause-and-effect relationships in the process and to discover the sources of variability. In Six Sigma projects, identifying root causes of variability and taking preventive or corrective actions are pivotal. The Analyze phase focuses on making data-driven decisions for the next phase of Six Sigma. Data-driven decision-making processes require Six Sigma practitioners to implement statistical analysis tools and techniques in the Analyze phase. This chapter presents how to conduct Analyze phase and incorporate these statistical analysis tools and techniques into the Analyze phase.

Practice and Discussion Questions

1. What is the main goal of "Analyze" phase in DMAIC?
2. What are the components of descriptive statistics?
3. What is the difference between statistic and parameter?
4. What descriptive statistics are used for continuous variable?
5. What descriptive statistics are used for categorical variable (discrete variable)?
6. What are the measures of central tendency?
7. What are the measures of variability?
8. When is the "median" preferred to the "mean" in descriptive analysis?
9. Calculate mean, mode, and median of 2, 4, 5, 7, 9, 10, 10, and 15, manually.
10. Calculate range, standard deviation, and variance of 2, 4, 5, 7, 9, 10, 10, and 15, manually.
11. Calculate the quartiles of 2, 4, 5, 7, 9, 10, 10, and 15, manually.
12. Calculate mean, mode, and median of 2, 4, 5, 7, 9, 10, 10, and 15 using Minitab.
13. Calculate range, standard deviation, and variance of 2, 4, 5, 7, 9, 10, 10, and 15 using Minitab.
14. Calculate the quartiles of 2, 4, 5, 7, 9, 10, 10, and 15 using Minitab.
15. Identify five measures of 2, 4, 5, 7, 9, 10, 10, and 15 manually.
16. Identify five measures of 2, 4, 5, 7, 9, 10, 10, and 15 using Minitab.
17. Discuss the impact of "variation" in processes?
18. What are the types of variation?
19. Suppose a coin will come up head with probability 0.6. This coin is tossed three times. Assume the random variable X defines the number of heads that appear in the three tosses. Determine the PMF of X.
20. On a Six Sigma green belt certification exam with four possible answers for each of the ten questions, what is the probability that a person who takes the exam would get five or more correct answers just by guessing?
21. A doctor's office knows that 5 percent of the patients making reservations on a certain day will not show up for their appointments. Consequently, their policy is to schedule ten appointments per day while only eight time slots are available per day

(i.e., the doctor can visit only 8 patient per day). What is the probability that there will be a time slot to see a doctor for every patient who shows up?

22. If you buy a lottery ticket in 40 lotteries where your chance of winning in each lottery is 1/1000, (a) calculate the probability that you will win a prize at least once.
23. (b) Calculate the probability that you will win a prize exactly once.
24. (c) Calculate the probability that you will win a prize at least twice.
25. In a manufacturing company, there is an average of 3.2 defective items produced per day. What is the probability that there are four or more defects produced in a day?
26. A process produces nonconformities according to a Poisson distribution. If the mean of nonconformities is 20, what is the standard deviation?
27. A farmer has an average of eight weeds per acre on his farm during the summer months. What is the probability that the farmer will have less than two weeds per acre in the summer?
28. If a process follows an exponential distribution with mean of 35, what is the standard deviation of the process?
29. Suppose that customers' arrivals to the ticket counter at a movie theater can be modeled as an exponential distribution with an average arrival rate of 10 per hour.
30. (a) what is the probability that time between arrivals of two customers will be less than 10 minutes?
31. (b) Given that the last customer arrived at 9:00 pm, and it is now 9:20 pm, what is the probability that the next customer will arrive sometime after 9:25 pm?
32. Assume that an exponential random variable X is defined to be the time required to repair a machine with mean 1/3 hours.
33. What is the probability that a repair time exceeds 1/3 hour?
34. What is the probability that a repair takes at least 10 hours given that its duration exceeds 9 hours?
35. For a normal distribution, two standard deviations on each side of the mean would include what percentage of the total population?
36. What is inferential statistics?
37. What kinds of problems a Six Sigma team can solve by using inferential statistics?
38. Discuss the importance of sampling in a Six Sigma project.
39. What is the sampling distribution of the mean?
40. What are the properties of the sampling distribution of the mean?
41. What is the standard error of the mean?
42. For a given population, which is larger, the standard deviation of the population or the standard error of the mean?
43. Discuss the following statement: regardless of the distributional shape of a data set, it is reasonable to use the normal distribution to solve probability questions regarding the sample mean, as long as the sample size is adequate.
44. What does the central limit theorem imply?
45. What is a point estimate?
46. A random sample is drawn from a population, and the observations are 6, 12, 13, 15, 18, and 20. Compute the following point estimates: (a) the estimate of the population mean; (b) the estimate of the population standard deviation; and (c) the estimate of the standard error of the mean.
47. A random sample of $n = 125$ observations is selected from a population with $\mu = 60$ and $\sigma = 20$. Calculate each of the following probabilities: (a) $P(\bar{x} > 62.5)$; (b) $P(\bar{x} < 61.8)$; (c) $P(59.1 < \bar{x} < 64.2)$; and (d) $P(\bar{x} < 55.0)$. Get the results also using Minitab.
48. Write the symbol for (a) the point estimator of a population proportion and (b) the point estimator of the standard error of the proportion.
49. What are the properties a point estimator can possess?
50. What does an unbiased estimator mean? Why is this property important?
51. What is an interval estimate?
52. Which has greater chance of being correct, a point estimate or an interval estimate? Why?
53. Assume two interval estimates, $20 \le \mu \le 30$ and $15 \le \mu \le 35$, are made from the same sample. Which interval estimate has the greater chance of being correct? Why?

7

54. How many types of interval estimates can be made?
55. What does a confidence interval estimate mean?
56. What are the different types of confidence intervals that a Six Sigma team may construct? Provide a practical application for each.
57. What is the margin of error? How is it calculated?
58. In a large-sample confidence interval for a population proportion, what does the confidence coefficient represent?
59. Discuss how a confidence interval changes with changes in the confidence level and the sample size.
60. Suppose that a random sample of size 105 is selected from a population with unknown mean μ and standard deviation σ. Summary statistics are obtained as $\bar{x} = 26.2$ and $s = 14.7$. Calculate a 95% confidence interval for the population mean.
61. Determine the value of $t_{\alpha/2}$ that would be used in the construction of a confidence interval for a population mean for the confidence coefficient 0.99 and sample size of 15.
62. What are differences between the t-distribution and the z-distribution?
63. A random sample of eight measurements from a normally distributed process outputs yielded $\bar{x} = 7.2$ and $s = 1.3$. Calculate a 90% confidence interval for the population mean.
64. Determine the values of $\chi^2_{n-1,\left(\frac{\alpha}{2}\right)}$ and $\chi^2_{n-1,\left(1-\frac{\alpha}{2}\right)}$ that would be used in the construction of a confidence interval for a population standard deviation for the confidence coefficient 0.90 and sample size of 20.
65. What are the assumptions required for the confidence interval estimation procedure of a population variance?
66. A Six Sigma Master black belt (BB) wishes to check the time candidate Six Sigma green belts (GB) spend on the 10-minute coffee breaks allowed during their training program. The Master BB observes a random sample of six candidates and notes the time, in minutes, spent by each for the coffee break. The times are 14, 8, 12, 9, 13, and 11. Suppose times spent are normally distributed. Using Minitab, construct the 95 percent confidence interval for the mean time and the standard deviation of the time per employee spent by all GB candidates who take coffee breaks. Interpret these two confidence intervals together.
67. A random sample of size 170 is selected from a population, and the number of successes is 60. (a) Calculate a 95% confidence interval for the population proportion, and interpret the interval. (b) How would the width of the confidence interval change if the confidence coefficient was decreased from 0.95 to 0.90?
68. What is the required sample size for the confidence interval estimation procedure based on normal distribution of a population proportion to be valid?
69. A 40 random sample of the light bulbs was selected from the manufacturing process, and the average lifetime of these bulbs was measured as 1500 hours. The sample standard deviation was found 200 hours. Set up a 95% tolerance interval for the lifetime of light bulbs in manufacturing process that has confidence level 99%.
70. What is a tolerance interval and how does it differ from a confidence interval?
71. In order to be 99% confident that at least 95% of the population will be included between the extreme values of the sample, what sample size will be required?
72. What is the difference between a prediction interval and a confidence interval?
73. What does correlation analysis tell us in Six Sigma projects?
74. What does it mean if correlation coefficient (r) equals 1?
75. What does it mean if correlation coefficient (r) equals 0?
76. What does it mean if correlation coefficient (r) equals −1?
77. What does it mean if correlation coefficient (r) equals 0.7?
78. What does it mean if correlation coefficient (r) equals −0.7?
79. Explain how statistical significance of the correlation analysis is tested and used.

80. Explain what pre-tests should be done prior to running correlation analysis.
81. What does regression analysis tell us in Six Sigma projects?
82. What does the "coefficient of determination" tell us in Six Sigma projects?
83. Discuss the "coefficient of determination" if correlation coefficient (r) was found to be 0.90.
84. Briefly discuss the assumptions of regression analysis.
85. What does multicollinearity mean in regression analysis?
86. How is multicollinearity issue eliminated in regression analysis?
87. What type of regression analysis should be used if there is more than one independent variable involved in the analysis?
88. What is the function of scatter diagram in regression analysis?
89. What would you say about the relationship between dependent and independent variables if R^2-adjusted was found to be 0.225?
90. Discuss the ways of evaluation of forecasting accuracy when the regression analysis is implemented in data analysis.
91. Discuss when the forecast error is zero.
92. What are the forecast accuracy measures that can be used in Six Sigma projects?
93. Discuss how ANOVA can be implemented in data analysis in Six Sigma projects.
94. Discuss how process capability analysis can be implemented in data analysis in Six Sigma projects.
95. What is the main goal of process capability analysis?
96. What are the basic indices used in process capability analysis?
97. Discuss the acceptable intervals of process capability analysis indices.
98. If the process has two-sided specifications, what is the minimum acceptable value for process capability indices in new processes?
99. If the process has one-sided specifications, what is the minimum acceptable value for process capability indices in existing processes?
100. What does process capability ratio (*PCR*) mean in process capability analysis?
101. What does Taguchi's loss function recommend to manufacturers to minimize the loss occurred due to low quality?
102. What are the types of Taguchi's loss function?
103. If CTQ aims at minimizing the measurements, what type of loss function should be utilized in the analysis?
104. If CTQ aims at maximizing the measurements, what type of loss function should be utilized in the analysis?
105. If CTQ aims at keeping the measurements at a target value, what type of loss function should be utilized in the analysis?

7

References

Barnes, J. W. (1994). *Statistical analysis for engineers and scientists a computer based approach*. Singapore: McGraw-Hill.

Benneyan, J. C. (1998). Statistical quality control methods in infection control and hospital epidemiology, part I introduction and basic theory. *Infection Control & Hospital Epidemiology, 19*(3), 194–214.

Benneyan, J. C., Lloyd, R. C., & Plsek, P. E. (2003). Statistical process control as a tool for research and healthcare improvement. *Quality & Safety in Health Care, 12*(6), 458–464.

Boardman, T. J., & Boardeman, E. C (1990, December). Don't touch that funnel. Quality Progress, 65–69.

Brassard, M., Finn, L., Ginn, D., & Ritter, D. (2017). *The Six Sigma Memory Jogger II: A Pocket Guide Tools for Six Sigma Improvement Teams* (Spiral Bound Edition ed.). GOAL/QPC.

Bruhl, R. (2018). *Understanding statistical analysis and Modeling*. Los Angeles: Sage.

Deming, W. E. (1982). *Quality, productivity, and competitive position*. Cambridge, MA: MIT Center for Advanced Engineering Study.

Duncan, G. T., Elliot, M., & Salazar-González, J. J. (2011). Why statistical confidentiality? In *Statistical confidentiality* (pp. 1–26). New York: Springer.

Evans, J. R. (2012). *Statistics, data analysis, and decision modeling* (5th ed.). Pearson.

Finison, L. J., Finison, K. S., & Bliersbach, C. M. (1993). The use of control charts to improve healthcare quality. *Journal for Healthcare Quality, 15*(1), 9–23.

Fisher, R. A., & Yates, F. (1957). Student's t-test of significance. *Statistical tables for agricultural, biological and other research workers. Edinburgh: Oliver and Boyd*, 157.

GOAL/QPC. (2002). *The Six Sigma Memory Jogger II: A Pocket Guide of Tools for Six Sigma Improvement Teams.* GOAL/QPC.

Hart, M. K., & Hart, R. F. (2002). *Statistical process control for health care.* Duxbury: Thomson Learning.

Juran, J. M. (1992). *Juran on quality by design: The new steps for planning quality into goods and services.* Simon and Schuster.

Juran, J. M., & Gryna, F. M. (1980). *Quality planning and analysis.* New York: McGraw-Hill.

Knapp, H. (2016). *Practical statistics for nursing using SPSS.* Los Angeles: Sage Publications.

Levine, D. M. (2006). *Statistics for six sigma green belts with Minitab and JMP.* Upper Saddle River: Prentice-Hall.

Levine, D. M., Krehbiel, T. C., & Berenson, M. L. (2015). *Business statistics: A first course* (7th ed.). New Jersey: Prentice Hall.

Lind, D. A., Marchal, W. G., & Wathen, S. A. (2012). *Statistical techniques in business and economics.* New York: McGraw-Hill/Irwin.

McDonald, J. H. (2009). *Handbook of biological statistics.* Baltimore: Sparky House Publishing.

Merriam-Webster Dictionary. (n.d.). Retrieved 01 06, 2019, from Definition of Inference: https://www.merriam-webster.com/dictionary/inference

Montgomery, D. C. (2005). *Introduction to statistical quality control* (5th ed.). New York: Wiley.

Montgomery, D. C. (2013). *Introduction to statistical quality.* New York: Wiley.

Montgomery, D. C., & Runger, G. C. (2014). *Applied statistics and probability for engineers* (6th ed.). Hoboken: Wiley.

Newbold, P., Carlson, W. L., & Thorne, B. M. (2013). *Statistics for business and economics: Global edition* (8th ed.). Pearson.

Park, G-J., T-H. Lee, K.H. Lee, and K-H. Hwang (2006), Robust design: an overview. *AIAA Journal*, 44(1), 181–191.

Pyzdek, T., & Keller, P. A. (2018). *The six sigma handbook* (5th ed.). New York: Mc-Graw Hill.

Salkind NJ. 2016. Statistics for People Who (Think They) Hate Statistics. 4th edition, Thousand Oaks: CA, Sage.

Sincich, T. (1996). *Business statistics by example* (5th ed.). New Jersey: Prentice-Hall.

Six Sigma Academy. (2002). *The Black Belt Memory Jogger: A Pocket Guide for Six Sigma Success* (1st ed.). GOAL/QPC.

Spence, J. R., & Stanley, D. J. (2016). Prediction interval: What to expect when You're expecting . . . A replication. *PLoS One, 11*(9), 1–22.

Taguchi, G. (1986). *Introduction to quality engineering: Designing quality into products and processes* (No. 658.562 T3).

The Minitab Blog. (2017, 08 22). Retrieved 05 06, 2019, from What's the Difference between Confidence, Prediction, and Tolerance Intervals?: https://blog.minitab.com/blog/understanding-statistics/whats-the-difference-between-confidence-prediction-and-tolerance-intervals

Wang, J.X. 2005.Engineering Robust Designs, Prentice Hall. Westford, MS.

Wedgwood, I. (2007). *Lean Sigma.* U.S.: Prentice Hall.

Analyze Phase: Other Data Analysis Tools

Contents

F. Pakdil, *Six Sigma for Students*, https://doi.org/10.1007/978-3-030-40709-4_8

Learning Objectives

After careful study of this chapter, you should be able to:

1. Construct and interpret seven old tools in data analysis in the Analyze phase in DMAIC
2. Construct and interpret seven new tools in data analysis in the Analyze phase in DMAIC
3. Construct and interpret other tools in data analysis in the Analyze phase in DMAIC.

8.1 Introduction

This chapter examines some problem-solving, data analysis, and data visualization tools under three sections: seven old tools, seven new tools, and other tools. The seven old tools are basic tools used by the lower hierarchical levels of the organization. Most of these tools do not require a statistical background or excessive training in statistics. Seven new tools are more complex and strategic planning-oriented tools used by higher hierarchical levels of the organization. Seven old and new tools and other tools analyzed in this chapter have important role in data analysis and visualization. Data visualization is critical in Six Sigma projects since it enables practitioners analyze and interpret data through charts, diagrams, tables, and graphs. Visual representation of data may help Six Sigma teams see the problems in detail, identify root causes more accurately, and develop solution ways more effectively. Statisticians John Tukey and Edward Tufte's work on data visualization significantly impacted statistics world, quality practitioners, and decision-makers (Sue and Griffin 2016). Additionally, technological advancements in computer technology improved data visualization, especially drawing and interpreting charts. Sue and Griffin (2016: xii) state that "pictures enable complex information to be delivered quickly and effectively and can help to reveal patterns in data and uncover anomalies that may not have been spotted otherwise."

8.2 Seven Old Tools

Seven old tools are an important part of statistical process control (SPC) and also known as the "magnificent seven tools" (Montgomery 2013). The proper use of seven old tools can solve 75% of quality problems when management is effectively involved in problem-solving processes. Therefore, the use of seven old tools throughout DMAIC process is effective, beneficial, and practical for Six Sigma project members. These seven old tools are:

1. Check sheet
2. Histogram
3. Fishbone diagram – cause-and-effect diagram
4. Pareto analysis and diagram
5. Scatter diagram
6. Stratification analysis
7. Control charts.

The control charts are the only one of these tools requiring specific statistical knowledge. The other seven old tools can easily be used by operators and employees without advanced statistical knowledge. These seven old tools should be taught to the majority of the organization and integrated into regular problem-solving processes throughout the organization. Seven old tools are detailed in the following sections.

8.2.1 Check Sheet

Check sheets are well-structured data collection sheets that simplify data collection and analysis. Since Big Data and data warehouses are implemented in decision-making processes, electronic and automatically generated check sheets are more functional and popular in data analytics. Computer software can create check sheets, but manually drawn check sheets can also be used in problem-solving processes. Check sheets are needed in the "Measure" phase of the DMAIC process. Historical data can be categorized, or current data collected in cheek sheets. Check sheets are used to categorize qualitative and quantitative data into several groups. They can also address problematic areas depending upon the type of variables included in check sheets. Montgomery (2013) suggests that time-oriented summary of data in check sheets may reveal trends and patterns. Check sheets function as a way to create frequency distributions which are a first step for further analyses. If the data are a set of observations from a population, the sample frequency distributions can be generated through check sheets.

Before collecting data using check sheets, a number of variables should be specified. For example, the type of data, date, process number, operation number, operator, season, etc. should be determined and used in the format of check sheets. If the structure of the cheek sheet does not assist in the analysis of the problem in Six Sigma projects, it may cause serious problems in the next phases of DMAIC.

► Example 1

A team that works in an endoscopy clinic collected data in the ultrasound process for 2 weeks. The team analyzes delays and generates a table, including delay reasons and frequencies as presented in ◘ Table 8.1. Analyze the table and interpret findings. ◄

▪ Solution

◘ Table 8.1 shows the frequencies of the delay reasons quarter by quarter and the total frequency in the last two columns. Six Sigma practitioners can use the check sheet to identify and differentiate relatively more important delay reasons. Since check sheets can also function as a tool to identify root causes of the problems, as seen from ◘ Table 8.1, the Six Sigma team first needs to focus on the details of "lack of supply." That particular item accounts for the highest frequency among the other reasons in this check sheet.

In ◘ Table 8.1, the "delay reasons" column represents the attribute data categories. The "total frequency" column shows how many times each category occurred in ultrasound process, and the "total frequency (tally)" column displays visual representation of frequencies in each category. After finalizing data collection on the check sheet, the Six Sigma project team can start working on the next phases of DMAIC.

8.2.2 Histogram

Histograms graphically depict the frequency distribution of a data set. Histograms and box plots provide information on skewness, outliers, behavior of data, and type of data distribution. A histogram is a graphical representation of a quantitative variable. It is also a special bar chart representing the relative frequency of the data.

In a histogram, the distribution of data is presented by grouping data in predefined intervals, based on frequencies and the center of the process. The amount of variation is highlighted. Even though a histogram is very similar to a bar graph, it differentiates from

◘ Table 8.1 Delay reasons and frequencies

Delay reasons	2018			Total frequency
	1st Q	2nd Q	3rd Q	
Lack of supply	25	12	17	54
Power loss and ultrasound down time	12	14	19	45
Lack of medical assistance	12	9	2	23
Physician absence	16	10	11	37

Source: Author's creation

bar graphs because there are no gaps between bars of histogram on the X-axis. Juran and Gryna (1980) state that many problems in quality control can be solved with the use of histograms. Analyzing a histogram does not require advanced statistical training. A basic knowledge of reading frequency distributions is sufficient to draw conclusions on histograms. Histograms are used for the following purposes (Amsden et al. 1998: 8):

1. Evaluating or checking processes
2. Indicating the need to take corrective action
3. Measuring the effects of corrective actions
4. Comparing machine performances and materials.

The following questions are answered by histograms:

1. Does the process produce parts that have a bell-shaped curve?
2. Where is the process centered?
3. Is the process capable of meeting the design specifications?
4. Is it expected to have any defective products that are located out of specification limits?
5. Is the variability of process in control?

These questions can be answered by analyzing (1) the center of the histogram; (2) the width of the histogram; and (3) the shape of the histogram (Juran and Gryna 1980: 44). The center of the histogram allows decision-makers to compare the target of the process with the actual center of the process, whereas the width of the process indicates the variation of the process. The shape of the histogram represents the distribution of the data.

In case of normality assumptions, at least 50 measurements are expected to show the pattern of the data set in a histogram (Juran and Gryna 1980: 46). Some other resources state that at least 30 measurements would be enough to check the normality of data set. The drawing of a histogram can be detailed step-by-step as follows:

- Step 1: Determine the CTQ characteristic to analyze based on internal (rework, scrap, and defect rates) and external issues or expectations (customer dissatisfactions, returns, warranty claims, etc.).
- Step 2: Design and detail how to collect data.
- Step 3: Collect data.
- Step 4: Check the accuracy, reliability, and validity of data, if necessary.
- Step 5: Calculate Range

$$\left(R = X_{\max} - X_{\min}\right) \tag{8.1}$$

- Step 6: Determine sample size (n).
- Step 7: Determine the class intervals, in other words the number of bars (K), based on sample size.
- Step 8: Calculate the interval width

$$\left(I = \frac{R}{K}\right) \tag{8.2}$$

- Step 9: Create intervals by starting from the minimum observation and adding internal (I) up to calculated cumulative observations.
- Step 10: Tally and count the number of observations that falls into each interval and draw a frequency table.
- Step 11: Draw the bar chart to show the frequency of each interval.
- Step 12: Interpret the findings.

► Example 2

A time and motion study in an assembly line generates a data set including 48 observations about cycle time (minutes) in six time periods as given in ▫ Table 8.2. Draw the histogram of data set. Interpret your findings. ◄

Let's follow the process flow given above for drawing histogram step-by-step.

- Step 1: Determine the CTQ characteristic to analyze based on internal (rework, scrap, and defect rates) and external issues (customer dissatisfactions, returns, warranty claims, etc.).
  The CTQ characteristic given in this example is "cycle time (minutes)" in the assembly line.
- Step 2: Design and detail how to collect data.

Table 8.2 Data set including cycle time (minutes) measurements in six time periods

	Period 1	Period 2	Period 3	Period 4	Period 5	Period 6
X1	40	25	13	31	20	35
X2	33	25	12	32	21	34
X3	30	27	10	32	22	33
X4	25	26	20	29	23	30
X5	20	29	19	28	25	21
X6	19	29	19	22	26	22
X7	14	30	18	23	28	19
X8	30	30	17	22	27	22

Source: Author's creation

The data were designed to be collected in six time periods with eight observations per time period.

- Step 3: Collect data.
  The data were already collected from the designated assembly line in this example.
- Step 4: Check the *accuracy*, *reliability*, and validity of data, if necessary.
  In this step, the data collectors should make sure that the 48 observations were measured by following the same process.
- Step 5: Calculate Range ($R = X_{max} - X_{min}$).
  The range of the data set is calculated as follows as given in Eq. 8.1:

$$R = 40 - 10 = 30$$

- Step 6: Determine sample size (n).
  The sample size is 48 observations.
- Step 7: Determine the class intervals or the number of bars (K) based on sample size.
  To determine the number of bars in histogram, there are simply two basic approaches used by practitioners. First, as presented in Juran's Quality Control Handbook (2017) and Amsden et al.'s book (1998), the number of bars can be determined based on the number of observations without following a rigid formula or calculation method, as given in Table 8.3. Even though there are no very strict rules for determining the number of bars, the histogram should demonstrate the density of the frequencies and the shape of the distribution. The higher the number of bars, the more spread the distributions.

Table 8.3 Number of bars in histograms

Number of observations	Number of bars[a]	Number of bars[b]
<50		5–7
20–50	6	
51–100	7	6–10
101–200	8	7–12
201–500	9	10–12
501–1000	10	
1000+	11–20	

Source: Author's creation based on [a]Defeo 2017; [b]Amsden et al. 1998

In the second approach, the number of observations is employed in order to determine the number of bars in histograms based on Eq. 8.3, where K is the number of bars and n is the number of observations:

$$2^{K-1} < n < 2^{K} \tag{8.3}$$

In the example, n = 48 and the associated K is 6. $2^{K-1} < n < 2^{K} \Rightarrow 2^{6-1} < 48 < 2^{6}$. As an outcome of this approach, the number of bars in this histogram can vary between 5 and 6. Let's determine K as 6.

- Step 8: Calculate the interval width $\left(I = \frac{R}{K}\right)$.
  In the example, the interval between bars should be 5, as the dividend of 30 (R) by 6 (K) as presented in Eq. 8.2 $\left(I = \frac{R}{K} = \frac{30}{6} = 5\right)$.
- Step 9: Create intervals by starting from the minimum observation and adding internal (I) up to calculated cumulative observations.
  The intervals are created, as presented in ◻ Table 8.4, starting from 10 minutes, which is the minimum observation of the data set. Each interval is created by adding 5 minutes, which is the interval width calculated in the previous step. It also stops at 40 minutes, which is the maximum observation in the data set.

The Six Sigma team needs to make sure that there are no gaps in between the intervals. For example, the first interval covers observations that fall into 10–15 minutes, whereas second interval carries observations that are between 15 and 20 minutes. Six Sigma practitioners may need to determine where cycle times of 15 minutes should be counted. Technically, a 15-minute cycle time can count in the first or second intervals. Once the decision is made and 15-minute cycle time is counted either in the first or second interval, the other cut points such as 20-minute and 25-minute cycle times will be counted in the same way.

- Step 10: Tally and count the number of observations that falls into each interval and draw a frequency table.
  All individual observations fall into certain intervals, as presented in ◻ Table 8.4. The frequency column shows how many individual observations are counted in each interval.
- Step 11: Draw the bar chart to show the frequency of each interval.
  Histogram can be drawn manually using check sheet generated in the previous step in ◻ Table 8.4. Histograms can also be drawn using software such as Minitab, JMP, or STATA. To draw histogram in Minitab, after transferring data set (◻ Table 8.2) in one column to a Minitab worksheet, click on Graph→Histogram. In the input screen, select the "With Fit" to see the distribution of the data. Click on OK. In the next input screen, enter the variable (cycle time) from left box to Graph Variables. Click on OK. The histogram will appear on Minitab output as shown in ◻ Image 8.1.
- Step 12: Interpret findings.

As seen in ◻ Image 8.1, the data are distributed into eight columns, and the type of the distribution fitted to data is shown by the distribution line. The information box at the upper right corner in histogram also shows the mean (24.73 minutes) and standard deviation of the data (6.427 minutes) and number of observations (48).

Amsden et al. (1998) suggest that the following points should be double checked before finalizing a histogram:

1. Use equal width intervals.
2. Do not use open intervals.
3. Do not make any breaks in the vertical or horizontal scales.
4. Do not have too few or too many intervals.
5. Do not put too much information on one histogram.
6. Give everything needed to identify all the information completely and to make the graph understandable.

◻ Image 8.1 complies with those points and does not create any issues with meeting the expectations of those suggestions.

◻ Table 8.4 Intervals of the histogram

Intervals	Frequencies	Tallies
10–15 minutes	4	////
15–20 minutes	9	~~////~~ ////
20–25 minutes	13	~~////~~ ~~////~~ ///
25–30 minutes	14	~~////~~ ~~////~~ ////
30–35 minutes	7	~~////~~ //
35–40 minutes	1	/

Source: Author's creation

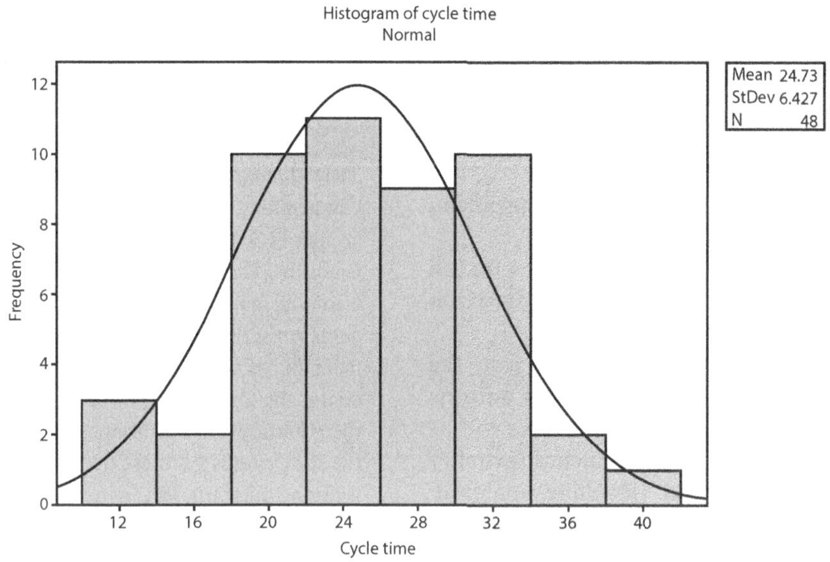

Image 8.1 Histogram in Minitab. (Source: Author's creation based on Minitab)

8.2.3 Fishbone Diagram Cause-and-Effect Diagram

Fishbone diagram, namely, cause-and-effect diagram, is a problem-solving tool developed by Japanese Quality Guru Kaoru Ishikawa to identify and categorize the potential causes of a quality problem (Ishikawa, 1985). The fishbone diagram is a powerful tool to identify the root causes, if used effectively and effectively. It is a pictorial demonstration of the relationships between a problem and its causes. Once a problem is identified, the potential root causes of the problem can be analyzed in depth by Six Sigma teams. The problem should be investigated with all possible aspects to discover the root causes. In the DMAIC process, the fishbone diagram is mostly used in the "Analyze" and the "Improve" phases. Teams that work on a particular defect or problem should construct a well-structured fishbone diagram.

The American Society for Quality (ASQ) states that the fishbone diagram identifies many possible causes for an effect or problem based on brainstorming outcomes.[1] In the fishbone diagram, the problem is put on the right side of the diagram. All other components of the diagram located on the left side are possible causes of the effect. These come from brainstorming sessions. Causes generated by team members are noted down on flip charts or sticky note cards and made visible to team members. After making sure that all causes are identified and articulated by the members, the causes are categorized into major causes, in other words primary reasons. The major causes in fishbone diagrams are:

1. Methods
2. Machines
3. Manpower
4. Materials
5. Measurement
6. Environment.

In some problems and cases, *communication* is considered another major category. The structure of the problem may also lead to problem-specific major primary reasons. Fishbone diagrams can also be constructed along with a five Whys analysis. Asking "Why" and "Why does this happen?" questions five times can identify relevant cause branches and root causes. In brainstorming sessions, when the team runs out of ideas, the ideas are trans-

1 ► https://asq.org/quality-resources/fishbone.

ferred to a fishbone diagram framework. The process flow of constructing fishbone diagram follows these work steps:

- *Construct*: Construct the team including all members from relevant functions or departments.
- *Define*: Define the particular problem, issue, defect, or error in detail.
- *Analyze*: Analyze the problem with all relevant aspects in several brainstorm sessions in the team.
- *Draw*: Draw a flat line and put the problem box at the right end to address the problem.
- *Add*: Place main cause branches (primary causes) such as 5M (machine, material, method, measurement, manpower) and environment; create new branches, if necessary, at an angle of about 70 degree on the line.
- *Identify*: Identify possible causes and place them under main branches as secondary causes.
- *Detail*: Detail secondary causes into third or fourth level causes, if necessary.
- *Rank*: Rank all causes in all levels and *identify* relatively more or most impactful one/s based on data and brainstorm sessions.
- *Focus*: Focus on corrective actions.

Fishbone diagrams are preferred by practitioners to categorize and organize the ideas of the team members about the causes of the problem. They also show relationships between and among the ideas generated by team members. Fishbone diagrams speed up the pace of problem-solving processes by focusing on identifying root cause/s of the problem. Furthermore, fishbone diagrams are effective tools to get employees and managers involved in problem-solving processes to work collaboratively. Well-structured brainstorming sessions prepare a platform for drawing fishbone diagrams. By integrating fishbone diagrams into problem-solving and DMAIC processes, Six Sigma team members find a formal platform to analyze and solve problems in a supportive and participative work environment, rather than searching for someone to blame for the problems. It is mostly used only to identify the potential X variables, not to prioritize those variables (Wedgwood 2007).

▶ Example 3

The Department of Transportation in Connecticut investigates the last 150 traffic accidents that occurred on I-91 over the last 6 months. The Director of Quality Management wants to map the potential causes of the accidents on a fishbone diagram. After forming a team that focuses on the reasons of the accidents, the director asks the team to present their findings in a fishbone diagram. Analyze the root causes of these accidents and draw a fishbone diagram. ◀

▪ **Solution**

After having several brainstorm sessions with team members and following the process given above, the director with the team drew the fishbone diagram. Software may be also used for drawing fishbone diagrams. Let's use Minitab and show the process step-by-step. To draw fishbone diagram in Minitab, first enter data in columns in a Minitab worksheet. Click on Stat→Quality tools→Cause-and-effect. In the input screen, transfer each branch to "Causes" column. Click on OK. In the next screen, the fishbone diagram will appear on Minitab output as shown in ◘ Image 8.2.

Under each primary reason, the team may also present detailed *secondary*- and *third-level reasons*. The team can explore the problem and each primary reason in more depth by asking "Why" question five times back to back. Asking five whys may help discover the root causes of the problem in the example. A fishbone diagram would be the best tool to picture all these root causes linked with primary reasons for determining the most effective solution ways for the relevant problem.

8.2.4 Pareto Analysis and Diagram

As well as the simplicity and practicality of Pareto diagrams, they are very helpful for analyzing the problems and processes in Six Sigma projects. Attribute data can be visualized

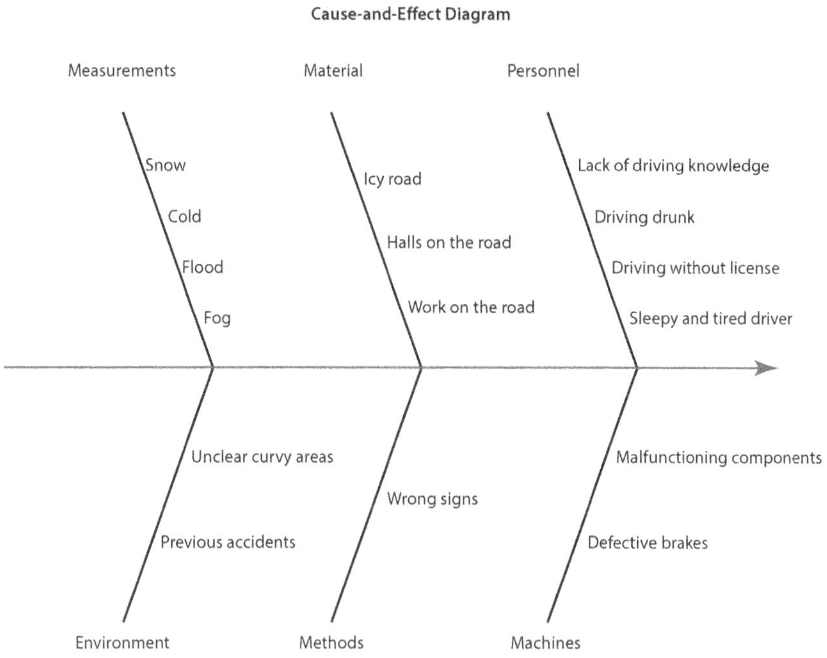

Image 8.2 Drawing fishbone diagram in Minitab. (Source: Author's creation based on Minitab)

based on the Pareto principle to differentiate "many trivial" from "vital a few" items in data set, which is the main idea behind the Pareto diagrams. In Pareto diagram, the *X*-axis represents the attribute variables or categories given in the data set while the *Y*-axis represents the cumulative percentages of the variables or categories used in the data set. The Pareto principle is also called the 80/20 or 70/30 rule.

The Italian Economist Vilfredo Pareto (1848–1923), the founder of Pareto Principle, stated that 80% of a nation's wealth was disproportionally owned by 20% of the population, as presented in his study of wealth distribution. Similarly, 20% of the customers may account for 80% of total sales. The Pareto diagram is applicable in any parts of Six Sigma projects whenever there is a need to focus on a specific part of the problem analyzed. It is also an effective tool to narrow the scope of the project and prioritize factors contributing to problems. The Pareto diagrams may also help teams prioritize the problems in a problem pool. The Pareto diagrams are mostly used in the "Measure" and the "Analyze" phases of DMAIC process.

The process of drawing a Pareto diagram is as follows:

- Step 1: Determine the CTQ characteristic to analyze based on internal (rework, scrap, and defect rates) and external issues (customer dissatisfactions, returns, warranty claims, etc.).
- Step 2: Identify and articulate the problem.
- Step 3: Analyze and identify the reasons of the problem.
- Step 4: Collect data to represent the frequency of the reasons/items.
- Step 5: Check the accuracy, reliability, and validity of data, if necessary.
- Step 6: Calculate the percentage of each reason/item.
- Step 7: Sort the reasons/items in a descending order starting from the highest one.
- Step 8: Calculate cumulative percentages of reasons/items.
- Step 9: Draw and label vertical axis from zero to 100%.
- Step 10: Draw and label horizontal axis and add reasons/items on horizontal axis from the highest percentage to the lowest one.

- Step 11: Draw the Pareto diagram.
- Step 12: Determine what items are covered by 80% of the cumulative percentage to differentiate many trivial from vital a few.
- Step 13: Interpret findings.

▶ Example 4

The causes of patient dissatisfaction in a microbiology lab were analyzed and identified, as demonstrated in the first two columns of ◘ Table 8.5, in a Six Sigma project. The total number of patients is 454 in 3 weeks. What reasons should receive more attention from the team, based on the Pareto principle, to diagnose the root causes? Draw a Pareto diagram and interpret your findings. ◀

Let's follow the process of drawing a Pareto diagram step-by-step as follows:

- Step 1: Determine the CTQ characteristic to analyze based on internal (rework, scrap, and defect rates) and external issues (customer dissatisfactions, returns, warranty claims, etc.).
  The CTQ characteristic analyzed in this example is patient dissatisfaction.
- Step 2: Identify and articulate the problem.
  The problem given in the example is about patient dissatisfaction in a microbiology lab.
- Step 3: Analyze and identify the reasons of the problem.
  The problem was already analyzed in this example and potential reasons were listed in the frequency table in the first column, based on the data collected in 3 weeks.
- Step 4: Collect data to represent the frequency of the reasons/items.
  The team collected data from the patients who declared a dissatisfaction reason about the lab, as represented in ◘ Table 8.5.
- Step 5: Check the accuracy, reliability, and validity of data, if necessary.
  After completing data collection, the team may want to analyze the data and ensure the reliability and validity by focusing on the amount of missing data or other potential issues with the data.
- Step 6: Calculate the percentage of each reason/item.
  The percentage of each patient dissatisfaction reason in the total is calculated and presented in the data set (◘ Table 8.5) in the third column.
- Step 7: Sort the reasons/items in a descending order starting from the highest one.
  The highest percentage belongs to "missing report" reason with 49.8% in the example. The second highest reason is "excessive waiting time" with 15.6%. The other reasons are ordered accordingly in ◘ Table 8.5.
- Step 8: Calculate cumulative percentages of reasons/items.

◘ Table 8.5 Microbiology lab patient dissatisfaction reasons and percentages

Dissatisfaction reasons	Frequency	Percentage	Cumulative percentage
Missing report	226	49.8%	49.8%
Excessive waiting time	71	15.6%	65.4%
Missing information in report	58	12.8%	78.2%
Wrong patient name in report	44	9.7%	87.9%
Lack of courtesy	39	8.6%	96.5%
Lack of communication	16	3.5%	100.0%
Total	454	100.0%	

Source: Author's creation

The cumulative percentages are calculated in the fourth column in ◘ Table 8.5 by adding up each percentage on the next percentage until the total cumulative percentage reaches 100%.

- Step 9: Draw and label vertical axis from zero to 100%.
  Y-axis is drawn, and percentages between 0 and 100% are added on *Y*-axis.
- Step 10: Draw and label horizontal axis by adding reasons/items on horizontal axis from the highest percentage to the lowest one.
  X-axis is drawn, and dissatisfaction reasons are added one by one on *X*-axis, starting from the item that has the highest percentage. Then, the second highest item is added on *X*-axis. All items are added one by one until the cumulative percentage goes up to 100%.
- Step 11: Draw the Pareto diagram.
  The Pareto diagram can be drawn manually or using software. At the end of Step 10, the diagram is complete and drawn manually. Customer dissatisfaction reasons from the highest to the lowest percentage are represented on *X*-axis of the diagram whereas each reason's percentage is demonstrated on *Y*-axis (◘ Image 8.3).
- Step 12: Determine what items are covered by 80% of the cumulative percentage to differentiate "many trivial" from "vital a few."
  The Six Sigma team identifies the reasons that comprise of 80% of the cumulative percentages of the dissatisfaction reasons. The Pareto Diagram given in ◘ Image 8.3 points out what causes are relatively more important based on the frequency of the occurrence in this lab.
- Step 13: Interpret findings.
  The *Y*-axis represents that 78.2% of the frequency accounts for the first three reasons: (1) missing report; (2) excessive waiting time; and (3) missing information in report. The Pareto diagram prioritizes these three reasons (vital a few) and differentiates them than the others (many trivial).

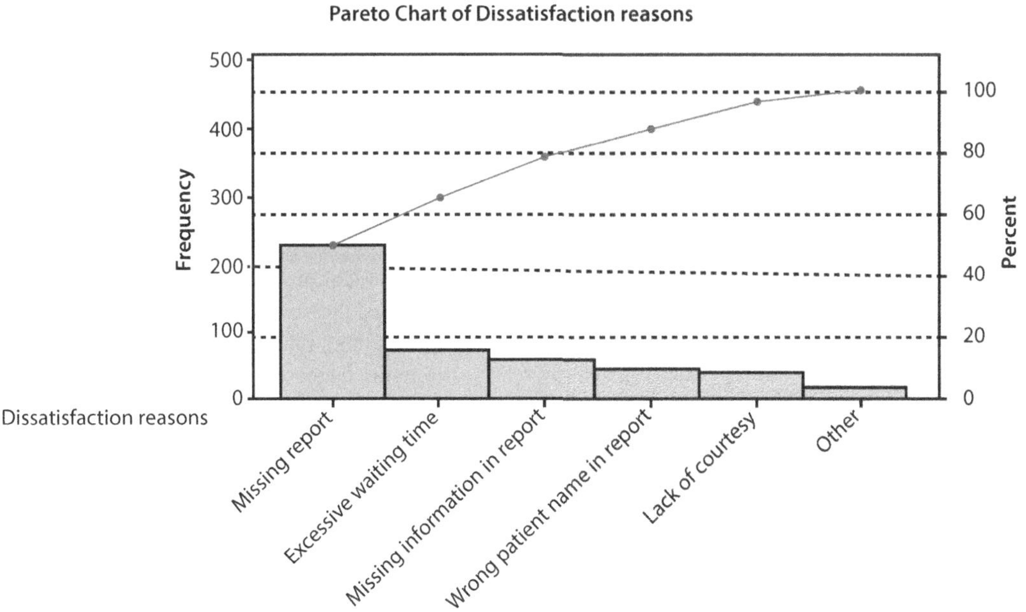

Frequency	226	71	58	44	39	16
Percent	49.8	15.6	12.8	9.7	8.6	3.5
Cum %	49.8	65.4	78.2	87.9	96.5	100.0

◘ **Image 8.3** Pareto diagram of the customer dissatisfactions in microbiology lab. (Source: Author's creation based on Minitab)

Pareto diagrams can also be drawn using software such as Minitab. To draw a Pareto diagram in Minitab, after transferring data set (first two columns of ◘ Table 8.5) to a Minitab worksheet, click on Stat→Quality Tools→Pareto Chart. In the input screen, select the Defects or attribute data in. Enter the variable (dissatisfaction reasons) from left box to defects or attribute data in. Enter frequency variable from left box to Frequencies in. Click on OK. The Pareto diagram will appear on Minitab output as shown in ◘ Image 8.3.

8.2.5 Scatter Diagram

A scatter diagram is used when the researchers and decision-makers are interested in finding whether there is an association between two variables. Scatter diagrams are used to analyze cause-and-effect relationships in root cause analysis process in regression analysis, as presented in ► Chap. 7. Scatter diagrams also show if there are any linear or nonlinear relationships based on the pairs of variables. A scatter diagram is a prelude to regression analysis. Using correlation and regression analysis results, it may be that one of the two variables is impactful on the other variable. These two variables are named *dependent variable* and *independent variable*. In a scatter diagram, the horizontal X-axis represents the scale for the independent variable, while the dependent variable is demonstrated on the vertical Y-axis. Individual observations are plotted at the points where the values of dependent and independent variables meet on the diagram. Each independent variable's individual value is represented as a dot according to its corresponding dependent variable's value on X-Y coordinate plane. In scatter diagrams, it is considered that dependent variables are statistically affected by independent variables. Scatter diagrams can be drawn with interval, ratio, or ordinal data collected in pairs (Scott and Mazhindu 2014).

Scatter diagrams are useful for demonstrating how one variable may relate to another one. For example, in a manufacturing setting, long work hours may influence the defect rates. To see the impact of work hours on defect rates, a scatter diagram of these two variables is drawn and analyzed. Each point plotted on scatter diagram represents how two variables are located on X-Y plane. The variables located in scatter diagrams are mostly continuous (measurable) ones. For example, time, weight, length, temperature, and pressure are good examples for continuous variables. But, depending on the type of the variable, discrete variables can be plotted on scatter diagrams, too. For example, customer dissatisfaction and defect rates or numbers can be used on scatter diagrams as discrete variables.

The scatter diagrams can be drawn using the process as follows:

- Step 1: Determine at least two CTQ characteristics, dependent and independent ones.
- Step 2: Collect data for each CTQ characteristic. Pay attention on pairs of CTQ characteristics. Each CTQ characteristic should have a corresponding CTQ to match on scatter diagram.
- Step 3: Determine the scales for each CTQ characteristic.
- Step 4: Draw the X-axis and Y-axis.
- Step 5: Plot the data pairs on X-Y plane on scatter diagram.
- Step 6: Analyze scatter diagram and interpret findings.

► Example 5

Sunshine Co. produces and sells umbrellas in six states in New England. The sales team assumes that "rainfall in one meter squared" is the main factor affecting the total umbrella sales in this region. Sunshine Co. wants to see the relationship between rainfall received in each state and umbrella sales in 2019. Using data given in ◘ Table 8.6, draw scatter diagram and interpret the findings. ◄

- Step 1: Determine at least two CTQ characteristics, dependent and independent ones. The amount of rainfall (independent) and total sales of umbrella (dependent) are the variables or CTQ characteristics used in this example.

- Step 2: Collect data for each CTQ characteristic.
  If the data needed are already collected, this step can be skipped. In case there are no data available, the data need to be collected by the team in Six Sigma project. The data are provided by the company in this example.
- Step 3: Determine the scales for each CTQ characteristic.
  The intervals indicating the range of data are determined in this step. For example, what is the expected highest and lowest total sales and rainfall? This question helps identify the interval of CTQ characteristics. In the example, rainfall varies between 30 liters and 40 liters, while total sales are in between 2,500 units and 3,100 units.
- Steps 4 and 5: Draw the *X*-axis and *Y*-axis and plot the data pairs on *X*-*Y* plane on scatter diagram.
  The points where the pairs of rainfall and total sales meet are plotted on *X*-*Y* axes. Drawing a scatter diagram can be done manually or using software. Since we have six states' information in ▣ Table 8.6, we will have six points plotted in scatter diagram. How to draw a scatter diagram in Minitab was discussed in ▸ Sect. 7.12 in ▸ Chap. 7. The scatter diagram will appear on Minitab output as shown in ▣ Image 8.4.

▣ Table 8.6 Data set used in Sunshine Co. in 2019

State	Rainfall (annual-liter)	Sales (annual)
Connecticut	30	2,500
Maine	35	2,800
Massachusetts	33	2,750
Rhode Island	32	2,650
New Hampshire	45	3,100
Vermont	40	2,950

Source: Author's creation

- Step 6: Analyze the scatter diagram and interpret findings.
  Scatter diagram in ▣ Image 8.4 display a positive relationship between the amount of rainfall and total umbrella sales in six states located in New England area. As rainfall increases, the total sales of umbrella rise, too. The general shape of the dots indicates whether the correlation is positive or negative. When the dots create a cluster from the lower left-hand corner through the upper right-hand corner, that cluster refers to a positive correlation. When the dots are accumulated from the upper left-hand corner through the lower right-hand corner, the correlation is considered negative.

8.2.6 Stratification Analysis

Stratification analysis focuses on sorting data into distinct groups, clusters, or layers. It is the separation of data into categories to diagnose which causes of a quality problem contribute most to the problem (Defeo 2017). According to ASQ, "it is a technique used in combination with other data analysis tools. When data from a variety of sources or categories have been lumped together, the meaning of the data can be difficult to see. This data collection and analysis technique separates the data so that patterns can be seen and is considered one of the seven basic quality tools."[2] A great variety of factors can be used to stratify the data. For example, days of the week, hours, quarters in a year, daily shifts, service types and lines, and types of customers or product lines are used to stratify the data. The following process is used to do stratification analysis:

- Step 1: Collect data that include stratification variables with the corresponding variables.
- Step 2: Establish stratification variables.
- Step 3: Establish frequency table to stratify the data.

2 ▸ https://asq.org/quality-resources/stratification.

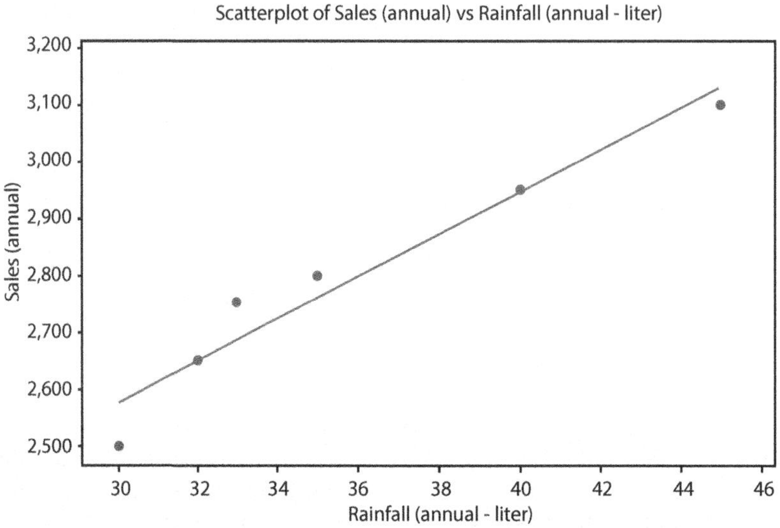

8

Image 8.4 Scatter diagram drawn in Minitab. (Source: Author's creation based on Minitab)

- Step 4: Count number of observations or occurrences for each variable.
- Step 5: Display total counts in bar graphs for each variable.
- Step 6: Interpret results.

▶ Example 6

The Big Moon Co. is a well-known local supermarket that sells a great variety of dairy products, bread, fresh fruits, vegetables, and meat. Over the last 3 months, Big Moon Co. received excessive customer complaints about certain products. The quality manager wants to see if there are any specific time periods in which customers complained or any specific products that caused customers to complaint about. The customer complaints were collected month by month, and a total of 583 item-related complaints in Store 1 and 271 item-related complaints in Store 2 were received over the last 3 months. The quality manager detailed and classified customer complaints in a systematic way so that the stratification analysis method can be implemented. Use stratification analysis to help quality manager in this example. ◀

Solution

We can follow the process given in stratification analysis step-by-step to help quality manager.

- Step 1: Collect data that include stratification variables with the corresponding variables.
  The data were collected by the quality manager from two stores: 583 complaints in Store 1 and 271 complaints in Store 2. At a first glance, there is no other detailed information. However, the stratification analysis will help us detail the data.
- Step 2: Establish stratification variables.
  The quality manager aimed at discovering root cause/s of the customer complaints and determined (1) *products*, (2) *months*, (3) *weeks*, and (4) *stores* as the stratification variables. The more detailed variables the analysis utilizes, the more detailed information the quality manager can generate in stratification analysis, depending on the availability of data.
- Step 3: Establish frequency table to stratify the data.
  The frequency table framework was established for the data as presented in Table 8.7 by the quality manager. Each *month* is divided into *weeks* in the columns and each *product* is presented in the rows for each *store*.
- Step 4: Count the number of observations or occurrences for each variable.
  The quality manager counted number of customer complaints received in each *week*

and *month*, in each *store*, and for each *product* as presented in ◘ Table 8.7.

- Step 5: Display total counts in bar graphs for each variable.
  To exemplify how to display total counts in bar graphs, ◘ Fig. 8.1 can be used. ◘ Figure 8.1 shows that customers do not complain about milk D vitamin 1 gallon in Store 2 as much as they do in Store 1. There is a significant difference between the stores, and that fact may help point out a specific root cause after analyzing the process related to milk D vitamin 1 gallon.
- Step 6: Interpret results.

As presented in ◘ Table 8.7 and bar graphs, some particular products cause more customer complaints compared to other products in this example. For instance, while milk (vitamin D ½ gallon) caused the majority of the customer complaints in two stores, buttermilk did not receive that level of complaints in the same time period. Similarly, mozzarella cheese does not have any complaints in two stores, while cheddar cheese brings substantial customer complaints, specifically in January 2019. Then, it decreases in the next months. At the

◘ Table 8.7 Stratified data

	January 2019				February 2019				March 2019				Total	Total
Store 1	**1**	**2**	**3**	**4**	**1**	**2**	**3**	**4**	**1**	**2**	**3**	**4**		**583**
Milk (vitamin D 1 gallon)	15	12	13	15	22	26	27	30	11	11	10	9	201	
Milk (vitamin D ½ gallon)	30	32	29	25	14	14	12	13	22	23	24	22	260	
Butter	1	3	4	2	2	2	3	2	1	2	2	3	27	
Buttermilk	0	0	1	0	0	0	0	0	1	1	1	1	5	
Cheddar cheese	14	12	13	12	4	2	4	3	1	1	2	4	72	
Mozzarella cheese	0	1	0	0	0	0	0	0	1	1	1	0	4	
Gourmet cheese	5	2	3	0	0	0	1	1	0	0	2	0	14	
Total	65	62	63	54	42	44	47	49	37	39	42	39	583	
Store 2	**1**	**2**	**3**	**4**	**1**	**2**	**3**	**4**	**1**	**2**	**3**	**4**		**271**
Milk (vitamin D 1 gallon)	0	2	1	2	0	0	0	0	1	1	1	1	9	
Milk (vitamin D ½ gallon)	2	2	3	2	17	18	14	22	22	20	20	19	161	
Butter	1	1	0	0	0	0	1	1	0	0	0	0	4	
Buttermilk	0	0	0	0	0	0	0	0	0	0	0	0	0	
Cheddar cheese	10	11	11	9	10	10	12	12	1	1	0	0	87	
Mozzarella cheese	0	0	0	0	0	0	0	0	0	0	0	0	0	
Gourmet cheese	5	0	0	0	1	1	1	1	0	0	0	1	10	
Total	18	16	15	13	28	29	28	36	24	22	21	21	271	854

Source: Author's creation

beginning of data analysis process, the quality manager reported the total number of complaints for each store. However, there were no specific details referring to time period or product-specific complaints. Using the stratification analysis method, the quality manager is able to detail the general data into various categories and identify root cause/s or specific products that cause more customer complaints. That information would be helpful for the quality manager to distinguish relatively more important products from less important ones to identify root causes of the customer complaints and find a solution for the problem.

8.2.7 Control Charts

Simply, control charts demonstrate the nature of the process. Control charts are constructed based on time series analysis and monitor the process using various performance indicators. Comparing aggregate data with summary statistics can be misleading and time-consuming unless the processes that produced the data are stable (Carey 2003: 7; Benneyan et al. 2003). SPC is a branch of statistics that combines rigorous time series analysis methods with graphical presentation of data (Benneyan et al. 2003). As a primary component of SPC, control charts plot the average (*CL*), upper control limit (*UCL*), and lower control limit (*LCL*) of the measurements of a CTQ characteristic based on a sample collected from the relevant process versus time (◘ Fig. 8.2). Shewhart proposed 3-sigma variation as *control limits* and two-sigma variation as *warning limits* in-control chart methodology, which is considered one of the critical parts of SPC.

Essentially, control charts, based on probability theory, are employed in process monitoring to detect whether there are any abnormalities or statistically out-of-control situations in the process. Control charts are ultimately used for improving the process. Control charts can only detect "special/assignable causes of variation" and graphically display how much variation exists in the process. However, the charts do not identify any improvement methods or reasons to eliminate abnormalities or the out-of-control situations. Since control charts represent retrospective data, it may not be easy to identify corrective

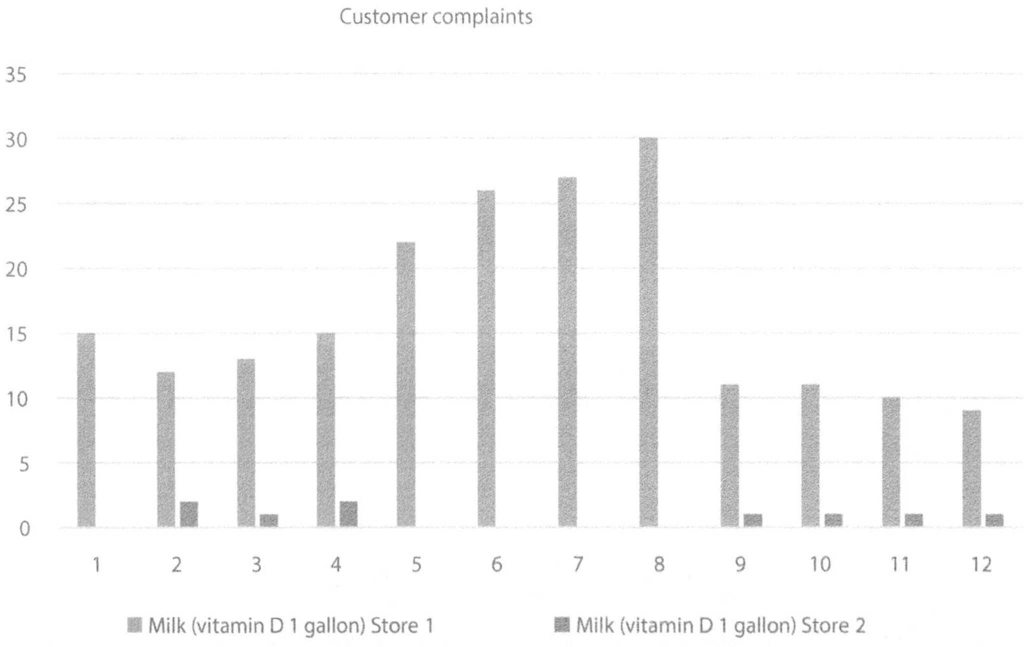

◘ **Fig. 8.1** Milk (Vitamin D 1 gallon) complaints bar graph. (Source: Author's creation)

actions and root causes of abnormalities by analyzing control charts. Control charts are also used to monitor the stability and statistical consistency of the variables and utilized as an alert to identify the abnormalities caused by either *common or assignable causes of variation* in the process. Deming (1982: 112) states that "a statistical chart detects the existence of a cause of variation that lies outside the system. It does not find the cause. Upon identifying out of control situation, the root causes of assignable cause/s should be identified to eliminate it." Hart and Hart (2002: 58) assert that "the control chart tells when to take action on the process and when to leave it alone." In the absence of special/assignable causes in control charts, the quality improvement efforts focus on identifying of effective and efficient ways of improving the process (Hart and Hart 2002). Juran (1945) states that "the control chart technique is admirably suited to executive review. This executive review carries with it the responsibility to see to it that corrective action is taken where such action is indicated. In the absence of corrective action, the control chart technique deteriorates into a sterile paper work procedure." Additionally, corrective actions include diagnosis of cause as well as provision of a remedy (Juran 1992: 277).

Shewhart established control charts by integrating the Tchebysheff's theorem into his experiments (Mohammed et al. 2008). They suggested that control charts could be used in a wide range of scientific disciplines. Control charts can tell us either to let the process go or correct mistakes found in the process. A control chart is seen as a record of small inspections (Amsden et al. 1998). Systematic use of a control chart is an excellent way to reduce variability (Montgomery 2005: 14), which is at the very heart of control charts (Kelley 2000). Control charts are perceived as the most convenient tools to differentiate assignable causes (variation occurring because of external or undefined factors) from common causes of variation (variation occurring due to systemic factors) in processes, since they generate easy-to-read data that displays the performance. Differentiating common causes from assignable causes is critical, since assignable ones are the causes of the variation driving the process or variable out-of-control status, and these causes cannot be diagnosed without a detail analysis. If no special causes are identified and only common causes take

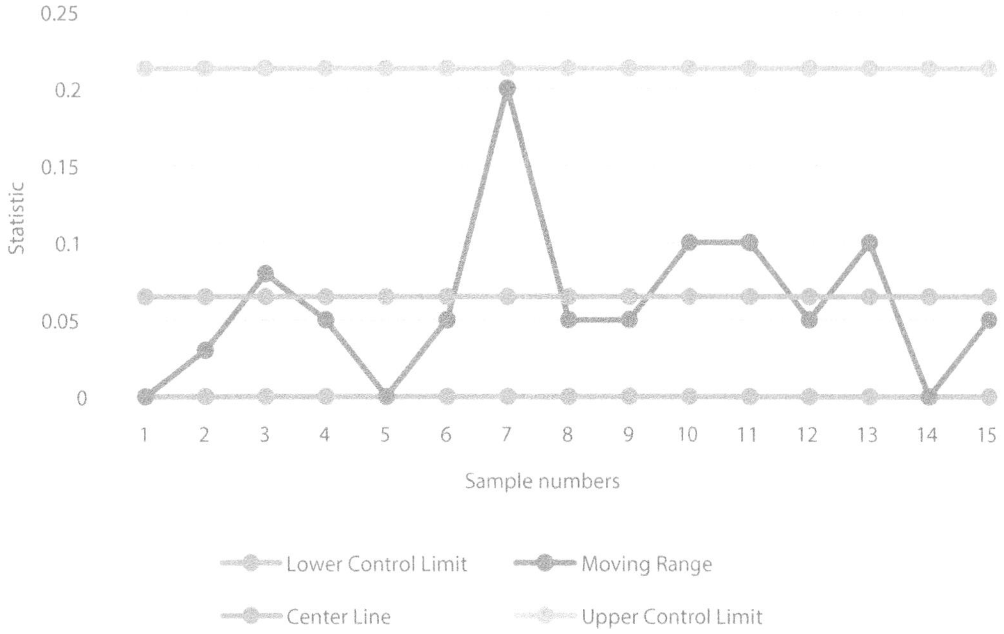

Fig. 8.2 A typical control chart. (Source: Author's creation)

place in the process, then the process is considered to be "in-control." Assignable causes are considered responsibility of the employees. However, common causes can be identified easily, since these causes are predictable, visual, and generated through the variability of the process itself. When the actual variation exceeds upper and lower control limits, it shows that assignable causes are present in the process and that it should be analyzed to discover and eliminate those assignable causes.

The data sets to draw control charts are gathered from a population within subgroups or sample of the items and categorized into several subgroups depending on the structure of data in control charts. The number of observations in each group is called "subgroup size." Subgrouping data may also be helpful to identify common and/or assignable causes. If a particular subgrouping method is not considered statistically significant, the control chart methodology will not be able to generate any significant evidence to differentiate common causes from assignable causes. Decision-makers should make sure that subgroup sizes are large enough to apply decision rules on control charts and to make correct and timely decisions. According to Hart and Hart (2002: 59), conventional 3-sigma control limits are usually set for data that have 25 subgroups. The 3-sigma control limits do not demonstrate good performance when the subgroups are way different from 25 subgroups. Furthermore, ongoing quality improvement projects may have larger numbers of subgroups. If the number of subgroups is greater than 20–25 and a common cause of variation is detected in the process, that process is considered stable. According to Mohammed et al. (2008), the control limits are regarded as "soft" or "provisional" with fewer subgroups. The control charts with 25 subgroups generate 6.5% false signals within 3-sigma limits that identify UCL and LCL, while 2-sigma limits produce 27.7% false signals in control charts (Montgomery 2005). Subgroups should be as homogeneous as possible. Specifically, for X charts, time-ordered values should be as close together in time as possible (Hart and Hart 2002: 286). The details of control charts will be discussed in ▶ Chap. 9.

8.3 Seven New Tools

8.3.1 Affinity Diagram

The affinity diagram is categorized as one of the new seven problem-solving tools. Juran (1992: 456) describes affinity diagram as "a method of converting vague concepts into specifics through use of language and diagrams." It collects and categorizes ideas and thoughts brainstormed by Six Sigma teams into several related subgroups. An affinity diagram clusters together similar type of items and helps see the relationships between cause and effects (Defeo 2017). It is a systematic way to sort items in meaningful groups or categories. When the topic analyzed is comprehensive and has a broad nature, related factors or items are put together under the same category. The affinity diagram requires teams to follow a procedure to categorize ideas and thoughts. Affinity diagrams can be used in various hierarchical levels of the organization. Either upper levels or operational levels can use affinity diagram to analyze the topics in the teams. "Problem definition" or "root cause analysis" steps may easily integrate affinity diagrams into problem-solving processes. Affinity diagrams effectively function while analyzing the problem in the "Analyze" phase of DMAIC. Affinity diagram is drawn by following these steps:

- Step 1: Conduct a brainstorm session to analyze the topic by team.
- Step 2: Note the ideas and thoughts on a visual board or platform.
- Step 3: Analyze all notes as a team.
- Step 4: Group individual ideas into sections.
- Step 5: Name each group.
- Step 6: Draw the affinity diagram.

As presented in process flow, in the brainstorm session, team members share their ideas and thoughts with the other team members and make each idea visible to everyone by posting their ideas on a visual board or platform. After all members run out of ideas, they come together around the ideas made visual to see any categories under which these ideas can be grouped. When the team makes sure that all ideas have been grouped under particular categories, the team focuses on naming each category. After that step, the affinity diagram is ready. The affinity diagram drawing process lets a group move beyond its habitual thinking and preconceived categories (ASQ, Dec 28, 2017). The following link presents ASQ's web site about affinity diagram ► https://asq.org/quality-resources/affinity.

► Example 7

Assume that you are the branch manager of a local fast-food restaurant and that customer complaints have been on the rise over the last 2 months. As the branch manager, you are expected to decrease customer complaints. You plan on analyzing customer complaint statements with your store team and drawing an affinity diagram to categorize those complaints. ◄

- **Solution**

Figure 8.3 shows the affinity diagram drawn by the team that analyzes customer complaints. The reasons of the problem are categorized into three main groups such as (1) cleaning up, (2) quality of service, and (3) quality of food, after going through the process given above. Each group is comprised of several detailed reasons that help the team identify the root causes of the customer complaints. For example, quality of service group includes (1) speed of order time, (2) speed of delivery, (3) courtesy of service associates, and (4) ambiance of the restaurant.

8.3.2 Systematic Diagram

A systematic diagram is also known as tree diagram, hierarchy diagram, or tree analysis. As one of the seven new problem-solving tools, systematic diagram is used to detail and break down the topic analyzed into categories. When the topic analyzed needs to be more specific, systematic diagrams are employed in the "Analyze" phase in DMAIC. A systematic diagram is useful to understand the overall picture of topics analyzed by the Six Sigma team and to identify all activities needed to achieve a goal or objective. The diagram begins with one item that breaks down into several branches. After the systematic diagram is drawn, decision-makers see the relationship paths between the branches and the main topic. If the team works on root cause analysis, systematic diagrams are helpful to identify root causes based on the relationship paths. Systematic diagrams start with an end result to be

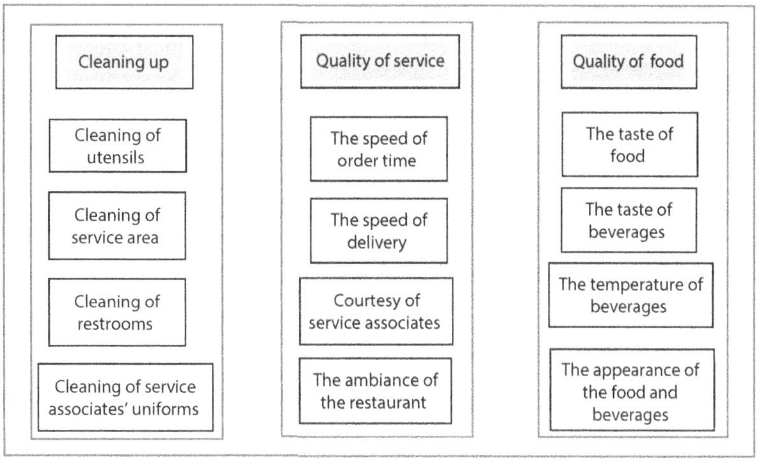

Fig. 8.3 An example of an affinity diagram. (Source: Author's creation)

attained or avoided and identify potential contributors to that result (Juran 1992: 457). The Six Sigma teams can also add "what," "who," and "when" components to systematic diagrams. Systematic diagram drawing process can be detailed as follows:

- Step 1: Determine a problem statement or topic to analyze.
- Step 2: Brainstorm the problem/topic by asking "What must be achieved?" or "What causes that problem?" until the team runs out of ideas.
- Step 3: Place the problem/topic on the left side of a paper or board.
- Step 4: Place each idea generated through brainstorm on a sticky note paper.
- Step 5: Put and group sticky note papers on the right side of the paper or board in a way to show the relationship between the next and previous ideas. Create a new branch representing new ideas if necessary.
- Step 6: Answer "who" and "when" questions for each branch.
- Step 7: The outcome of the brainstorm session should seem as a tree breaking the branches into new and detailed ideas in systematic diagram.

Specifically, systematic diagrams detail strategic plans to show what tactical goals and means are required to achieve strategic goals and objectives. The necessary tactics to achieve the goals and objectives in strategic planning can be virtually shown using systematic diagrams.

► Example 8

Assume that the marketing department aims to increase customer satisfaction and identify what tactical goals and means should be targeted in the following 5-year strategic plans. Draw a systematic diagram to help them identify the details of goals and means. ◄

- **Solution**

- Step 1: Determine a problem statement or topic to analyze.
  The topic analyzed in this example a detail of the goals and means of 5-year strategic plan for improving customer satisfaction.
- Step 2: Brainstorm the problem/topic by asking "What must be achieved?" or "What causes that problem?" until the team runs out of ideas.
  The Six Sigma team asks these two questions about increasing customer satisfaction in the marketplace. The team lists the answers one by one.
- Step 3: Place the problem/topic on the left side of a paper or board.
  The topic is articulated as "increasing customer satisfaction" and placed on the left side of the board.
- Step 4: Place each idea generated through brainstorm on a sticky note paper.
  The Six Sigma team goes back to answers given in Step 2 and puts the answers on sticky notes.
- Step 5: Put and group sticky note papers on the right side of the paper or board in a way to show the relationship between the next and previous ideas. Create a new branch representing new ideas if necessary.
  Starting from the topic articulated on the left side of the board, team members group ideas and thoughts about the goals and means of increasing customer satisfaction. This is the step where the team is expected to successfully link the goals and means related to increasing customer satisfaction.
- Step 6: Answer "who" and "when" questions for each branch.
  Depending on the topic analyzed by Six Sigma team, it may be possible to identify "who" will be working on "what" activity or branch by "when" in systematic diagram.
- Step 7: The outcome of the brainstorm session should seem as a tree breaking the branches into new and detailed ideas in systematic diagram.
  ◘ Figure 8.4 represents a systematic diagram that can be drawn in this example.

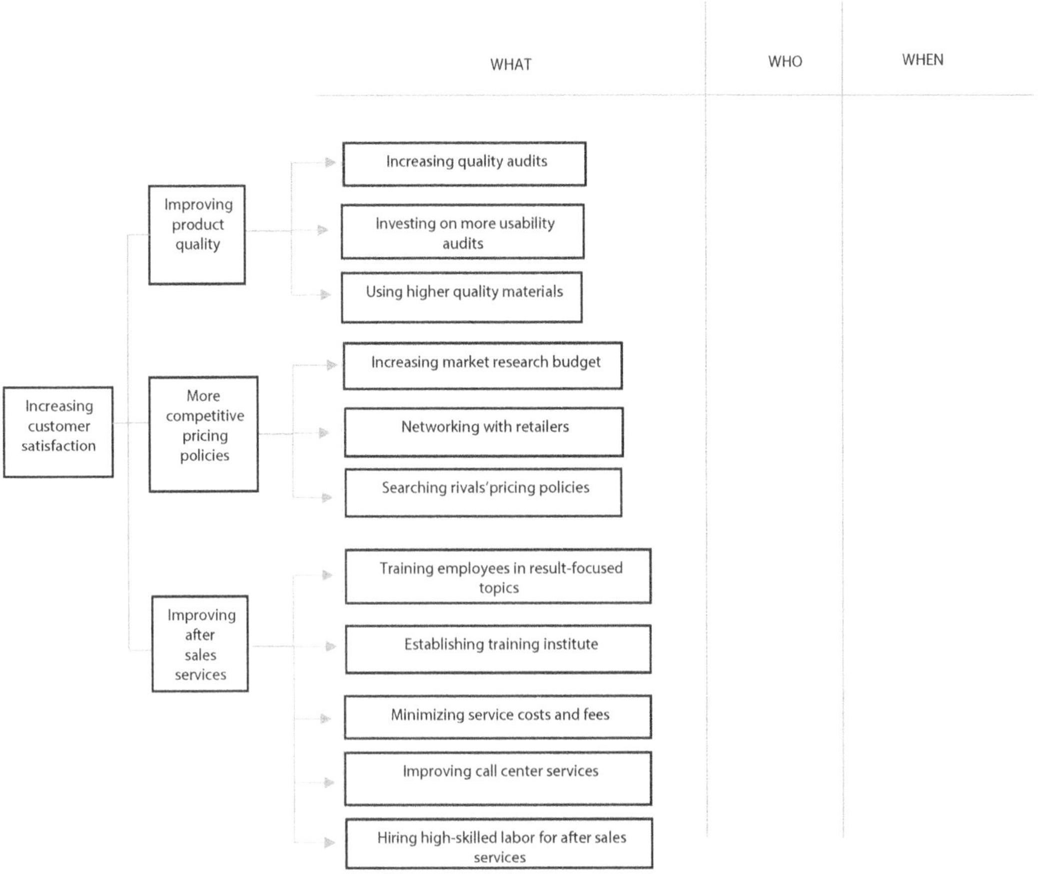

Fig. 8.4 An example of a systematic diagram. (Source: Author's creation)

8.3.3 Arrow Diagram

The arrow diagram is also known as a Gantt chart, activity on node diagram, network diagram, activity network diagram, or node diagram. Arrow diagrams are used not only in quality improvement activities but also in broader management activities, such as project management and strategic management, to picture the necessary activities, tasks, and the relationships between the activities in a project or process. Using arrow diagrams, decision-makers can see the predecessors of the relevant activities, estimate project completion time, and identify critical paths, bottlenecks, resource allocation problems, and some critical performance indicators such as mean flow time of the activities.

The arrow or Gantt chart was developed by Henry Laurence Gantt in the 1910s, a mechanical engineer and management consultant contributing to the Scientific Management era. As scheduling tools, arrow charts show the activities or tasks of a project, the activity duration times, and when each activity or task starts and ends in the project. As the timeline continues and the project progresses, the completed parts of the project are also shown on arrow charts. There are several types of information that can be generated on arrow charts, such as planned timetable of a project, actual activity durations, responsible positions, and project completion time. In some types of arrow charts, resources and skills needed to complete the activities and tasks are presented. Basic arrow charts can be drawn manually, whereas professional software is preferred in large-size cases for several reasons. Using software makes it easier to construct and modify the arrow charts. MS Excel spreadsheets are also helpful for building arrow charts. An example of an arrow or Gantt chart drawn in MS Excel is shown in ◘ Table 8.8.

Table 8.8 An example of arrow chart (Gantt chart)

	Project Teammates:		Justen, Lynn, Josh																						
	Project Start Date:		9/25/2019 (Wednesday)																						
	Today's Date:		8/24/2019 (Saturday)		23	24	25	26	27	28	29	30	1	2	3	4	5	6	7	8	9	10	11	12	13
	Display Week:		1		Week 1							Week 2							Week 3						
					9/23/19							9/30/19							10/7/19						
	Task	**Lead**	**Start**	**End**	M	T	W	Th	F	Sa	Su	M	T	W	Th	F	Sa	Su	M	T	W	Th	F	Sa	Su
1	**Brainstorm**	[Name]	Wed 9/25/19	Sat 9/28/19																					
1.1	Gantt Chart Write Up/ Gather Info	Justen	Wed 9/25/19	Sat 9/28/19																					
1.2	Gantt Chart Write Up/ Gather Info	Lynn	Wed 9/25/19	Sat 9/28/19																					
1.3	Gantt Chart Write Up/ Gather Info	Josh	Wed 9/25/19	Sat 9/28/19																					
2	**Gather Info**		Wed 10/02/19	Thu 10/10/19																					
2.1	Research Business Plan	Justen	Wed 10/02/19	Thu 10/10/19																					
2.2	Gather Theories from Book	Josh	Wed 10/02/19	Fri 10/11/19																					
2.3	Study Product Lines	Lynn	Wed 10/02/19	Sat 10/12/19																					
2.4	Meetup	All	Thu 10/03/19	Sun 10/13/19																					
3	**Analyze**		Wed 10/09/19	Tue 10/15/19																					
3.1	Collaborate Info	All	Wed 10/09/19	Tue 10/15/19																					
3.2	Evaluate Gaps in Info	All	Wed 10/09/19	Tue 10/15/19																					
3.3	Meetup	All	Thu 10/10/19	Thu 10/10/19																					
4	**Write Up/Evaluate Gantt Chart**		Wed 10/16/19	Tue 10/22/19																					
4.1	Intro Summary	Josh	Wed 10/16/19	Mon 10/21/19																					
4.2	Conclusion Summary	Lynn	Tue 10/22/19	Tue 10/22/19																					
4.3	Evaluate Gantt Chart	Justen	Wed 10/23/19	Wed 10/23/19																					
4.4	Meetup	All	Thu 10/17/19	Thu 10/17/19																					
5	**Evaluate and Practice**		Wed 10/23/19	Fri 11/29/19																					
5.1	Practice	All	Thu 10/24/19	Thu 10/24/19																					
5.2	Practice	All	Thu 10/31/19	Thu 10/31/19																					
5.3	Practice	All	Thu 11/07/19	Thu 11/07/19																					
5.4	Practice	All	Sun 11/07/19	Sun 11/07/19																					

The members of a Six Sigma team clearly think about the answers of *what*, *when*, *who*, and *how* questions. What must be accomplished by when should be answered in arrow chart drawing process. Updating the chart also helps the team manage the project more effectively. Critical points and/or accomplishments may be colored on arrow charts to let the team members know those points. Additionally, comparison between planned and actual timetable generates beneficial information for Six Sigma teams to see how effectively the plans were executed by the team.

8.3.4 Relations Diagram

The relations diagram is similar to a fishbone diagram (cause-and-effect diagram) and helps identify the root causes of the problems analyzed in the team. It is also known as interrelationships diagram and network diagram. The relations diagram demonstrates cause-and-effect relationships visually. Complex problems and topics can be analyzed in relations diagrams to depict potential complex relations between causes and effects of the topic analyzed and differentiates major from trivial causes. The other tools that are helpful to draw a relations diagram are brainstorming and checklists. The ASQ's web site in the following link may be used for further information: ▶ https://asq.org/quality-resources/relations-diagram. The basic procedure for drawing the relations diagram is detailed below:

- Step 1: Articulate the problem/topic in a way to make it visual to all team members. Place the problem/topic statement on a paper or board.
- Step 2: Comprehensively brainstorm the ideas about the problem/topic in the team, and put those ideas on sticky note papers or cards to add on the board. Fishbone diagram, affinity diagram, or systematic diagram previously drawn may be helpful in this step to generate the ideas.
- Step 3: Analyze one idea at a time and figure out if the ideas are related to this idea. If the team finds any related ideas, they put those ideas together on the paper or board.
- Step 4: Draw arrows from relevant ideas to other relevant ones. By doing so, the team will see all relationships between and among the ideas.
- Step 5: Count how many arrows come in and go out of each idea. The ideas that have the highest number of arrows are the key ideas.
- Step 6: Identify what ideas have the highest outgoing arrows. These ideas are considered "causes."
- Step 7: Identify what ideas have the highest incoming arrows. These ideas are considered "effects."

In ▶ Example 9, let's use the same scenario given in ▶ Example 7, and draw the relations diagram for this example.

▶ Example 9

Assume that you are the branch manager of a local fast-food restaurant and that customer complaints have been on the rise over the last 2 months. As the branch manager, you are expected to decrease customer complaints. You plan on analyzing customer complaint statements with your store team and drawing a relations diagram to categorize those complaints. Follow the steps given above and draw the relations diagram. ◀

- Step 1: Articulate the problem/topic in a way to make it visual to all team members. Place the problem/topic statement on a paper or board.
  The main problem presented in this example is increasing customer complaints.
- Step 2: Comprehensively brainstorm the ideas about the problem/topic in the team, and put those ideas on sticky note papers or cards to add on the board.

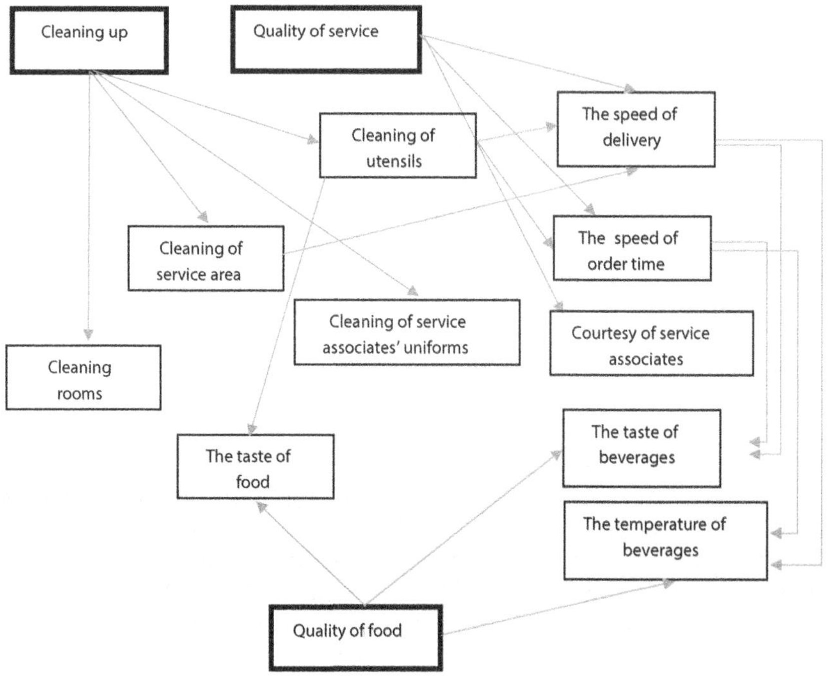

Fig. 8.5 An example of relations diagram. (Source: Author's creation)

As presented in affinity diagram above (Fig. 8.5), the problem is analyzed in three main branches such as cleaning up, quality of service, and quality of food.

- Step 3: Analyze one idea at a time and figure out if the ideas are related to this idea.
  After analyzing the problem based on three branches, the ideas for each branch are articulated by team members.
- Step 4: Draw arrows from relevant ideas to other relevant ones.
  After analyzing the problem based on three branches, the team links related ideas to each other as presented in Fig. 8.5 below. Certain ideas may cause and affect some of the other ideas.
- Step 5: Count how many arrows come in and go out of each idea. The ideas that have the highest number of arrows are the key ideas.
  The team identifies numbers of incoming and outgoing arrows generated in the relations diagram. The speed of delivery and the speed of order time have the highest total number of arrows in this example.
- Step 6: Identify what ideas have the highest outgoing arrows. These ideas are considered "causes."
  The cleaning of utensils has the highest number of outgoing arrows in the relations diagram. This idea represents the "cause" of the customer complaints.
- Step 7: Identify what ideas have the highest incoming arrows. These ideas are considered "effects."
  The speed of delivery, the speed of order time, the taste of the beverages, and the temperature of beverages have the highest number of incoming arrows. These ideas are considered the "effects" of the customer complaints.

Using the relations diagram, the Six Sigma team can analyze the causes and effects of the problem to identify root causes. The structure of the relations diagram visualizes the components of the problem and makes it easy for the teams to understand the problem comprehensively.

8.3.5 Matrix Diagram

Fundamentally, the matrix diagram demonstrates the relationships between more than two groups of data. Matrix diagrams are grouped into six types: L, T, Y, X, C, and roof-shaped matrix diagrams. L-shaped diagram shows the interactions of two groups of variables. T-shaped diagram includes the interrelationships between two groups out of three groups of variables. Y-shaped diagram shows the interactions among three groups of variables. X-shaped diagram demonstrates the relationships between and among four groups of variables. C-shaped diagram shows all the relationships between three or four groups of variables. Roof-shaped diagram depicts the relationship between two groups.

To exemplify, L-shaped matrix diagrams are used to show the details of customer requirements. For example, if the process is about baking a cake and the end product is a variety of cakes, a matrix diagram can be constructed as presented in ◘ Table 8.9. The types of CTQ characteristic and target values of technical requirements are presented in matrix diagram. More information about matrix diagrams are found at ASQ's web site ► http://asq.org/learn-about-quality/new-management-planning-tools/overview/matrix-diagram.html.

The House of Quality (HOQ) in quality function deployment (QFD) is a roof-shaped matrix diagram that shows the relationships between "customer needs" and "technical requirements." The details of QFD were previously presented in ► Chap. 4.

◘ **Table 8.9** L-shaped matrix diagram

	Customer 1	Customer 2	Customer 3
Thickness	<5 cm	>7 cm	=5.5 cm
External texture	Crispy	Soft	Crispy
Internal texture	Soft	Soft	Crunchy
Type	Cocoa	Coconut	Dried fruit

Source: Author's creation

8.3.6 Matrix Data Analysis

Matrix data analysis generalizes the data arrays (horizontal and vertical) widely used to facilitate evaluating composite relationships (Juran 1992: 457). When the data analysis contains more than one variable, matrix data analysis allows decision-makers to see the relationship between and among the variables in one platform. For example, the HOQ is a basic matrix data analysis where customer needs are matched with technical requirements of the products or services. Similarly, ◘ Table 8.7, shown in stratification analysis in ► Example 6, is another matrix data analysis where stores, products, weeks, and months are analyzed simultaneously in one table.

8.3.7 Process Decision Program Chart (PDPC)

PDPC is similar to the familiar flow diagram but modified to include unpredictable outcomes and thereby to assist in anticipating future events (Juran 1992: 457). PDPC is used to identify and anticipate what may be unpredictable in plans. First, as a preventative action, after building plans or systematic diagrams, Six Sigma team may focus on what kind of unexpected situations may occur in each branch of systematic diagram. PDPC allows Six Sigma practitioners to analyze and assess potential risks of plans and costs of unexpected situations. Second, Six Sigma teams may also develop countermeasures for those identified unpredictable outcomes or unexpected situations. Those countermeasures are categorized into "practical" and

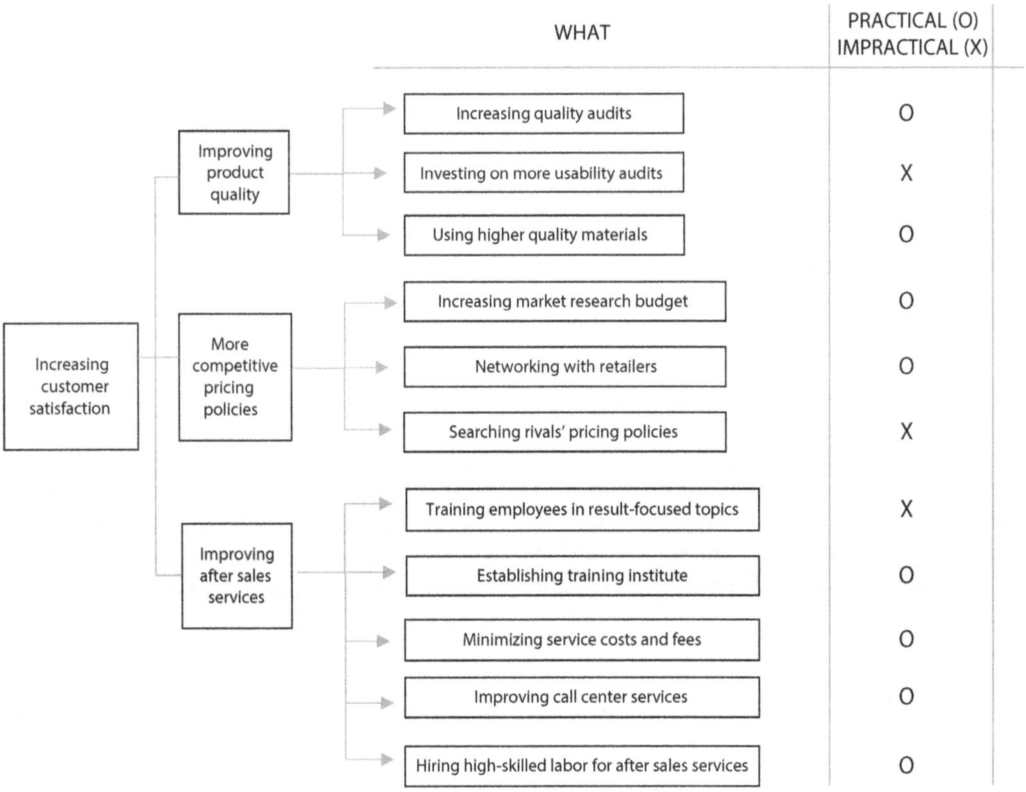

Fig. 8.6 Process decision program chart. (Source: Author's creation)

"impractical" to decide what kind of resources will be allocated to those countermeasures. Time availability, cost, and feasibility may be taken into consideration by the Six Sigma team to identify practical and impractical countermeasures. ASQ's web site details PDPC in this link: ► https://asq.org/quality-resources/process-decision-program-chart. An example of PDPC is shown in Fig. 8.6.

8.4 Other Tools

Graphical representation of data makes decision-making processes easier and more meaningful in Six Sigma projects. Rather than dealing with raw data and information, it is more helpful for decision-makers to work with graphics, tables, and charts. As Napoleon Bonaparte said, "War is 90% information."

According to Scott and Mazhindu (2014), "less is more" should be the guiding principle behind using graphical techniques. When too much information is presented in the graph, the target audience may get lost in complicated presentations. However, simple ways of data representation give decision-makers more information to make more accurate and appropriate decisions. The following sections will present other tools that are used in the "Analyze" phase of DMAIC.

8.4.1 Brainstorming

Brainstorming sessions are held in many different settings. For example, Six Sigma teams can identify problems and potential causes of the problems through brainstorming sessions in the DMAIC process. They can list and prioritize the problems using brainstorming sessions. They can also analyze the problem/s or project/s in depth as a team through brainstorming.

Brainstorming can function as a way to build a team, and it is a group problem-solving tool, which inspires team members with other members' contributions and statements in brainstorming process. What is stated by a member may help and inspire another member develop a parallel or different idea. It basically encourages members to contribute to the team. There are a few critical rules for an effective brainstorming session. These rules can be listed as follows:

1. In each session, one topic is analyzed by the team.
2. The team should organize meetings where the members cannot be interrupted because of routine daily work processes.
3. The team should have flipcharts, board markers, if necessary recorders, and sticky note cards to write all the ideas down and make them visual to everyone.
4. Everyone contributes to the session by participating in conversations in each phase of session. If the member does not have any other ideas about the topic analyzed by the team, he or she says "pass" for that turn.
5. A recorder may be assigned to type or record the statements of the members.
6. Criticism is not accepted if it is personal and not supportive for the topic analyzed by the team. Positive and constructive feedbacks are acceptable for inspiring and stimulating the other members at the end of the session.
7. The rules should be announced to all members ahead of time to get them familiar with the rules.
8. The leader of the team may run the brainstorming sessions to manage the team effectively.
9. Each team member should have equal time and opportunity to speak up and state their ideas in brainstorming sessions.

8.4.2 5 Whys Analysis

The 5 Whys analysis is a tool used to determine the root cause/s of the problem, often hidden causes of the problems, which is crucial for identifying the solution ways in problem-solving processes. The main idea behind 5 Whys analysis is to understand contributing factors of a problem or failure and develop improvement plans based on the root causes. Asking "why" five times back to back drills down into the details of the problem.

This analysis tool was developed by Sakichi Toyoda, who was the founder of Toyota Industries Corporation. Toyoda stated that "by repeating 'why' five times, the nature of the problem as well as its solution becomes clear."[3] Taiichi Ohno, the founder of lean production systems at Toyota, emphasized using 5 Whys questions to identify root causes rather than dealing with the surface of the problems. It is also a qualitative analysis tool that focuses on the ideas and opinions of the experts and team members. In complicated cases, it may not even be possible to go further second or third Why questions. In general, reasons identified without performing a detailed and systematic analysis may not be enough to identify root causes of the problem, where the successful problem-solving process lies down. One of the common mistakes made by Six Sigma project teams is to just focus on reasons that are seen on the surface of the iceberg. However, the teams are supposed to identify the root causes that result in the problems, and 5 Whys analysis is considered one of the effective tools used in Six Sigma projects. A step-by-step procedure to perform 5 Whys analysis is presented as follows:

- Step 1: Articulate the problem.
- Step 2: Focus and brainstorm on the problem and causes.
- Step 3: Ask, as a team, "why" this problem occurred.
- Step 4: Try to find an answer for the first why question.
- Step 5: Ask one more time "why" the statement given in the previous step occurs.
- Step 6: Ask three times more "why" question and try to articulate detailed answers.
- Step 7: Identify root causes of the problem, based on the answers given in the analysis.

3 ▶ http://asq.org/healthcare-use/why-quality/five-whys.html.

The 5 Whys is also a lean tool. A framework and example for performing 5 Whys is presented below:

Problem: There are missing screws on the finished goods.

- **Question 1:** Why does the finish good have missing screws?
- **Answer 1:** Because the operator forgets to assemble the screws.
- **Question 2:** Why does the operator forget to assemble the screws?
- **Answer 2:** Because the operator picks up more screws than he needs.
- **Question 3:** Why does he take more screws than what he needs?
- **Answer 3:** Because he doesn't take them by counting.
- **Question 4:** Why doesn't he count the screws?
- **Answer 4:** Because there isn't such a step in the process.
- **Question 5:** Why isn't there such a step in the process?
- **Answer 5:** Because no one is aware of this need and such step was never implemented in this station.

Using step-by-step procedure given above, the problem statement is phrased at the beginning of the process. The "why" questions take place in the process, and each why question is answered one by one. Then, the team identifies the root cause/s resulting in the problem analyzed. Wedgwood (2007) states that 5 Whys method analyzes various aspect of the process as presented below:

1. First why – symptom
2. Second why – excuse
3. Third why – blame
4. Fourth why – cause
5. Fifth why – root cause.

8

8.4.3 Dot Plot

A dot plot is a primitive data analysis tool. It is a graph showing where individual observations of the variable or CTQ characteristic are located at the number line on the horizontal axis. The dot plot is a graph of measurement data where dots are stacked vertically on the horizontal axis for each value of the variable of interest. The appearance of dot plot is similar to histogram and generally demonstrates the type of the distribution of data where each dot displays a value. Dot plot shows the behavior of data based on the frequencies of each observation. Whether the data are skewed, multimodal, spread, or contains any outliers is identified on dot plots. If the data set is small, dot plots can either be drawn manually or computer software can be used for drawing dot plots.

The sample size of the data set may affect the structure and appearance of the dot plot. When the sample size is less than 50 observations, each dot in dot plot represents one individual observation. When the sample size is greater than 50, each dot may represent more than one observation. The drawing of a dot plot can be detailed step-by-step as follows:

- Step 1: Determine the CTQ characteristic to analyze.
- Step 2: Design and detail how to collect data.
- Step 3: Collect data.
- Step 4: Check the accuracy, reliability, and validity of data, if necessary.
- Step 5: Locate individual observations on a continuum and draw dot plot.
- Step 6: Interpret findings.

▶ Example 10

An assembly line is monitored by a Six Sigma team, and the data about cycle time (seconds) of the line are collected for 8 days, as presented in Table 8.10. Draw the dot plot and interpret the findings. ◀

Solution

- Step 1: Determine the CTQ characteristic to analyze.

  In the example, CTQ characteristic is the cycle time of the activities in the assembly line.
- Step 2: Design and detail how to collect data.

  The Six Sigma team identifies how to collect data that represent the population,

■ **Table 8.10** Data set

Date	Time	X1	X2	X3	X4	X5
01-14-2020	4:49:22	74.030	74.002	74.019	73.992	74.008
01-15-2020	4:49:37	73.995	73.992	74.001	74.011	
01-16-2020	4:49:50	73.988	74.024	74.021		
01-17-2020	4:50:01	74.002	73.996	73.993	74.015	74.009
01-18-2020	4:50:10	73.992	74.007	74.015	73.989	
01-19-2020	4:50:21	74.009	73.994	73.997	73.985	
01-21-2020	4:50:30	73.995	74.006	73.994		
01-22-2020	4:50:39	73.985	74.003	73.993	74.015	

Source: Author's creation

in terms of sample size and time of data collection. In this example, there are varying sample sizes for the days.

- Step 3: Collect data.
  In this example, the data about the cycle time are collected from the assembly line and represent individual observations in varying sample sizes in each day.
- Step 4: Check the accuracy, reliability, and validity of data, if necessary.
  The Six Sigma team members need to make sure that the data statistically represent the population. If the data are reliable, it will be used in the further data analysis steps. The team may check the normality of data if they plan on using data analysis tools that require normally distributed data sets. If the data need to have another type of distribution such as Poisson, binomial, or beta, the team needs to make sure that the data set has that required distribution.
- Step 5: Locate individual observations on a continuum and draw dot plot.
  Individual observations are located on a continuum that varies between minimum and maximum observations of the data. To draw dot plot in Minitab, after transferring data set (■ Table 8.10) to a Minitab worksheet, click on Graphs→Dot Plot. Select the type of the dot plot in the following screen. If there is one Y variable in the question, you can select "One Y" and "Simple." If there are multiple Y variables, you can select one of four alternatives. Click on OK. Then, enter the data set from left box to the right box. Click on OK. The dot plot of the data set is presented in ■ Image 8.5 as Minitab output.
- Step 6: Interpret findings.
  To interpret the dot plots, first, the peaks with the most common observations are identified. Then, the spread of data is analyzed to see the variation of the data set. Outliers are also detected on dot plots. The dot plot in ■ Image 8.5 shows that the majority of the observations are located in between 73.990 and 73.996 seconds. The outliers of the data set are 73.985 and 74.030 seconds. The dot plot does not provide enough information to say something about the distribution of the data set in this example.

8.4.4 Run Chart

A run chart is utilized in problem-solving or data analysis processes when the Six Sigma team needs to visualize the data set over time. In *X*-*Y* plane, *Y*-axis represents the corresponding values of the relevant variable, while *X*-axis represents the time sequence. A bar chart displays the distribution of a categorical variable, showing the counts for each

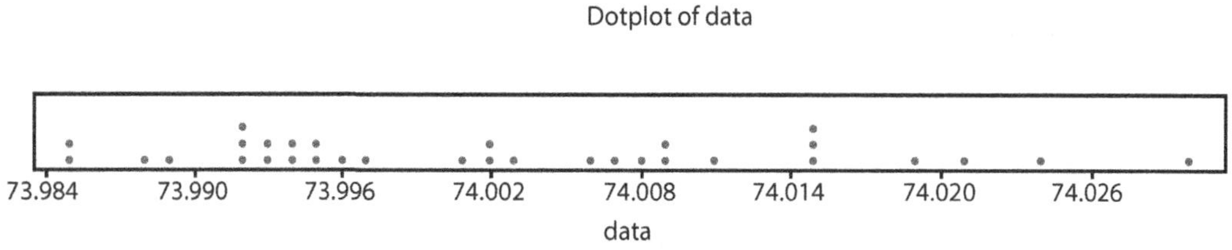

Image 8.5 Dot plot. (Source: Author's creation based on Minitab)

category next to each other for easy comparison. The run chart does not use control limits as control charts do. But, they contain median and/or predetermined threshold to analyze the data set. There are two interpretation ways for the run charts. First, any individual observation located above the threshold represents a statistically out-of-control or abnormal condition. Second, eight consecutive points located either above or below the median address another statistically out-of-control or abnormal condition. To establish randomness and minimizing special causes of the variation, the second way is commonly used on run charts. To draw a run chart, the following steps can be followed:

- Step 1: Determine the CTQ characteristic to analyze.
- Step 2: Design and detail how to collect data.
- Step 3: Collect data.
- Step 4: Check the accuracy, reliability, and validity of data, if necessary.
- Step 5: Locate individual observations consecutively one by one on a continuum and draw run chart.
- Step 6: Identify and draw "threshold" or "median."
- Step 7: Interpret findings.

▶ Example 11

An assembly line is monitored by a Six Sigma team, and the data about cycle time are collected for 8 days, as presented in Table 8.10. Draw the run chart and interpret the findings. ◀

Solution

Let's follow the steps given above one by one.

- Step 1: Determine the CTQ characteristic to analyze.
  The cycle time in the assembly line is the CTQ characteristic analyzed in this example.
- Step 2: Design and detail how to collect data.
  The data collection plan should focus on how to monitor, measure, and record the individual observations since individual observations will be important for drawing and interpreting run chart.
- Step 3: Collect data.
  The data set is demonstrated in Table 8.10.
- Step 4: Check the accuracy, reliability, and validity of data, if necessary.
  The Six Sigma team members need to make sure that the data statistically represents the population, are reliable, and are appropriate to use in the further data analysis steps. The team may check the normality of the data if they plan on using data analysis tools that require normally distributed data sets. If the data need to have another type of distribution such as Poisson, binomial, or beta, the team needs to make sure that the data set has that required distribution.
- Step 5: Locate individual observations consecutively one by one on a continuum.
  The individual consecutive observations are located on a continuum. To draw the run chart in Minitab, put all individual observations in one column. Click on Stat→Quality Tools→Run Chart B. Then, enter the data set from left box to Single column on the right box. Enter 1 for subgroup size. Click on OK. The run chart of the data set is presented in Image 8.6 as Minitab output.
- Step 6: Identify and draw "threshold" or "median."

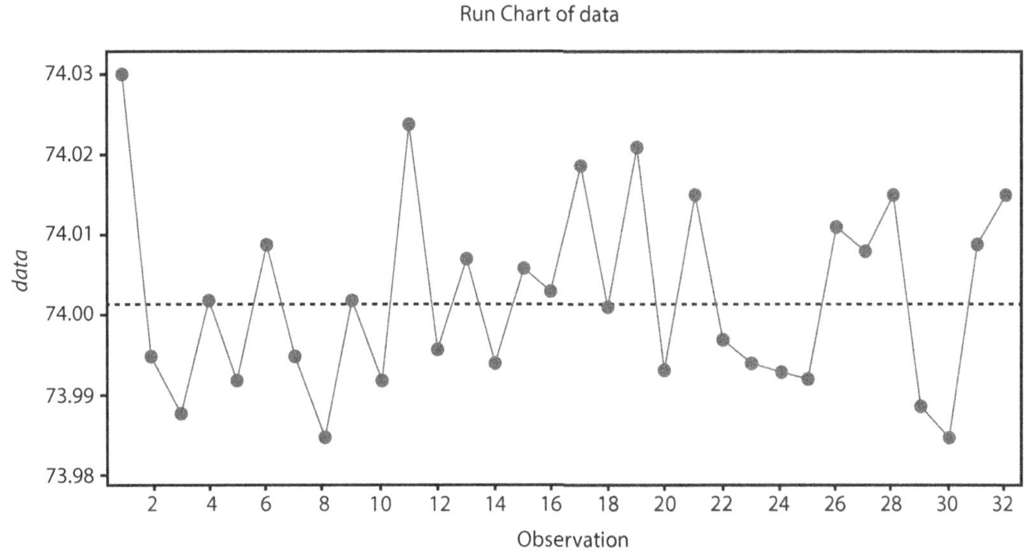

Number of runs about median:	21	Number of runs up or down:	24
Expected number of runs:	17.0	Expected number of runs:	21.0
Longest run about median:	4	Longest run up or down:	4
Approx P-Value for Clustering:	0.925	Approx P-Value for Trends:	0.902
Approx P-Value for Mixtures:	0.075	Approx P-Value for Oscillation:	0.098

Image 8.6 Run chart. (Source: Author's creation based on Minitab)

Depending on the performance expectations about the CTQ characteristic, the threshold to show the maximum acceptable cycle time can be identified by the Six Sigma team and other decision-makers. The Six Sigma team can also use "median" (74.0015 minutes) to analyze the performance of the CTQ characteristic.

- Step 7: Interpret findings.
  To interpret the run chart, we can follow the two ways as presented above. First, let's assume that the decision-makers previously identified the threshold as 74.025 minutes. In this case, any individual observation located above the threshold should represent a statistically out-of-control or abnormal condition, and there is one individual point that is considered abnormal in Image 8.6. According to second rule, eight consecutive points located either above or below the median given with dotted line (74.0015 minutes) are not detected in this run chart, and no statistically out-of-control or abnormal points are seen in this process. The "longest run about median" was found to be four in the run chart. Since it is less than eight, the process does not show any abnormalities. As a conclusion, depending on the threshold, the process can be considered statistically out of control since the first observation (74.030 minutes) is great than threshold (74.025 minutes).

8.4.5 Box-and-Whisker Plot

According to ASQ,[4] a box-and-whisker plot, called a "box plot" for simplicity, is a graphical method of displaying variation in a set of data. It allows comparison of data from different categories for easier and more effective decision-making. In addition to information generated through a histogram, a box plot also provides decision-makers with more details, allowing them to see the multiple data sets in one graph. It rapidly and simply represents a visual impression of the data by demonstrating both central tendency and the

4 ► http://asq.org/learn-about-quality/data-collection-analysis-tools/overview/box-whisker-plot.html.

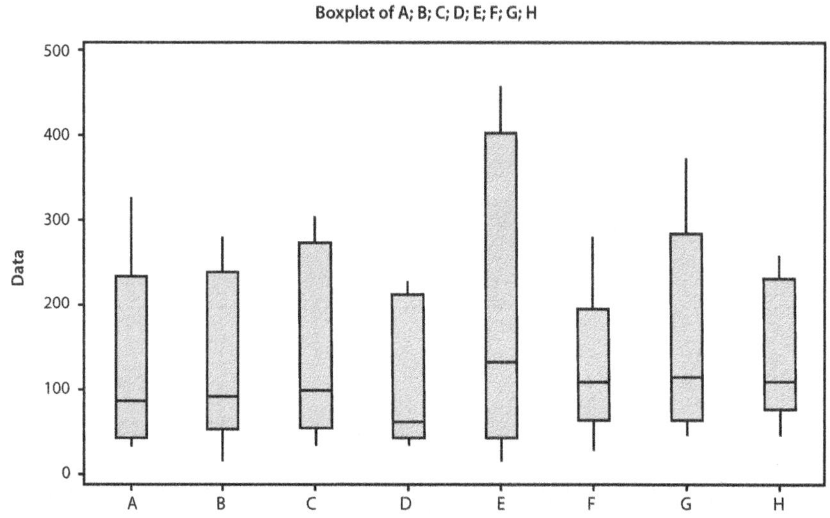

Image 8.7 The box plots of the magnitudes of over speed violations by drivers A through H. (Source: Author's creation based on Minitab)

variability of the data. It contains elements showing the distribution of the data set. The details of box plots are shown in ▶ Sect. 7.3.1 in ▶ Chap. 7.

For comparison purposes, box plots generate a good amount of information. Box plots may show whether there are any differences between levels of X values. ◘ Image 8.7 demonstrates eight drivers' speed magnitudes in a period of time. In the box-and-whisker plot, decision-makers can easily see who violates speed limits, who is the fastest and slowest drivers, and whose speed magnitude has the highest or lowest variability. Multiple box plots in ◘ Image 8.7 help identify the differences between the drivers. As seen in box plots, driver E has the fastest driving speed, whereas driver F drives with the lowest speed. Compared to the other box plots in ◘ Image 8.7, driver E has the highest difference between Q1 and Q3 which shows the largest variation among the drivers.

8.4.6 Probability Plot

A probability plot graphically represents cumulative frequencies of the data. It helps identify the distribution type of the data. It provides the best estimate for the percentages in the tails of the distribution and of the process capability (Hart and Hart 2002). If the distribution is moderately right skewed, the data yields a probability plot slightly convex toward the upper side. If the distribution is highly skewed to the right, the probability plot shows severely curved plot.

Compared to histograms, the probability plot is a better way of showing center of the data and type and spread of the distribution. The probability plot also better represents the structure of data for small samples (Montgomery 2013). When the data are near-normally distributed, the line in the probability plot is expected to be straight, gathering individual observations around it (Hart and Hart 2002). According to Shapiro (1990), individual observations shown on the points are gathered on a straight line when the model is appropriate. When the model is not appropriate, the points systematically deviate and move away from the line.

The normal probability plots are used to estimate process yields and fallouts. However, if there is not an objective procedure to determine if the process is statistically in-control, probability plotting-based analysis may be best and is supported by additional statistical analysis, such as goodness-of-fit tests (Montgomery 2013).

Whether the data are normally distributed can be checked by the probability plot. The advantage of the probability plot is that in cases where the fit to the normal distribution is not satisfactory, the probability plot may give the idea of a transformation of the data such that the transformed data approximately follow a normal distribution (DIS-16269). To draw the probability plot, the following steps can be followed:

- Step 1: Determine the CTQ characteristic to analyze.
- Step 2: Design and detail how to collect data.
- Step 3: Collect data.
- Step 4: Check the accuracy, reliability, and validity of data, if necessary.
- Step 5: Draw a probability plot on Minitab.
- Step 6: Interpret findings.

▶ Example 12

Moonlight Inc. produces various metal sheets on make-to-order system. One of the CTQ characteristics monitored in manufacturing processes is the diagonal length of the end products. ◘ Table 8.11 displays a group of measurement collected in 10 days. USL is 30 mm and LSL is 50 mm in this process. Draw the probability plot and analyze the distribution of the data. ◀

Solution

Let's follow the steps given above for drawing probability plot using Minitab.

- Step 1: Determine the CTQ characteristic to analyze.
  The CTQ characteristic analyzed in this example is the diagonal length of the metal sheets.
- Step 2: Design and detail how to collect data.
  The data are collected from seven batches, and each batch includes five observations.
- Step 3: Collect data.
  The data set is presented in ◘ Table 8.11.
- Step 4: Check the accuracy, reliability, and validity of data, if necessary.
  The Six Sigma team members need to make sure that the data statistically represents the population and that the data are reliable and can be used in the further data analysis steps.
- Step 5: Draw probability plot on Minitab.
  Let's draw the probability plot of the data using Minitab. As exemplified in ▶ Chap. 7 previously, first, transfer individual observations to Minitab in one column. Then, click on Graph→Probability Plot. In the next screen, click on Single if there is one Y variable or click on Multiple if there is more than one variable. Then, click on OK. Enter the data from left box to Graph variables. Click on OK. ◘ Image 8.8 below demonstrates the probability plot of the data.
- Step 6: Interpret findings.
  The probability plot information box on the right upper corner shows that there are

◘ Table 8.11 Diagonal length of metal sheets ordered in descending order (mm) ($n = 35$)

Order	Length (mm)	Order	Length (mm)	Order	Length (mm)	Order	Length (mm)	Order	Length (mm)
1	30	8	32	15	33	22	36	29	39
2	30	9	32	16	33	23	37	30	39
3	30	10	32	17	34	24	37	31	40
4	31	11	32	18	34	25	37	32	41
5	31	12	32	19	35	26	37	33	41
6	31	13	33	20	35	27	37	34	41
7	31	14	33	21	36	28	39	35	41

Source: Author's creation

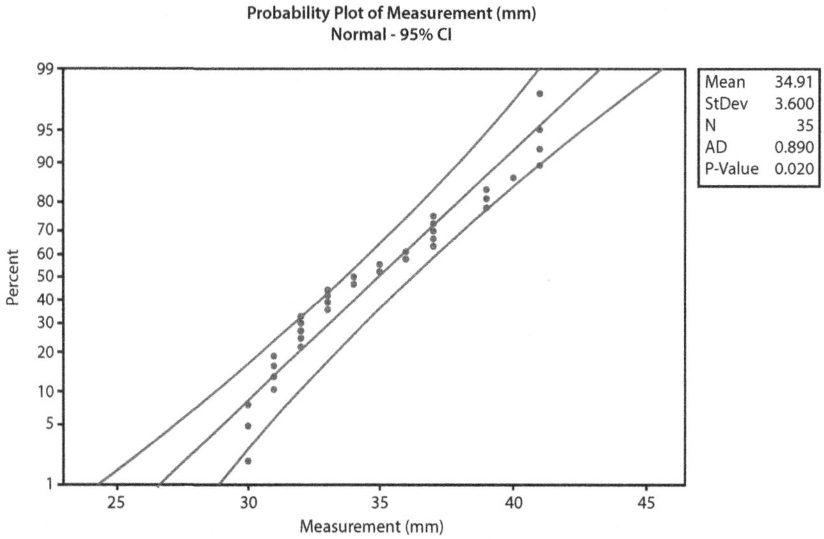

8

Image 8.8 The probability plot. (Source: Author's creation based on Minitab)

35 individual observations and that the mean of the data is 34.91 mm with 3.6 mm standard deviation. *P*-value (0.02) shows that the data set is not normally distributed, since α is greater than *p*-value, which results in rejecting the Ho hypothesis stating that the data set is normally distributed. If α is 0.01, it is accepted that the data are normally distributed.

8.4.7 Bar Chart

Bar charts are used to categorize either continuous or discrete data and demonstrate the general structure and distribution of the raw data based on frequencies of CTQ characteristics. Bar charts help decision-makers and practitioners find out significant and critical information and knowledge generated through data analysis and visualization. They are effectively used for displaying categorical data. When the data are divided into categories, we can generate frequencies for each category. Bar charts are also known as frequency charts. A bar chart is constructed along two orthogonal axes. The *X*-axis contains the range or category of the data, while the *Y*-axis displays the frequency of each category or range. Each frequency is displayed with a bar on the *X*-axis in bar chart.

The process of bar graph drawing can be detailed as follows:

- Step 1: Determine CTQ characteristic.
- Step 2: Generate categories under which the data are grouped.
- Step 3: Collect data by counting each occurrence in each category.
- Step 4: Plot a horizontal line on the *X*-axis and a vertical line on the *Y*-axis, crossing the *X*-axis.
- Step 5: Put the categories on the *X*-axis.
- Step 6: Put the frequency of each category on the *Y*-axis, and show the bars in bar graph.

Since bars have visual weight, they make it easy to see the individual importance of items given in data set. The following example shows how to draw a bar chart and to picture the impact of individual weights in data sets.

► Example 13

A doctor's office analyzes the causes of data entry errors and generates Table 8.12 showing the type of errors and frequency of each error. Draw a bar chart to show the frequency and distribution of the errors. ◄

- **Solution**

To draw a bar chart for this data set, we can use either MS Excel or Minitab. If we don't have any electronic tools, we can draw bar chart manually using the data set. Let's follow the process of drawing a bar chart as follows:

- Step 1: Determine the CTQ characteristic. The CTQ characteristic given in this data set is "errors made in data entry process in the doctor's office."
- Step 2: Generate categories under which the data are grouped.
  The categories are presented in the first column of ◘ Table 8.12: missing SSN, missing last name, missing date of birth, missing address, missing zip code, and others.
- Step 3: Collect data by counting each occurrence in each category.
  ◘ Table 8.12 shows how many times each error occurred in the system. The second column of the table displays frequencies of each error category.
- Step 4: Plot a horizontal line on X-axis and vertical line on Y-axis crossing X-axis. If you prefer to draw a bar chart manually, you can draw the X-axis and the Y-axis on a coordinate system.
- Step 5: Put the categories on X-axis.
  Let's put the "type of error" on the X-axis.

◘ **Table 8.12** Check sheet of causes of data entry errors in customer database

Type of error	Frequency	Percentage (%)
Missing SSN	25	16.67
Missing last name	33	22.00
Missing date of birth	12	8.00
Missing address	29	19.33
Missing zip code	42	28.00
Others	9	6.00
Total	150	100.00

Source: Author's creation

- Step 6: Put the frequency of each category on Y-axis and show the bars in bar graph. Let's put the "frequencies" on the Y-axis numerically for each category of type of error. ◘ Figure 8.7 shows the bar chart for this data set. Based on the bar chart, the most frequent occurrence is seen in "missing zip code" category, whereas the least frequent occurrence appears on "others" category.

To draw the bar graph in Minitab, the data set is entered in one column in Minitab. Click on Graph→Bar Chart. In the next screen, click on Bars represent→Select Values from a table. Click on Simple if there is one variable in the data set and click on OK. Enter data from left box to the right box titled Graph variables. For Categorical variable, enter the attribute variable from the left box. Click on OK. The bar graph of the data set is presented in ◘ Image 8.9 as Minitab output.

8.4.8 Line Graph

A line graph is an easy way to show the trend of the data. It demonstrates data in a time series of observations, linked by a line, and shows the trend of data and changes on data over time. When the data include more than one variable, multiple line graphs on the same X-Y axes enable comparisons between variables. In line graphs, similar to scatter diagrams, the X-axis carries the scale for the variable, and the Y-axis represents the individual observations for the corresponding X variable.

The process of drawing a line graph can be detailed as follows:

- Step 1: Determine the CTQ characteristic.
- Step 2: Generate categories under which the data are grouped.
- Step 3: Collect data by counting each occurrence in each category.
- Step 4: Plot a horizontal line on X-axis and vertical line on Y-axis crossing X-axis.
- Step 5: Put the categories on X-axis.
- Step 6: Plot each data point for each category.
- Step 7: Connect the data points with a line.

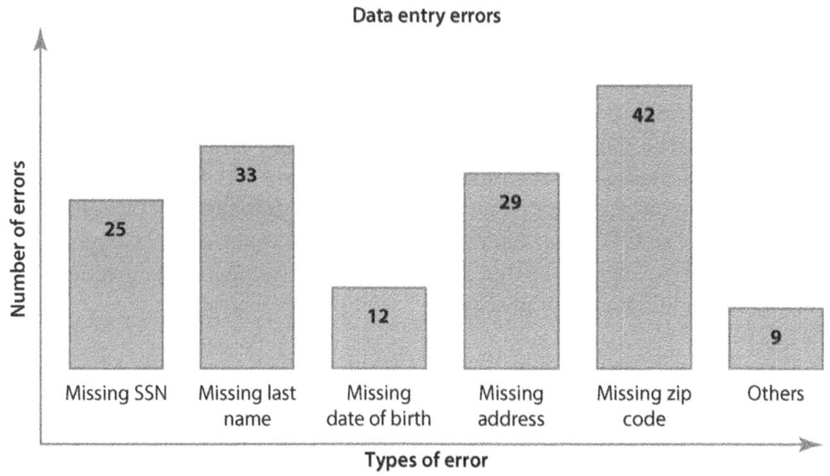

Fig. 8.7 Bar chart of the type of data-entry errors. (Source: Author's creation)

8

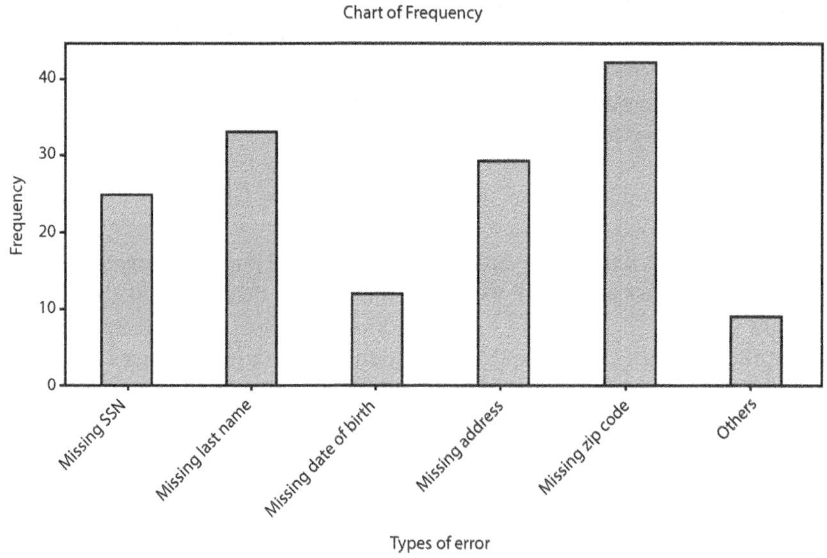

Image 8.9 Bar chart of Example 13. (Source: Author's creation based on Minitab)

► Example 14

The number of customers per day at a branch of Sunset Bank located in Boston is presented in Table 8.13. The decision-makers would like to see the trend of the number of customers. Draw the line graph representing the general trend of the weekly data and interpret your findings. ◄

Using the process flow, we can draw a line graph as follows:

- Step 1: Determine the CTQ characteristic. The CTQ characteristic used in this example is the number of customers of Sunset Bank in a branch in Boston.
- Step 2: Generate categories under which the data are grouped.

Table 8.13 The weekly number of customers at a branch in Boston Sunset Bank

Days	The number of customers
Monday	529
Tuesday	415
Wednesday	420
Thursday	475
Friday	502
Saturday	120
Sunday	0

Source: Author's creation

The categories, days of the week, are given in the data set in Table 8.13.

- Step 3: Collect data by counting each occurrence in each category.
  The number of occurrence for each day is presented in the second column of the data in Table 8.13.
- Step 4: Plot a horizontal line on *X*-axis and vertical line on *Y*-axis crossing *X*-axis. If you prefer to draw the bar chart manually, you can draw the *X*-*Y* axis on a coordinate system.
- Step 5: Put the categories on *X*-axis.
  The categories are added on the *X*-axis as presented in Fig. 8.8.
- Step 6: Plot each data point for each category.
  The number of customer for each day is added on the *Y*-axis as presented in Fig. 8.8.
- Step 7: Connect the data points with a line.
  Figure 8.8 shows the line graph demonstrating the trend of the customers of Sunset Bank located in Boston branch. The branch has more customers during weekdays as opposed to weekend. The decision-makers may establish scheduling processes based on the number of customers using line graphs, where they can easily see the customer trends.

8.4.9 Stem-and-Leaf Plot

The stem-and-leaf plot is a useful graphical tool to analyze and summarize a quantitative data set. It is used to display important features and aspects of the data, such as variability, central tendency, and the shape of the distribution. The individual observations in the data set are represented as a two-digit number. Each number, x_i, is divided into two parts: stem and leaf. The stem consists of the leading digits, and the leaf includes the remaining ones. For example, if the individual observation is 37, the stem is identified as 3, and 7 is considered leaf. The stem-and-leaf plot drawing process can be given as follows:

- Step 1: Collect the data.
- Step 2: Order the individual observations from smallest to the greatest.
- Step 3: Draw a vertical bar.
- Step 4: Build stems on the left side of the vertical bar.
- Step 5: Add leaf-leaves on the right side.
- Step 6: Interpret your findings.

▶ Example 15

Suppose that the data collected below represent lead times (minutes) in an assembly line. Draw the stem-and-leaf plot and interpret the findings.

36	24	33	36	37	38	20	14	15	17	30	31	33
33	35											

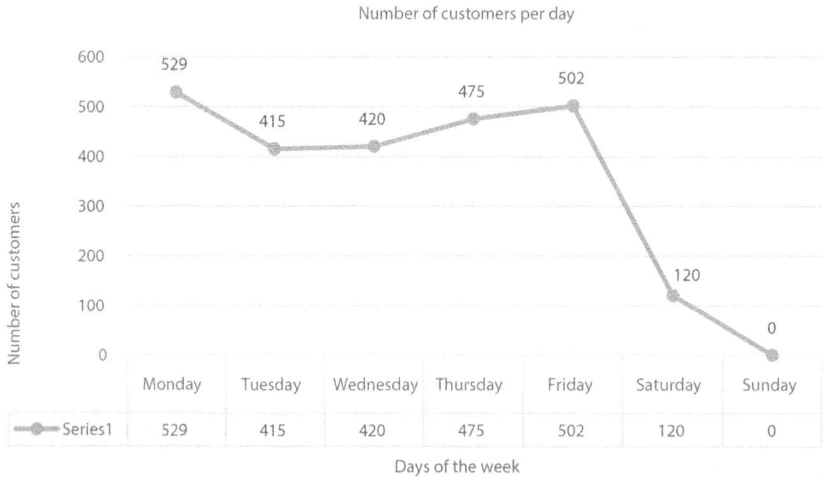

Fig. 8.8 Line graph of number of customers per day at a branch in Boston Sunset Bank. (Source: Author's creation)

8

- **Solution**

Let's follow the stem-and-leaf plot drawing process of and detail each step.

- Step 1: Collect the data.
  We assume that the data are already collected in the example. If there are no historical data collected, the Six Sigma team needs to work on the details of the data collection plan and collect the data.
- Step 2: Order the individual observations from smallest to the greatest.
  The collected data are put in ascending order as follows:

14
15
17
20
24
30
31
33
33
33
35
36
36
37
38

- Step 3: Draw a vertical bar.
  If the Six Sigma team wants to draw stem-and-leaf plot manually, a vertical bar is drawn first.
- Step 4: Build stems on the left side of the vertical bar.
  Then, stems are presented on the left side of the vertical bar to show the leading digits of the individual observations. The data include 10s, 20s, and 30s in the example. Therefore, stems will be 1, 2, and 3 to represent the tens digits of the numbers.

1 |
2 |
3 |

- Step 5: Add leaf-leaves on the right side.
  The remaining digits (ones digits) are located on the right side of the vertical bar as shown below:

1	457
2	04
3	0133356678

For example, we have 14, 15, and 17 minutes in the data set. These observations are shown in the first row of stem-and-leaf. The stem-and-leaf plot helps decision-makers understand the statistical structure of the data set. As seen in the stem-and-leaf plot, the assembly line process has a right-skewed distribution that has more observations between 30 and 38 minutes. It also helps draw and interpret a histogram in the next data analysis steps. There are 15 observations, and the first column of the plot represents the frequency of the observations that fall into each leaf. For example, there are three observations (x_i = 14, 15, and 17) in the first stem of the plot, whereas there are two observations such as 20 and 24 minutes in the second stem. Totally, there are 15 observations in the stem-and-leaf plot.

To draw the stem-and-leaf plot in Minitab, the data set is entered in one column in Minitab. Then, click on Graph→ stem-and-leaf. In the next screen, click on Graph variables→enter data column from left box to Graph variables→OK. The plot of the data set is presented in Session Window (◘ Image 8.10). In the plot, the leaf unit is 1.0; in other words, it is one-digit whole numbers.

The first column of the plot represents a cumulative count of the number of observations. The second column shows the stems for each observation, while the last column indicates the leaf, the second digit of each observation. The number 5 in the parenthesis in the first column shows where the mean of the process is located.

Stem-and-leaf of Lead time N = 15

```
1   1 4
3   1 57
5   2 04
5   2
(5) 3 01333
5   3 56678

    Leaf Unit = 1
```

◘ Image 8.10 Stem-and-leaf plot

Key Concepts

DMAIC, Analyze phase, statistical process control, big data, seven old tools (magnificent seven), check sheet, histogram, interval width, number of bars, range, fishbone diagram – cause-and-effect diagram, Kaoru Ishikawa, 5 Whys analysis, Pareto analysis and diagram, 80/20 rule, 70/30 rule, many trivial, vital a few, scatter diagram, dependent variable, independent variable, stratification analysis, control charts, 3σ control limits, upper control limit (UCL), lower control limit (LCL), 2σ warning limits, common causes of variation, assignable causes of variation, affinity diagram, systematic diagram, arrow diagram, Gantt chart, relations diagram, matrix diagram, House of Quality (HOQ), quality function deployment (QFD), matrix data analysis, process decision program chart (PDPC), brainstorming, dot plot, run chart, box-and-whisker plot, probability plot, bar chart, line graph, and stem-and-leaf plot.

Summary

This chapter examines seven old tools, seven new tools, and other tools to solve problems, analyze and visualize the data used in Six Sigma teams. The seven old tools are check sheets, histograms, fishbone diagrams – cause-and-effect diagrams, Pareto analysis and diagrams, scatter diagrams, stratification analysis, and control charts. The seven new tools are affinity diagrams, systematic diagrams, arrow diagrams, relations diagrams, matrix diagrams, matrix data analysis, and process decision program charts. The other tools are brainstorming, 5 Whys analysis, Gantt charts, dot plots, run charts, box-and-whisker plots, probability plots, bar charts, line graphs, and stem-and-leaf plots.

Practice and Discussion Questions

1. What are the seven old tools used in Six Sigma methodology?
2. Discuss when and why check sheets are needed and used in Six Sigma methodology.
3. What are the functions of histogram in Six Sigma methodology?
4. Discuss the benefits of fishbone diagrams in Six Sigma methodology.
5. Draw a fishbone diagram to analyze possibly why your order at a fast-food restaurant was incorrectly fulfilled and late.
6. What are the "primary reasons" used in fishbone diagrams?
7. What are the functions of Pareto analysis in Six Sigma methodology?
8. What is the main function of scatter diagram in Six Sigma methodology?
9. Discuss how to use stratification analysis in Six Sigma methodology.
10. What is the main function of control charts in Six Sigma methodology?
11. What type of variation is identified in control charts?
12. What are the seven new tools?
13. What is the main function of affinity diagram in Six Sigma methodology?
14. What is the main function of systematic diagram in Six Sigma methodology?
15. What is the main function of arrow diagram in Six Sigma methodology?
16. What is the main function of relations diagram in Six Sigma methodology?
17. What is the main function of matrix diagram in Six Sigma methodology?
18. What is the main function of matrix data analysis in Six Sigma methodology?
19. What is the main function of process decision program chart in Six Sigma methodology?
20. Discuss how to use brainstorming effectively in Six Sigma methodology.
21. Discuss how to use 5 Whys analysis effectively in Six Sigma methodology.
22. Assume that you are the project champion of a Six Sigma project and that the team analyzes the root cause of the low customer satisfaction in a movie theater. The movie theater employs 15 associates and is open with 8 saloons between 10:00 am and 1:00 am 7 days in a week. Apply 5 Whys on this problem and identify potential root causes.
23. Discuss what kind of information is possibly generated through dot plots.
24. Discuss how to use run charts effectively in Six Sigma methodology.
25. Discuss how to use box plots effectively in Six Sigma methodology.
26. Discuss how to use probability plot effectively in Six Sigma methodology.
27. Discuss what kind of information is generated through probability plots.
28. Discuss what kind of information bar charts produce in Six Sigma methodology.
29. Discuss what kind of information line graphs produce in Six Sigma methodology.
30. Discuss what kind of information stem-and-leaf plots produce in Six Sigma methodology.

References

Amsden, D. M., Amsden, R. T., & Butler, H. E. (1998). *SPC simplified workbook: Practical steps to quality*. New York: SteinerBooks.

Benneyan, J. C., Lloyd, R. C., & Plsek, P. E. (2003). Statistical process control as a tool for research and healthcare improvement. *BMJ Quality and Safety, 12*(6), 458–464.

Carey, R. G. (2003). *Improving healthcare with control charts: Basics and advanced SPC methods and case studies*. Milwaukee: ASQ Quality Press.

Defeo, J. (2017). *Juran's quality management and analysis*. Singapore: McGraw-Hill Higher Education.

Deming, W. E. (1982). *Quality, productivity, and competitive position*. Cambridge, MA: MIT Center for Advanced Engineering Study.

Hart, M. K., & Hart, R. F. (2002). *Statistical process control for health care*. Duxbury: Thomson Learning.

Ishikawa, K. (1985). *What is total quality control? The Japanese way*. Englewood Cliffs, NJ: Prentice Hall.

Juran, J. M. (1945). *Management of inspection and quality control*. New York: Harper and Brothers.

Juran, J. M. (1992). *Juran on quality by design: The new steps for planning quality into goods and services*. New York, The Free Press: Simon and Schuster.

Juran, J. M., & Gryna, F. M. (1980). *Quality planning and analysis*. New York: McGraw-Hill.

Kelley, L. (2000, January). *How to use control charts in healthcare. ASQ World Conference on Quality and Improvement Proceedings (p. 696)*. American Society for Quality.

Mohammed, M. A., Worthington, P., & Woodall, W. H. (2008). Plotting basic control charts: Tutorial notes for healthcare practitioners. *BMJ Quality and Safety, 17*(2), 137–145.

Montgomery, D. C. (2005). *Introduction to statistical quality control* (5th ed.). New York: John Wiley & Sons.

Montgomery, D. C. (2013). *Introduction to statistical quality control*. New York: Wiley.

Scott, I., &Mazhindu, D. (2014). *Statistics for healthcare professionals: An introduction*. Los Angeles, CA: Sage.

Shapiro, S. (1990). *How to test normality and other distribution assumptions* (p. 1990). Milwaukee: American Society for Quality.

Sue, V. M., & Griffin, M. T. (2016). *Data visualization & presentation with Microsoft office*. Thousand Oaks, CA: Sage.

Wedgwood, I. (2007). *Lean sigma a practitioner's guide*. Upper Saddle River, NJ: Prentice Hall.

Control Charts

Contents

F. Pakdil, *Six Sigma for Students*, https://doi.org/10.1007/978-3-030-40709-4_9

Learning Objectives

After careful study of this chapter, you should be able to:

- Understand common and assignable causes of variation
- Understand and describe the statistical background of control charts
- Understand the impact of subgroup size in control charts
- Apply control charts in the Analyze and Control phases of DMAIC
- Draw and interpret control charts for variables
- Draw and interpret control charts for attributes.

9.1 Introduction

9

Control charts are used in the "Analyze" and "Control" phases of DMAIC process. The plotted values demonstrate successive performance over time of the relevant CTQ characteristics. The data represented in control charts are historical data. The *CL* indicates where, generally, the average of CTQ characteristics is located, while *UCL* and *LCL* show the acceptable limits for the individual measurements. The control charts differentiate common causes of variation from assignable causes of variation through control limits. When the representative data points fall within the control limits, it is accepted that the variability of the data stems from common causes. The control limits are calculated based on probability theory, and actual variation exceeds control limits when an assignable cause is signaled. When random causes are present in the process, the observations are located in between *UCL* and *LCL*. Where at least one point falls out of the control limits, the process is considered statistically out-of-control and contains an assignable cause resulting in this status. After constructing control charts, Six Sigma teams assess the patterns and trends on the charts. Non-random trends and behaviors or any kind of abnormalities are expected to be detected on control charts. A control chart-based analysis is also required to eliminate assignable causes of variation for process capability analysis in the "Analyze" phase of DMAIC. This chapter analyzes control charts for variables and control charts for attributes.

9.2 Elements of Control Charts

Control charts are used in the "Analyze" and "Control" phases of DMAIC process. Control charts contain five main components: (1) upper control limit (*UCL*), (2) central line (*CL*), (3) lower control limit (*LCL*), (4) representative data points, and (5) rules in the control charts. As demonstrated in ◘ Fig. 9.1, *UCL* and *LCL* are expected to cover *CL*, and these control limits are symmetric to *CL*. Control charts demonstrate trends of time-ordered data as an ongoing process monitoring procedure. Since control charts are constructed based on time series analysis, the representative data are added in the control chart on the *Y*-axis in a time flow. The horizontal scale (*X*-axis) generally represents the time variable, whereas the vertical scale (*Y*-axis) shows the performance of CTQ characteristic analyzed in control charts. The plotted values demonstrate successive performance over time of the relevant CTQ characteristic. The data represented in control charts are historical data. The *CL* indicates where, generally, the average of CTQ characteristic is located, while *UCL* and *LCL* show the acceptable limits for the individual measurements. In skewed data sets, the median can be used as *CL* in control charts.

To calculate *UCL, CL,* and *LCL*, a general framework can be used, as presented in Eqs. 9.1, 9.2, and 9.3, respectively.

$$UCL = \mu_w + L\sigma_w \tag{9.1}$$

$$CL = \mu_w \tag{9.2}$$

Fig. 9.1 The structure of a control chart. (Source: Adapted from Montgomery, (2013))

$$LCL = \mu_w - L\sigma_w \tag{9.3}$$

where

- w = a statistic of CTQ characteristic,
- μ_w = the mean of w,
- σ_w = the standard deviation of w,
- L = the distance between control limits and CL.

In each chart, UCL, LCL, and CL are calculated based on relevant equations as presented in the following sections. L is generally accepted as 3 in control charts. UCL and LCL represent the ± 3-sigma (3σ) distances from the CL. If warning limits are needed in calculations, L is accepted as 2. The warning limits on control charts show a 2-sigma distance from the CL. To decrease the risk of error, it is also suggested to use ± 2-sigma (2σ) control limits as warning limits in control charts. If at least one point falls in between warning limits and control limits, the process should be investigated for assignable and/or common causes of the variation. Control charts can be plotted by increasing sample size and frequency. Montgomery (2005) states that warning limits can improve the sensitivity of control charts even if using these limits may be confusing for operators.

The structure of control charts is divided into three sections as shown in Fig. 9.1: Zone A, Zone B, and Zone C. While Zone A covers ± 3-sigma (3σ) control limits, Zone B refers to warning limits with ± 2-sigma (2σ) distance. Zone C contains ± 1-sigma (1σ) distance in control charts.

The control charts differentiate common causes of variation from assignable causes of variation through control limits. When the representative data points fall within the control limits, it is accepted that the variability of the data stems from common causes. The control limits are calculated based on probability theory, and actual variation exceeds control limits when an assignable cause is signaled. When random causes are present in the process, the observations are located in between UCL and LCL. Where at least one point falls out of the control limits, the process is considered statistically out-of-control and contains an assignable cause resulting in this status. Even where at least one point is determined to be out-of-control limits, the system shows a non-random behavior and abnormality. In this case, the decision-makers should analyze whether the process is statistically out-of-control. In cases where the process is statistically in-control, the process demonstrates a random pattern. Either common or assignable causes result in non-random patterns, abnormalities, and out-of-control conditions that indicate improvement needs in the process.

Control charts are also associated with hypothesis testing. For instance, in $\bar{X}$ charts, null hypothesis claims that the process mean(μ_1) is located in between control limits (μ_0), while alternative hypothesis claims the opposite as follows:

$$H_0 : \mu_0 = \mu_1$$

$$H_1 : \mu_0 \neq \mu_1$$

If the points fall into the control limits, the null hypothesis fails to be rejected. At that point, the probability of type I error and type II error should be considered. Type I error may occur if the process is identified as out-of-control when it is actually in-control. Type II error may arise if the process is identified as in-control when it is actually out-of-control. When the control limits are located further from the *CL*, the risk of type I error is reduced along with an increasing risk of type II error. If control limits are closer to *CL*, the risk of type I error increases, whereas the risk of type II error decreases (e.g., Montgomery 2005; Benneyan 2008; Mohammed et al. 2008; Matthes et al. 2007).

Control limits vary based on sample size, in other words, subgroup size. The higher the sample size, the narrower the control limits. Benneyan et al. (2003) state that using the rule of one data point beyond three standard deviations is likely to generate more type I errors when using more than 30 data points. Koetsier et al. (2012) suggest using 10–35 data points, while Carey and Stake (2001) and Carey (2002) propose using 20–30 data points in control charts. Less than 20–25 points should not be used (Benneyan 2008; Matthes et al. 2007) because a higher chance of type II error occurs in the control charts. Benneyan et al. (2003) and Benneyan (2008) state that 25–35 data points are advised when using control limits set at three standard deviations from the *CL*. Finally, Matthes et al. (2007) suggest using 10–34 data points in control charts.

Sampling methods and sizes for subgroups and batches should be determined prior to collecting data for control charts using various sampling methods in control charting. The sequence of the occurrence is one of the common ways for collecting data in the process. The sampling and data collection processes need to focus on revealing a pattern or trend about the process or CTQ characteristic to be analyzed in control charting.

Control charts are categorized into two main groups:

1. Control charts for variables
2. Control charts for attributes.

Control charts for variables are utilized to monitor the performance of the CTQ characteristic when the CTQ characteristic is measurable, in other words, where the characteristic is a continuous variable, such as dimension, temperature, weight, length, or volume. Measurement data are variables. Control charts for variables are used to monitor two statistics: (1) the mean ($\bar{X}$), as the measure of central tendency, and (2) the variability of the CTQ characteristic. The measure of the central tendency is monitored through the mean $(\bar{X})$ of the process, whereas the variability of the process is measured through standard deviation (*S*) or range (*R*). *S* charts for standard deviation and *R* charts for range are used in control charts for variables. Control charts for variables include

1. $\bar{X} - R$,
2. $\bar{X} - S$, and
3. $I - MR$ control charts.

When the CTQ characteristic is a countable one, that is, a discrete variable, control charts for attributes are employed in deciding whether the CTQ characteristic is statistically in-control or out-of-control. *P* charts, *np* charts, *u* charts, and *c* charts are the control charts used for discrete CTQ characteristics. The decision tree that shows the most appropriate control chart is represented in ◘ Fig. 9.2. The process

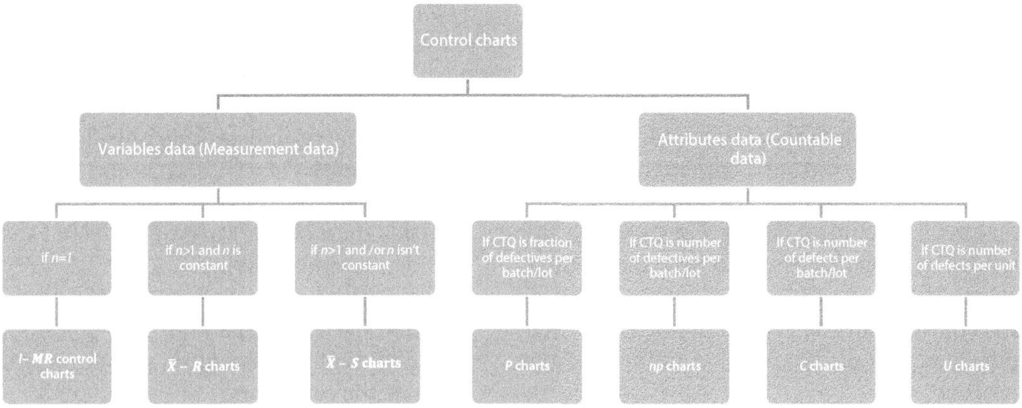

Fig. 9.2 Decision tree for control charts. (Source: Author's creation)

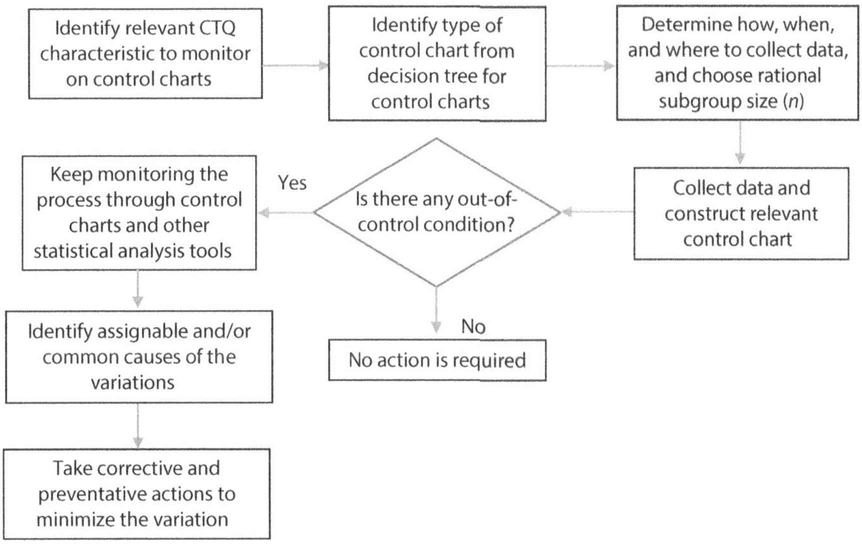

Fig. 9.3 Implementation of control charts. (Source: Author's creation)

types and data collection methods determine the type of control chart, based on underlying appropriate statistical distribution.

9.3 Implementation of Control Charts

Six Sigma teams need a well-structured process to establish control charts in the "Analyze" and "Control" phases of DMAIC process. The details of that process are presented in **Fig. 9.3**.

9.4 Decision-Making on Control Charts

After constructing control charts, Six Sigma teams assess the patterns and trends on the charts. Non-random trends and behaviors or any kind of abnormalities are expected to be detected on control charts. A "state-of-control" condition in control charts means minimum possible process variation, whereas an "out-of-control" condition indicates that some rules are violated by the process or the process exhibits a nonrandom behavior created by an assignable cause. A control chart-based analysis is also required to eliminate assignable causes of variation for process capability analysis in the "Analyze" phase of DMAIC.

There are various approaches implemented by researchers for decision-making processes on control charts. For example, the following rules are used to test the control charts. If at least one of those rules is violated, the control chart detects an abnormality that indicates an out-of-control condition in the process.

- Rule 1. One point located beyond 3-sigma zones
- Rule 2. Nine points in a row in 1-sigma zones
- Rule 3. Six points in a row constantly increasing or decreasing
- Rule 4. Fourteen points in a row constantly going up and down
- Rule 5. Two out of three points in a row in between 2-sigma and 3-sigma zones
- Rule 6. Four out of five points in a row beyond 1 sigma zone
- Rule 7. Fifteen points in a row in 1-sigma zone
- Rule 8. Eight points in a row on both sides of the *CL* and none of the points are located in 1 sigma zone.

As presented by Montgomery (2013), a group of points located in a systematic sequence is called "run" that points out a potential out-of-control condition. For example, constantly increasing or decreasing group of points show a *run* in control charts. A run may be perceived as a signal of an out-of-control condition. In addition to runs, other types of non-random behavior may display an out-of-control condition. For example, where there is a cyclic behavior on control chart, even if the points are located between control limits, the Six Sigma team should be able to analyze the process to find any hidden out-of-control conditions caused by assignable or common causes in the process. Six Sigma team members are considered competent to analyze and discover those causes.

In the *Western Electric Statistical Quality Control Handbook* (1956), several rules are presented for detecting nonrandom trends or abnormalities on control charts. These rules are:

- Rule 1: One or more points outside of the control limits
- Rule 2: Two of three consecutive points outside the 2-sigma warning limits but still inside the control limits
- Rule 3: Four of five consecutive points beyond the one-sigma limits
- Rule 4: Eight consecutive points on one side of the center line
- Rule 5: Consecutive six points constantly increasing or decreasing
- Rule 6: Fifteen consecutive points in zone C
- Rule 7: Fourteen consecutive points going up and down
- Rule 8: Eight consecutive points in a row on both sides of the *CL* with no points in zone C
- Rule 9: Any unusual trend in the data set
- Rule 10: One or more points near a warning or control limit.

The first four rules are known as the Western Electric Rules. Additionally, Shewhart's rules are generally accepted in control chart applications. All of these rules stem from Shewhart's studies from the 1920s. For example, in rule 1, all points represented in control charts should be located in between 3-sigma control limits, namely, *UCL* and *LCL*. To detect abnormalities, all of the rules are tested and controlled.

9.5 Control Charts for Variables

Control charts for variables generate useful information about the performance of the process, where the CTQ characteristic is a continuous variable. The information on mean and variability and the location of points plotted on control charts makes the decision-making processes more meaningful and easier for Six Sigma practitioners. Root cause analysis and process capability analysis are more easily constructed after analyzing the process through control charts for variables.

Control charts for variables assume that the data are normally distributed and require a smaller sample size per batch than do control charts for attributes. When the inspection method is destructive, a smaller sample size requirement may be very helpful to organize the data collection in Six Sigma projects (Montgomery 2013). $\bar{X}-R$, $\bar{X}-S$, and *I-MR* charts will be detailed in this section.

9.5.1 $\bar{X}-R$ Charts

$\bar{X}-R$ charts assume that the data used in the charts are normally distributed and signals to decision-makers and operators before the defectives are produced. $\bar{X}-R$ charts are convenient for use, particularly when the decision-makers want to monitor the central tendency and variability of the CTQ characteristic and the sample taken from each batch is a constant value, greater than one ($n > 1$). To construct the control limits, we first need to calculate $\bar{\bar{X}}$ and $\bar{R}$. If $\bar{x}_1, \bar{x}_2, \ldots, \bar{x}_m$ are the averages of the batches of m, then, the average of the averages ($\bar{\bar{X}}$), as the best estimator of the population mean (μ), is

$$\bar{\bar{X}} = \frac{\bar{x}_1 + \bar{x}_2 + \ldots + \bar{x}_m}{m} = \frac{\sum_{i=1}^{m} \bar{x}_i}{m} \quad (9.4)$$

where $\bar{\bar{X}}$ is the center line of $\bar{X}$ chart, in other words, the overall average of the observations, and m is the number of the batches. If $x_1, x_2, \ldots, x_n$ are the individual observations in a sample size of n, the range of the ith batch is

$$R_i = x_{i\_\max} - x_{i\_\min} \quad (9.5)$$

where $x_{i\_\max}$ is the maximum value of the ith batch and $x_{i\_\min}$ is the minimum value of the ith batch in data set. Let $R_1, R_2, \ldots, R_m$ be the ranges of batches of m. Then, the average of ranges is

$$\bar{R} = \frac{R_1 + R_1 + \ldots + R_m}{m} = \frac{\sum_{i=1}^{m} R_i}{m} \quad (9.6)$$

If the standard deviation of the population is known, *UCL*s and *LCL*s of $\bar{X}$ charts and R charts are separately calculated as given in Eqs. 9.7, 9.8, 9.9, and 9.10, respectively.

$$UCL_{\bar{X}} = \bar{\bar{X}} + z\sigma_{\bar{X}} \quad (9.7)$$

$$LCL_{\bar{X}} = \bar{\bar{X}} - z\sigma_{\bar{X}} \quad (9.8)$$

$$UCL_R = D_4\bar{R} \quad (9.9)$$

$$LCL_R = D_3\bar{R} \quad (9.10)$$

where

- $UCL_{\bar{X}}$ = the upper control limit of $\bar{X}$ chart
- $LCL_{\bar{X}}$ = the lower control limit of $\bar{X}$ chart
- $\bar{\bar{X}}$ = the average of the averages of batches
- z = the corresponding standard normal distribution value, which is always 3 for control limits in control charts
- $\sigma_{\bar{X}}$ = the standard deviation of the batches
- UCL_R = the upper control limit of R chart
- LCL_R = the lower control limit of R chart
- $\bar{R}$ = the average of ranges (R) of the batches
- D_4 and D_3 are the 3-sigma limit coefficients for control charts.

An unbiased estimator of σ is $\hat{\sigma} = \frac{\bar{R}}{d_2}$. When the true standard deviation of the population is unknown, $\hat{\sigma}$ can be used as the best estimator. If the true standard deviation of the population is not known, *UCL*s and *LCL*s of $\bar{X}$ and R charts are separately calculated as given in Eqs. 9.11, 9.12, 9.13, and 9.14, respectively.

$$UCL_{\bar{X}} = \bar{\bar{X}} + A_2\bar{R} \quad (9.11)$$

$$CL_{\bar{X}} = \bar{\bar{X}}$$

$$LCL_{\bar{X}} = \bar{\bar{X}} - A_2\bar{R} \quad (9.12)$$

$$UCL_R = D_4\bar{R} \quad (9.13)$$

$$CL_R = \bar{R}$$

$$LCL_R = D_3\bar{R} \quad (9.14)$$

where

- $UCL_{\bar{X}}$ = the upper control limit of $\bar{X}$ chart
- $LCL_{\bar{X}}$ = the lower control limit of $\bar{X}$ chart
- $\bar{\bar{X}} = CL_{\bar{X}}$ = the average of the averages of the batches
- UCL_R= the upper control limit of R chart
- LCL_R= the lower control limit of R chart
- $\bar{R} = CL_R$ = the average of ranges (R) of the batches
- D_4, D_3, and A_2 are the 3-sigma limit coefficients for control charts.

► Example 1

Green Light Pub Restaurant wants to make sure that customers are served their order within a range of 15 and 20 minutes. The restaurant management monitors the process and collects data from four customer orders/day for 10 consecutive days. Using the data set in ◘ Table 9.1, develop an $\bar{X} - R$ chart for this process and decide whether the process mean and range are statistically in-control. ◄

9

▪ Solution

The CTQ characteristic is order processing time at Green Light Pub Restaurant. Let's get started by calculating mean ($\bar{x}$) and range (R) of the CTQ characteristic for each day (batch). The last two columns in ◘ Table 9.1 show the mean and range of order processing time for each day.

Now we can calculate $\bar{\bar{X}}$ and $\bar{R}$ as follows:

$$\bar{\bar{X}} = \frac{\sum_{i=1}^{10} \bar{x}_i}{10} = \frac{18.08 + 16.96 + \ldots + 18.93}{10}$$

$$= 17.484 \text{ minutes}$$

$$\bar{R} = \frac{\sum_{i=1}^{m} R_i}{m} = \frac{3.60 + 3.95 + \ldots + 1.80}{10}$$

$$= 2.68 \text{ minutes}$$

$A_2 = 0.73$ ($n = 4$) from Table A.11.

Control limits for $\bar{X}$ chart:

$$UCL_{\bar{X}} = \bar{\bar{X}} + A_2\bar{R} = 17.484 + 0.73(2.68)$$

$$= 19.433 \text{ minutes}$$

◘ **Table 9.1** Data of order processing time for 10 days at Green Light Pub Restaurant

Day	Customer 1 (minutes) X1	Customer 2 (minutes) X2	Customer 3 (minutes) X3	Customer 4 (minutes) X4	Mean ($\bar{x}_i$)	Range (R_i)
1	16.40	17.40	18.50	20.00	18.08	3.60
2	15.45	15.50	17.50	19.40	16.96	3.95
3	17.20	18.30	20.10	19.10	18.68	2.90
4	13.50	14.40	15.25	25.00	17.04	11.50
5	17.25	16.40	16.50	17.40	16.89	1.00
6	18.20	18.00	18.30	18.10	18.15	0.30
7	19.00	19.00	18.50	19.10	18.90	0.60
8	16.50	17.00	17.10	17.10	16.93	0.60
9	14.00	14.20	14.50	14.50	14.30	0.50
10	18.20	18.25	19.25	20.00	18.93	1.80
				Mean	$\bar{\bar{X}}$ = 17.484	$\bar{R}$ = 2.68

Source: Author's creation

$$LCL_{\bar{X}} = \bar{\bar{X}} - A_2\bar{R} = 17.484 - 0.73(2.68)$$
$$= 15.535 \text{ minutes}$$
$$CL_{\bar{X}} = \bar{\bar{X}} = 17.484 \text{ minutes}$$

After calculating *UCL* and *LCL* for $\overline{X}$ chart, we can draw the chart as shown in ◘ Image 9.1.The averages of order processing time of 10-day data are time series data in $\overline{X}$ chart. When the sample means for 10 days are plotted on the chart, the process is diagnosed as statistically out-of-control, because the mean of the 9th batch with 14.30 minutes is less than *LCL*, which is 15.53 minutes. This batch alerts the manager of an assignable cause of variation for order processing time. Additionally, the process does not exceed *USL* in the chart, considering the upper specification limit (*USL*) = 20 minutes and the lower specification limit (*LSL*) = 15 minutes. However, the 9th batch's average value is less than *LSL*, which shows that the process does not flow in between specification limits.

$\bar{R}$ is 2.68 minutes as the average ranges of 10 batches. D_3 is 0 and D_4 is 2.28 as shown in Table A.11 for sample size of four ($n = 4$).

Control limits and *CL* for *R* chart:

$$UCL_R = D_4\bar{R} = 2.28(2.68) = 6.11 \text{ minutes}$$

$$LCL_R = D_3\bar{R} = 0(2.68) = 0 \text{ minutes}$$

$$CL_R = \bar{R} = 2.68 \text{ minutes}$$

As seen in *R* chart (◘ Image 9.1), when 10 days' sample ranges are placed on *R* chart, the process is identified as statistically out-of-control because the range of the 4th batch with 11.50 minutes is greater than *UCL*, which is 6.11 minutes.

To draw $\bar{X} - R$ charts on Minitab, first, the data set is entered in Minitab worksheet. Then, click on Stat→control charts→ Variables charts for subgroups→ Xbar & R, transfer "Measurements" variable from left box to data box and "day" to subgroup sizes box. Click on Xbar & R options→ Tests→Perform all tests for special causes→

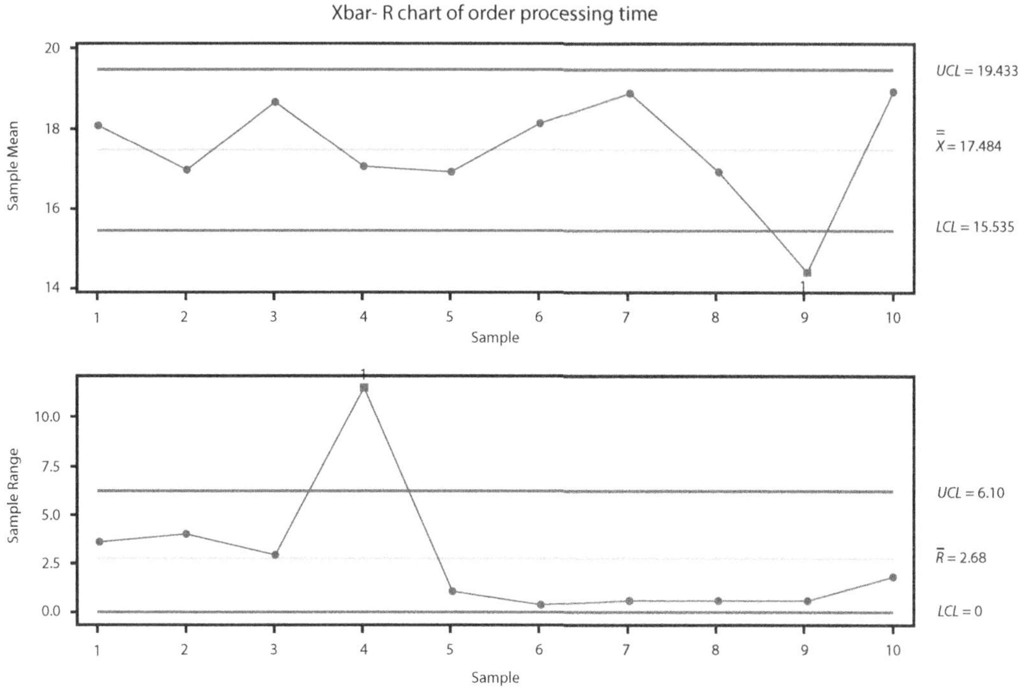

◘ **Image 9.1** $\bar{X} - R$ chart of order processing time in Minitab. (Source: Author's creation based on Minitab)

OK→OK. The $\bar{X}-R$ charts are presented in ◘ Image 9.1 as Minitab outputs.

Since both $\bar{X}$ and R charts show out-of-control points (9th day on $\bar{X}$ chart and 4th day on R chart), we can conclude that the process is not statistically in-control. After diagnosing these out-of-control points, Six Sigma teams analyze and identify assignable and/or common causes of the variations causing these abnormalities.

Let's also compute the lower natural tolerance limit ($LNTL$) and the upper natural tolerance limit ($UNTL$) of order processing time in this question. Since we have a data set that has four samples per day for 10 days, we have 40 individual observations. Since the data set does not include the population, we need to estimate mean of the population ($\hat{\mu}$) and the standard deviation of the population ($\hat{\sigma}$) to calculate $LNTL$ and $UNTL$. As given in the example, the restaurant wants to serve the orders between 15 and 20 minutes after the customer orders, and we identify LSL = 15 minutes and USL = 20 minutes. Before calculating $LNTL$ and $UNTL$, let's compute $\hat{\mu}$ and $\hat{\sigma}$. As we mentioned previously, the best estimator of μ is $\bar{\bar{X}}$. Therefore, $\hat{\mu}=\bar{\bar{X}}=17.484$. To estimate σ, we use d_2 = 2.059 for n = 2.

$$\hat{\sigma}=\frac{\bar{R}}{d_2}=\frac{2.68}{2.059}=1.301$$

$$UNTL=\hat{\mu}+3\hat{\sigma}=17.484+3(1.301)$$
$$=21.387 \text{ minutes}$$

$$LNTL=\hat{\mu}-3\hat{\sigma}=17.484-3(1.301)$$
$$=13.581 \text{ minutes}$$

As shown in ◘ Image 9.2, $UNTL$ is located far beyond USL, which indicates that the process is likely to produce defective outcomes between 20 and 21.387 minutes. The same trend appears on the other end of the process, since $LNTL$ = 13.581 is less than LSL = 15. Let's calculate the probability that order processing time is (1) greater than USL and (2) lower than LSL.

1. $$P(X>USL)=P(X>20)=P\left(Z>\frac{X-\mu}{\sigma}\right)$$
$$=P\left(Z>\frac{20-17.484}{1.301}\right)$$
$$=P(Z>1.9339)=2.68\%$$

From the standard normal distribution table (Table A.2), we find the probability of 1.93 Z value, that is, the probability that order processing time is greater than USL is 2.68%.

2. $$P(X<LSL)=P(X<15)=P\left(Z<\frac{X-\mu}{\sigma}\right)$$
$$=P\left(Z<\frac{15-17.484}{1.301}\right)$$
$$=P(Z<-1.9093)$$
$$=P(Z>1.9093$$
$$=1-0.9719=2.81\%$$

From the standard normal distribution table (Table A.2), we find the probability of 1.91 Z value; therefore, the probability that order processing time is lower than LSL is 2.81%.

9.5.2 $\bar{X}-S$ Charts

When decision-makers seek to monitor the central tendency and variability of the CTQ characteristic, and the sample taken from each batch has a constant and/or variable value and greater than one (n > 1), $\bar{X}-S$ charts are convenient for use. $\bar{X}-S$ charts are preferred where (a) the sample size per batch is not constant, (b) has a relatively higher number of observations (n >*10*), and (c) the decision-makers aim to analyze variability in a deeper way than R. We will set up this section based on two divisions. First, we will analyze $\bar{X}-S$ when n is constant, in other words, when each batch has the same number of samples. Second, we will analyze $\bar{X}-S$ when n is a variable value, in other words, when each batch has a varying number of samples.

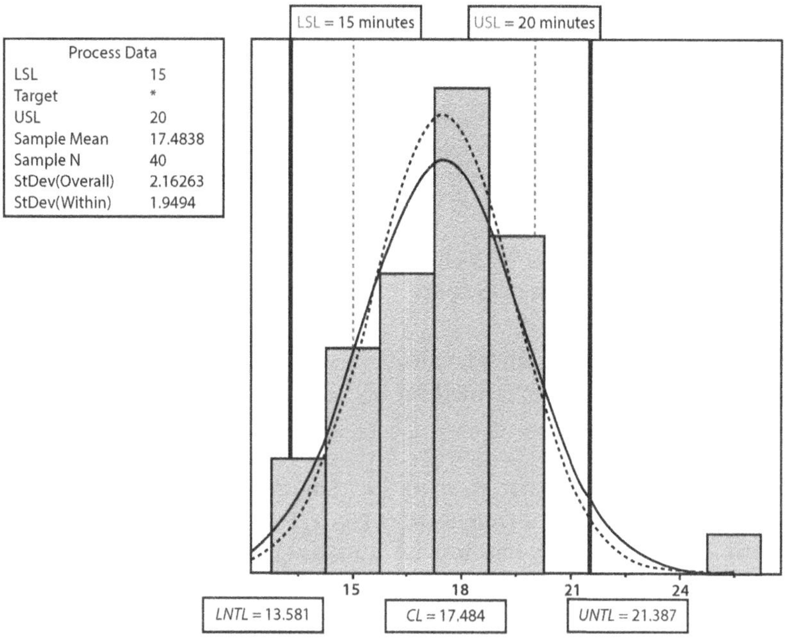

Image 9.2 The distribution of order processing time at Green Light Pub Restaurant. (Source: Author's creation based on Minitab)

9.5.2.1 The $\bar{X}$ and *S* Charts When the Sample Size Is Constant

According to central limit theorem, the distribution of the averages becomes closer to the normal distribution (e.g., Montgomery 2005) when the number of individual observations or subgroup size increase in data set. Similar to $\bar{X}-R$ charts, the control limits on *S* charts assume that the data are normally distributed. In $\bar{X}$ charts, when the data are normally distributed, it is expected that 99.73% of the averages will fall within ± 3-sigma on the distribution. If σ^2 is the unknown variance of a probability distribution, then an unbiased estimator of σ^2 is the sample variance (Montgomery 2013) and is shown in Eq. 9.15.

$$s^2 = \frac{\sum_{i=1}^{n}(x_i - \bar{x})^2}{n-1} \tag{9.15}$$

However, the sample standard deviation *s* is not an unbiased estimator of σ. The best estimator of σ can be calculated as follows (Eq. 9.16):

$$\hat{\sigma} = \frac{\bar{s}}{c_4} \tag{9.16}$$

where $\bar{s}$ is the mean of the sample standard deviation and c_4 is a coefficient that depends on the sample size *n*. The mean of the standard deviations of the batches is shown in Eq. 9.17.

$$\bar{s} = \frac{\sum_{i=1}^{m} s_i}{m} \tag{9.17}$$

If the standard deviation of the population is not known, *UCL*s and *LCL*s of $\bar{X}$ and *S* charts are separately calculated as given in Eqs. 9.18, 9.19, 9.20, and 9.21, respectively.

$$UCL_{\bar{X}} = \bar{\bar{X}} + A_3\bar{s} \tag{9.18}$$

$$CL_{\bar{X}} = \bar{\bar{X}}$$

$$LCL_{\bar{X}} = \bar{\bar{X}} - A_3\bar{s} \tag{9.19}$$

$$UCL_S = B_4\bar{s} \tag{9.20}$$

$$CL_S = \bar{s}$$

$$LCL_S = B_3\bar{s} \qquad (9.21)$$

where
- $UCL_{\bar{X}}$ = the upper control limit of $\bar{X}$ chart
- $LCL_{\bar{X}}$ = the lower control limit of $\bar{X}$ chart
- $\bar{\bar{X}} = CL_{\bar{X}}$ = the average of the averages of the batches
- UCL_S = the upper control limit of S chart
- LCL_S = the lower control limit of S chart
- $\bar{s}$ = the average of S variable
- B_4, B_3, and A_3 are the 3-sigma limit coefficients for control charts and presented in ◘ Table A.11.

The same rules presented in the *Western Electric Statistical Quality Control Handbook* (1956) apply to $\bar{X}$ charts, while one test is applicable to S charts: one or more points are located beyond control limits. Although drawing and interpreting $\bar{X}-R$ charts seems easier, $\bar{X}-S$ charts are a statistically better alternative to monitor the process. $\bar{X}-S$ charts for batches with at least subgroup sizes of 24 time-ordered samples generate more accurate and effective results (Hart and Hart 2002). For varying sample size, manually drawing $\bar{X}-S$ charts may be challenging and open to errors. Where the sample size varies in the batches, drawing $\bar{X}-S$ charts using software would be a better option to minimize calculation errors.

▶ Example 2

Let's use the question given in ▶ Example 1 and construct $\bar{X}-S$ charts for order processing time in the service process during dinner time (◘ Table 9.1). ◀

▪ Solution

The CTQ characteristic is order processing time at Green Light Pub Restaurant. Let's get started by calculating mean ($\bar{x}$) and standard deviation of batches (s) of the CTQ characteristic for each day (batch). The last two columns in ◘ Table 9.2 show the mean and standard deviation of order processing time for each day.

9

◘ Table 9.2 Data of order processing time for 10 days at Green Light Pub Restaurant

Day	X1	X2	X3	X4	Mean ($\bar{x}_i$)	Standard deviation (s_i)
1	16.40	17.40	18.50	20.00	18.08	1.544
2	15.45	15.50	17.50	19.40	16.96	1.885
3	17.20	18.30	20.10	19.10	18.68	1.228
4	13.50	14.40	15.25	25.00	17.04	5.356
5	17.25	16.40	16.50	17.40	16.89	0.511
6	18.20	18.00	18.30	18.10	18.15	0.129
7	19.00	19.00	18.50	19.10	18.90	0.271
8	16.50	17.00	17.10	17.10	16.93	0.287
9	14.00	14.20	14.50	14.50	14.30	0.245
10	18.20	18.25	19.25	20.00	18.93	0.865
				Mean	$\bar{\bar{X}} = 17.484$	$\bar{s} = 1.232$

Source: Author's creation

$$\bar{\bar{X}} = \frac{\sum_{i=1}^{10} \bar{x}_i}{10} = \frac{18.08 + 16.96 + \ldots + 18.93}{10}$$
$$= 17.484 \text{ minutes}$$

$$\bar{s} = \frac{\sum_{i=1}^{10} s_i}{10} = \frac{1.544 + 1.885 + \ldots + 0.865}{10}$$
$$= 1.232 \text{ minutes}$$

$A_3 = 1.63$ ($n = 4$)
Control limits for $\bar{X}$ chart:

$$UCL_{\bar{X}} = \bar{\bar{X}} + A_3\bar{s} = 17.484 + 1.63(1.232)$$
$$= 19.490$$

$$LCL_{\bar{X}} = \bar{\bar{X}} - A_3\bar{s} = 17.484 - 1.63(1.232)$$
$$= 15.478$$

$$CL_{\bar{X}} = \bar{\bar{X}} = 17.484$$

As seen in $\bar{X}$ chart in ◘ Image, 9.3 when the sample averages for 10 days are plotted on the chart, the process is diagnosed as statistically out-of-control because the average of the 9th sample with 14.30 minutes is less than *LCL*, which is 15.478 minutes. This alerts the decision-maker that there is an assignable cause of variation for order processing time at the restaurant. Additionally, considering the USL = 20 minutes and LSL = 15 minutes, the process does not exceed *USL* in the chart. However, the 9th batch's average value is less than *LSL,* which shows that the process does not flow in between specification limits.

The $\bar{s}$ is 1.232 minutes as the average of standard deviations of ten batches. B_3 is 0 and B_4 is 2.27 as shown in Table A.11 for sample size of four ($n = 4$).

Control limits for *S* chart:

$$UCL_S = B_4\bar{s} = 2.27(1.232) = 2.796$$

$$LCL_S = B_3\bar{s} = 0(1.232) = 0$$

$$CL_S = \bar{s} = 1.232$$

As seen in *S* chart in ◘ Image 9.3, when 10 days of sample standard deviations are placed on the *S* chart, the process is statistically out-of-control, because the standard deviation of the 4th batch with 5.356 minutes is greater than *UCL,* which is 2.796 minutes.

To draw $\bar{X} - S$ charts on Minitab, the data set is entered in Minitab worksheet. Then, click on Stat→control charts→Variables charts for subgroups→Xbar & S, and transfer data "Measurements" from left box to data box and "day" to "subgroup sizes". Next, click on Xbar & S options→Tests→Perform all tests for special causes→OK→OK. The $\bar{X} - S$ charts are presented in ◘ Image 9.3 as Minitab outputs.

Since both $\bar{X}$ and *S* charts detect out-of-control points, we can conclude that the process is not statistically in-control. After diagnosing this, Six Sigma teams are required to analyze and identify assignable and common causes of the variability. As presented on *R* and *S* charts, measures of variability display a similar trend.

We can also estimate the process standard deviation by using unbiased estimator of σ. The estimated process standard deviation is

$$\hat{\sigma} = \frac{\bar{s}}{c_4} = \frac{1.232}{0.9213} = 1.337$$

▶ Example 3

A large-scale dishwasher producer wants to monitor K11 (mm) CTQ characteristic through $\bar{X} - S$ charts. The data collected in 19 days are presented in ◘ Table 9.3. The number of observations per day is four ($n = 4$). For K11, target value is 299.9 mm and toler-

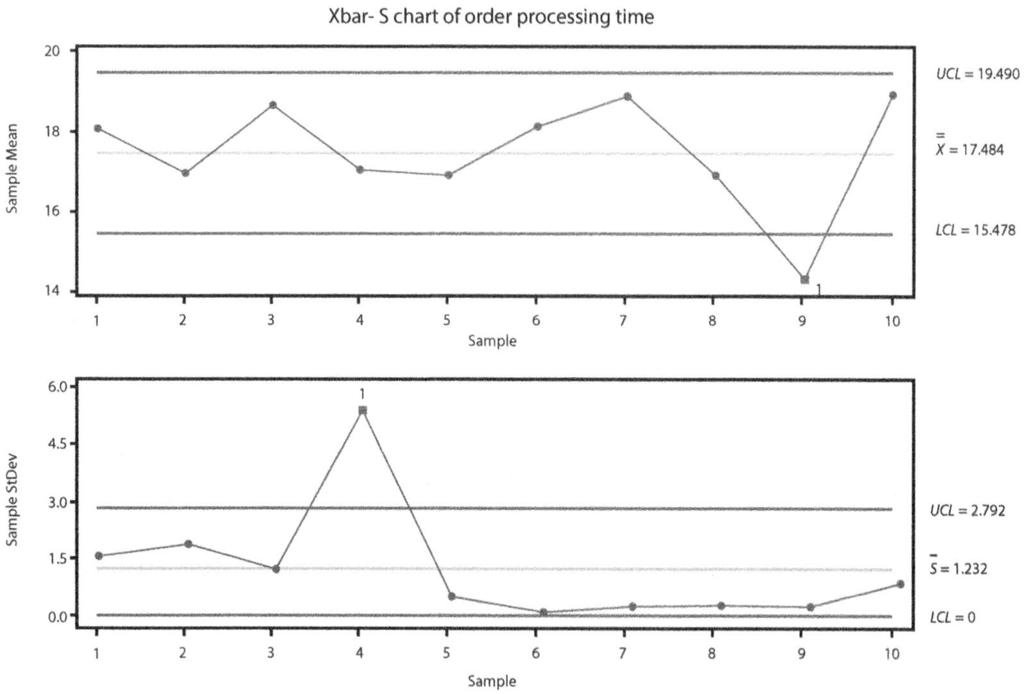

Image 9.3 $\overline{X}-S$ chart on Minitab. (Source: Author's creation based on Minitab)

ance interval is ±0.8 mm. Construct $\overline{X}-S$ charts and interpret your findings. ◂

- **Solution**

Let's get started by calculating means $(\overline{x_i})$ and standard deviations (s_i) of each day. The columns in Table 9.3 represents mean (x_i) and standard deviation (s_i) of K11 CTQ characteristic for each day, respectively.

$$\overline{\overline{X}} = \frac{\sum_{i=1}^{19} \overline{x}_i}{19} = \frac{299.98 + 299.92 + \ldots + 300.03}{19}$$

$$= 299.9387$$

$$\overline{s} = \frac{\sum_{i=1}^{19} s_i}{19} = \frac{0.22 + 0.14 + \ldots + 0.08}{19}$$

$$= 0.1606$$

$A_3 = 1.63$ $(n = 4)$

Control limits for $\overline{X}$ chart:

$$UCL_{\overline{X}} = \overline{\overline{X}} + A_3\overline{s} = 299.9387 + 1.63(0.1606)$$

$$= 300.2002$$

$$LCL_{\overline{X}} = \overline{\overline{X}} - A_3\overline{s} = 299.9387 - 1.63(0.1606)$$

$$= 299.6772$$

$$CL_{\overline{X}} = \overline{\overline{X}} = 299.9387$$

Using Minitab, as explained in the previous example, $\overline{X}-S$ chart can be drawn as shown in Image 9.4. When the sample means for 19 days are plotted on the chart, the process is diagnosed as statistically in-control because no sample means are greater than $UCL = 300.2002$ mm or lower than $LCL = 299.6772$ mm. Additionally, the process does not exceed

Table 9.3 Data of K11 CTQ characteristic and means and standard deviations of measurements

Date	K11	$\bar{x}_i$	*si*	Date	K11	$\bar{x}_i$	*si*	Date	K11	$\bar{x}_i$	*si*
9/27/18	299.76	299.98	0.22	10/16/18	299.98			11/8/18	300	299.92	0.08
9/27/18	299.91			10/16/18	299.67			11/8/18	299.95		
9/27/18	300.28			10/18/18	299.87	299.92	0.23	11/8/18	299.81		
9/27/18	299.98			10/18/18	300.11			11/8/18	299.92		
9/28/18	299.84	299.92	0.14	10/18/18	299.62			11/18/18	299.91	300.10	0.14
9/28/18	299.86			10/18/18	300.08			11/18/18	300.13		
9/28/18	300.13			10/19/18	299.85	299.81	0.06	11/18/18	300.25		
9/28/18	299.86			10/19/18	299.82			11/18/18	300.12		
9/29/18	299.59	299.93	0.26	10/19/18	299.84			11/19/18	299.74	299.83	0.20
9/29/18	300.17			10/19/18	299.72			11/19/18	299.62		
9/29/18	299.87			10/20/18	300.04	300.00	0.13	11/19/18	299.86		
9/29/18	300.08			10/20/18	300.15			11/19/18	300.1		
9/30/18	300.08	299.98	0.15	10/20/18	299.85			11/22/18	299.81	300.03	0.15
9/30/18	300.06			10/20/18	299.95			11/22/18	300.04		
9/30/18	300			10/23/18	299.98	300.02	0.14	11/22/18	300.15		
9/30/18	299.76			10/23/18	300.1			11/22/18	300.12		
10/14/18	300.21	299.94	0.21	10/23/18	299.83			11/25/18	299.84	299.89	0.13
10/14/18	299.74			10/23/18	300.15			11/25/18	300.06		
10/14/18	300			10/27/18	300.14	299.98	0.12	11/25/18	299.92		
10/14/18	299.81			10/27/18	299.87			11/25/18	299.75		
10/15/18	299.77	299.76	0.07	10/27/18	299.89			11/26/18	299.97	300.03	0.08
10/15/18	299.75			10/27/18	300			11/26/18	300.08		
10/15/18	299.85			11/1/18	299.65	299.83	0.31	11/26/18	299.96		
10/15/18	299.68			11/1/18	299.87			11/26/18	300.12		
10/16/18	300.2	299.98	0.22	11/1/18	299.54						
10/16/18	300.05			11/1/18	300.25						

USL = 300.7 mm (299.9 + 0.8 mm) or LSL = 299.1 mm (299.9 – 0.8 mm), which shows that the process flows in between specification limits.

The $\bar{s}$ is 0.1606 mm as the average of standard deviations of 19 days. B_3 is 0 and B_4 is 2.27 as shown in Table A.11 for sample size of four ($n = 4$).

Control limits for S chart:

$$UCL_S = B_4\bar{s} = 2.27(0.1606) = 0.3640$$

$$LCL_S = B_3\bar{s} = 0(0.1606) = 0$$

$$CL_S = \bar{s} = 0.1606$$

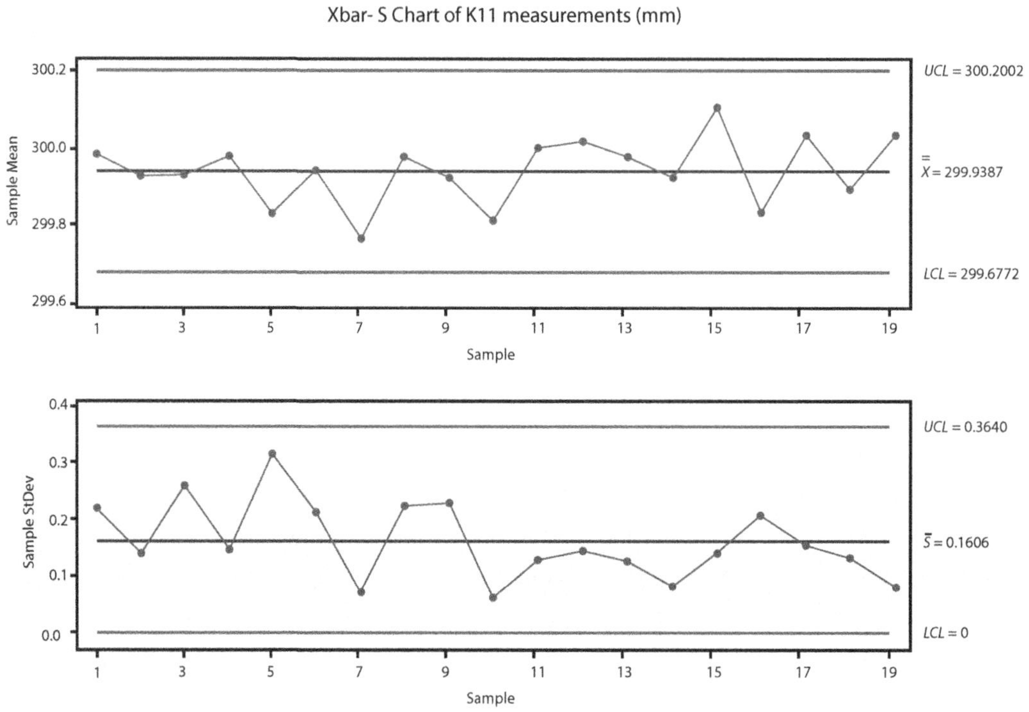

Image 9.4 $\overline{X}-S$ charts of K11. (Source: Author's creation based on Minitab)

As seen in the S chart in Image 9.4, when 19 days of sample standard deviations are placed on the chart, the process is statistically in-control because no sample standards are located beyond the control limits. Since both $\overline{X}$ and S charts do not show any out-of-control point, we can conclude that the process is statistically in-control.

9.5.2.2 The $\overline{X}$ and S Charts When the Sample Size Is Not Constant

If sample size is not constant in each batch or subgroup, we need to compute weighted averages of $\overline{\overline{X}}$ and $\overline{s}$, as shown in Eqs. 9.22 and 9.23, respectively.

$$\overline{\overline{X}} = \frac{\sum_{i=1}^{m} n_i \overline{x_i}}{\sum_{i=1}^{m} n_i} \tag{9.22}$$

$$\overline{s} = \left[\frac{\sum_{i=1}^{m} (n_i - 1) s_i^2}{\sum_{i=1}^{m} n_i - m} \right]^{1/2} \tag{9.23}$$

UCLs and LCLs for $\overline{X}-S$ charts are calculated as presented in Eqs. 9.18, 9.19, 9.20, and 9.21 in the previous sections. Note that A_3, B_3, and B_4 coefficients vary based on the sample size used in each batch or subgroup.

► Example 4

The same large-scale dishwasher producer used in ► Example 3 wants to monitor K11 CTQ characteristic through $\overline{X}-S$ charts based on another data set where sample sizes vary. The data collected in 13 days are presented in Table 9.4. For K11, target value of the CTQ characteristic is 299.9 mm, and tolerance interval is ±0.8 mm. Construct $\overline{X}-S$ charts and interpret your findings. ◄

Table 9.4 Data of K11 CTQ characteristic for varying sample size

Date	K11 (mm)	Day	$\overline{x}_i$	s_i	Date	K11 (mm)	Day	$\overline{x}_i$	s_i
9/27/18	300.2	1	299.9773	0.269892	11/1/18	299.65	5	299.89	0.305205
9/27/18	299.81	1			11/1/18	299.87	5		
9/27/18	300.09	1			11/1/18	299.54	5		
9/27/18	299.76	1			11/1/18	300.25	5		
9/27/18	300.51	1			11/1/18	300.14	5		
9/27/18	299.59	1			10/4/18	299.75	6	299.9733	0.207445
9/27/18	299.86	1			10/4/18	300.16	6		
9/27/18	299.76	1			10/4/18	300.01	6		
9/27/18	299.91	1			10/5/18	299.97	7	299.9433	0.211266
9/27/18	300.28	1			10/5/18	299.72	7		
9/27/18	299.98	1			10/5/18	300.14	7		
9/28/18	299.84	2	299.89	0.140357	10/6/18	299.82	8	299.79	0.036056
9/28/18	299.86	2			10/6/18	299.75	8		
9/28/18	300.13	2			10/6/18	299.8	8		
9/28/18	299.86	2			10/7/18	299.96	9	299.86	0.088882
9/28/18	299.76	2			10/7/18	299.83	9		
9/29/18	300.05	3	299.9867	0.222411	10/7/18	299.79	9		
9/29/18	300.16	3			10/9/18	300.19	10	300.0133	0.185562
9/29/18	299.59	3			10/9/18	300.03	10		
9/29/18	300.17	3			10/9/18	299.82	10		
9/29/18	299.87	3			10/11/18	299.82	11	300.05	0.325269
9/29/18	300.08	3			10/11/18	300.28	11		
9/30/18	300.08	4	299.975	0.147309	10/12/18	299.98	12	299.97	0.014142
9/30/18	300.06	4			10/12/18	299.96	12		
9/30/18	300	4			10/13/18	300.11	13	300.1833	0.087369
9/30/18	299.76	4			10/13/18	300.28	13		
					10/13/18	300.16	13		
							Mean	$\overline{\overline{X}} = 299.9585$	$\overline{s} = 0.2166$

Source: Author's creation

■ **Solution**

First, the average of averages and average of standard deviations are calculated as presented in the last two columns in Table 9.4 for each day (batch). To calculate the average of the averages ($\overline{\overline{X}}$) and the average of standard deviations ($\overline{s}$) for the entire data, we need to work on weighted $\overline{\overline{X}}$ and $\overline{s}$ as shown in below, using Eqs. 9.22 and 9.23.

$$\overline{\overline{X}} = \frac{\sum_{i=1}^{m} n_i \overline{x_i}}{\sum_{i=1}^{m} n_i} = \frac{11(299.9773)+5(299.89)+\ldots+3(300.1833)}{11+5+\ldots+3} = 299.9585$$

$$\overline{s} = \left[\frac{\sum_{i=1}^{m}(n_i-1)s_i^2}{\sum_{i=1}^{m} n_i - m}\right]^{1/2} = \frac{10(0.269892)^2+4(0.140357)^2+\ldots+2(0.087369)^2}{10+4\ldots+3-13} = 0.2166$$

Afterward, we can compute *UCL*s and *LCL*s for each day in $\overline{X}-S$ charts, respectively. Since sample sizes vary from day to day, we need to calculate *UCL*s and *LCL*s for each day. For example, for the first day A_3 coefficient is 0.93 for $n = 11$ in Table A.11, and *UCL* and *LCL* for the first day are calculated as follows:

$$UCL_{\overline{X_1}} = \overline{\overline{X}} + A_3\overline{s}$$
$$= 299.9585 + 0.93(0.2166)$$
$$= 300.1599$$

$$LCL_{\overline{X_1}} = \overline{\overline{X}} - A_3\overline{s}$$
$$= 299.9585 - 0.93(0.2166)$$
$$= 299.7571$$

$$CL_{\overline{X}} = \overline{\overline{X}} = 299.9585$$

After calculating *UCL*s and *LCL*s for the other 12 days as presented in ◘ Table 9.5, we can draw $\overline{X}$ chart as shown in ◘ Image 9.5. The center line of the $\overline{X}$ chart is 299.9585 mm, and the center line of the *S* chart is 0.2166 mm, respectively.

The $\overline{X}$ chart (◘ Image 9.5) shows that the process is statistically in-control because all means of the days are located in between *UCL*s and *LCL*s in each subgroup. In the *S* chart, *UCL*s and *LCL*s are separately calculated for each subgroup (◘ Table 9.5). The center line is located at 0.2166 ($CL_S = \overline{s} = 0.2166$). For the first day $n = 11$, $B_4=1.68$, $B_3=0.32$ (◘ Table A.11), and

$$UCL_{S1} = B_4\overline{s} = 1.68(0.2166) = 0.3638$$

and

$$LCL_{S1} = B_3\overline{s} = 0.32(0.2166) = 0.0693.$$

After calculating *UCL*s and *LCL*s for the other 12 days, as presented in ◘ Table 9.5, we can draw *S* chart as shown in ◘ Image 9.5. To draw $\overline{X}-S$ charts on Minitab, the steps given in ► Example 2 can be followed. The $\overline{X}-S$ charts are presented in ◘ Image 9.5 as Minitab outputs.

As seen in ◘ Table 9.5 and $\overline{X}-S$ charts in ◘ Image 9.5, the process is statistically under control since none of $\overline{X}$ or *S* variables are beyond either *UCL*s or *LCL*s. Note that Minitab calculates $\overline{s} = 0.1912$, while our weighted standard deviation is found to be 0.2166 using Eq. 9.23. As stated by Minitab, when the subgroup size is not constant, process standard deviation is used as the average standard deviation $\overline{s} = \sigma$.

For reducing the calculations and using an approximate approach, an alternative approach is to calculate the control limits on average sample size as $\overline{n}$ or most common sample size. In our example, the most common sample size is 3 in 6 days. To calculate $\overline{s}$, the average of all values of s_i for which $n_i = 3$, the following formula is used:

Table 9.5 *UCLs and LCLs in $\bar{X}-S$ charts for each day in data*

Date	Day	n	A_3	B_3	B_4	$\bar{X}_i$	s_i	$UCL_{\bar{X}}$	$LCL_{\bar{X}}$	UCL_s	LCL_s
9/27/2004	1	11	0.93	0.32	1.68	299.9773	0.269892	300.1599	299.7571	0.363839	0.069303
9/28/2004	2	5	1.43	0	2.09	299.8900	0.140357	300.2682	299.6488	0.452633	0
9/29/2004	3	6	1.29	0.03	1.97	299.9867	0.222411	300.2379	299.6791	0.426645	0.006497
9/30/2004	4	4	1.63	0	2.27	299.9750	0.147309	300.3115	299.6055	0.491616	0
11/1/2004	5	5	1.43	0	2.09	299.8900	0.305205	300.2682	299.6488	0.452633	0
10/4/2004	6	3	1.95	0	2.57	299.9733	0.207445	300.3808	299.5362	0.556587	0
10/5/2004	7	3	1.95	0	2.57	299.9433	0.211266	300.3808	299.5362	0.556587	0
10/6/2004	8	3	1.95	0	2.57	299.7900	0.036056	300.3808	299.5362	0.556587	0
10/7/2004	9	3	1.95	0	2.57	299.8600	0.088882	300.3808	299.5362	0.556587	0
10/9/2004	10	3	1.95	0	2.57	300.0133	0.185562	300.3808	299.5362	0.556587	0
10/11/2004	11	2	2.66	0	3.27	300.0500	0.325269	300.5346	299.3824	0.708187	0
10/12/2004	12	2	2.66	0	3.27	299.9700	0.014142	300.5346	299.3824	0.708187	0
10/13/2004	13	3	1.95	0	2.57	300.1833	0.087369	300.3808	299.5362	0.556587	0
						$\bar{\bar{X}} = 299.9585$	$\bar{s} = 0.2166$				

Source: Author's creation

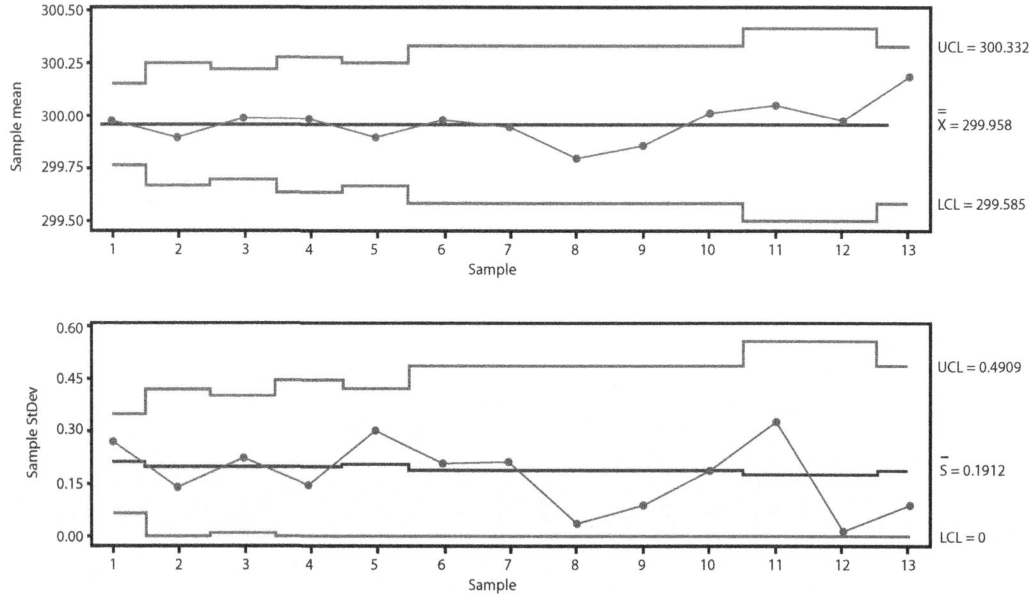

Image 9.5 $\bar{X}-S$ charts in Minitab. Tests are performed with unequal sample sizes. (Source: Author's creation based on Minitab)

$$\bar{s} = \frac{0.207445 + 0.211266 + 0.036056 + 0.088882 + 0.185562 + 0.087369}{6} = 0.065328$$

The UCL and LCL for the most common sample size can be calculated as follows, where $A_3 = 1.95$, $B_3 = 0$, and $B_4 = 2.57$ for $n = 3$.

$$UCL_X = \bar{\bar{X}} + A_3\bar{s} = 299.9585 + 1.95(0.0653) = 300.0858$$

$$LCL_X = \bar{\bar{X}} - A_3\bar{s} = 299.9585 - 1.95(0.0653) = 299.8311$$

$$CL_{\bar{X}} = \bar{\bar{X}} = 299.9585$$

$$UCL_S = B_4\bar{s} = 2.57(0.0653) = 0.1678 \text{ and}$$

$$LCL_S = B_3\bar{s} = 0(0.0653) = 0$$

$$CL_{\bar{s}} = \bar{s} = 0.0653$$

As seen in the data set, the process is statistically out-of-control since batch 13 is located beyond UCL in $\bar{X}$ chart and batches 1, 3, 5, 6, 7, 10, and 11 are located above UCL in S chart.

9.5.3 *X* – *MR* Charts

$X - MR$ charts, also known as *I-MR* charts, are utilized in SPC when each output of the process is required to be inspected, in other words, if the sample size is one ($n = 1$). X charts are also considered simply run charts with control limits in $X - MR$ charts. Where each CTQ characteristic of the outcome is critical in the process, $X - MR$ charts are employed in monitoring individual observations and variation of the process and outcome. X charts are used to detect shift on the observed raw data in time order while MR charts show the differences between current and previous observations of two consecutive parts, (x_i and x_{i-1}). When variation is low in the process, the MR variable is expected to be as low as possible, ideally zero. Especially, aerospace and healthcare industries are some of the industries where $X - MR$ charts are preferable. Automated testing and inspection methods make it easier to collect data from individual items for each CTQ characteristic. Any X or MR variable in out-of-control limits indicates abnormality, which needs to be eliminated in the process.

There are a few critical requirements to consider when drawing $X - MR$ control charts. First, if the data are not normally distributed, the statistical interpretations of data cannot be considered as accurate. Second, time-ordered data are assessed in terms of the independence of data. The data should demonstrate that there is no statistically significant relationship or autocorrelation between individual observations (Mohammed et al. 2008).

MR, as computed in Eq. 9.24, is a better measure of variability. It measures variation from point to point without paying attention to the average. Taking the average of the individual measures and moving ranges, as computed in Eqs. 9.25 and 9.26, allows decision-makers to construct $X - MR$ charts.

$$MR_i = |X_t - X_{t-1}| \tag{9.24}$$

where

- MR_i = moving range of the ith item
- X_t = current measurement
- X_{t-1} = previous measurement

$$\bar{X} = \frac{\sum_{i=1}^{m} x_i}{m} \tag{9.25}$$

where

- m = subgroup size
- x_i = ith measurement
- $\bar{X}$ = average of the measurements

$$\overline{MR} = \frac{\sum_{i=1}^{m} MR_i}{m-1} \tag{9.26}$$

where $\overline{MR}$ is the average of MRs.

The UCLs and LCLs of $X - MR$ charts are calculated as presented in Eqs. 9.27, 9.28, 9.29, and 9.30, respectively.

$$UCL_x = \bar{X} + 2.66\left(\overline{MR}\right) = \bar{X} + 3\frac{\overline{MR}}{d_2} \tag{9.27}$$

$$CL_x = \bar{X}$$

$$LCL_x = \bar{X} - 2.66\left(\overline{MR}\right) = \bar{X} - 3\frac{\overline{MR}}{d_2} \tag{9.28}$$

$$UCL_{MR} = D_4\overline{MR} \tag{9.29}$$

$$CL_{MR} = \overline{MR}$$

$$LCL_{MR} = D_3\overline{MR} \tag{9.30}$$

where

- UCL_X = the upper control limit of X chart
- LCL_X = the lower control limit of X chart
- $\bar{X} = CL_X$ = the average of the samples
- UCL_{MR} = the upper control limit of MR chart
- LCL_{MR} = the lower control limit of MR chart
- $\overline{MR} = CL_R$ = the average of MR variable
- D_4, D_3, and d_2 are the 3-sigma limit coefficients for control charts.

Technically LCL can be below zero in MR charts. In this case, it is rounded up and accepted that LCL can be reset to zero. X charts are generally used with no standard given at the beginning of the process.

► Example 5

At a manufacturing company that produces metal sticks in aerospace industry, quality inspector in assembly line 1 takes measures for the diameter of the consecutive 15 products. Ideally, the products are expected to be in between 11.50 mm and 11.80 mm. The target value is 11.65 mm. Develop control limits of 3 standard deviations for the process. Decide whether the process is statistically in-control, based on the data set. The diameter data collected by the inspector are shown in ◘ Table 9.6. ◄

Table 9.6 Measurements and moving ranges of 15 consecutive products

# of Observation	Measurements of diameter (mm)	Moving Range (MR)
1	11.55	-
2	11.58	0.03
3	11.5	0.08
4	11.55	0.05
5	11.55	0
6	11.6	0.05
7	11.8	0.2
8	11.75	0.05
9	11.7	0.05
10	11.8	0.1
11	11.7	0.1
12	11.65	0.05
13	11.75	0.1
14	11.75	0
15	11.7	0.05
Total	$\sum_{i=1}^{15} x = 174.93$	$\sum_{i=1}^{15} MR = 0.91$
Mean	$\overline{X} = 11.662$	$\overline{MR} = 0.065$

Source: Author's creation

▪ **Solution**

Let's, first, compute mean of individual observations($\overline{X}$) and MR values of the data set as presented in Table 9.6.

$$\overline{X} = \frac{\sum_{i=1}^{15} x_i}{15} = \frac{11.55 + 11.58 + \ldots + 11.7}{15}$$

$$= \frac{174.93}{15} = 11.662\,\text{mm}$$

$$\overline{MR} = \frac{\sum_{i=1}^{m} MR_i}{m-1} = \frac{0.03 + 0.08 + \ldots + 0.05}{14}$$

$$= \frac{0.91}{14} = 0.065\,\text{mm}$$

In $X - MR$ charts, d_2 coefficient is 1.128 for $n = 2$, since there are two variables used in MR, such as x_i and x_{i-1}. D_4 is 3.27 and D_3 is 0 for $n = 2$ as shown in Table A.11.UCLs, LCLs, and CL for X and MR charts are calculated as follows.

$$UCL_x = \overline{X} + 3\frac{\overline{MR}}{d_2} = 11.662 + 3\frac{0.065}{1.128}$$

$$= 11.8349\,\text{mm}$$

$$CL_x = \overline{X} = 11.662\,\text{mm}$$

$$LCL_x = \overline{X} - 3\frac{\overline{MR}}{d_2} = 11.662 - 3\frac{0.065}{1.128}$$

$$= 11.4891\,\text{mm}$$

$$UCL_{MR} = D_4\overline{MR} = 3.27(0.065)$$

$$= 0.2124\,\text{mm}$$

$$LCL_{MR} = D_3\overline{MR} = 0(0.0065) = 0$$

$$CL_{MR} = \overline{MR} = 0.065\,\text{mm}$$

$X - MR$ control charts can be drawn using Minitab or MS Excel. To draw $X - MR$ charts on Minitab, the data set is entered in a Minitab worksheet. Then, click on Stat→Control Charts→Variables Charts for Individuals→I-MR, then transfer data in "Measurements" variable from left box to variables box. Next, click on I-MR options→Tests→Perform all tests for special causes→OK→OK. The $X - MR$ charts are presented in Image 9.6 as Minitab outputs.

As demonstrated in the control charts, multiple pieces are detected as abnormalities in the process. The diameters of the pieces numbered 4, 5, 6, and 8 are not statistically in-control in this example. Minitab outputs also identify which rules are broken by pieces with red colors in the control charts. Pieces numbered 4, 5, and 6 violate test rule number 6, whereas piece numbered 8 violates test rule number 8. As presented in Minitab, test rule 6 is "4 out of 4 + 1 points > 1 standard deviation from center line (same side)." Test rule 8 is "8 points in a row > 1 standard deviation from center line (either side)."

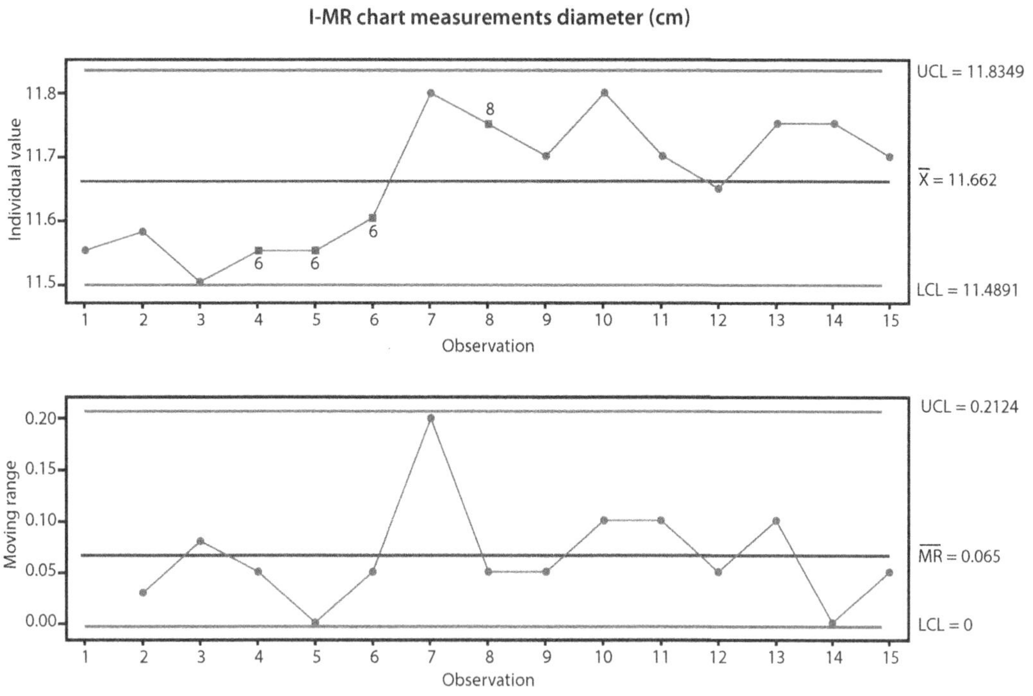

Image 9.6 $X - MR$ control charts in Minitab. (Source: Author's creation based on Minitab)

9.6 Control Charts for Attributes

Control charts for attributes monitor the process and outcomes when the CTQ characteristic is a countable/discrete variable. The number of defectives in a batch or lot, the number of customer complaints in a store, the number of defects of a product/service, and the number of conformities/nonconformities are some of the CTQ characteristic examples for attributes data. CTQ characteristics in control charts for attributes are considered a binary variable, such as one or zero (1–0), since there are only two probabilities in each case, such as, true/false, go/no-go, conformities/nonconformities, or defective/nondefective products. The control charts for attributes can be categorized into two main sections:

1. Control charts for fraction nonconforming
2. Control charts for nonconformities.

Attribute data have just one statistic, the average. Control charts for attributes are useful and beneficial in both manufacturing and non-manufacturing industries and in business processes. Control charts for attributes have several advantages. For example, classifying a product/service as conforming or nonconforming takes into consideration at least one or more CTQ characteristics together. Rather than focusing on individual CTQ characteristics, using a joint approach may ease the decisions for the inspectors and minimize inspection time and cost. Therefore, attribute data can be utilized to decide if the product/service is acceptable. In the following sections, we will analyze each control chart for attribute data.

9.6.1 Control Charts for Fraction Nonconforming

9.6.1.1 *P* Charts

P charts, also known as control charts for fraction nonconforming, are used where the decision-makers monitor the percentage or fraction of detectives per batch/lot. The fraction nonconforming is a ratio of number of nonconforming parts to the total number of parts inspected in a batch/lot. If a part does not conform to the specifications of the relevant CTQ characteristics, it is categorized as "nonconforming." For example, if a manager wants to see the percentage of customer complaints in a time period or a manufacturing engineer needs to determine the percentage of defectives of the can bottles, *P* charts would be the best tool to use in decision-making processes. The attributes data used in *P* charts are assumed to be distributed binomially with parameters *n* and *p* as follows in Eq. 9.31:

$$P\{D=x\}=\binom{n}{x}p^x(1-p)^{n-x} \quad x=0,1,\ldots,n \tag{9.31}$$

where *D* is the number of nonconforming, *p* is the fraction defective/nonconforming, *n* is the sample size, and *x* is the number of nonconforming products found in a random sample of *n*. The mean of the random variable *D* is *np* and the variance is *np(1-p)*. In case fraction nonconforming is unknown, *p* is estimated from the collected data. The average of estimated fraction nonconforming $\hat{p}$ is calculated as presented in Eq. 9.32.

$$\bar{p}=\hat{p}=\frac{\sum_{i=1}^{m} D_i}{nm}=\frac{\sum_{i=1}^{m}\hat{p}_i}{m} \tag{9.32}$$

where

- $\bar{p}$ = the average of the nonconforming of the batches
- $\hat{p}_i$ = the estimated fraction nonconforming of the *i*th batch
- D_i = the number of nonconforming of the *i*th batch
- n = the number of sample per batch
- m = the number of batches.

The fraction nonconforming for the *i*th sample (p_i) is calculated as follows in Eq. 9.33:

$$\hat{p}_i=\frac{D_i}{n} \quad i=1,2,\ldots,m \tag{9.33}$$

The mean and variance of $\hat{p}$ are calculated as follows in Eqs. 9.34 and 9.35:

$$\mu_{\hat{p}}=p \tag{9.34}$$

$$\sigma_{\hat{p}}^2=\frac{p(1-p)}{n} \tag{9.35}$$

As a reminder, the structure of control charts is based on a general model as follows (Eqs. 9.36, 9.37, and 9.38), where *w* is a statistic of CQT, μ_w is the mean of statistic *w*, σ_w is the standard deviation of statistic *w*, *L* is the distance of the control limits from center line:

$$UCL=\mu_w+L\sigma_w \tag{9.36}$$

$$CL=\mu_w \tag{9.37}$$

$$LCL=\mu_w-L\sigma_w \tag{9.38}$$

L is always considered 3 for control limits in control charts. If the sample sizes are constant per batch and true fraction nonconforming is unknown, *UCL, CL,* and *LCL* are calculated as presented in Eqs. 9.39, 9.40, and 9.41, respectively.

$$UCL=\bar{p}+3\sqrt{\frac{\bar{p}(1-\bar{p})}{n}} \tag{9.39}$$

$$CL=\bar{p} \tag{9.40}$$

$$LCL=\bar{p}-3\sqrt{\frac{\bar{p}(1-\bar{p})}{n}} \tag{9.41}$$

where

- *LCL* = the lower control limit
- *UCL* = the upper control limit
- $\bar{p}$ = *CL* = the average of the fraction nonconforming
- n = the sample size per batch.

If the true fraction nonconforming, *p,* is known, *UCL, CL,* and *LCL* are calculated as presented in Eqs. 9.42, 9.43, and 9.44, respectively.

$$UCL=p+3\sqrt{\frac{p(1-p)}{n}} \tag{9.42}$$

$$CL=p \tag{9.43}$$

$$LCL=p-3\sqrt{\frac{p(1-p)}{n}} \tag{9.44}$$

After computing *UCL, CL,* and *LCL*, the fraction nonconforming for each batch (p_i) is plotted on *P* chart and the process is monitored and evaluated through the chart with regard to the rules presented in section "Decision-Making on Control Charts for Variables."

► Example 6

A tire manufacturing company is concerned with the number of defective tires returned by the customers over the last months. To analyze the production processes, 50 randomly selected units per batch were inspected in 30 batches. The numbers of defective tires found in each batch in the inspection process are presented in ◘ Table 9.7. Monitor the fraction nonconforming on the data set through a *P* chart and decide whether the process is statistically in-control in terms of the fraction nonconforming. ◄

▪ Solution

Let's calculate percentage of nonconforming (p_i), in other words, the fraction nonconforming, for each batch. The sample size per batch is $n = 50$. D_i variable represents the number of nonconforming products per batch. For example, for the first batch, percentage of nonconforming ($\hat{p}_1$) is 0.3 as presented in Eq. 9.45. The $\hat{p}_i$ values for the other batches are presented in ◘ Table 9.7.

◘ Table 9.7 The numbers and percentages of nonconforming per batch

Batch number (i)	Di	$\hat{p}_i = \frac{D_i}{n}$	Batch number (i)	Di	$\hat{p}_i = \frac{D_i}{n}$
1	15	0.3	16	9	0.18
2	15	0.3	17	12	0.24
3	12	0.24	18	7	0.14
4	7	0.14	19	8	0.16
5	6	0.12	20	15	0.3
6	1	0.02	21	15	0.3
7	12	0.24	22	18	0.36
8	13	0.26	23	22	0.44
9	4	0.08	24	17	0.34
10	10	0.2	25	10	0.2
11	7	0.14	26	10	0.2
12	7	0.14	27	3	0.06
13	12	0.24	28	15	0.3
14	13	0.26	29	10	0.2
15	10	0.2	30	3	0.06
				$\sum_{i=1}^{m=30} D_i = 318$	$\hat{p} = \overline{p} = 0.212$

Source: Author's creation

$$\hat{p}_1 = \frac{D_i}{n} = \frac{D_1}{n} = \frac{15}{50} = 0.3 \quad (9.45)$$

To calculate the average of estimated fraction nonconforming, $\hat{p}$, for 30 batches and 50 samples per batch, total number of nonconforming products are $\sum_{i=1}^{30} D_i = 318$ and

$$\bar{p} = \hat{p} = \frac{\sum_{i=1}^{m} D_i}{nm} = \frac{\sum_{i=1}^{25} D_i}{nm}$$

$$= \frac{318}{(50)(30)} = \frac{318}{1500} = 0.212 \quad \text{or}$$

$$\bar{p} = \hat{p} = \frac{\sum_{i=1}^{m} \hat{p}_i}{m} = \frac{6.36}{30} = 0.212$$

9

To calculate the *UCL, CL,* and *LCL* based on estimated true fraction nonconforming for *P* chart,

$$UCL = \bar{p} + 3\sqrt{\frac{\bar{p}(1-\bar{p})}{n}}$$

$$= 0.212 + 3\sqrt{\frac{0.212(1-0.212)}{50}}$$

$$= 0.212 + 3(0.0578) = 0.3854$$

$$CL = \bar{p} = 0.212$$

$$LCL = \bar{p} - 3\sqrt{\frac{\bar{p}(1-\bar{p})}{n}}$$

$$= 0.212 - 3\sqrt{\frac{0.212(1-0.212)}{50}}$$

$$= 0.212 - 3(0.0578) = 0.0386$$

After calculating *CL, UCL,* and *LCL*, *P* chart can be drawn as shown in ◘ Image 9.7. The fraction nonconforming products for each batch are represented on *P* chart.

As well as MS Excel, Minitab can also be used to draw *P* charts. To draw *P* charts in Minitab, after transferring data set (◘ Table 9.7) to a Minitab worksheet, click on Stat→Control charts→Attributes charts→p charts, then transfer "number of defectives" data from left box to variables box and enter "50" in the subgroup sizes. Click on p charts options →Tests →Perform all tests for special causes→OK→OK. The *P* chart is presented in ◘ Image 9.7 as Minitab output.

We note that the fractions nonconforming on batch 6 ($\hat{p}_6 = 0.02$)and batch 23 ($\hat{p}_{23} = 0.44$) are located beyond $UCL = 0.3854$ and LCL = 0.0386, respectively. Batches 6 and 23 indicate abnormalities and out-of-control points in terms of fraction nonconforming. The number "1" on batch 6 and batch 23 in ◘ Image 9.7 indicate that these batches violate Rule 1 of control charts. After this diagnostic step, *P* charts will allow decision-makers in Six Sigma projects to detect the causes of abnormalities and out-of-control conditions. In other words, *P* charts enables them to discover assignable or common causes of variation resulting in those conditions.

9.6.1.2 *np* Charts

The *np* charts are utilized in SPC when CTQ characteristic is the number of nonconforming items per batch as an attribute variable. These charts are also known as control charts for number of nonconforming items. Similar to *P* charts, if the true fraction nonconforming is unknown, $\hat{p}$ is used as the unbiased estimator of p as shown in Eq. 9.46.

$$\bar{p} = \hat{p} = \frac{\sum_{i=1}^{m} D_i}{nm} = \frac{\sum_{i=1}^{m} \hat{p}_i}{m} \quad (9.46)$$

where

- D_i = the number of nonconforming of the *i*th batch

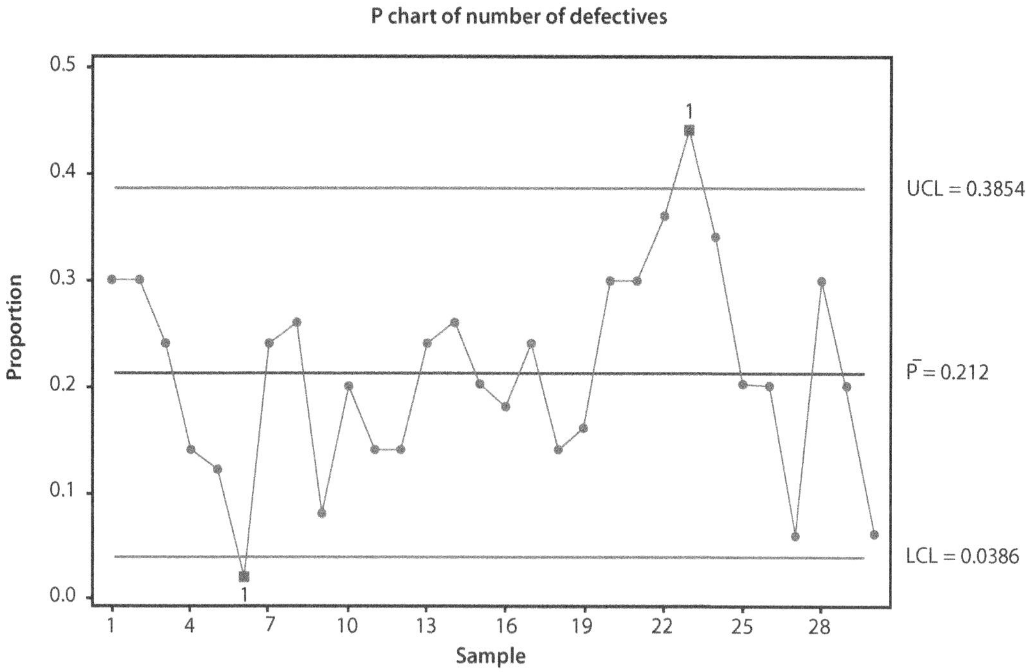

Image 9.7 *P* chart for tire manufacturing firm drawn in Minitab. (Source: Author's creation based on Minitab)

- n = the number of sample
- m = the number of batches.

The *UCL, CL,* and *LCL* are calculated using equations presented in Eqs. 9.47, 9.48, and 9.49, respectively.

$$UCL = np + 3\sqrt{np(1-p)} \quad (9.47)$$

$$CL = np \quad (9.48)$$

$$LCL = np - 3\sqrt{np(1-p)} \quad (9.49)$$

where
- *LCL*= the lower control limit
- *UCL*= the upper control limit
- *CL*=*np*=the average of the number of the nonconforming
- *n*= the sample size.

The *np* charts can be used in two conditions: 1) a constant subgroup size or 2) varying subgroup size. When the subgroup size is constant, *UCL* and *LCL* are calculated as shown above in Eqs. 9.47 and 9.49. When the subgroup size varies from batch to batch, the *np* chart turns into *P* chart, and *UCL* and *LCL* are calculated as shown above in *P* charts in Eqs. 9.42 and 9.44.

▶ Example 7

Using the data set from ▶ Example 6, monitor the number of nonconforming using an *np* chart and decide whether the process is statistically in-control. ◀

Solution

To calculate the average of estimated fraction nonconforming, $\hat{p}$, for 30 batches and 50 samples per batch (subgroup size is constant), total number of nonconforming products is $\sum_{i=1}^{30} D_i = 318$ and

$$\bar{p} = \hat{p} = \frac{\sum_{i=1}^{m} D_i}{nm} = \frac{\sum_{i=1}^{25} D_i}{nm}$$

$$= \frac{318}{(50)(30)} = \frac{318}{1500} = 0.212 \quad \text{or}$$

$$\bar{p} = \hat{p} = \frac{\sum_{i=1}^{m} \hat{p}_i}{m} = \frac{6.36}{30} = 0.212$$

The *UCL, LCL,* and *CL* for *np* chart are

$$UCL = np + 3\sqrt{np(1-p)}$$
$$= 50(0.212) + 3\sqrt{50(0.212)(1-0.212)}$$
$$= 19.2703$$

$$CL = np = 50(0.212) = 10.6$$

$$LCL = np - 3\sqrt{np(1-p)}$$
$$= 50(0.212) - 3\sqrt{50(0.212)(1-0.212)}$$
$$= 1.9296$$

The *np* chart is built as shown in ◘ Image 9.8. Since the sample size per batch is constant ($n = 50$), there will be constant *UCL* and *LCL* for the entire data set. The number of nonconforming products for each batch is represented on the *np* chart along with *UCL, CL,* and *LCL*. Batch 23 with 22 nonconforming products is located beyond $UCL = 19.2703$, and batch 6 with one nonconforming product is below $LCL = 1.9296$. These two batches on the *np* chart shows an out-of-control condition in the process. Similar to our interpretation on *P* chart, Six Sigma project members can analyze out-of-control conditions to identify assignable and/or common causes of the variation that occur in the tire production process.

9

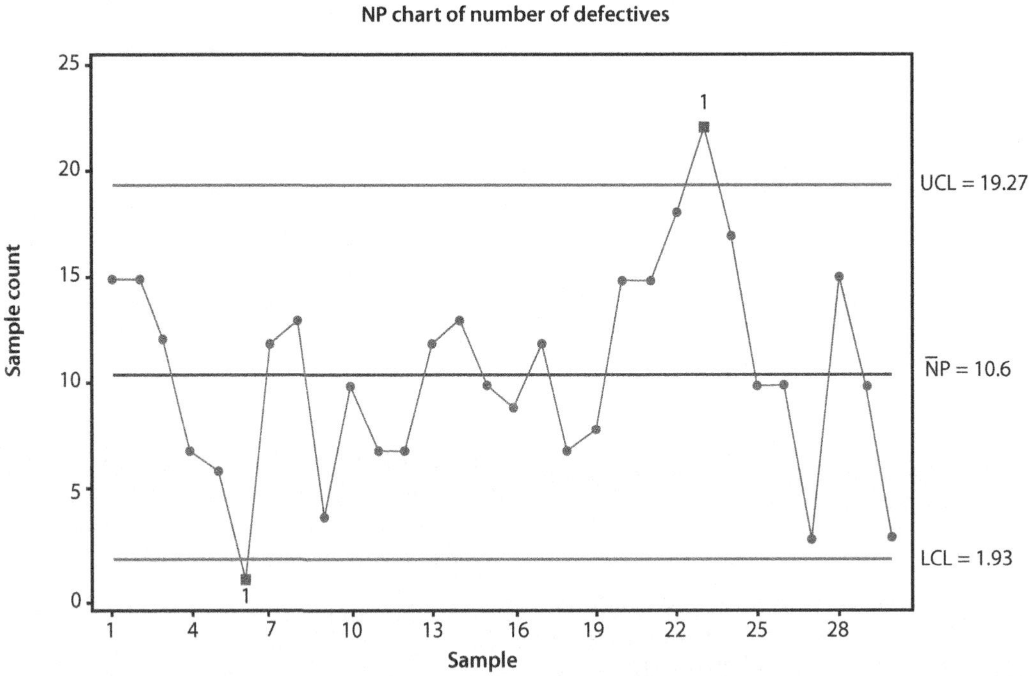

◘ **Image 9.8** *np* chart drawn in Minitab for tire manufacturing firm. (Source: Author's creation based on Minitab)

To draw *np* charts, as well as MS Excel, we can also use Minitab. To draw *np* charts in Minitab, after transferring data set (◘ Table 9.7) to a Minitab worksheet, click on Stat→Control Charts→Attributes Charts→np charts, then transfer "number of defectives" data from left box to variables box and enter "50" in subgroup sizes. Click on np charts options→Tests→Perform all tests for special causes→OK→OK. The *np* chart is presented in ◘ Image 9.8 as Minitab output.

When the sample size per batch (n) is not constant, there are several methods to monitor and evaluate the process through *np* charts. Where the sample size varies, as stated above, the *np* chart automatically turns into *P* chart. In the first approach, *UCL*s and *LCL*s are individually computed for each batch. As a reminder, $\overline{p}$, *UCL, CL,* and *LCL* in *P* charts are calculated as shown in Eqs. 9.50, 9.51, 9.52, and 9.53, respectively.

$$\overline{p} = \frac{\sum_{i=1}^{m} D_i}{\sum_{i=1}^{m} n_i} \tag{9.50}$$

$$UCL = \overline{p} + 3\sqrt{\frac{\overline{p}(1-\overline{p})}{n}} \tag{9.51}$$

$$CL = \overline{p} \tag{9.52}$$

$$LCL = \overline{p} - 3\sqrt{\frac{\overline{p}(1-\overline{p})}{n}} \tag{9.53}$$

where
- D_i = the number of nonconforming of the *i*th batch
- $\overline{p} = CL$ = the average of the fraction nonconforming
- n = the number of sample
- m = the number of batches
- UCL = the upper control limit
- LCL = the lower control limit.

► Example 8 demonstrates how to construct *P* charts when the subgroup size varies.

► Example 8

The tire manufacturing company given in ► Example 6 is still concerned about the number of defective tires returned by the customers. In this analysis, the quality inspectors select varying numbers of samples from each batch, as presented in the first three columns of ◘ Table 9.8. The numbers of defective tires found in the inspection process and the sample size per batch are also presented. Monitor the process using the number of nonconforming products with varying sample sizes. ◄

▪ **Solution**

To solve this problem, we can work on two options. In the first option, we will focus on *P* charts. In the second option, we will take an approximate approach and accept an average of sample size for in *P* charts.

In the first option, although the data set represents the number of nonconforming products per batch, the varying sample size per batch leads the Six Sigma team to use *P* charts. The *np* charts cannot be used since it is not possible to compute a constant *UCL* and *LCL* for the entire data set. Instead, each batch will have different *UCL* and *LCL* depending on the sample size, as shown in ◘ Table 9.8. To calculate the average of fraction nonconforming, we first determine the total number of nonconforming products ($\sum_{i=1}^{30} D_i = 318$) and the total number of products inspected ($\sum_{i=1}^{30} n_i = 1950$).

$$\overline{p} = CL = \frac{\sum_{i=1}^{m} D_i}{\sum_{i=1}^{m} n_i} = \frac{\sum_{i=1}^{30} D_i}{\sum_{i=1}^{30} n_i} = \frac{318}{1950} = 0.163$$

As an exemplary, the *UCL* and *LCL* for the first batch can be calculated as follows:

$$UCL_1 = \overline{p} + 3\sqrt{\frac{\overline{p}(1-\overline{p})}{n_1}}$$
$$= 0.163 + 3\sqrt{\frac{0.163(1-0.163)}{50}} = 0.320$$

$$LCL_1 = \overline{p} - 3\sqrt{\frac{\overline{p}(1-\overline{p})}{n_1}}$$
$$= 0.163 - 3\sqrt{\frac{0.163(1-0.163)}{50}} = 0.006$$

Table 9.8 *UCL*s and *LCL*s for 30 batches in ► Example 8

Batch number (i)	Number of defectives (D_i)	Sample size (n_i)	Sample fraction nonconforming	UCL_i	LCL_i
1	15	50	0.300	0.320	0.006
2	15	60	0.250	0.306	0.020
3	12	50	0.240	0.320	0.006
4	7	80	0.088	0.287	0.039
5	6	80	0.075	0.287	0.039
6	1	70	0.014	0.296	0.031
7	12	70	0.171	0.296	0.031
8	13	60	0.217	0.306	0.020
9	4	60	0.067	0.306	0.020
10	10	70	0.143	0.296	0.031
11	7	80	0.088	0.287	0.039
12	7	80	0.088	0.287	0.039
13	12	50	0.240	0.320	0.006
14	13	50	0.260	0.320	0.006
15	10	50	0.200	0.320	0.006
16	9	60	0.150	0.306	0.020
17	12	60	0.200	0.306	0.020
18	7	60	0.117	0.306	0.020
19	8	60	0.133	0.306	0.020
20	15	70	0.214	0.296	0.031
21	15	80	0.188	0.287	0.039
22	18	70	0.257	0.296	0.031
23	22	80	0.275	0.287	0.039
24	17	60	0.283	0.306	0.020
25	10	60	0.167	0.306	0.020
26	10	60	0.167	0.306	0.020
27	3	60	0.050	0.306	0.020
28	15	70	0.214	0.296	0.031
29	10	70	0.143	0.296	0.031
30	3	70	0.043	0.296	0.031

Source: Author's creation

9

The *UCLs* and *LCLs* for the following batches are computed as presented in the last two columns of ◘ Table 9.8.

P charts can be drawn in MS Excel. To draw *P* charts in Minitab, after transferring data set (◘ Table 9.7) to a Minitab worksheet, click on Stat→Control Charts→Attributes Charts→p charts, then transfer "number of defectives (*Di*)" data from left box to variables box and "Sample size(n_i)" to subgroup sizes. Click on p charts options→Tests→Perform all tests for special causes→OK→OK. The *P* chart is presented in ◘ Image 9.9 as Minitab output.

The main rule in *P* chart for varying sample size is that the fraction nonconforming for each batch should locate in between *UCL* and *LCL* of the batch to consider the process statistically in-control. As presented in ◘ Image 9.9, the *P* chart demonstrates several abnormalities in the fraction nonconforming products. First, batch 6 indicates an out-of-control point since its sample fraction nonconforming ($\hat{p}_6 = 0.014$) is lower than $LCL = 0.031$. The number "1" on batch 6 in ◘ Image 9.9 shows that it violates Rule 1 of control charts. Second, the batches between 21 and 24 show a run with an increasing trend. After diagnosing abnormalities in the control chart, the Six Sigma team should analyze the reasons and root causes for the out-of-control points.

In the second option, an approximate approach is taken for sample size (n), and *P* charts are constructed based on an average sample size. In this case, *UCL* and *LCL* are computed for the overall data and become constant values. For this example, average sample size ($\overline{n}$) and average of fraction nonconforming ($\overline{p}$) are calculated as follows and used to compute *UCL* and *LCL*:

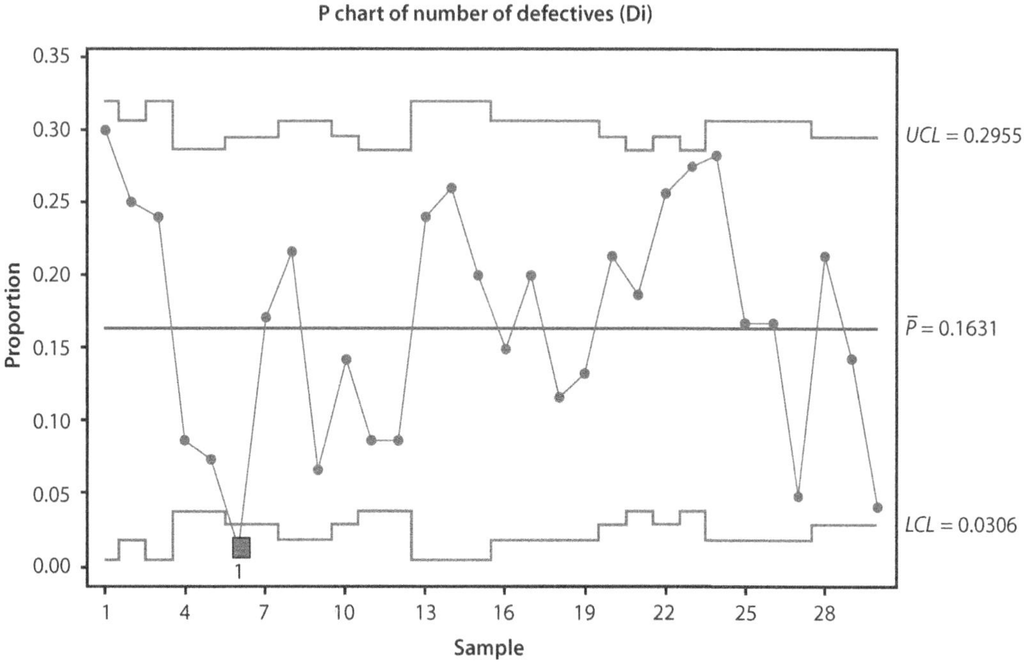

◘ **Image 9.9** *P* chart drawn in Minitab for varying sample size at tire manufacturing firm. Tests are performed with unequal sample sizes. (Source: Author's creation based on Minitab)

$$\bar{n} = \frac{\sum_{i=1}^{m} n_i}{m} = \frac{1950}{30} = 65$$

$$\bar{p} = CL = \frac{\sum_{i=1}^{m} D_i}{\sum_{i=1}^{m} n_i} = \frac{\sum_{i=1}^{30} D_i}{\sum_{i=1}^{30} n_i} = \frac{318}{1950} = 0.163$$

$$UCL = \bar{p} + 3\sqrt{\frac{\bar{p}(1-\bar{p})}{n}}$$
$$= 0.163 + 3\sqrt{\frac{0.163(1-0.163)}{65}} = 0.3005$$

$$LCL = \bar{p} - 3\sqrt{\frac{\bar{p}(1-\bar{p})}{n}}$$
$$= 0.163 - 3\sqrt{\frac{0.163(1-0.163)}{65}} = 0.0256$$

The *P* chart drawn in Minitab based on this approach is presented in ◘ Image 9.10. As seen in the charts, again, the fraction nonconforming in batch 6 is located beyond the *LCL*, causing the process to be in a statistically out-of-control condition.

The main difference is that there are varying *UCL* and *LCL* for each batch in the first approach, whereas the second (approximate) approach generates constant *UCL* and *LCL* for the entire data set. To draw *P* charts in Minitab using this approach, we can follow the same steps given in the discussion of *P* charts. The only difference is that subgroup sizes are entered as "65," in Minitab. The *P* chart is presented in ◘ Image 9.10 as Minitab output.

9.6.2 Control Charts for Nonconformities

A nonconforming product/service may have more than one nonconformity, depending on the number of CTQ characteristics. Each CTQ characteristic specified may result in a nonconformity when the product/service does not meet the specifications of the CTQ characteristics. From that perspective, a nonconforming item may have at least one nonconformity. A product, theoretically, may have as many nonconformities as the number of CTQ characteristics.

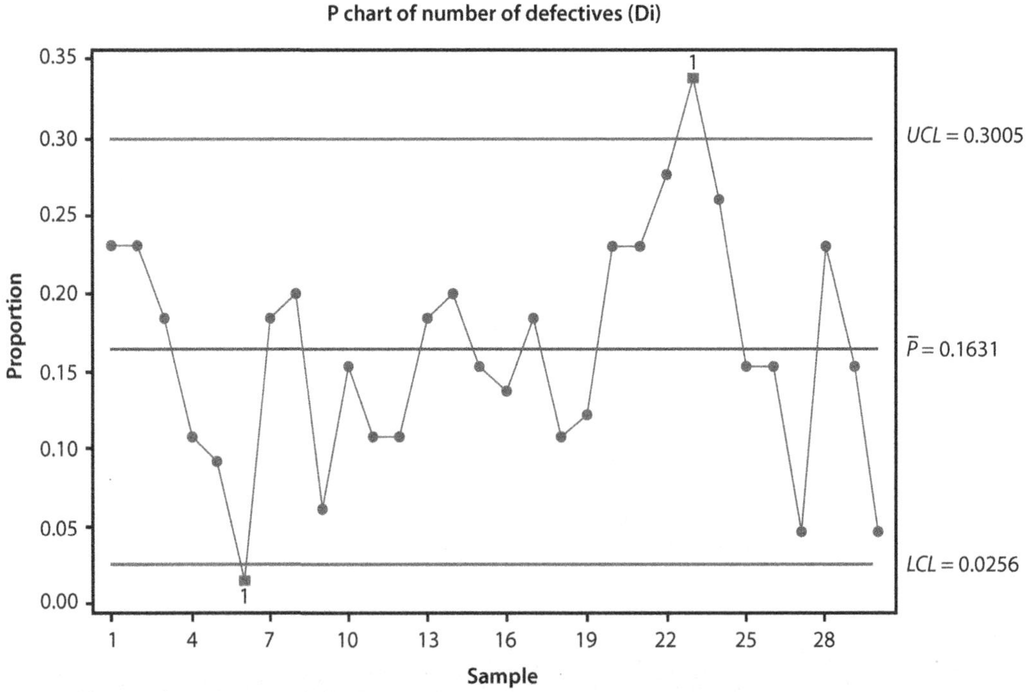

◘ **Image 9.10** *P* chart for average sample size drawn in Minitab. (Source: Author's creation based on Minitab)

Six Sigma teams may develop control charts for (1) the number of nonconformities per unit or (2) the number of nonconformities per batch. Poisson distribution models the occurrence of nonconformities per batch or unit as follows in Eq. 9.54:

$$p(x) = \frac{e^{-c} c^x}{x!} \quad x = 0,1,2,\ldots \tag{9.54}$$

where x is the number of nonconformities and c is the parameter of the Poisson distribution. As a reminder, the mean and variance of the Poisson distribution are c. It is assumed that the inspection unit is the same for each sample. As Montgomery (2013: 318) stated, "each inspection unit must always represent an identical area of opportunity for the occurrence of nonconformities." Montgomery (2013) mentions that Poisson distribution is not the only probability model for the count of nonconformities. When the mean and variance of the counts differ from each other, other types of distributions, such as negative binomial distribution and compound Poisson distribution, should be taken into consideration. Control charts for nonconformities are categorized into two types of charts: c charts and u charts. C charts analyze the number of nonconformities per batch, while u charts represent the number of nonconformities per unit.

9.6.2.1 c Charts

The c charts are implemented in monitoring the process when CTQ characteristic is the number of nonconformities or defects per batch. It is assumed that the data have a Poisson distribution with parameter c and average of nonconformities is calculated as shown in Eq. 9.55.

$$\bar{c} = \frac{\sum_{i=1}^{m} d_i}{m} \tag{9.55}$$

where

- $\bar{c}$ is the estimated average of nonconformities or defects per batch
- d_i is the number of nonconformities or defects per batch
- m is the number of batches.

If parameter c is known, UCL, CL, and LCL are calculated as shown in Eqs. 9.56, 9.57, and 9.58, respectively.

$$UCL = c + 3\sqrt{c} \tag{9.56}$$

$$CL = c \tag{9.57}$$

$$LCL = c - 3\sqrt{c} \tag{9.58}$$

Where there is no standard value for c, UCL, CL, and LCL are calculated as shown in Eqs. 9.59, 9.60, and 9.61, respectively. If the subgroup size is constant, the c chart is used to monitor the process. When subgroup sizes vary, u charts are preferred to monitor attribute variables. If the distribution is too skewed, the usual control charts cannot be used in decision-making processes, making u charts very effective. If UCL and/or LCL are negative values, these values are considered zero.

$$UCL = \bar{c} + 3\sqrt{\bar{c}} \tag{9.59}$$

$$CL = \bar{c} \tag{9.60}$$

$$LCL = \bar{c} - 3\sqrt{\bar{c}} \tag{9.61}$$

where

- LCL = the lower control limit
- UCL = the upper control limit
- $\bar{c}$ = CL = the average of the nonconformities or defects per batch.

► Example 9

A production manager wants to monitor labeling CTQ characteristic in a local beverage bottling company. The labeling CTQ characteristic monitors the number of labels that do not meet the specifications of the designated labels. The production manager collects data from samples of 50 bottles for 20 consecutive days, as presented in ◘ Table 9.9. Construct a c chart and analyze the performance of labeling process using the c chart ◄

Table 9.9 The number of nonconformities for 20 days

Batch number (i)	Number of nonconformities (d_i)
1	13
2	8
3	16
4	9
5	10
6	15
7	17
8	16
9	12
10	11
11	20
12	15
13	12
14	13
15	14
16	17
17	12
18	20
19	18
20	20
	$\sum_{i=1}^{m=20} d_i = 288$

Source: Author's creation

9

Solution

First, we can estimate an average of c, based on the data provided in the question, as follows:

$$\bar{c} = \frac{\sum_{i=1}^{m} d_i}{m} = \frac{\sum_{i=1}^{20} d_i}{20} = \frac{288}{20} = 14.4$$

That means average number of defects per batch is 14.4. Then, we can calculate *UCL*, *CL*, and *LCL* as presented below:

$$UCL = \bar{c} + 3\sqrt{\bar{c}} = 14.4 + 3\sqrt{14.4} = 25.7842$$

$$CL = \bar{c} = 14.4$$

$$LCL = \bar{c} - 3\sqrt{\bar{c}} = 14.4 - 3\sqrt{14.4} = 3.0158$$

After calculating the elements of the c control chart and drawing the chart, as given in Image 9.11, the product manager can monitor and assess the product labeling process using the c chart, and abnormalities and out-of-control conditions may be detected. Looking at the c chart, none of the rules are violated by the process in terms of number of nonconformities per batch. The process can be considered statistically in-control since no abnormalities are detected in the control chart.

To draw c charts in Minitab, after transferring data set (Table 9.9) to a Minitab worksheet, click on Stat→ Control Charts→ Attributes Charts→C charts, then transfer "number of nonconformities (di)" data from left box to variables box. Click on c charts options→Tests→Perform all tests for special causes→OK→OK. The c chart is presented in Image 9.11 as Minitab output.

9.6.2.2 *u* Charts

U charts are convenient to monitor the process when decision-makers detect abnormalities based on the number of defects or nonconformities per unit. In u charts, the number of defects detected (x_i) in a batch is divided by sample size (n) of the relevant batch in order to calculate relevant number of defects per unit (u_i) for each batch (Eq. 9.62). Similar to c charts, u charts do not require minimum or maximum number of observations.

$$u_i = \frac{x_i}{n} \tag{9.62}$$

The average of nonconformities per unit ($\bar{u}$), *UCL*, *CL*, and *LCL* values are calculated as presented in Eqs. 9.63, 9.64, 9.65, and 9.66, respectively.

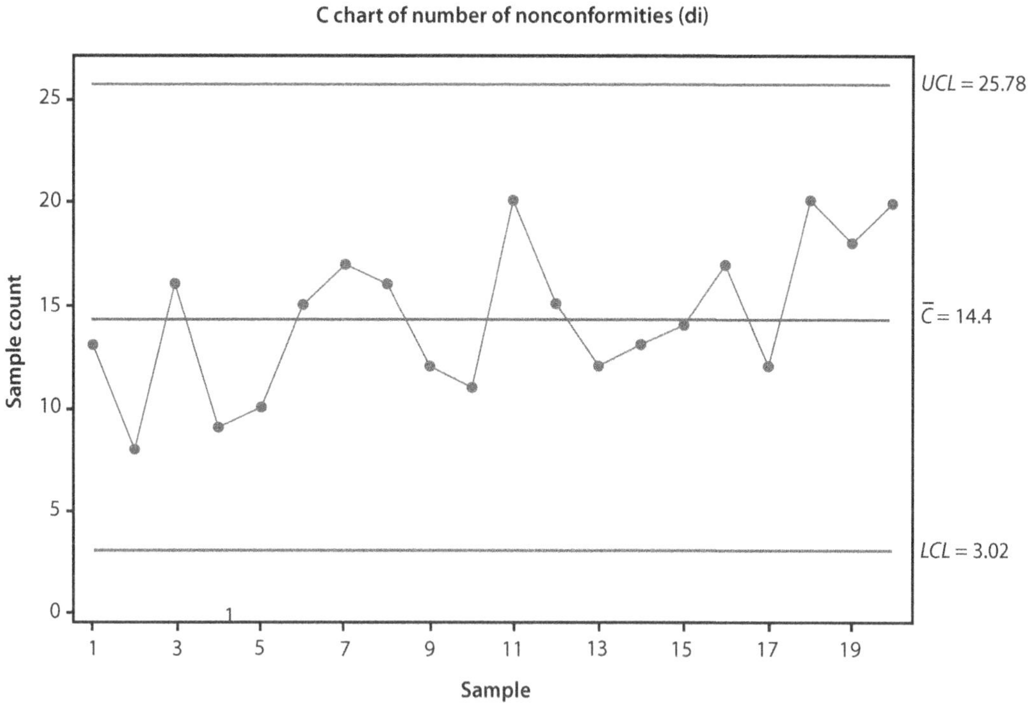

Image 9.11 *C* chart of the labeling process drawn in Minitab. (Source: Author's creation based on Minitab)

$$\bar{u} = \frac{\sum_{i=1}^{m} u_i}{m} \tag{9.63}$$

$$UCL = \bar{u} + 3\sqrt{\bar{u}/n} \tag{9.64}$$

$$CL = \bar{u} \tag{9.65}$$

$$LCL = \bar{u} - 3\sqrt{\bar{u}/n} \tag{9.66}$$

where

- $\bar{u}$ $=CL=$ the average of the nonconformities or defects per unit
- UCL = the upper control limit
- LCL = the lower control limit
- n = the number of sample size
- m = the number of batches.

If the sample size is not constant, UCL and LCL are calculated separately for each batch. For varying sample sizes, u_i is the variable that is in the u chart.

► Example 10

A restaurant manager in a popular tourist town measures customer satisfaction based on the number of customer complaints received in 25 days in June 2019. Each day, the restaurant conducts a survey of 25 customers and tabulates data as presented in the first three columns of Table 9.10. Develop a u chart and analyze the customer complaints using the u chart. ◄

■ Solution

Let's get started answering the question by calculating u_i variable for each day. As presented in Eq. 9.62, the number of customer complaints per customer for the first day (u_1) is

$$u_1 = \frac{x_i}{n} = \frac{3}{25} = 0.12$$

Table 9.10 The data of customer complaints for 25 days

Days (i)	Number of customer complaints (x_i)	Sample size (n_i)	$u_i = x_i/n$
1	3	25	3/25 = 0.120
2	3	25	0.120
3	4	25	0.160
4	4	25	0.160
5	5	25	0.200
6	6	25	0.240
7	1	25	0.040
8	2	25	0.080
9	2	25	0.080
10	5	25	0.200
11	3	25	0.120
12	2	25	0.080
13	4	25	0.160
14	5	25	0.200
15	6	25	0.240
16	7	25	0.280
17	8	25	0.320
18	3	25	0.120
19	5	25	0.200
20	3	25	0.120
21	2	25	0.080
22	3	25	0.120
23	4	25	0.160
24	5	25	0.200
25	3	25	0.120
	$\sum_{i=1}^{m=25} X_i = 98$		$\sum_{i=1}^{m=25} u_i = 3.92$

Source: Author's creation

The number of customer complaints per customer for the following days is presented in the last column of Table 9.10.

Then, we can now calculate the average number of customer complaints by

$$\bar{u} = \frac{\sum_{i=1}^{m} u_i}{m} = \frac{\sum_{i=1}^{25} u_i}{25} = \frac{3.92}{25} = 0.1568$$

Afterward, *UCL, CL,* and *LCL* can be computed for *u* chart as follows:

$$UCL = \bar{u} + 3\sqrt{\bar{u}/n}$$
$$= 0.1568 + 3\sqrt{0.1568/25} = 0.3944$$

$$CL = \bar{u} = 0.1568$$

$$LCL = \bar{u} - 3\sqrt{\bar{u}/n}$$
$$= 0.1568 - 3\sqrt{0.1568/25} = -0.0807 \cong 0$$

After calculating the elements of *u* chart, it can be drawn as shown in Image 9.12 in Minitab. To draw *u* charts in Minitab, after transferring data set (Table 9.10) to a Minitab worksheet, click on Stat→ Control Charts→ Attributes Charts→U charts, then transfer "number of nonconformities (di)" data from left box to variables box and enter "25" for "subgroup sizes". Click on u charts options→Tests→Perform all tests for special causes→OK→OK. The *u* chart is presented in Image 9.12 as Minitab output.

The *u* chart (Image 9.12) shows that the number of customer complaints per day is statistically in-control. However, $u_{17} = 0.32$ is relatively high compared to other days. The process should be analyzed, specifically on day 17 at the restaurant, to determine what is different about the process on that day, i.e., different employees, different food supplies used, weather, cleanliness.

If the decision-makers prefer to use varying sample sizes per unit in the data set, the procedure for constructing the *u* chart significantly changes. In many actual cases, the CTQ characteristics, specifications, standards, and the dynamics of processes may lead decision-makers to monitor the process with varying sample sizes. Similar to *np* charts, if the sample size per unit varies, separate *UCL*s and *LCL*s are computed for each unit. Let's focus on ▶ Example 10 and change the sample size

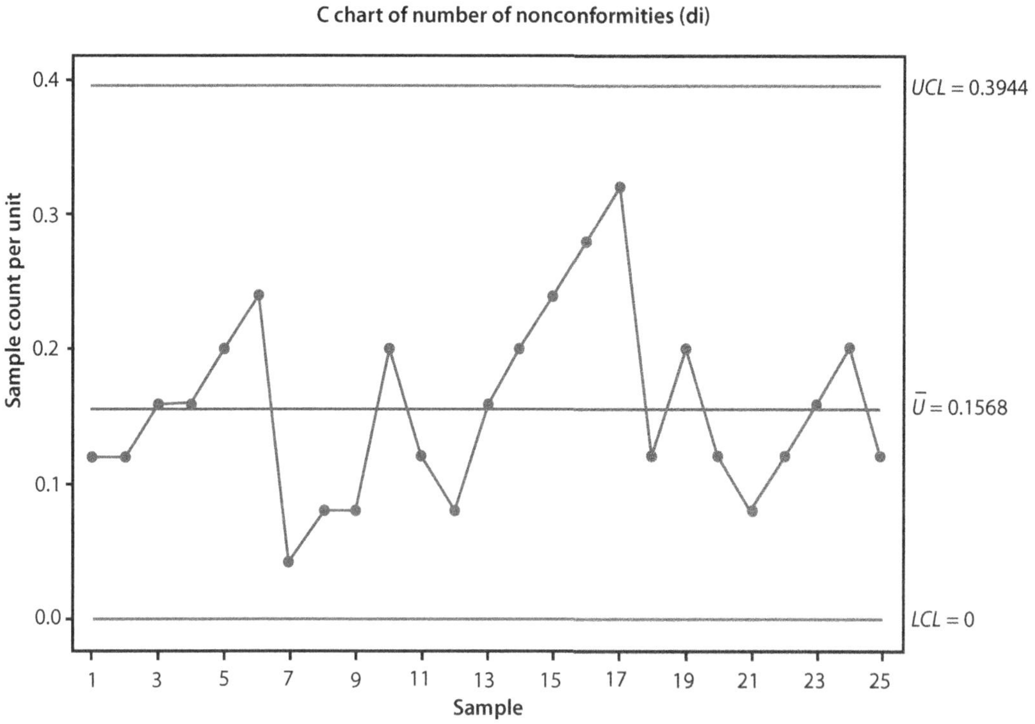

Image 9.12 *U* chart for the number of customer complaints per customer for 25 days drawn in Minitab. (Source: Author's creation based on Minitab)

per day. Then, it is required to compute u_i variable for each day, as presented in Table 9.11. Afterward, *UCL*s and *LCL*s are computed for each day. To exemplify, let's calculate $\bar{u}$, and *UCL* and *LCL* for the first day as follows:

$$\bar{u} = \frac{\sum_{i=1}^{m} u_i}{m} = \frac{\sum_{i=1}^{m} d_i}{\sum_{i=1}^{m} n} = \frac{\sum_{i=1}^{25} d_i}{785} = \frac{98}{785} = 0.1248$$

$$UCL_1 = \bar{u} + 3\sqrt{\bar{u}/n_1} = 0.1248 + 3\sqrt{0.1248/30} = 0.318$$

$$CL = \bar{u} = 0.1248$$

$$LCL_1 = \bar{u} - 3\sqrt{\bar{u}/n_1} = 0.1248 - 3\sqrt{0.1248/30} = -0.068 \cong 0$$

The following days' *UCL*s and *LCL*s are presented in Table 9.11.

As presented in Image 9.13, the *u* chart demonstrates that each day has its own *UCL* and *LCL*, and the process is statistically in-control with regard to the number of customer complaints per day. None of the days violate the rules of control charts.

Table 9.11 Customer complaints for varying sample size for 25 days

Days (i)	Number of customer complaints (x_i)	Sample size (n_i)	$u_i = x_i/n$	LCL_i	UCL_i
1	3	30	0.100	0	0.318
2	3	30	0.100	0	0.318
3	4	20	0.200	0	0.362
4	4	25	0.160	0	0.337
5	5	35	0.143	0	0.304
6	6	40	0.150	0	0.292
7	1	40	0.025	0	0.292
8	2	25	0.080	0	0.337
9	2	35	0.057	0	0.304
10	5	35	0.143	0	0.304
11	3	20	0.150	0	0.362
12	2	25	0.080	0	0.337
13	4	25	0.160	0	0.337
14	5	35	0.143	0	0.304
15	6	35	0.171	0	0.304
16	7	40	0.175	0	0.292
17	8	45	0.178	0	0.283
18	3	45	0.067	0	0.283
19	5	20	0.250	0	0.362
20	3	25	0.120	0	0.337
21	2	25	0.080	0	0.337
22	3	30	0.100	0	0.318
23	4	30	0.133	0	0.318
24	5	35	0.143	0	0.304
25	3	35	0.086	0	0.304

Source: Author's creation

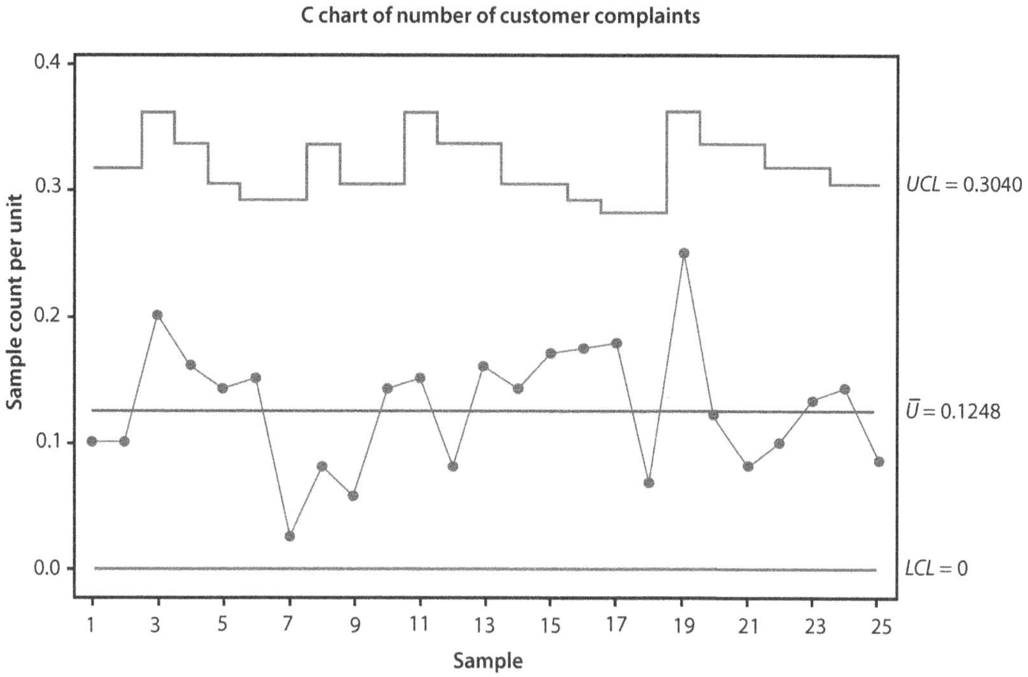

Image 9.13 *U* chart for the number of customer complaints per customer with varying sample size for 25 days drawn in Minitab. Tests are performed with unequal sample sizes. (Source: Author's creation based on Minitab)

Key Concepts

DMAIC, analyze phase, defect, defective, control phase, 3σ control limits, 2σ warning limits, upper control limit (UCL), central line (CL), lower control limit (LCL), Zone A, Zone B, Zone C, non-random patterns, abnormality, out-of-control process, in-control process, subgroup size, type I error, type II error, control charts for variables, control charts for attributes, measure of central tendency, measure of variability, $\bar{X}-R$ charts, $\bar{X}-S$ charts, $I - MR$ charts, P charts, np charts, u charts, and c charts, normal distribution, Poisson distribution, binomial distribution, lower natural tolerance limit ($LNTL$), upper natural tolerance limit ($UNTL$), average, range, standard deviation, moving range, control charts for fraction nonconforming, control charts for number nonconforming, control charts for nonconformities, the fraction of defectives per batch/lot, the number of defectives per batch/lot, the number of defects per unit, the number of defects per batch/lot, the number of nonconforming, the fraction defective/nonconforming, the estimated average of nonconformities or defects per batch, the number of nonconformities or defects per batch, and the number of defects or nonconformities per unit.

Summary

A control chart-based analysis is required to monitor the process and eliminate assignable causes of variation for in the "Analyze" phase of DMAIC. Either common or assignable causes result in non-random patterns, abnormalities, and out-of-control conditions that indicate improvement needs in the process. This chapter analyzes control charts for variables and control charts for attributes.

Practice and Discussion Questions

1. What are the main components of a control chart?
2. Where is the ideal place of *CL* against *LCL* and *UCL* in control charts?
3. Explain what variables are represented on *X*-*Y* axes of control charts.
4. Discuss what variables are used in the general framework to calculate *UCL, CL,* and *LCL* in control charts.
5. What is the distance of *UCL* or *LCL* from *CL* in control charts?
6. What is the distance of warning limits from CL in control charts?
7. Discuss Zones A, B, and C in control charts. What areas are covered in each zone?
8. How do you interpret the signals of control charts when at least one point is out of the control limits?
9. What is the ideal sub-group size in control charts?
10. Statistically, what hypothesis is tested using control charts?
11. What are the two main groups of control charts?
12. List the types of the control charts for variables.
13. List the types of the control charts for attributes.
14. Explain what in-control and out-of-control conditions mean in control charts.
15. What are the two statistics monitored in $\overline{X}-R$ charts?
16. What are the two statistics monitored in $\overline{X}-S$ charts?
17. What are the two statistics monitored in $\overline{I}-MR$ charts?
18. What is the statistic monitored in *P* charts?
19. What is the statistic monitored in *np* charts?
20. What is the statistic monitored in *u* charts?
21. What is the statistic monitored in *c* charts?
22. List and discuss the rules of control charts.
23. What are the unbiased estimators of σ when *R* and/or *s* are known?
24. When the CTQ characteristic is a variable and sample taken from each batch is a constant value and greater than one, what control chart can be used for process monitoring?
25. When the CTQ characteristic is a variable and sample taken from each batch is a constant value and equals one, what control chart can be used for process monitoring?
26. When the CTQ characteristic is a variable and sample taken from each batch is a constant value and greater than 10, what control chart can be used for process monitoring?
27. When the CTQ characteristic is a variable and sample taken from each batch is not a constant value, what control chart can be used for process monitoring?
28. When the sample taken from each batch is a constant/varying value and CTQ characteristic is the number nonconforming (defectives), what control chart can be used for process monitoring?
29. When the sample taken from each batch is a constant/varying value and CTQ characteristic is the fraction nonconforming (defectives), what control chart can be used for process monitoring?
30. When the sample taken from each batch is a constant/varying value and CTQ characteristic is the number of nonconformities (defects) per batch/lot, what control chart can be used for process monitoring?
31. When the sample taken from each batch is a constant/varying value and CTQ characteristic is the number of nonconformities (defects) per unit, what control chart can be used for process monitoring?
32. After diagnosing out-of-control points on control charts, what do you expect from Six Sigma teams?
33. In control charts for variables, what type of distribution is assumed that the data was distributed?
34. In $\overline{X}$ charts, when the data are normally distributed, what percent of the averages does fall within ± 3-sigma on the distribution?
35. To reduce manual calculations when sample size is not a constant value in $\overline{X}-S$ charts, what approach would you suggest to researchers to calculate control limits?
36. What actions should be taken in control charts for variables if the data are not normally distributed?

37. When the sample size per batch/lot is not a constant, what approaches can be taken to monitor the process using fraction or number of nonconforming in control charts for attributes?
38. In control charts for attributes, what type of distribution is assumed that the data was distributed?

References

Benneyan, J. C. (2008). The design, selection, and performance of statistical control charts for healthcare process improvement. *International Journal of Six Sigma and Competitive Advantage, 4*(3), 209–239.

Benneyan, J. C., Lloyd, R. C., & Plsek, P. E. (2003). Statistical process control as a tool for research and healthcare improvement. *BMJ Quality & Safety, 12*(6), 458–464.

Carey, R. G. (2002). How do you know that your care is improving? Part II: Using control charts to learn from your data. *The Journal of Ambulatory Care Management, 25*(2), 78–88.

Carey, R. G., & Stake, L. V. (2001). *Improving healthcare with control charts: Basic and advanced SPC methods and case studies*. Milwaukee: ASQ Quality Press.

Hart, M. K., & Hart, R. F. (2002). *Statistical process control for health care*. Duxbury: Thomson Learning.

Koetsier, A. V. d. V., Jager, S. N. K. J., Peek, N., & de Keizer, N. F. (2012). *Methods of Information in Medicine, 51*(3), 189–198.

Matthes, N., Ogunbo, S., Pennington, G., Wood, N., Hart, M. K., & Hart, R. F. (2007). Statistical process control for hospitals: Methodology, user education, and challenges. *Quality Management in Healthcare, 16*(3), 205–214.

Mohammed, M. A., Worthington, P., & Woodall, W. H. (2008). Plotting basic control charts: Tutorial notes for healthcare practitioners. *BMJ Quality & Safety, 17*(2), 137–145.

Montgomery, D. C. (2005). *Introduction to statistical quality control* (5th ed.). New York: Wiley.

Montgomery, D. C. (2013). *Introduction to statistical quality*. Wiley. NYC.

Western Electric Company. (1956). *Statistical quality control handbook*. Indianapolis: Western Electric Company.

Improve Phase: I Is for Improve

Contents

F. Pakdil, *Six Sigma for Students*, https://doi.org/10.1007/978-3-030-40709-4_10

Learning Objectives

After careful study of this chapter, you should be able to:

- Explain the Improve phase of DMAIC
- Understand the basics of the Improve phase
- Understand the basic concepts of Design of Experiment (DOE)
- Explain how DOE is used in the Improve phase
- Conduct 2^k factorial designs and interpret the effect of interactions
- Conduct response surface designs and interpret the effect of interactions
- Understand the basic concepts of simulation
- Explain how simulation is used in the Improve phase
- Understand the basic concepts of lean
- Understand the basic concepts of Failure Modes and Effects analysis (FMEA).

10.1 Introduction

Improving the processes and systems is the ultimate goal of Six Sigma projects. Along with all activities done in the previous phases of DMAIC process, the Improve phase aims to identify ways to improve the outcomes of the process and system and minimize the variation throughout the system. In other words, the Improve phase aims for the identification and development of multiple alternatives for increasing performance and for selecting and implementing best alternative/s for improvement.

In the Improve phase, based on what has been discussed so far in previous phases of the DMAIC (Define-Measure-Analyze) process, the Six Sigma team focuses on specific changes that may have the desired impacts on the relevant processes by redesigning the process, eliminating NVA activities and wastes, and testing them using such methods as simulation, optimization, Design of Experiment (DOE), lean implementation, and Failure Modes and Effects Analysis (FMEA).

The Improve phase includes the following tasks:

1. Develop alternative solutions to improve sigma level of the CTQ characteristics.
2. Select the optimal solution/s.
3. Map the new potential process based on the selected solution/s (future-state map).
4. Analyze the potential risks in the new process.
5. Pilot test the new process by collecting data and analyzing the data and process.
6. Implement the new process based on Go and No-Go decision.

- Step 1. Develop alternative solutions to improve sigma level of the CTQ characteristics.
  First and foremost, the Improve phase includes developing alternative solutions for the CTQ characteristics and conditions analyzed in the Six Sigma team. Alternative solutions are developed either by using advanced methods or by doing small-scale modifications. Typically, the levels of resources, such as materials, equipment, workforce, technology, processes, and work instructions, are modified and changed to develop alternative solutions. In this step, all inputs and transformation processes are considered to make those changes and modifications. Considering the previous outcomes, performance indicators should be identified to quantify expected change on performance. The Improve phase answers this question: What changes should be done to processes, products, or systems for improvement? Technical and technological advancements and developments, technology transfer, and breakthrough improvements that occur within or outside of the industry may also be considered in this step. The outcomes of this step impact the effectiveness of the Improve phase. If alternative solutions do not solve the problem or achieve the goal identified in the project, the next steps of the Improve phase may not flow as intended.

- Step 2: Select the optimal solution/s.
  After completing step 1, the Six Sigma team analyzes alternative solutions and focuses on determining the optimal solution/s for the problem under consideration. *Optimal solution* refers to the most effective, efficient, feasible, and acceptable solution, considering the decision-making criteria, constraints, resources, and capabilities of the organization. In this step, the Six Sigma team may utilize qualitative and/or quantitative tools. For example, optimization, simulation, DOE, lean principles, and FMEA are some of the methods used in this step and are discussed in the next sections in this chapter. These methods help Six Sigma teams analyze the technical feasibility of the alternatives. Additionally, alternative solutions may provide different types of benefits and necessitate different resources. While analyzing alternative solutions, their *costs and benefits* are calculated and considered in the Improve phase. The types of costs and benefits of the alternative solutions may vary depending on the scope of the Six Sigma project. An evaluation matrix shown in ◘ Table 10.1 is used for this analysis. If the potential benefits exceed the resources that need to be utilized, this alternative can be considered by the Six Sigma team for the next evaluations. If the resources do not produce expected benefits, this alternative should be excluded from alternative solutions list. To identify the potential benefits in the Improve phase, the improvement goal/s developed in the project charter in the Define phase are helpful. The theoretical best solutions may not always be the optimal solution for the organization due to various factors such as organizational culture, constraints, available resources, and capabilities.

  After analyzing technical feasibility, costs, and benefits of the alternative solutions, if the alternative solution contains high benefit and low cost, that means these solutions may be implemented with "Go" decision. If the solution has high cost and brings low benefit, this alternative is not considered for implementation and should be given a "No-Go" decision. A cost-benefit matrix can be used in this step.

  As stated by Montgomery (2013), there are two questions that determine what type of improvement methodologies would be more convenient to evaluate improvement activities. These questions are (1) is the process statistically in-control? and (2) is the process capable? These questions are asked in the Analyze phase and identify the direction of the Improve phase of DMAIC process. If the two questions are answered with "Yes," the best method for improvement activities is SPC. If one of the two questions is answered with "No," either SPC or DOE is the best alternative to shape the improvement activities in Six Sigma projects.

◘ **Table 10.1** Evaluation of costs and benefits of alternative solutions

	Costs		**Benefits**	
	Implementation cost	**Maintenance cost**	**Return on investment**	**Improvement on CTQ characteristic**
Alternative 1	$130,000	$25,000	100 days	20%
Alternative 2	$120,000	$12,000	90 days	25%
Alternative 3	$130,000	$5,000	60 days	15%
Alternative 4	$150,000	$25,000	180 days	25%

Source: Author's creation

After developing alternative solutions and analyzing the technical aspects of these solutions through simulation, optimization, DOE, and FMEA, Six Sigma teams may want to take a multi-criteria decision (MCDM) approach to select the optimal solution. Due to the structure of the Improve phase, selecting the optimal solution among all potential solutions is an MCDM problem. Given the complex nature of Six Sigma projects, MCDM methods with multiple objectives may help increase the probability of selecting correct solutions. MCDM methods are utilized when critical decisions cannot be made based upon one dimension. All potential solutions cannot be implemented in the same time period, and limited resources cannot be equally allocated to potential solutions. Prioritizing and selecting appropriate solutions and allocating resources to the correct solution are the two critical success factors in the Improve phase of the process. Having a systematic optimal solution selection method in place impacts the allocation of limited resources of the organization in the direction aligned with the strategic direction and competitive advantages of the organization. Poor optimal solution selection processes result in wasting resources, decreasing organizational performance, diminishing belief in the benefits of Six Sigma, and lowering the long-term success of Six Sigma efforts. When the number of factors and potential solutions that are prioritized and selected increases, the optimal solution selection problem becomes much more complex. In this case, running this process without systematic and well-structured methods may put the process at risk. The Analytic Hierarchy Process (AHP), goal programming, Analytic Network Process (ANP), Delphi, Decision-Making Trial and Evaluation Laboratory (DEMATEL), linear and nonlinear programming, fuzzy logic, project prioritization matrix, FMEA, and hybrid methods are used for selection optimal solutions. Combining multiple decision-making approaches and integrating them in a unique hybrid methodology may help decision-makers prioritize and select optimal solutions, depending on the complexity of the organizational structure, market dynamics, resource availability, and potential solution scopes.

- Step 3: Map the new potential process based on the selected solution/s (future-state map).
  After selecting the optimal solution/s in step 2, the future-state map is developed, based on the outcomes of the optimal solution/s by the Six Sigma team. Using process mapping tools, such as value stream maps (VSM) or process flow charts, the Six Sigma team integrates the chosen alternative solution into the existing process or system. Process mapping tools were previously discussed in ▶ Chap. 4. Using process mapping tools, the team can also determine the benefits and savings to be gained after the implementation of the optimal solution/s. Total lead time, VAT, NVAT, and other similar performance indicators are analyzed and compared on current-state and future-state maps. Planning some changes and integrating those changes into real processes may not always have the expected result. Therefore, drawing future-state maps shows Six Sigma teams potential benefits and obstacles. An example of a future-state map is shown in ◘ Image 10.1. The flow of a CT scan laboratory at a university hospital after implementing changes in the process is demonstrated in the VSM.
- Step 4: Analyze the potential risks in the new process.
  By using process mapping tools with basic or advanced risk analysis tools, the Six Sigma team also analyzes potential risks of the optimal solution/s in the future-state map. Some potential risks may not be easily discovered in the previous cost and benefit analysis. After mapping the future state, potential technical issues or risks are identified by the team. If there are no risks anticipated, the optimal solution is ready for the pilot test.

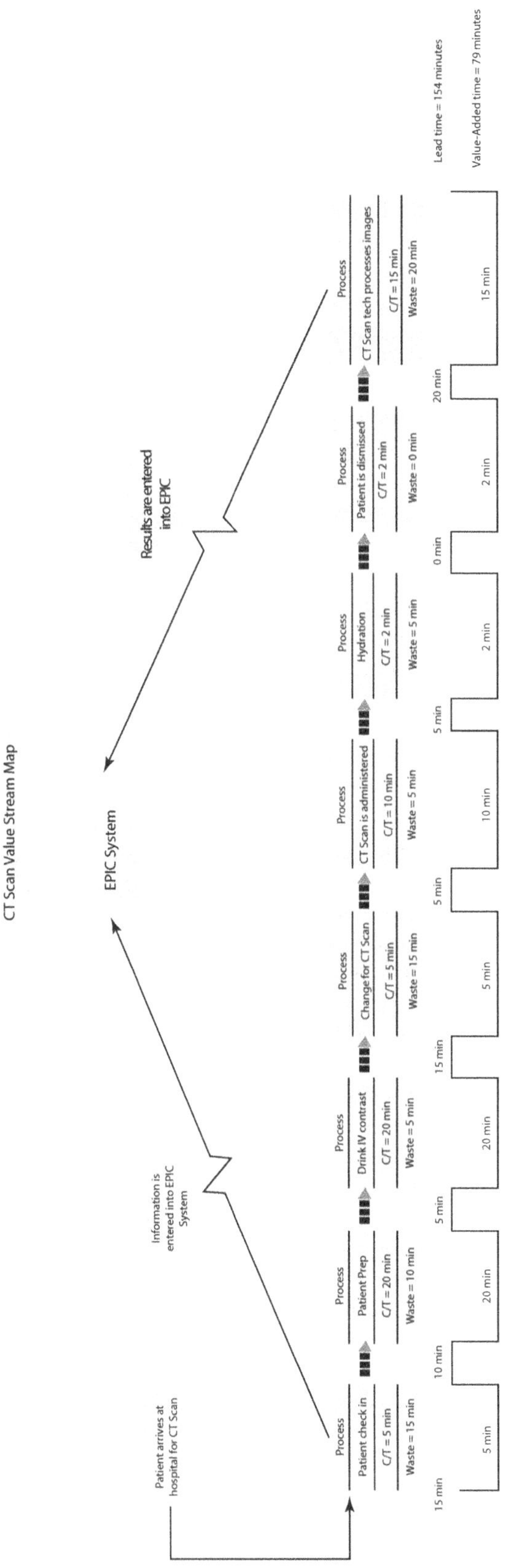

Image 10.1 An example of a future-state map. (Source: Author's creation)

- Step 5: Pilot test the new process by collecting data and analyzing the data and process. It is beneficial to pilot test the selected optimal solution/s as much as the system allows. A pilot test allows the Six Sigma team to see how the implementation of optimal solution/s works and what outputs and values are generated through optimal solution/s. The pilot test allows Six Sigma teams to decide whether the proposed optimal solutions are the best solutions for the system analyzed. Simulation or DOE results may not always anticipate some of the technical issues or obstacles that may arise during implementation. Pilot test results help decision-makers with Go and No-Go decisions.
- Step 6: Implement the new process based on Go and No-Go decision.
  Based on the findings from the previous steps, the Six Sigma team can decide whether these optimum solutions will be implemented in the system in the Improve phase. If a "Go" decision is made by the team, the optimal solution is implemented in the process.
  Implementation of the optimal solution/s is a unique task, not only for the Six Sigma team but also for the relevant departments, functions, and employees. As part of the Improve phase, the optimal solution/s is presented to higher-level decision-makers for approval. If the decision-makers are being continuously informed during the project period, they will be aware of the performance and expected outcomes of the Six Sigma team. Therefore, keeping decision-makers in the communication loop during Six Sigma projects is a critical success factor. Additionally, implementing the optimum solution in the system may start a change management process throughout the organization. Organizational development and change, although a major topic analyzed in the organizational behavior discipline, is not within the scope of this textbook and will not be discussed here.

The next sections will discuss DOE, simulation, lean principles, and FMEA, respectively. These methods are helpful while developing alternative solutions in the Improve phase of DMAIC. In addition, Six Sigma teams may consider other basic or advanced qualitative and quantitative methods to integrate into the Improve phase especially while developing alternative solutions. For example, in service organizations, queuing theory and methods may be helpful to develop improvement ways for optimizing time-oriented CTQ characteristics such as wait time, order fulfill time, total lead time, and mean time between arrivals. Advanced decision-making methods can also be utilized in the Improve phase.

10.2 Experimental Design – Design of Experiment (DOE)

Quality control activities are classified into two types: (1) *online* quality control and (2) *off-line* quality control. *Off-line* quality control section involves hearing the voice of customer from the marketplace; learning customer needs and expectations, product and process design, and development processes; determining product/process specifications; and quality improvement activities. DOE is a critical part of off-line quality control activities. DOE aims to maximize the performance of the product or service to meet customers' needs and expectations. It allows the analysis of the relationships between independent variables and their interactions on the dependent variables. DOE also helps reduce the variability in manufacturing and service delivery processes. DOE helps identify the best levels of the design characteristics to maximize the performance of the product, service, or processes. Various design characteristics are tested at various levels to identify the optimum levels of the design characteristics. DOE has a critical role in DMAIC process, specifically in the Analyze and Improve phases. Experiments are performed in manufacturing and service organizations at various levels to understand the behavior of processes, products, and services, which, in turn, allows variation to be minimized and performance to be maximized.

As presented in ISO 3534-3 standard (p. vii), "Design of Experiments (DOE) catalyzes innovation, problem solving and discovery. DOE comprises a strategy and a body of methods that are instrumental in achieving quality improvement in products, services and processes. Although statistical quality control, management resolve, inspection and other quality tools also serve this goal, experimental design represents the methodology of choice in complex, variable and interactive settings."

DOE was developed by Ronald Fisher at the Rothamsted Agricultural Field Research Station in London in the 1920s. In his experiments, Fisher investigated the effects of various fertilizers on crops in multiple lands. He discovered that the performance of the crop was dependent on a group of variables such as type of soil, soil condition, level of moisture, and type and amount of fertilizers. After Fisher's experiments, biological and agricultural disciplines implemented DOE in research and development. Since then, DOE has been used in a great variety of industries such as aerospace, electronics, automotive, chemical, and pharmaceutical industries.

DOE is an effective component of continuous quality improvement efforts. For example, if control charts in SPC detect any abnormalities, DOE is used to identify what factors result in that out-of-control condition. Since detailing DOE is beyond the scope of this chapter, DOE is presented at introductory level in this chapter.

DOE assumes that Six Sigma team already developed alternative solutions to be considered in Improve phase. Process changes and alternative solutions are already planned, and data collection plans are developed prior to DOE. A DOE study needs to identify some variables at the planning stage. These variables are:

1. Factors
2. Levels of the factors
3. Response variable (Y variable)
4. Number of experiments (runs)
5. Treatments
6. Repetitions
7. Experiment conditions.

Factors are the independent variables that are expected to create variability and affect the performance of the product or process. Each process or output has multiple factors. The factors are also called inputs and categorized into controllable and uncontrollable factors. Uncontrollable factors are also known as noise factors. For example, oven temperature; amounts of egg, sugar; and flour; and baking time are factors in a cake baking process (◘ Fig. 10.1). *Levels of factors* are the experiment levels chosen for each factor. For example, Six Sigma team may run the experiments with the following factors and levels:

- Oven temperature with two levels: 350 °F(level 1) and 380 °F (level 2)
- Amount of egg with two levels: two eggs (level 1) and three eggs (level 2)
- Amount of sugar with two levels: 1 cup (level 1) and 1 ½ cups (level 2)

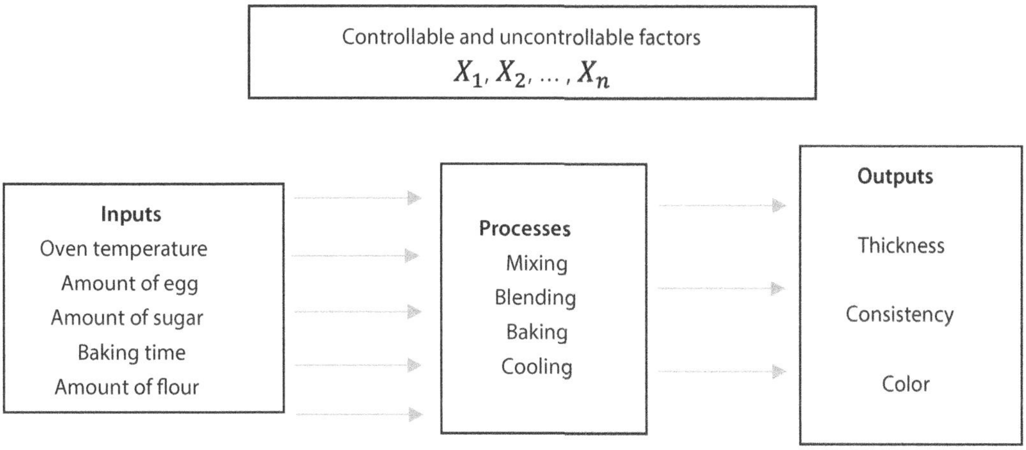

◘ **Fig. 10.1** Illustration of cake baking process. (Source: Author's creation)

- Baking time with two levels: 25 minutes (level 1) and 30 minutes (level 2)
- Amount of flour with two levels: 1 cup (level 1) and 1 ¼ cups (level 2).

A *response variable* is the one that needs to be optimized. For the cake example, the thickness, consistency, and color of the cake may be used as response variables. One product may have multiple response variables. They are also called outputs or CTQ characteristic. Before designing the experiments, important factors, appropriate number of levels, and units of measurements should be identified by the Six Sigma team. Although not known completely at the beginning of DOE, as the studies go on, the details are clarified and determined in some cases. *The number of experiments*, in other words *runs*, shows how many experiments will be conducted in DOE. The budget, time, features of the product, and DOE method selected determine the number of experiments. If the components of the product, or the product itself, are destroyed in experiments, the number of the experiments may be, by necessity, limited. The method chosen by the Six Sigma team is a critical factor for identifying the number of experiments as well. The *treatment* represents combination of levels of factors in each run. The number of factors and levels determine the size of the treatment in experiment. For example, if the experiment contains two factors and two levels for each factor, the number of treatments is four. *Repetition* refers to multiple runs of the same treatment. The details of the methods are discussed in the following sections.

10.2.1 DOE Steps

To conduct a DOE study, Six Sigma teams usually follow these steps:

- Step 1. Thoroughly analyze and understand the process and product/service.
  Prior to starting DOE studies, Six Sigma teams need to understand the structure of the process and product/service in depth. Understanding the process or product allows team members to identify significant response variables and the factors affecting those response variables. A good understanding of process and product enables an effective DOE.
- Step 2: Identify the goals and objectives of the experiment.
  The goals and objectives of DOE should be determined by Six Sigma teams to decide what DOE method/s would be more practical and valuable. After analyzing the product/process and project charter, Six Sigma team can identifies the goals and objectives of DOE. Generally, process optimization, product optimization, and identifying the optimal levels of factors are the main goals of DOE.
- Step 3: Identify the response variable/s.
  Depending on the structure and design of the product or process, Six Sigma team will identify what response variables (dependent variables) will be measured and optimized in DOE. As well as single response variable, multiple response variables are simultaneously utilized in DOE. For example, tensile strength of a metal cable or fabric, reaction or response time of a chemical test, cycle time of a process, and thickness of a cake may be utilized as a response variable in DOE.
- Step 4: Identify the factors affecting the response variable/s.
  In this step, design outcomes, VSM, bills of material, fishbone diagrams, affinity diagrams, and product trees may be used by Six Sigma teams to identify the factors affecting the response variables. The technical details of a product or process will help identify these factors in this step. For instance, the factors affecting "tensile strength of metal cable" may be temperature, operation time, pressure, and angle of cut in a manufacturing setting. The number of factors is pivotal for identifying DOE method.
- Step 5: Identify the levels of the factors.
  In this step, the Six Sigma team identifies the levels of the factors. Each factor may be applied at different values and levels on the product or process. As exemplified above, in the metal cable manufacturing process, temperature may vary between two levels, such as 350 °F and 380 °F. In operation

time, three different time periods, such as 10 minutes, 12 minutes, and 14 minutes, may be tested in the experiments.

- Step 6: Determine DOE method.
  The numbers of factors, levels of factors, and response variables identified above help determine what DOE method/s are more useful, beneficial, and practical for Six Sigma teams. Single factor experiments, full factorial experiments, fractional factorial experiments, response surface, and Taguchi designs are considered in this step.
  While planning experiments, potential biases that may stem from the conditions of the experiments must be minimized. Randomization, blocking, and replication minimize those biases. Some specific blocking strategies are randomized block designs, Latin square designs, and balanced incomplete block designs. Additionally, mixture designs, nested designs, graphical methods, and regression analysis are used in DOE.
- Step 7: Identify the number of experiments or runs.
  The number of experiments or runs is determined by the DOE method selected, availability of the resources, and the likelihood of discarding the tested resources. If the Six Sigma team has limited budget and time, some methods that need more experiments may not be considered by the team.
- Step 8: Identify the structure of the experiments.
  The technical aspect of the product or process, the numbers of factors and levels, the type of DOE method, and the complexity of the experiments assist in identifying the structure of the experiments. Decisions about which experiments to run will be affected by whether they are technically detailed, time-consuming, and costly. A well-designed experiment in DOE should be simple and time-, and cost-effective, although this is dependent on the product, service, or process being studied.
- Step 9: Conduct experiments.
  The experiments are conducted as required by the DOE method in this step.
- Step 10: Collect data.
  A well-structured data collection manual and sheets are needed in this step. The quality of data collection and keeping records is likely to affect the results of DOE. Clear steps are identified, and the experiment results are clearly recorded.
- Step 11: Analyze data.
  As detailed in the next sections, the data regarding factors, levels, and response variables are analyzed manually or by using statistical software packages. Minitab and JMP are the two software packages that are usually selected by Six Sigma teams.
- Step 12: Determine significant factors and levels on response variable/s.
  Based on the DOE results, Six Sigma teams will identify statistically significant factors and the levels for the response variable/s. This step will generate the most important information for the team. After identifying significant factors and levels, Six Sigma teams may consider modifying the design of the product or process.

10.2.2 DOE Methods

In this section, several DOE methods are discussed. These methods are single factor experiments, two-factor factorial designs, full factorial experiments, fractional factorial experiments, screening experiments, and response surface design methods.

10.2.2.1 Single Factor Experiments

If the team prefers to focus on one factor of the response variable and keep the others fixed in the experiments, they use single factor experiments. Using one factor in experiments takes the approach called One-variable-at-a-Time. The single factor experiment isolates its impact on the response variable in the experiments by focusing on the effects of one factor. Single factor experiments may have some drawbacks. There is a need for a high number of repetitions of the experiments, and it ignores the interactions between various factors that might create different effects. In manufacturing and service settings, it is more likely that more than one factor would impact the response variable/s.

▶ **Example 1**

A Six Sigma team studies the effects of temperature, operation time, pressure, and angle of cut on "tensile strength of metal cable" in a manufacturing process. It is assumed that each factor is applied at two levels in the manufacturing process as follows:

- Temperature: 510 °F (level 1) and 550 °F (level 2)
- Operation time: 125 seconds (level 1) and 128 seconds (level 2)
- Pressure: 300 Pa (level 1) and 400 Pa (level 2)
- Angel of cut: 0° (level 1) and 2° (level 2).

Create an experiment plan for the first level of the first factor using the single factor experiment method. ◀

■ **Solution**

In the example, the response variable is tensile strength of metal cable. Using the single factor experiment method, in each trial, only one of the factors will change each time. The other factors and their levels will alter in each experiment. Let's use temperature set at 510 °F as one factor and list all possible combinations of the experiments in a decision tree, as shown in ◘ Fig. 10.2. As seen in the experiment plan, only one factor is focused upon, and the others are ignored in the single factor experiment. (To take all factors and levels into account, a full factorial experiment needs to be used in DOE.) Keeping the temperature factor constant at 510 °F, we create experiments that include other factors: operation time, pressure, and angle of cut.

10.2.2.2 Two-Factor Factorial Designs

Two-factor factorial experiments are also known as 2^2 factorial designs, since they assess the effects of two factors on the response variables. In this section, we only focus on situations that have equal numbers of repetitions for each levels of the factors. When the experiment has two factors and two levels, total variation (SST) is generated by three factors: factor A, factor B, and the interactions of factors A and B. Total variation (SST) is comprised of sum of squares among groups (SSA) and sum of squares within groups (SSW) as presented in Eq. 10.1.

$$SST = SSA + SSW \quad (10.1)$$

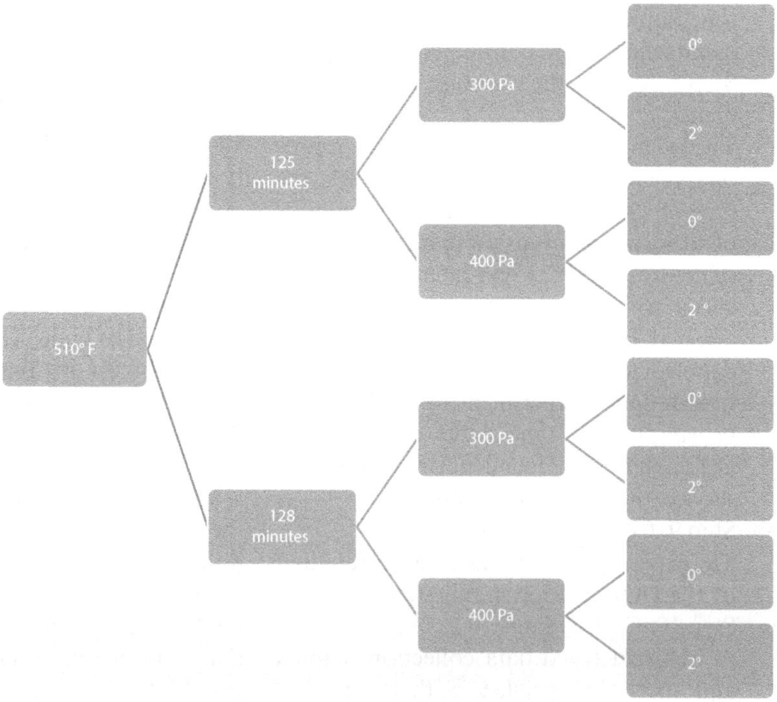

◘ **Fig. 10.2** Decision tree of experiment plan for level 1 (510° F) of temperature factor in ▶ Example 1. (Source: Author's creation)

In two-factor factorial designs, the total variation (SST) is also calculated using Eq. 10.2, where SSA is sum of squares of factor A, SSB is sum of squares of factor B, SSAB is sum of squares of interaction of A and B, and SSE is sum of squares of random error.

$$\mathrm{SST} = \mathrm{SSA} + \mathrm{SSB} + \mathrm{SSAB} + \mathrm{SSE} \tag{10.2}$$

SSA calculates the differences among the various mean levels of factor A and overall mean of the response variable (Eq. 10.3). SSB computes the differences among the various mean levels of factor B and overall mean of the response variable (Eq. 10.4). SSAB represents the combined impact of factor A and factor B on response variable (Eq. 10.5). Finally, SSE refers to the differences among the individual observations of response variable. After calculating these variables, we can calculate degrees of freedom, variances, and F test statistics of those variables as presented in ◘ Table 10.2, where r is number of levels of factor A, c is the number of levels of factor B, k is the number of replications for each experiment, and n is the total number of observations in the experiment.

$$\mathrm{SSA} = \frac{\left[a + ab - b - (1)\right]^2}{4n} \tag{10.3}$$

$$\mathrm{SSB} = \frac{\left[b + ab - a - (1)\right]^2}{4n} \tag{10.4}$$

$$\mathrm{SSAB} = \frac{\left[ab + (1) - a - b\right]^2}{4n} \tag{10.5}$$

In two-factor factorial designs, the levels of factors in runs are shown in a design matrix by 2 * 2 using signs as low (−) and high (+) (◘ Table 10.3). Runs are shown in the rows in ◘ Table 10.3. For example, the first run with notation *(1)* shows that factors A and B are at the low levels (−). The second run with notation a shows that factor A is at the high level (+) and factor B is at the low level (−). The signs of the runs in the column called AB show the product of signs from columns A and B.

The main effects of factors in a two-factor factorial design are (1) the effect of factor A, (2) the effect of factor B, and (3) the effect of the interactions of factor A and factor B. The main effects of factor A, factor B, and interaction of A and B are calculated using Eqs. 10.6, 10.7, and 10.8, where n is the number of runs, *(1)* is the sum of the measurements of the response variable in the first run, a is the sum of the measurements of the response variable in the second

◘ **Table 10.2** The framework of Analysis of Variance (ANOVA) for two-factor factorial design

Source of variation	Degree of freedom	Sum of squares	Variance (mean square)	*F* test statistic
Factor A	$r-1$	SSA	$\mathrm{MSA} = \frac{\mathrm{SSA}}{r-1}$	$F = \frac{\mathrm{MSA}}{\mathrm{MSE}}$
Factor B	$c-1$	SSB	$\mathrm{MSB} = \frac{\mathrm{SSB}}{c-1}$	$F = \frac{\mathrm{MSB}}{\mathrm{MSE}}$
Interaction of A and B	$(r-1)(c-1)$	SSAB	$\mathrm{MSAB} = \frac{\mathrm{SSAB}}{(r-1)(c-1)}$	$F = \frac{\mathrm{MSAB}}{\mathrm{MSE}}$
Error	$rc(k-1)$	SSE	$\mathrm{MSE} = \frac{\mathrm{SSE}}{rc(k-1)}$	
Total	$n-1$	SST		

Source: Adapted from Montgomery (2013)

Table 10.3 Signs of the levels in the 2^2 factor design

Runs	Notations	Factors A	B	AB
1	(1)	-	-	+
2	a	+	-	-
3	b	-	+	-
4	ab	+	+	+

Source: Author's creation

run, *b* is the sum of the measurements of the response variable in third run, *ab* is the sum of the measurements of the response variable in the fourth run.

$$A = \frac{1}{2n}\left(a + ab - b - (1)\right) \quad (10.6)$$

$$B = \frac{1}{2n}\left(b + ab - a - (1)\right) \quad (10.7)$$

$$AB = \frac{1}{2n}\left(ab + (1) - a - b\right) \quad (10.8)$$

In two-way ANOVA, three tests of hypotheses are conducted as follows:

1. In the first test of hypothesis, the effect of factor A is investigated.

   $H_o : \mu_1 = \mu_2$
   (Means of factor A equal means of response variable.)

   $H_1 : \mu_1 \neq \mu_2$
   (Means of factor A do not equal means of response variable.)

2. In the second test of hypothesis, the effect of factor B is investigated.

   $H_o : \mu_1 = \mu_2$
   (Means of factor B equal means of response variable.)

   $H_1 : \mu_1 \neq \mu_2$
   (Means of factor B do not equal means of response variable.)

3. In the third test of hypothesis, the effect of the interaction of factors A and B is investigated.

   $H_o : \mu_1 = \mu_2$
   (Means of interaction of factors A and B equal means of response variable.)

   $H_1 : \mu_1 \neq \mu_2$
   (Means of interaction of factors A and B do not equal means of response variable.)

► Example 2

A Six Sigma team in a manufacturing firm wants to use DOE to see the impacts of "temperature" and "pressure" on the "tensile strength" of the metal cables. Two levels are considered for each factor. Temperature will be tested at 510 °F and 550 °F, and pressure will be applied at 300 Pa and 400 Pa. Using a 2^2 factorial design, four tests are conducted, and each test is repeated four times. The tensile strength measurements are shown in Table 10.4. Analyze the manufacturing process based on the data set and develop improvements for this process. ◄

Solution

The design used in this example is a $2 * 2$ design or 2^2 design with four replications, where the exponent number represents the number of levels and the base number represents the number of factors. The data presented in Table 10.4 are converted into the structure that represents each variable in a single column for temperature, pressure, and tensile strength, respectively.

Since we have two factors involved in this example, let's use two-way ANOVA in our analysis. To run two-way ANOVA in Minitab, after transferring data set (Table 10.4) to a Minitab worksheet, click on Stat→ANOVA→General

Table 10.4 Data collected on tensile strength in Example 2

Factors	Notations	Temperature (° F)	Pressure (Pa)	Tensile strength (psi)				Total (psi)
Runs				1	2	3	4	
1	(1)	–	–	135	125	127	130	517
2	a	+	–	125	122	120	118	485
3	b	–	+	140	142	141	140	563
4	ab	+	+	138	139	137	135	549

Source: Author's creation

Linear Model→Fit General Linear Model. In the input screen, first select "tensile strength" for "responses" box, and then select "temperature" and "pressure" variables for "factors" box. Click on OK. The ANOVA results will appear in the report section in Minitab as represented in Table 10.5.

As presented in the "factor information" section in Minitab output in Table 10.5, temperature and pressure variables have two levels: 510 °F and 550 °F and 300 Pa and 400 Pa, respectively. Let's interpret findings of "Analysis of variance" in Table 10.5 and test our hypotheses. In two-way ANOVA, three hypotheses are tested as follows.

- **Test for Temperature**

In the first test of hypothesis, we will analyze the effect of factor of temperature on tensile strength.

$H_o : \mu_1 = \mu_2$
(Means of temperature equal means of tensile strength.)

$H_1 : \mu_1 \neq \mu_2$
(Means of temperature do not equal means of tensile strength.)

The F statistic for temperature is 16.71 and the p-value is 0.002. Since $\alpha > p$ (0.05>0.002), therefore, the null hypothesis is rejected, meaning that the means are not equal, and there is statistically significant evidence of the effect of temperature on the tensile strength.

- **Test for Pressure**

In the second test of hypothesis, we will analyze the effect of factor of pressure on tensile strength.

$H_o : \mu_1 = \mu_2$
(Means of pressure equal means of tensile strength.)

$H_1 : \mu_1 \neq \mu_2$
(Means of pressure do not equal means of tensile strength.)

The F statistic for temperature is 95.53 and the p-value is 0.000. Since $\alpha > p$ (0.05>0.000), the null hypothesis is rejected, meaning that the means are not equal, and there is statistically significant evidence of effect of pressure on the tensile strength.

- **Test for Interaction of Temperature and Pressure**

In the third test of hypothesis, we will analyze the effect of the interaction of two factors, temperature and pressure, on tensile strength.

$H_o : \mu_1 = \mu_2$

(Means of interaction of temperature and pressure equal means of tensile strength.)

$H_1 : \mu_1 \neq \mu_2$

(Means of interaction of temperature and pressure do not equal means of tensile strength.)

Table 10.5 Two-way ANOVA output of Minitab in Example 2

General Linear Model: Tensile Strength Versus Temperature, Pressure					
Method					
Factor coding (−1, 0, +1)					
Factor information					
Factor	**Type**	**Levels**	**Values**		
Temperature	Fixed	2	510, 550		
Pressure	Fixed	2	300, 400		
Analysis of variance					
Source	**DF**	**Adj SS**	**Adj MS**	***F*-value**	***P*-value**
Temperature	1	132.25	132.250	16.71	0.002
Pressure	1	756.25	756.250	95.53	0.000
Temperature*pressure	1	20.25	20.250	2.56	0.136
Error	12	95.00	7.917		
Total	15	1003.75			
Model summary					
S / **R-sq**	**R-sq(adj)**	**R-sq(pred)**			
2.81366 / 90.54%	88.17%	83.17%			
Coefficients					
Term	**Coef**	**SE Coef**	***T*-value**	***P*-value**	**VIF**
Constant	132.125	0.703	187.83	0.000	
Temperature 510	2.875	0.703	4.09	0.002	1.00
Pressure 300	−6.875	0.703	−9.77	0.000	1.00
Temperature*pressure 510 300	1.125	0.703	1.60	0.136	1.00
Regression equation					
Tensile strength	= 132.125 + 2.875 temperature_510 – 2.875 temperature_550 – 6.875 pressure_300 + 6.875 pressure_400 + 1.125 temperature*pressure_510 300 – 1.125 temperature*pressure_510 400 – 1.125 temperature*pressure_550 300 + 1.125 temperature*pressure_550 400				
Fits and diagnostics for unusual observations					
Term	**Coef**	**SE coef**	***T*-value**	***P*-value**	
1	135.00	129.25	5.75	2.36	R
R Large residual					
Source: Author's creation based on Minitab					

10

The F statistic for the interaction of temperature and pressure is 2.56 and the p-value is 0.136. Since $\alpha < p$ ($0.05<0.136$), we fail to reject null hypothesis. There is no statistically significant evidence to accept the alternative hypothesis. In other words, there is no sufficient evidence of the effect of the interaction between temperature and pressure on the tensile strength.

Using regression analysis, "Model Summary," "Coefficients," and "Regression Equation" sections in Minitab output (◘ Table 10.5) present how well the regression model generated through the analysis fits the data. As the measure that shows how close the data are to the regression line, R^2-adjusted= 88.17% shows that the regression model fits the data and that 88.17% of the variation in the tensile strength is explained by the interaction of "temperature" and "pressure" factors. The regression equation generated through ANOVA is shown in Eq. 10.9. Since the interactions of factor A and factor B are not statistically significant, we will not keep the interaction factors in the regression equation.

$$\begin{aligned}\text{Tensile strength} &= 132.125 + 2.875\\ &\text{Temperature\_510} - 2.875\\ &\text{Temperature\_550} - 6.875\\ &\text{Pressure\_300} + 6.875\,\text{Pressure\_400}\end{aligned} \quad (10.9)$$

The main effects plot in ◘ Image 10.2 demonstrates the means of tensile strength for two levels of pressure and temperature. When the temperature is 510 °F and the pressure is 400 Pa, the mean of the tensile strength is higher. When the temperature is 550 °F and the pressure is 300 Pa, the mean of the tensile strength is lower. As presented in ◘ Image 10.2, the mean of tensile strength is higher when the temperature is 510 °F and pressure is 400 Pa. One of the important findings of DOE is that the decision-makers should keep the temperature at 510 °F and pressure at 400 Pa to maximize the level of tensile strength.

After analyzing the effects of factors (temperature and pressure) on the tensile strength, we can also study the interaction between factors. The interaction between factors is demonstrated in an interaction

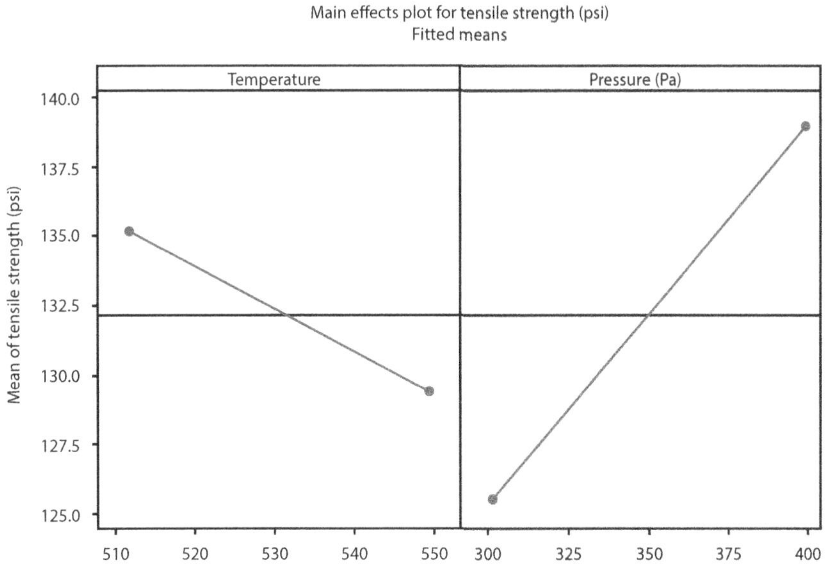

◘ **Image 10.2** Main effects plot for tensile strength. (Source: Author's creation based on Minitab)

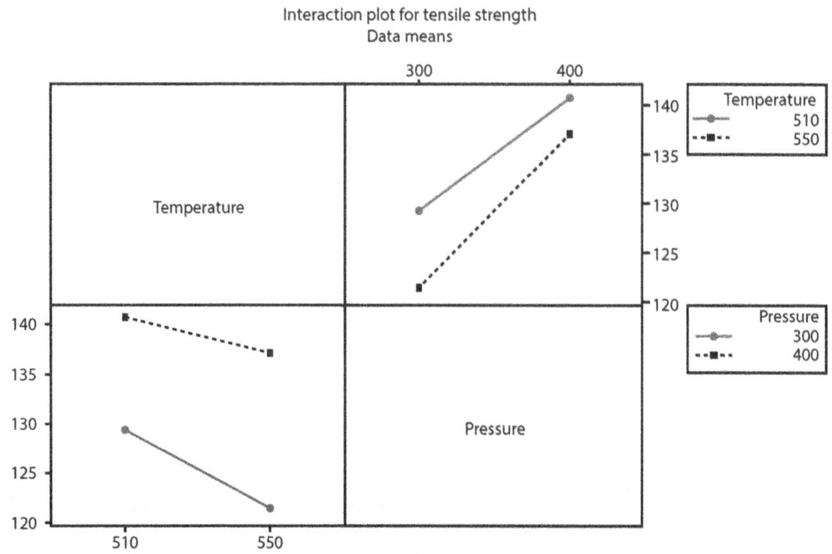

▣ **Image 10.3** Interaction plot for tensile strength. (Source: Author's creation based on Minitab)

10

plot (▣ Image 10.3) in Minitab. If there is not any interaction between the factors (temperature and pressure), the lines are parallel. ▣ Image 10.3 shows that, when temperature is at 510 °F or pressure is at 400 Pa, the mean of the tensile strength increases. Also, the interaction plot presents the results for temperature in the upper right diagram and for pressure in the lower left diagram. In both diagrams, it is seen that the lines are roughly parallel, which shows that the interaction between temperature and pressure is not significant enough to affect the tensile strength. That finding is also supported with the *p*-value (0.136) of the hypothesis testing presented above for the interaction of these two factors.

To analyze the factor effects and their interaction, let's recall the individual observations of tensile strength with four repetitions in each experiment shown in ▣ Table 10.4. After assigning notations of *(1)*, *a*, *b*, and *ab* to each combination of experiment, we can calculate the mean effect for each factor and interaction as presented in Eqs. 10.6, 10.7 and 10.8 above for two-factor designs.

$$A = \frac{1}{2(4)}(485 + 549 - 563 - 517) = -5.75$$

$$B = \frac{1}{2(4)}(563 + 549 - 485 - 517) = 13.75$$

$$AB = \frac{1}{2(4)}(549 + 517 - 485 - 563) = 2.25$$

Variable A shows that the mean difference in tensile strength for temperature 550 °F is 5.75 psi less than the mean for temperature 510 °F. In other words, when we change temperature from low level to high level, the average tensile strength will decrease by 5.75 psi. Variable B shows that when we change pressure from low level to the high level, the mean tensile strength increases by 13.75 psi. We can confirm the significance of these effects with the analysis of variance, in ▣ Table 10.5 above.

The ANOVA results show that the "temperature" and "pressure" factors are statistically significant but their interactions are not (p = 0.136). Therefore, variable AB in Eq. 10.3 is not statistically significant. If the interaction between two factors was statistically significant, AB = 2.25 would mean that the mean effect of combining temperature and pressure would increase tensile strength 2.25 psi. The interaction effect demonstrates the combined effects of factors on the response variable. When an interaction effect is statistically significant in ANOVA, the impact of one factor depends on the level of the other factor. When the interaction lines in the "interaction plot" do not cross, that is called ordinal interaction. If the lines are not parallel in an ordinal interaction, the interaction effect is considered significant. If the lines are parallel, there is no interaction effect of factors on the response variable, as shown in ◘ Image 10.3.

10.2.2.3 Full Factorial Experiments

A full factorial experiment includes all factors and levels of the experiment in the analysis. All possible combinations of factors, levels, and interactions are mapped. To calculate the potential number of combinations, the number of levels for each factor is multiplied by the number of factors. For instance, if we have four factors and two levels, the total number of combinations is $2 * 2 * 2 * 2$. We can also show it in an exponential notation as $2^4 = 16$. We can generalize the calculations, as shown in Eq. 10.10.

$$\prod f^{l} = n \quad (10.10)$$

where f is the number of levels, l is the number of factors, and n is the number of combinations in full factorial experiment. A full factorial experiment requires time and resources to run all of the necessary experiments. If the Six Sigma team has enough time and budget to work on DOE, a full factorial experiment is implemented in the team's work plan. Using a full factorial experiment, the factors and levels that optimize the response variable are determined.

▶ Example 3

Let's focus on the same scenario given in ▶ Example 1. The Six Sigma team studies the effects of (1) temperature, (2) operation time, (3) pressure, and (4) angle of cut on "tensile strength of metal cable" in a manufacturing process. It is assumed that each factor has two levels in DOE. Create an experiment plan using all factors and levels in full factorial experiment method. ◀

- **Solution**

Similar to our solution in ▶ Example 1, we will use the same factors and levels of factors as shown in ◘ Fig. 10.3. The response variable is tensile strength of metal cable. Using full factorial experiment method, let's list all possible combinations of the experiments in a decision tree as shown in ◘ Fig. 10.3. The experiment includes 16 runs, eight for 510 °F and eight for 550 °F.

$$\prod f^{l} = n \Rightarrow 2^4 = 16$$

▶ Example 4

The Six Sigma team uses the same case given in the previous example. Tensile strength variable is measured on 16 pieces, as reported in ◘ Fig. 10.3. Using the data, run a full factorial design in Minitab and interpret your findings. ◀

- **Solution**

Since we have multiple factors involved in this example, let's use factorial regression analysis in our analysis. To run factorial regression analysis in Minitab, after transferring data set to a Minitab worksheet, click on Stat→DOE→Factorial→Analyze Factorial Design. In the next screen, click on Yes to enter factors and response variable manually. Then, in the input screen, first select four factors from left box for "factors" box, and then click on "Low/High..." button. Mark "uncoded" option at the bottom for "worksheet data are" section since we use raw data collected. The levels of factors will appear on the right screen. Click on OK and OK. In the next screen, select "tensile strength" from left box to "response" box. Click on "terms" and exclude "ABCD" from the "selected terms" list to prevent degrees of freedom being zero in the analysis. Click on OK. Click on OK and OK. The ANOVA results will appear in

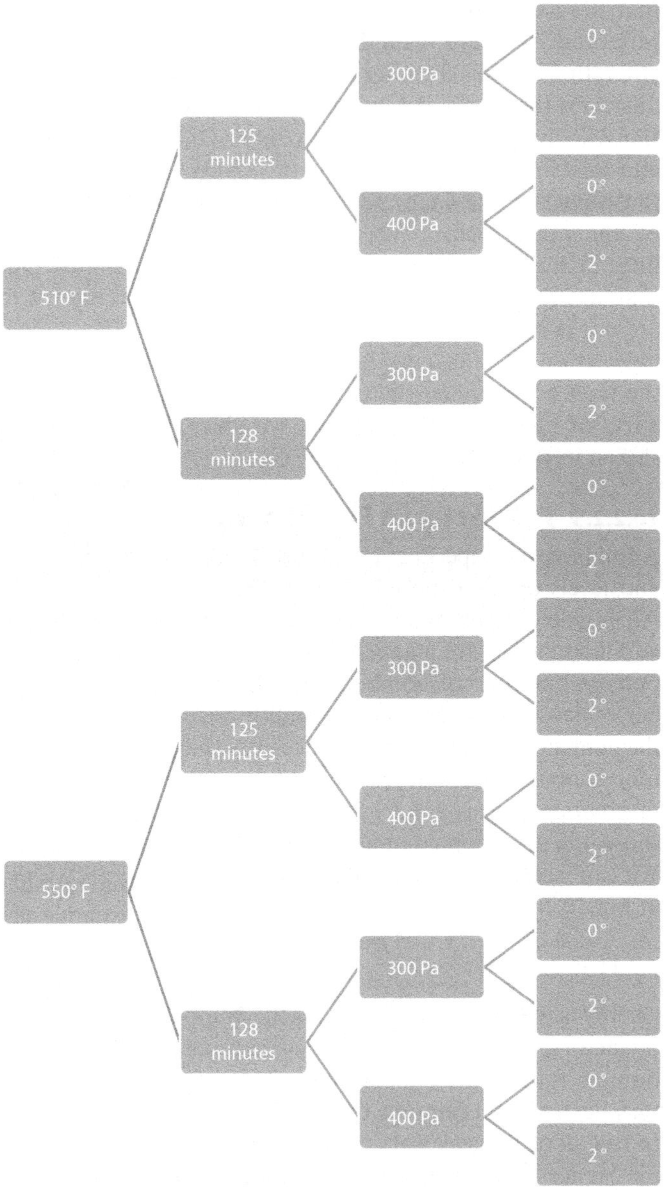

Fig. 10.3 Decision tree of experiment plan for all factors and levels. (Source: Author's creation)

the report section in Minitab as represented in Table 10.6.

As presented in "Coded Coefficients" section in Table 10.6, *p*-values of each individual factor and factor pairs are greater than $\alpha = 0.05$. Additionally, the variance inflation factor (VIF = 1) values in this section demonstrate that there is a low multicollinearity among the factors. The ANOVA results in Table 10.6 show that none of the factors are statistically significant since *p*-value of each factor is greater than level of significance ($\alpha = 0.05$). Therefore, we can assume that these four factors and factor pairs are not statistically significant to predict the tensile strength in this full factorial experiment.

Table 10.6 Minitab output for Example 4

Factorial Regression: Tensile Strength Versus Temperature, Operation Time, Pressure, Angle of Cut

Coded Coefficients

Term	Effect	Coef	SE coef	*T*-value	*P*-value	VIF
Constant		132.125	0.750	176.17	0.004	
Temperature	13.750	6.875	0.750	9.17	0.069	1.00
Operation time	−5.750	−2.875	0.750	−3.83	0.162	1.00
Pressure	−2.250	−1.125	0.750	−1.50	0.374	1.00
Angle of cut	−1.500	−0.750	0.750	−1.00	0.500	1.00
Temperature*operation time	2.250	1.125	0.750	1.50	0.374	1.00
Temperature*pressure	0.750	0.375	0.750	0.50	0.705	1.00
Temperature*angle of cut	1.500	0.750	0.750	1.00	0.500	1.00
Operation time*pressure	−1.250	-0.625	0.750	−0.83	0.558	1.00
Operation time*angle of cut	0.000	0.000	0.750	0.00	1.000	1.00
Pressure*angle of cut	1.000	0.500	0.750	0.67	0.626	1.00
Temperature*operation time*pressure	0.250	0.125	0.750	0.17	0.895	1.00
Temperature*operation time*angle of cut	−0.500	−0.250	0.750	−0.33	0.795	1.00
Temperature*pressure*angle of cut	−2.500	−1.250	0.750	−1.67	0.344	1.00
Operation time*pressure*angle of cut	−1.500	−0.750	0.750	−1.00	0.500	1.00

Model summary

S	R-sq	R-sq(adj)	R-sq(pred)
3	99.10%	86.55%	0.00%

Analysis of variance

Source	DF	Adj SS	Adj MS	*F*-value	*P*-value
Model	14	994.75	71.054	7.89	0.273
Linear	4	917.75	229.438	25.49	0.147
Temperature	1	756.25	756.250	84.03	0.069
Operation time	1	132.25	132.250	14.69	0.162
Pressure	1	20.25	20.250	2.25	0.374
Angle of cut	1	9.00	9.000	1.00	0.500
2-way interactions	6	41.75	6.958	0.77	0.701
Temperature*operation time	1	20.25	20.250	2.25	0.374
Temperature*pressure	1	2.25	2.250	0.25	0.705
Temperature*angle of cut	1	9.00	9.000	1.00	0.500

(continued)

Table 10.6 (continued)

Operation time*pressure	1	6.25	6.250	0.69	0.558
Operation time*angle of cut	1	0.00	0.000	0.00	1.000
Pressure*angle of cut	1	4.00	4.000	0.44	0.626
3-way interactions	4	35.25	8.812	0.98	0.631
Temperature*operation time*pressure	1	0.25	0.250	0.03	0.895
Temperature*operation time*angle of cut	1	1.00	1.000	0.11	0.795
Temperature*pressure*angle of cut	1	25.00	25.000	2.78	0.344
Operation time*pressure*angle of cut	1	9.00	9.000	1.00	0.500
Error	1	9.00	9.000		
Total	15	1003.75			
Regression equation in uncoded units					
Tensile strength	1717 – 2.4 temperature – 11.3 operation time + 4.5 pressure– 1257 angle of cut + 0.017 temperature*operation time – 0.0089 temperature*pressure + 1.53 temperature*angle of cut – 0.043 operation time*pressure + 7.9 operation time*angle of cut + 1.94 pressure*angle of cut + 0.000083 temperature*operation time*pressure – 0.0083 temperature*operation time*angle of cut – 0.001250 temperature*pressure*angle of cut – 0.0100 operation time*pressure*angle of cut				

Source: Author's creation based on Minitab

10

10.2.2.4 Fractional Factorial Experiment

Fractional factorial experiment helps Six Sigma teams reduce the number of runs, as only a limited part of the full factorial experiments is used. Selection of the partial groups should be done carefully. Several types of fractional factorial experiments are used in DOE, such as screening designs and Taguchi designs. Fractional factorial experiments enable Six Sigma teams to analyze the interrelationships among multiple factors. These methods generate more effective results than single factor experiments. When the experiment includes a large number of factors, fractional factorial experiments are preferred to full factorial experiments mainly because of the excessive number of runs required in full factorial experiments.

10.2.2.5 Screening Experiments

When full factorial experiments develop a large number of tests and the experiment needs a relatively high number of factors, screening experiments are utilized to minimize the number of tests. If the goal of DOE is to identify what factors need more analysis, screening designs may be helpful. In screening experiments, + and – signs are used to depict the levels of the factors, level 1 and level 2. Screening experiment can also be called 2^k experiments. An experiment plan for screening experiment using 2^3 experiments is shown in Tables 10.7.

10.2.2.6 Response Surface Designs

If the main goal is to optimize a response variable, the response surface method (RSM) may be preferred, while full or frac-

Table 10.7 Experiment plans for screening experiments using 2^3 experiments

Factors	F1	F2	F3	F4	F5	F6	F7
Runs							
1	–	–	–	–	–	–	–
2	+	–	+	–	+	–	+
3	+	+	–	+	–	–	+
4	+	–	+	+	–	+	–
5	–	–	–	+	+	+	+
6	–	+	+	+	+	–	–
7	–	+	+	–	–	+	+
8	+	–	+	+	–	+	–

Source: Author's creation

tional factorial experiments are utilized to determine what factors behave differently from the other factors at which levels. RSM is used to optimize the response variable (dependent variable) that is affected by several factors (independent variables). RSM has been used in a great variety of industries since the 1950s. Mostly manufacturing industries such as electronics, chemistry, pharmacy, metal processing, and aerospace industries have implemented RSM in decision-making processes. To exemplify the method, let's recall the process given in ► Example 1. The tensile strength of metal cable is response variable and a function of the levels of (1) temperature (x_1), (2) operation time (x_2), (3) pressure (x_3), and (4) angle of cut (x_4) in metal cable manufacturing process, where ε is the error term in the response variable, (Eq. 10.11).

$$y = f\left(x_1, x_2, x_3, x_4\right) + \varepsilon \tag{10.11}$$

To minimize the size of the analysis using RSM, let's focus on two factors: temperature (x_1) and operation time (x_2). If there are two factors involved in the analysis, the expected value of y is denoted as response variable, $E(y) = f(x_1, x_2) = f$ (*temperature, operation time*). Six Sigma teams may see graphically the location of the response variable versus the levels of these factors in a three-dimensional diagram, surface plot (Image 10.4). The surface plots demonstrate the location of response variable, in the z-axis against two factors in the X- and Y-axes. The contours of the response surface are presented in area and line contour plots (Image 10.5) using lines and areas. Surface plots and contour plots enable Six Sigma team to see the behavior of the response variable against the factors.

In RSM, the Six Sigma team determines whether the data set fits in a first-order model that is built by a linear regression model (Eq. 10.12). If the data set does not fit in a first-order model, a second-order model that is based on polynomial or higher-degree functions is tested (Eq. 10.13) (Montgomery 2013). First, it is suggested to explore the presence of first-order model in RSM. If the response variable is explained by factors (independent variables) in a first-order model, there is no need to test second-order models. But, if the first-order model does not model the response variable, then second-order models are tested to represent response variable,. The parameters ($\beta_0, \beta_1, \beta_2, \beta_k, \beta_i, \beta_{ii}, \beta_{ij}$) of the models are estimated using regression analysis. Then, a response surface analysis is performed by Six Sigma team.

$$y = \beta_0 + \beta_1 x_1 + \beta_2 x_2 + \ldots + \beta_k x_k + \varepsilon \tag{10.12}$$

$$y = \beta_0 + \sum_{i=1}^{k} \beta_i x_i + \sum_{i=1}^{k} \beta_{ii} x_i^2 + \Sigma \sum_{i<j=2}^{k} \beta_{ij} x_i x_j + \varepsilon \tag{10.13}$$

► Example 5

Recall the example used in two-factor factorial design. Assume that in a metal cutting process, the response variable is tensile strength, and the factors are temperature and pressure. Sixteen measures are presented in Table 10.4 above. Using RSM, determine optimal operating conditions for this process. ◄

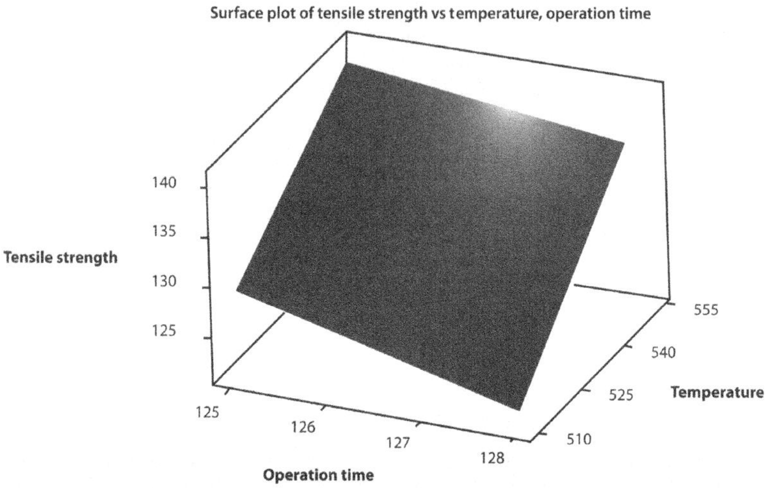

Image 10.4 An example of surface plot of response variable and two factors. (Source: Author's creation based on Minitab)

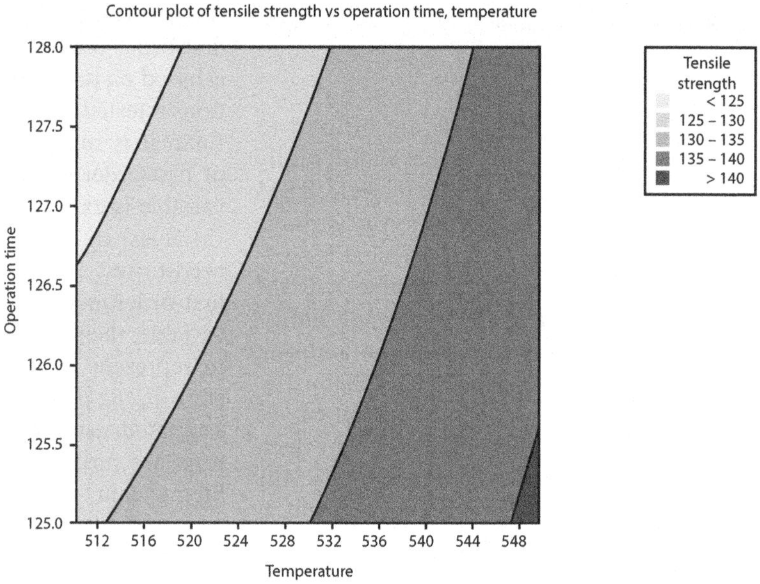

Image 10.5 Area contour plot of response variable and two factors. (Source: Author's creation based on Minitab)

10

Solution

To run response surface analysis in Minitab, after transferring data set (Table 10.4) to a Minitab worksheet, click on Stat→ DOE→Response Surface→Analyze Response Surface Design. Click on Yes in the next screen. In the input screen, first select "temperature" and "pressure" variables for "continuous factors" box. Then, click on "Low/High…" button, and see factors entered with the levels in the box. Click on OK and OK. In the next screen, select "tensile strength" from left box to "responses" box on the right. Click on OK. The response surface analysis results are given in the report section in Minitab as shown in Table 10.8.

As presented in "Coded Coefficients" section in Table 10.8, the first-order model that fits to the tensile strength (y) is

$$y = 132.125 - 2.875\,\text{temperature} + 6.875\,\text{pressure} + 1.125\,\text{temperature} * \text{pressure}$$

However, the term of "temperature*pressure" is not statistically significant since p-value (0.136) is greater than $\alpha = 0.05$ in t-test. Therefore, the "temperature*pressure" term needs to be excluded from the first-order model. Final first-order model is

Table 10.8 Response surface analysis results for Example 5

Response Surface Regression: Tensile Strength (psi) Versus Temperature (°F), Pressure (Pa)

The following terms cannot be estimated and were removed:

Temperature (°F) *temperature (°F), pressure (Pa)*pressure (Pa)

Coded coefficients

Term	Coef	Adj SS	Adj MS	*F*-value	*P*-value
Constant	132.125	0.703	187.83	0.000	
Temperature (°F)	−2.875	0.703	−4.09	0.002	1.00
Pressure (Pa)	6.875	0.703	9.77	0.000	1.00
Temperature (°F)*pressure (Pa)	1.125	0.703	1.60	0.136	1.00

Model summary

S	R-sq	R-sq(adj)	R-sq(pred)
2.81366	90.54%	88.17%	83.17%

Analysis of variance

Source	DF	Adj SS	Adj MS	*F*-value	*P*-value
Model	3	908.75	302.917	38.26	0.000
Linear	2	888.50	444.250	56.12	0.000
Temperature (° F)	1	132.25	132.250	16.71	0.002
Pressure (Pa)	1	756.25	756.250	95.53	0.000
2-way interaction	1	20.25	20.250	2.56	0.136
Temperature (°F)*pressure (Pa)	1	20.25	20.250	2.56	0.136
Error	12	95.00	7.917		
Total	15	1003.75			

Regression equation in uncoded units

Tensile strength (psi) = 369 − 0.537 temperature (°F) − 0.459 pressure (Pa) + 0.001125 temperature (°F)*pressure (Pa)

Fits and diagnostics for unusual observations

Obs	Tensile strength (psi)	Fit	Resid	Std resid	Fit
1	135.00	129.25	5.75	2.36	R

R Large residual

Source: Author's creation based on Minitab

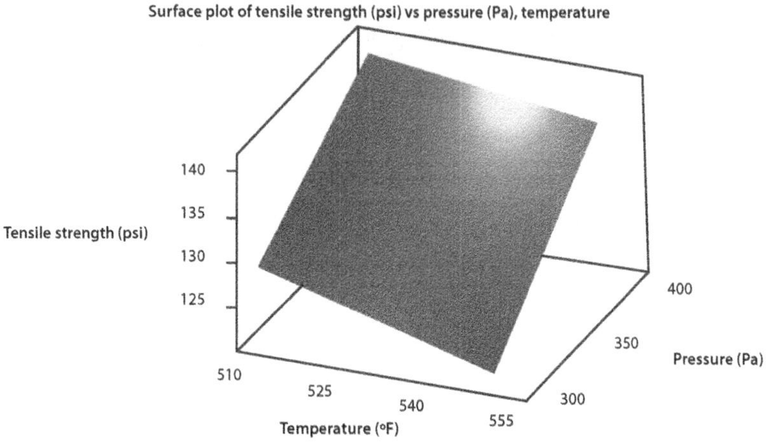

Image 10.6 Surface plot of tensile strength vs pressure and temperature. (Source: Author's creation based on Minitab)

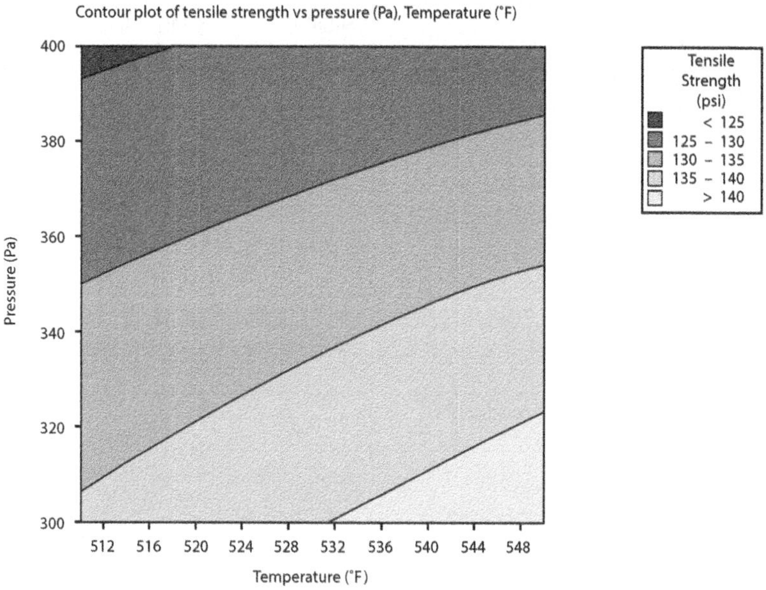

Image 10.7 Contour plot of tensile strength vs pressure and temperature. (Source: Author's creation based on Minitab)

$$y = 132.125 - 2.875\,\text{temperature} + 6.875\,\text{pressure}$$

In "Analysis of Variance" section in Table 10.8, temperature (p = 0.002) and pressure (p = 0.000) variables are statistically significant, whereas two-way interaction of these variables is not (p = 0.136). "Regression Equation in Uncoded Units" section in Table 10.8 shows that when the measurements are used as "uncoded" in the analysis, the tensile strength is estimated as follows:

$$y = 369 - 0.537\,\text{temperature} - 0.459\,\text{pressure}$$

Image 10.6 demonstrates the surface plot of tensile strength against factors of temperature and pressure, while Image 10.7 shows the contours of tensile strength in the first-order model. The surface plot indicates that tensile strength is maximized when temperature is low (510 °F) and pressure is high level (440 Pa). A similar trend is also seen in contour plot. The area covered in the left upper corner of the contour plot in Image 10.7

shows that tensile strength level is maximized when pressure is at 440 Pa and temperature is at 510 °F.

As presented in the main effects plot in ◘ Image 10.2, the mean of tensile strength is higher when temperature is 510 °F and pressure is 400 Pa. One of the important findings of DOE is that the decision-makers should keep the temperature at 510 °F and pressure at 400 Pa to maximize the level of tensile strength.

When we add all four factors, (1) temperature, (2) operation time, (3) pressure, and (4) angle of cut, in the analysis using the same steps as presented below, the response surface regression analysis presents the results shown in ◘ Table 10.9. According to the results presented in "Coded Coefficients" section, the relationship between the response variable and factors is modeled as follows:

$$y = 132.125 - 2.875\,\text{temperature} + 6.875\,\text{pressure}$$

As presented in ◘ Table 10.9, except for temperature and pressure factors, the factors and the interactions between factors are not statistically significant, since p-values are greater than $\alpha = 0.05$ in t-test. R^2-adjusted = 86.77% shows that the regression model given above fits the data, and 86.77% of the variation in the tensile strength is explained by "temperature" and "pressure" factors.

The Pareto chart in ◘ Image 10.8 shows the absolute values of the standardized effects. The standardized effects show t-statistics testing the null hypothesis that the effect is 0. The Pareto chart includes a vertical reference line that shows statistically significant effects. It also shows magnitude and significance of the effects. The bars on the chart that cross the vertical reference line show statistically significant effects at the 0.05 α level. According to the Pareto chart in ◘ Table 10.9, "pressure" and "temperature" pass the vertical reference line that is at 2.571.

As presented in the main effects plot in ◘ Image 10.9, the mean of tensile strength is maximized when:

1. Temperature is lower (510 °F)
2. Operation time is lower (125)
3. Pressure is higher (400 Pa)
4. Angle of cut is low (0).

The interaction plot in ◘ Image 10.10 demonstrates the impacts of interactions between factors on the response. Each end point in the interaction plot shows the means at different combinations of factor levels. For example, operation time has two levels, 125 minutes and 128 minutes, in the experiments. The midpoint of the operation time (126.5 minutes) is also presented in the upper right box titled "Operation ti." Similarly, pressure is presented with two levels, 300 Pa and 400 Pa. Additionally, the midpoint of pressure, 350 Pa, is also added in the interaction plot. As shown in ◘ Image 10.10, the lines are parallel in interaction plots 2, 3, and 4. The parallel lines in interaction plots show that there is no statistically significant interaction between the two factors analyzed. This result is also confirmed by the p-values in t-tests in ◘ Table 10.9. To analyze the interaction plots in depth, in the first interaction plot, it is seen that tensile strength is maximized when both operation time (125 minutes) and temperature are low (510 °F). Note that the relationship between these two factors is not statistically significant (p = 0.439). The second interaction plot including temperature and amount of pressure shows that tensile strength is maximized when temperature is high (400) and pressure is low (510 °F). Note that the relation ship between these two factors is not statistically significant (p = 0.191). The third interaction plot including temperature and angle of cut shows that tensile strength is maximized when temperature is low and angle cut is 0°. Note that the relationship between these two factors is not statistically significant (p = 1.000). The fourth interaction plot including operation time and amount of pressure indicates that

Table 10.9 Response surface regression analysis results for Example 5

Response Surface Regression: Tensile Strength Versus Temperature, Operation Time, Pressure, Angle of Cut

The following terms cannot be estimated and were removed:

Temperature*temperature, operation time*operation time, amount of pressure*amount of pressure

Coded coefficients					
Term	**Coef**	**SE coef**	***T*-value**	***P*-value**	**VIF**
Constant	132.125	0.744	177.65	0.000	
Temperature	−2.875	0.744	−3.87	0.012	1.00
Operation time	−1.125	0.744	−1.51	0.191	1.00
Pressure	6.875	0.744	9.24	0.000	1.00
Angle of cut					
0	0.750	0.744	1.01	0.360	1.00
Temperature*operation time	−0.625	0.744	−0.84	0.439	1.00
Temperature*pressure	1.125	0.744	1.51	0.191	1.00
Temperature*angle of cut					
0	−0.000	0.744	−0.00	1.000	1.00
Operation time*pressure	0.375	0.744	0.50	0.636	1.00
Operation time*angle of cut					
0	−0.500	0.744	--0.67	0.531	1.00
Pressure*angle of cut					
0	−0.750	0.744	−1.01	0.360	1.00
Model summary					
S	**R-sq**	**R-sq(adj)**	**R-sq(pred)**		
2.97489	95.59%	86.77%	54.86%		
Analysis of variance					
Source	**DF**	**Adj SS**	**Adj MS**	***F*-value**	***P*-value**
Model	10	959.50	95.950	10.84	0.008
Linear	4	917.75	229.437	25.93	0.002
Temperature	1	132.25	132.250	14.94	0.012
Operation time	1	20.25	20.250	2.29	0.191
Pressure	1	756.25	756.250	85.45	0.000
Angle of cut	1	9.00	9.000	1.02	0.360
2-way interaction	6	41.75	6.958	0.79	0.616
Temperature*operation time	1	6.25	6.250	0.71	0.439
Temperature*pressure	1	20.25	20.250	2.29	0.191
Temperature*angle of cut	1	0.00	0.000	0.00	1.000

Table 10.9 (continued)

Operation time*pressure	1	2.25	2.250	0.25	0.636
Operation time*angle of cut	1	4.00	4.000	0.45	0.531
Pressure*angle of cut	1	9.00	9.000	1.02	0.360
Error	5	44.25	8.850		
Total	15	1003.75			
Regression equation in uncoded units					
Angle of cut					
0	Tensile strength	= –663 + 2.10 temperature + 8.2 operation time – 1.11 pressure – 0.0208 temperature*operation time + 0.001125 temperature*pressure + 0.00500 operation time*pressure			
2	Tensile strength	= –760 + 2.10 temperature + 8.9 operation time – 1.08 pressure – 0.0208 temperature*operation time + 0.001125 temperature*pressure + 0.00500 operation time*pressure			

Source: Author's creation based on Minitab

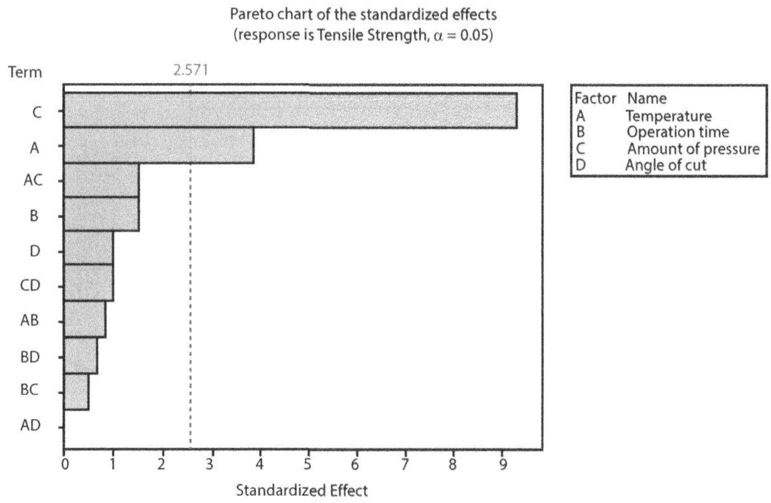

Image 10.8 Pareto chart in RSM. (Source: Author's creation based on Minitab)

tensile strength is maximized when operation time is low (125 minutes) and pressure is high (400 Pa). Note that the relationship between these two factors is not statistically significant ($p = 0.636$). In the next interaction plot including operation time and angle of cut, tensile strength is maximized when operation time is low (125 minutes) and angle of cut is 0°. Note that the relationship between these two factors is not statistically significant ($p = 0.531$). The last interaction plot shows that tensile strength is maximized when the amount of pressure is high (400 Pa) and angle of cut is 0°. Note that the relationship between these two factors is not statistically significant ($p = 0.360$). These interactions between the factors are also presented in surface plots for tensile strength in Image 10.11.

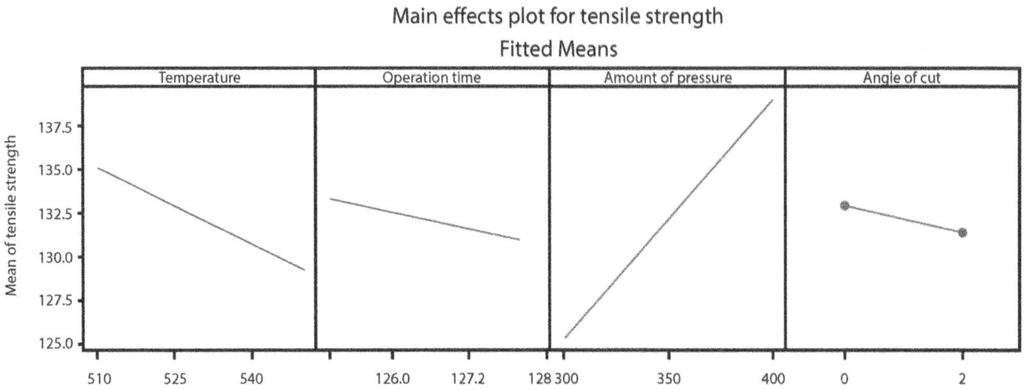

Image 10.9 Main effects plot for tensile strength. (Source: Author's creation based on Minitab)

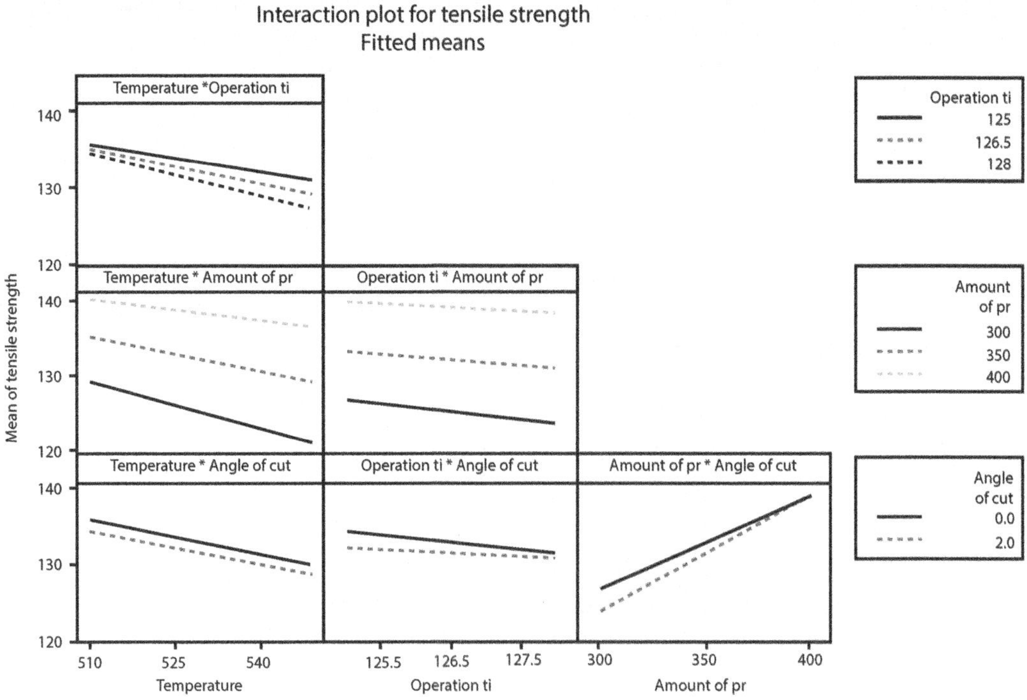

Image 10.10 Interaction plot for tensile strength. (Source: Author's creation based on Minitab)

10

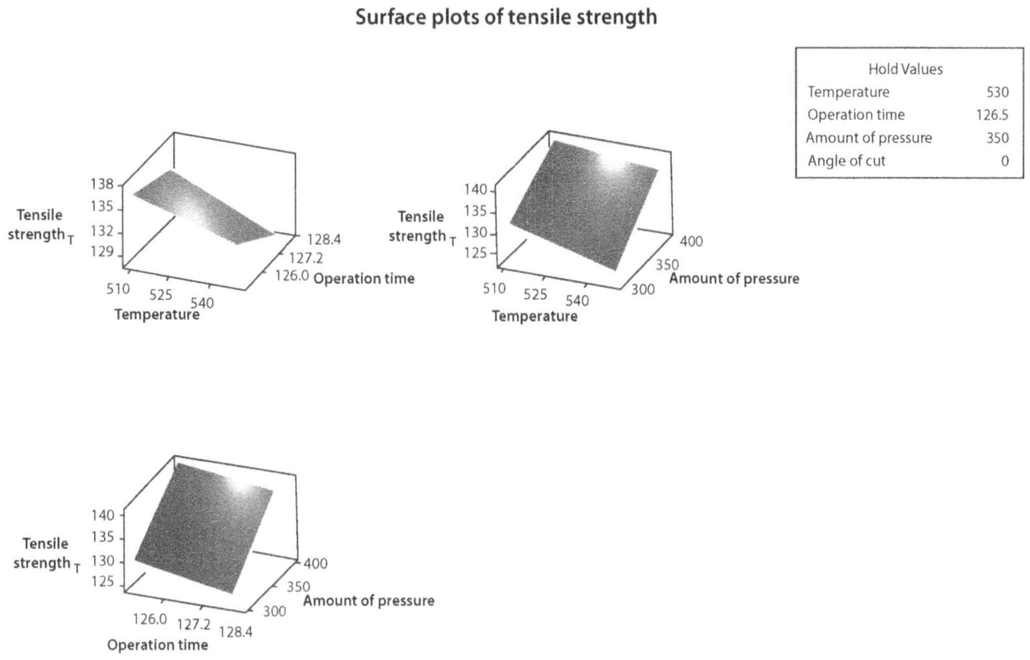

Image 10.11 Surface plots for tensile strength. (Source: Author's creation based on Minitab)

10.3 Simulation

Burçin Çakır Erdener
Assistant Professor
Başkent University

10.3.1 Introduction

Depending on the topic identified in the Define phase, alternative improvements and scenarios and process or product designs are developed, proposed, and tested using the simulation method by Six Sigma teams. The simulation method helps teams test various improvement alternatives in a simulated environment before implementing changes in the processes or products. Using simulation outcomes, the performance of the current and alternative systems is compared, and the optimum system is selected for implementation in the Improve phase.

Simulation techniques have been widely used by organizations to analyze their operations, generate process improvements, and compare alternative system performances for many years. Six Sigma has been developed as a disciplined, highly quantitative approach and introduced alternative ways of thinking with regard to process or product improvement. The concept of integrating simulation and Six Sigma approaches enhances the overall performance of the processes. The most significant challenge in Six Sigma projects is to measure the effects of the changes in a variable or process. To run a successful Six Sigma project, the variability of the system has to be correctly modeled. Here, simulation is an effective and practical tool since it can capture the impact of variability of the real-life problems in ways that traditional static and deterministic methods cannot. With a simulation model, it is possible to analyze the performance of an existing system or to test new alternatives (Altiparmak et al. 2002).

In DMAIC, the process considered within a Six Sigma project is stochastic in nature, and simulation provides a powerful platform to analyze dynamic and complex features. Simulation also predicts the consequences of potential changes through a series of steps designed to observe the system, analyze the input data, generate distributions, build a model, and analyze results using appropriate statistical output analyzers. Changes that are most effective in improving performance can be applied in the real processes after the benefits are confirmed with the simulation model (Hussein et al. 2017).

Despite the advantages of using simulation in Six Sigma projects, it should be used correctly and in appropriate areas. Simulation is more effective when the system being studied is complicated and difficult to visualize. For simple processes, it might be better to use traditional improvement methods. If the decision is made that simulation is necessary, the assumptions in a complicated system are determined very carefully. The simulation model focuses on critical processes that affect the quality of the whole system. Using the simulation model functionality, it becomes possible to determine the impact of the decisions within a Six Sigma project on the process outputs. It also allows Six Sigma teams to understand the interactions between system components together with their significance in the overall system (Taneja and Manchanda 2013). Therefore, the integration of Six Sigma modeling and simulation is an effective decision-making tool (Ahmed et al. 2017).

In the following sections, first, we introduce the basic concepts of the simulation modeling with a particular focus on process modeling and discrete event simulation. To this end, we provide an introduction of the concept, terminology, classification of simulation models, and utilizing simulation tools in real-life applications. The following section briefly explains the simulation, and the fundamental concepts are presented later. The next section contains the features of performing a simulation analyses, and final section includes manual simulation examples.

10.3.2 What Is Simulation?

Simulation is a powerful tool used to design and analyze a complicated system; it is a computer-based model that mimics the operation of an existing or proposed system (Shannon 1975). The simulation model is an abstract model of a real system to determine how the system will respond to changes in its structure, environment, or underlying assumptions. The assumptions in a simulation model are represented by the mathematical and logical relations between the elements and the entities in the system. Simulation works best when the system has medium-sized complication and variation. Simulation can also be a valuable candidate even for less complicated systems if real-world processing is costly, the process is hard to replicate, and it requires considerable resources for implementation.

While building a simulation model, the modeler must specify the scope of the model and the level of detail needed. Only those factors with a significant impact on the model's ability to serve its stated purpose should be included. The level of detail must be consistent with the purpose. The idea is to create, as economically as possible, a replica of the real-world system that can provide necessary information regarding important questions (Martha 1996). This is usually possible at a reasonable level of detail. Commonly, simulations provide data on a wide variety of systems metrics, such as throughput, resource utilization, waiting times, and production requirements. While useful in modeling and understanding existing systems, they are even better suited to evaluating proposed process changes. In essence, simulation is a tool for rapidly generating and evaluating ideas for process improvement (Ricki 2008).

In today's world, simulation is used in a wide range of areas, but especially for service and manufacturing systems. Some simulation applications in manufacturing and service systems are used for:

- Improving resource allocations in healthcare and hospital management
- Identifying and solving bottlenecks in airports and aviation
- Analyzing alternative work processes in logistics

- Increasing profitability in restaurants and food services
- Supporting decision-making in IT systems
- Testing the effect of an alternative process in the banking system
- Designing and testing alternative layouts in warehousing
- Improving the quality of production in manufacturing plant
- Addressing risk and vulnerabilities in an assembly line
- Forecasting demand and predicting performance in manufacturing planning
- Reducing the time for waste and rework in the production process.

10.3.3 Types of Simulation Models

Simulation models can be classified in many ways, but one important classification discussed here uses three dimensions: static-dynamic models, deterministic-stochastic models, and discrete-continuous models.

- **Static-Dynamic Models**

In a static model, time is not considered, meaning that the model is a snapshot. Monte-Carlo simulation is an example of static simulation, and it is used to model the probability of different outcomes that cannot be predicted due to random variables. In contrast, a dynamic model considers the system during a period of time. In a dynamic model, the time changes, and the simulation of the system is implemented while the system is evolving over time. The majority of the systems in practice are dynamic.

- **Discrete-Continuous Models**

In a discrete model, time changes in incremental steps over a period. For example, the number of the customers in a bank changes at discrete points in time when a new customer arrives or an existing one departs. Other examples for discrete models could be a manufacturing system with parts arriving and leaving at specific times and machines working or failing at specific times. Continuous models have a time that changes continuously and smoothly. For instance, the speed and status of an airplane change continuously over time.

- **Deterministic-Stochastic Models**

Deterministic models do not include any probabilistic components, and randomness does not affect the behavior of the system. Therefore, the outputs of a deterministic model are not random variables. A strict appointment-book operation with fixed service time could be an example. Stochastic or probabilistic models are affected by randomness, and they have at least some random input components. For instance, a queuing system in a bank with randomly arriving customers and varying service times can be modeled with stochastic models. The outputs of the stochastic models are also random.

An example of an emergency room in a hospital:

- *The emergency room of ABC hospital works 24 hours a day to serve the patients. The service has two doctors and a nurse. There is also an associate working at the registration desk. When the patients come in, they first have to register and then wait in a queue if the doctor is busy. The doctors and nurses serve the patients based on the severity level of illness or accident. Patients come into the service at different and random intervals. From 8.00 am to 5.00 pm, the arrivals per hour have a Poisson 5 arrival rate, and from 5.01 pm to 07.59 am, the arrivals per hour have a Poisson 8 arrival rate. The service time for the patients is normally distributed with a mean of 10 minutes and variance of 2 minutes. In such a system, the state variables change at discrete points over the time. Therefore, the system can be modeled using discrete structure. During the day, it is a dynamic model. Inter-arrival times and service times are random. Therefore, the system is a stochastic one. Eventually, we can model this emergency room as a discrete, dynamic, and stochastic simulation model.*

10.3.4 How Are Simulations Performed?

Once the Six Sigma team determines that simulation is the right tool for the team to solve the problem, the next step is to decide on how to carry it out. Below the options to run a simulation are discussed.

10.3.4.1 Simulation by Hand (Manual Simulation)

Simulation can be performed manually if the system is not very complicated. The common purpose in all simulations is to estimate a value that is hard to compute. This estimation will not be exactly right and will have some errors. To reduce the errors, the number of replications is increased. Therefore, although simulation by hand is easy to implement, it is limited because of the complexity of the systems.

10.3.4.2 Simulation with General Purpose Languages

Using general purpose languages such as C, C++, and Python in simulation requires a high level of programming skills, which makes simulating a system highly flexible and customizable. As the system gets more complicated, modeling may take longer. Some simulations can be run using Excel spreadsheets that provide random number generators that enable teams to model simple dynamic systems. The inherent limitations of spreadsheets make it difficult to use them for realistic, large, and dynamic systems.

10.3.4.3 Special Purpose Simulation Languages

Special purpose simulator languages such as GPSS, SIMSCRIPT, EXTEND, SLAM, SIMAN, and ARENA are very popular tools for modeling realistic, complex, and dynamic systems. Although the programming skills are easier, compared to simulation with general purpose languages, it is still necessary to learn the features of the simulators to use them effectively.

10.3.5 Concepts of the Simulation Model

In this section, the basic concepts of a simulation model are presented to identify essential terms used while building a model.

10.3.5.1 The System

A system in simulation is a combination of elements that interact with each other to accomplish a common purpose. For instance, a group of machines performing related manufacturing operations would constitute a system. These machines may be considered a group or element in a larger production system. The production system may be an element of a larger system involving design, delivery, etc. To properly model the system, the limits of the system should be determined carefully, considering the properties of the study. The elements of a system from a simulation perspective are shown in ◘ Fig. 10.4 and explained in detail below.

- **Entities**

A system in a simulation study is comprised of a group of entities interacting toward a set of goals. Entities are both inputs and outputs of the system, and they follow a series of activities defined for the system using resources and control plans. When the entities leave the system, the feedback mechanism gives the statistics of the study to draw conclusions on the key decisions. Entities are objects which are the subject of the study. Entities are the dynamic parts of the simulations that are created to move in the system for a while and are disposed when they leave. Without an entity there will be no action in the model. Entities can be human (customer, employees), product, project, etc.

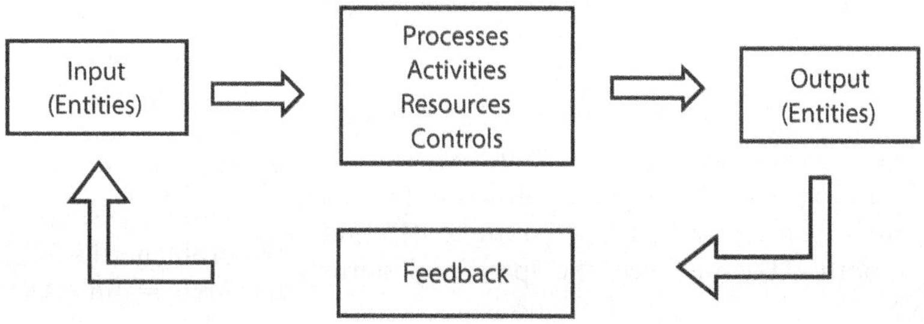

◘ **Fig. 10.4** Elements of a system in a simulation study. (Source: Author's creation)

▪ **Attributes**

Attributes are the properties that an entity can have. An attribute can be a common characteristic for all entities, but the value of entities can differ from one to another. For instance, if a product is the entity, then color, price, or due date could be its attributes. The definition of the attributes is decided based on the nature of the problem. The programmer defines the attributes that are needed, assign values to them, and change when necessary.

▪ **Activities**

Activities are actions performed by an entity over a period (repairing a machine, filling the order form, assembling the product, ordering food, waiting for the food to be cooked, etc.). A different type of an activity could be also queuing when the system is not available. Entities awaiting resources to be available are also in an activity for a simulation model.

▪ **Resources**

Resources are the required tools used to perform the activities (personnel, tool, space, energy, time, money, etc.). Entities get service from a resource by making it busy when available and release when finished. An entity could need simultaneous service from multiple resources.

▪ **Control**

Control is a process plan that represents the place, order, and ways of doing activities (process plans, production plan, maintenance policy, etc.).

▪ **System Statement**

The system statement is the sum of variables required to describe the system at a time, depending on the purpose of the study.

▪ **Event**

Event is an instantaneous occurrence that changes the system state. In simulation models, mainly three kinds of events are observed:

- Arrival event: A new entity enters the system.
- Departure event: An entity finishes its service and leaves the system.
- The end: Termination criteria for the simulation model.

The logic behind each event will be explained in detail later.

▪ **Performance Measures**

Performance measures are the outputs of the system that is of interest of the study (cycle time, utilization rate, waiting time, quality, cost, etc.). To obtain performance measures, the model has to track statistical accumulator variables while the simulation progresses. The possible statistical accumulators could be the number of produced products so far, the total waiting time in the system for a part so far, and numerous other variables. All of the accumulators are initialized to zero. In special purpose simulators, those accumulators are automatically recorded, but while implementing simulation by hand, it is done manually. Some real-life system element examples are given below. It should be noted that, depending on the purpose of the study, different elements might be considered for the same systems.

Banking Service Example ABC is a branch of an international bank. The analyst is interested in knowing the average total waiting time that a customer spends in the system in 1 month. In such a system, *entities* would be the customers of the bank. Customers arrive at the bank, request the service, receive the service, and depart. *Attributes* are properties of the entities, and some attributes of the customers in banking system could be prioritized customers or regular customers, customers that require individual services or standard services. *Activities* are operations that a customer performs such as applying for an individual credit, withdrawing money or transferring money to an international bank, and so on. *Resources* are personnel of the bank, the number-generating machine for queuing, and computer system. The *control* of the system is the order of the activities that a customer needs to follow. For example, a customer first obtains a number from the machine for the corresponding activity and follows the required process. The *statement of the system* is the collection of variables that defines the system at any given time. The system statement is the snapshot of the system at a given time such as the number of customers waiting in the queue, the idle and busy personnel, and the arrival time interval of a new customer (the arrival of a customer at a time would change the system state).

Events can be defined in three groups for this example: arrival event, departure event, and the ending event. An arrival event can change the system state in two different ways. Arrival of a new customer at the system will increase the number of customers in the queue if the service line is busy or will make an idle service busy if there is no customer receiving service. The second type of event is the departure event. It is similar to arrival event and can change the system statement in two different ways. Departure of a customer when completing the service can result in decreased number of customers waiting in the queue or changed status of personnel from busy to idle (if there is no customer waiting in the queue). The ending event for the system is 1 month. *Performance measure* is the average total waiting time of a customer spent in the system. Other performance measures that could be observed are utilization of personnel, average waiting time in the queue, and the total number of customers served in a day.

Manufacturing Example XYZ manufacturing plant produces 100 parts in 16 hours per day, of which 5% of the parts need to be reworked. The managers want to improve the production quality and have a lower number of defective parts. In this example, the *entities* of the system are the parts that are produced. The relevant *attributes* of the parts are defective parts or non-defective parts. *Activities* are the production steps that are required for the part such as drilling, assembling, and painting. *Resources* are the operators, raw material, components, and machines. *Control* is the production process order. For example, a control plan for the production of a plastic could be plastic extrusion, molding, cooling, and releasing. The variables needed to define the *system statement* are the number of parts waiting in the queues, the number of busy/idle resources, and the arrival time interval of the parts. *Events* are the arrival event of a part or the completion of a service (departure event). Finally, the *performance measure* is the number of defective parts produced in one working day.

10.3.5.2 Steps of Building a Simulation Model

Modeling is the most important stage of a simulation study. Indeed, the outputs are closely related to how the model is built. A simulation model comprises of the following steps given below:

- Step 1: Problem formulation
  A simulation study begins with clear identification of the problem and purpose. The bounds of the system and the overall objective are defined. The working plan is carefully determined, including identifying the alternatives and performance measures, assigning members of the team, time frame, cost, and so on. The problem must be formulated as precisely as possible.
- Step 2: Conceptual modeling
  Conceptual modeling is the transformation of the real-life problem's essential features, logical relations, and structure into an abstract model, defined as simulation representation. The representation can be a block diagram, flow chart, or process map depicting key characteristics of the real system, such as entities, parameters, logic, and outputs. The conceptual model is then transferred to a simulation model using simulation tools.
- Step 3: Data collection
  If the system exists, the required information and data are collected. Sources of randomness are identified and processed statistically to select the appropriate probability distributions. Software packages for distribution fitting and selection include ARENA input Analyzer, Minitab, ExpertFit, BestFit, and add-ons in some standard statistical packages. These aids combine goodness-of-fit tests, e.g., $\chi 2$ test, Kolmogorov-Smirnov test, Anderson-Darling test, and parameter estimation. Also, if possible, the performance measures of the system are recorded to validate the simulation model. The real and model outputs can be statistically compared for verification of the model. For instance, in a banking system, the inter-arrival time between two consecutive customers and

their service times are recorded for a certain time period to build the model. If the performance measure is the average waiting time in queue for a customer, the waiting times of each customer can be collected for verification.

Input Analyses: The collection and statistical analyses of the input data are defined as input data analyses or input data modeling. Especially while dealing with a stochastic system, the input data change over time randomly or according to certain probability distributions. Therefore, the collected data are analyzed carefully to find the probability function, and, once it is obtained, the distribution function is used to generate samples (Ungureanu et al. 2005).

- Step 4: Pre-model building
There is no standard procedure for building a simulation model. The procedure is often based on the modeler approach and the software used. However, a generic procedure for building a simulation model effectively includes key basic steps, such as constructing model components, developing the logic and flow, inserting data, and determining parameters. Once the model is generated, it is verified with experts who have knowledge of the system and/or end users. Early agreement with experts prevents waste of resources and enhances the reliability of the model.
- Step 5: Programming and validation
A model can be developed either by general purpose programming languages (C, C++, C sharp vb.) or appropriate simulation software (ARENA, GPSS, ExtendSim, SLAM, etc.) depending on the availability of the tool or modeling capability of the programmer. Validation is completed by traces, varying input parameters over their acceptable range, and checking the output, manual checking of outputs, and animation.
- Step 6: Verification
Once the model is developed in an acceptable form, pilot runs are executed for verification. The verification step determines whether the simulation model is a good representation of the real system. It is implemented step by step by identifying the differences using statistical analysis and correcting errors. At this stage, the performance measure data collected from the real system at Step 2 is used for comparing the outputs of the model. Statistical inference tests (see ▶ Chap. 7) are performed and examined considering the confidence level placed by the end users.
- Step 7: Model analysis
Having a verified and validated model provides a great platform to run experiments and apply various types of engineering analyses. Model analysis includes statistical analysis and experimental design. The objective of these methods is to evaluate the performance of the system and compare the performance of alternative scenarios. Statistical analyses include computation of numerical estimates (mean, variance, confidence intervals) for the desired performance measures. Experimental design with simulation includes conducting a partial or full factorial design of experiments to provide the best settings to model control variables. Also, before executing the runs, the analyst runs the model for a certain period to identify the input parameters that need to be changed, a warm-up period for non-terminating systems, the number of replications, and replication length (Karnon et al. 2012).
- Step 8: Study documentation
Following generation of results, outputs are documented based on the objectives of the project. If the main objective is to assess the performance of the system, the statistical analyses of the performance measures are summarized. If the main purpose is to compare the performance of the alternatives, the alternative that outperforms is highlighted.

▶ Example 6 Applying the Simulation Process to an Ambulatory Care Center

In this section, a simulation modeling example is applied on an ambulatory care center. The main goal of the example is to observe the overall performance of the system in terms of waiting times for patients in the queues and utilization of care providers. The example also evaluates the current status and suggests improvements if necessary. ◀

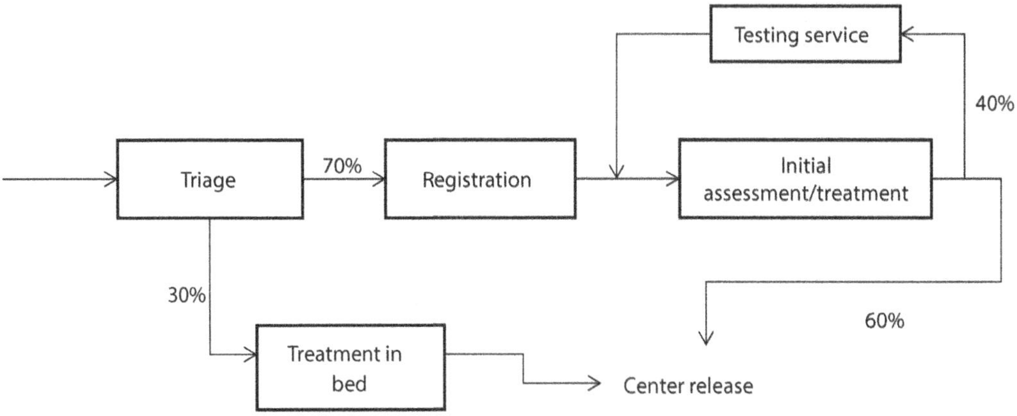

Fig. 10.5 Conceptual model for ambulatory care center example. (Source: Author's creation)

- Steps 1 and 2: Problem formulation and conceptual modeling
  Ambulatory care center process description:

  As shown in Fig. 10.5, the center has five stations: (1) the triage station where the patients are evaluated with regard to their severity levels, (2) treatment station where the severe patients get immediate treatment in bed, (3) the registration station where the station attendant enters the data for patients, (4) the initial assessment/treatment station where the patients receive the first treatment and get evaluated if they need laboratory examinations, and (5) the testing station where patients that need additional examinations are directed. Thirty percent of the patients arriving at the triage station are classified as severely ill, and they are immediately directed to the treatment in bed station. The rest of the patients wait in the registration station and then are directed to the initial assessment/treatment station. The 40% of patients in the initial assessment/treatment station are directed to the testing station for additional examinations, and 60% of the patients are released from the system. The patients that go to the testing station will go back to the initial assessment/treatment station to show the results of their examinations. Those patients leave the system after they finish their appointments with the doctor. There is one physician working in the treatment in bed station, one nurse in the registration, two physicians in initial assessment, and one physician in the testing service.

  This system is considered a queuing system with different properties. Generally, healthcare organizations are not typical first-in-first-out systems, since the severity of patient illness is more important than arrival time. Therefore, a priority-based queuing discipline can be used. The system described above can be effectively modeled by simulation, since it has an arrival rate of patients and service rates for stations, similar to a discrete event simulation model.

  The modeling steps of the system described are explained below. The simulation model is implemented using ARENA simulator. The steps of the model within the simulator are shown as screenshots. The training mode of the simulator can be downloaded at ▶ https://www.arenasimulation.com/.
- Steps 3 and 4. Data collection and pre-model modeling
  To collect data from this system, the analysts observe the system and collect data by counting the patients waiting in the queue, measuring the service times at each process stage for each type of patient, inter-arrival of patients, and so on. After collecting all available data, they are fitted to corresponding statistical distributions.
- Steps 5 and 6: Programming, validation, and verification

The system is modeled using simulation software to gain more accurate results in the outputs. The system given in the example is too complex to model by hand or using spreadsheets. Instead, it is quite simple when utilizing software, such as ARENA simulation package and EXTEND in discrete event simulation. The model logic first is verified, and the built model is validated. The components of the model are:

- Entity: patients
- Attributes: patients' health classification
- Activity: treatment, testing, and registration
- Events: arrival of patients and departure of patients
- State variables: number of patients waiting in queues, number of busy physicians and nurses.

The model is built for the care center in ARENA. Although the model looks simpler with ARENA features, the SIMAN blocks and elements of ARENA are used to give the fundamental steps of the model in detail. The model is developed using the data obtained in the input analyzing step. The data is comprised of distribution of inter-arrival times of patients, which is an exponential distribution with a mean of 4 minutes, and registration time, which is normally distributed with mean parameters of 4 minutes and variance of 2 minutes.

- Step 7: Model analysis

Let's assume that the care center works two shifts per day (960 min/day), and the analysts want to analyze the system behavior for a time period of 30 days. Therefore, the simulation model will terminate at the end of the 30th day or on the 28,800th minute (960*30). Since the system inputs are probabilistic, the outputs are probabilistic as well. Therefore, the simulation model is replicated n times for better confidence intervals. The results of the first run of the simulation model are shown in ◘ Image 10.12. The results are analyzed carefully to evaluate the performance, identify the bottlenecks, and make suggestions for improving the system.

As seen from the results, the average waiting times in the queues for the patients that go to "initial assessment" and "treatment in bed" stations are 56.241 minutes and 15.494 minutes, respectively. Let's assume the analyst wants to improve those times and observe how the results would improve if the number of physicians in these stations increases. In the alternative scenario, the number of physicians in "initial assessment" and "treatment in bed" stations is increased by 1, resulting in an increase in the number of physicians to 2 and 3, respectively.

The results of the alternative scenario are given in ◘ Image 10.13. In the alternative scenario, the average waiting times of the patients are 1.6373 minutes and 1.0019 minutes, respectively. The change of the number of physicians in those stations changed the values of the performance measures.

- Step 8: Study documentation

Once the simulation model is run for the current and alternative system, the analyst can report the results. In this example, the main aim is to analyze the improvement in waiting times when the number of physicians is increased. As seen in ◘ Image 10.13 (the results of the first run of the alternative system), the waiting times are decreased. In ◘ Table 10.10, the average waiting times (minutes) for initial assessment and treatment in bed stations for current and alternative system for ten runs are given.

As seen from the results of ◘ Table 10.10, the alternative scenario is much better than the current system in terms of waiting times. However, the cost of employing new physicians should also be considered by the decision-makers. This is a very simple and basic implementation of designing an alternative scenario. The results of the current and alternative scenarios usually need to be statistically analyzed using paired t-tests to draw conclusions for the final decision (see ► Chap. 7 for paired t-tests).

ARENA Simulation Results
Windows User

Summary for Replication 1 of 10

Project: care center example   Run execution date :12/12/2019
Analyst: BC   Model revision date:12/12/2019

Replication ended at time : 28800.0 Minutes
Base Time Units: Minutes

TALLY VARIABLES

Identifier	Average	Half Width	Minimum	Maximum	Observations
initial assesment_queue.WaitingTime	56.241	(Corr)	.00000	257.00	5560
treatment in bed_queue.WaitingTime	15.494	2.9908	.00000	96.747	1741
testing_queue.WaitingTime	7.4331	2.5325	.00000	108.34	1613
reg_queue.WaitingTime	3.1761	.41591	.00000	29.697	4002

DISCRETE-CHANGE VARIABLES

Identifier	Average	Half Width	Minimum	Maximum	Final Value
doctor_bed.NumberBusy	.72781	.03940	.00000	1.0000	.00000
doctor_bed.NumberScheduled	1.0000	(Insuf)	1.0000	1.0000	1.0000
doctor_bed.Utilization	.72781	.03940	.00000	1.0000	.00000
nurse.NumberBusy	.56477	.01126	.00000	1.0000	1.0000
nurse.NumberScheduled	1.0000	(Insuf)	1.0000	1.0000	1.0000
nurse.Utilization	.56477	.01126	.00000	1.0000	1.0000
doctor.NumberBusy	1.9319	.03474	.00000	2.0000	2.0000
doctor.NumberScheduled	2.0000	(Insuf)	2.0000	2.0000	2.0000
doctor.Utilization	.96597	.01737	.00000	1.0000	1.0000
doctor_test.NumberBusy	.48882	.03148	.00000	1.0000	.00000
doctor_test.NumberScheduled	1.0000	(Insuf)	1.0000	1.0000	1.0000
doctor_test.Utilization	.48882	.03148	.00000	1.0000	.00000
initial assesment_queue.NumberInQueue	11.082	(Corr)	.00000	60.000	54.000
treatment in bed_queue.NumberInQueue	.93669	.22183	.00000	8.0000	.00000
testing_queue.NumberInQueue	.41631	.12946	.00000	9.0000	.00000
reg_queue.NumberInQueue	.44136	.06250	.00000	7.0000	.00000

COUNTERS

Identifier	Count	Limit
patients_test	3945	Infinite

OUTPUTS

Identifier	Value

Image 10.12 The results of the first run of simulation for the current system. (Source: Author's creation based on ARENA)

ARENA Simulation Results
Windows User

Summary for Replication 1 of 10

Project: care center example
Analyst: BC

Run execution date :12/12/2019
Model revision date:12/12/2019

Replication ended at time : 28800.0 Minutes
Base Time Units: Minutes

TALLY VARIABLES

Identifier	Average	Half Width	Minimum	Maximum	Observations
initial assesment_queue.WaitingTime	1.6373	.24870	.00000	24.540	5551
treatment in bed queue.WaitingTime	1.0019	.17417	.00000	21.904	1710
testing_queue.WaitingTime	7.1154	1.9396	.00000	121.52	1543
reg_queue.WaitingTime	3.3959	.47739	.00000	42.081	4012

DISCRETE-CHANGE VARIABLES

Identifier	Average	Half Width	Minimum	Maximum	Final Value
doctor_bed.NumberBusy	.71599	.03416	.00000	2.0000	1.0000
doctor_bed.NumberScheduled	2.0000	(Insuf)	2.0000	2.0000	2.0000
doctor_bed.Utilization	.35799	.01708	.00000	1.0000	.50000
nurse.NumberBusy	.55710	.01899	.00000	1.0000	1.0000
nurse.NumberScheduled	1.0000	(Insuf)	1.0000	1.0000	1.0000
nurse.Utilization	.55710	.01899	.00000	1.0000	1.0000
doctor.NumberBusy	1.9343	.05577	.00000	3.0000	3.0000
doctor.NumberScheduled	3.0000	(Insuf)	3.0000	3.0000	3.0000
doctor.Utilization	.64478	.01859	.00000	1.0000	1.0000
doctor_test.NumberBusy	.46159	.03752	.00000	1.0000	1.0000
doctor_test.NumberScheduled	1.0000	(Insuf)	1.0000	1.0000	1.0000
doctor_test.Utilization	.46159	.03752	.00000	1.0000	1.0000
initial assesment_queue.NumberInQueue	.31572	.05678	.00000	9.0000	2.0000
treatment in bed_queue.NumberInQueue	.05949	.01288	.00000	4.0000	.00000
testing_queue.NumberInQueue	.38122	(Corr)	.00000	11.000	.00000
reg_queue.NumberInQueue	.47308	.07468	.00000	11.000	.00000

COUNTERS

Identifier	Count	Limit
patients_test	4005	Infinite

OUTPUTS

Identifier	Value

Image 10.13 The results of the first run of simulation for the alternative system. (Source: Author's creation based on ARENA)

Table 10.10 The average waiting times (minutes) in initial assessment and treatment in bed station queues for 10 replications

Current system		Alternative system	
Initial ass.sta.wait.time (minutes)	Treat.in bed.wait.time (minutes)	Initial ass.sta.wait. time (minutes)	Treat.in bed.wait.time (minutes)
56.2410	15.4940	1.6373	1.0019
52.8550	16.2480	1.7434	0.9955
66.3480	14.0270	1.9247	1.0866
103.2100	18.9800	1.548	0.99435
66.0650	21.3000	2.0063	0.79902
145.1900	14.9140	1.9429	1.0962
77.6910	18.0450	1.8294	1.0542
80.6920	13.6690	1.9386	1.0946
151.3000	15.4490	1.8032	1.1656
51.8950	15.3760	1.5243	1.2102

Source: Author's creation

10

10.3.6 Simulation Modeling Features

In this section, we describe an example system and discuss the features to simulate the system behavior and performance. The example is a simple case of a manufacturing process as given in Fig. 10.6. Parts arriving at a milling operation center are processed by a single milling-cutter and then leave the system. If a part arrives and milling-cutter is idle, its processing will start immediately. If the milling-cutter is busy, the part will wait in the queue with first-in-first-out principle. This is the main logic of the system. The other features of the system will be explained below.

10.3.6.1 Discrete Event Simulation (DES)

DES deals with modeling the systems which represent the evolution of variation of variables instantaneously in the dynamic system over a period of time (Karnon et al. 2012). The milling center example is a simple, but good, example for understanding DES. Parts (entities) arrive and leave the system resulting in the change of the state of the system. The state variables change in discrete steps over a time period. For example, the number of waiting parts in the queue over a time period is given in Fig. 10.7.

10.3.6.2 Start and Stop of Simulation

While building and running simulation studies, the Six Sigma team specifies parameters for when the simulation starts and stops. First, the time unit is determined depending on the system properties. In the milling center example, minutes are used as time representations. The system starts at a time zero with no parts processing, and the milling-cutter is idle. This initial condition can be realistic if the system does not continue its operation at the end of the day and the new day starts with an idle service. But for most manufacturing processes, this is not the case; if the process is an ongoing one, then the initial conditions should be determined accordingly. Together with the starting time and initial conditions of the simulation, the stopping condition (termination criteria) is decided. In this example, termination of the simulation model is given as 25 minutes.

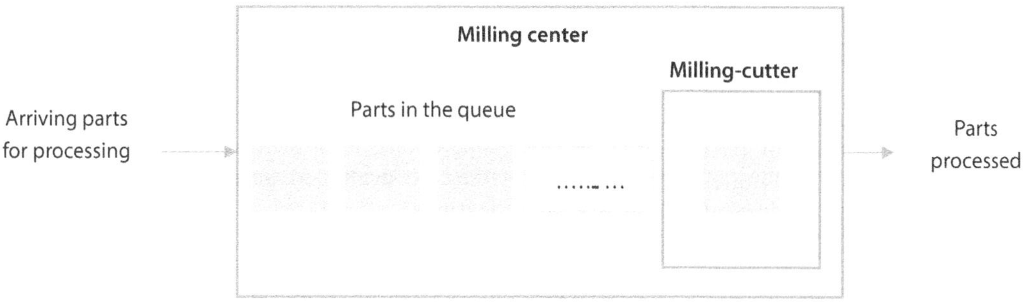

■ **Fig. 10.6** An example of a single server manufacturing process. (Source: Author's creation)

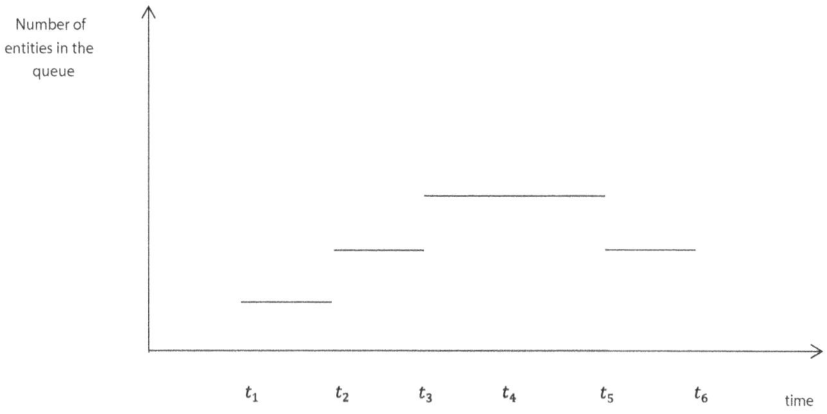

■ **Fig. 10.7** The number of entities in the queue over a time period t for a discrete system. (Source: Author's creation)

10.3.6.3 Queueing Theory

A queueing system is described by population, arrival rate, service mechanism, system capacity, and queueing discipline. The milling center example is a typical representation of single server queue systems in discrete event simulation. Calling population is infinite, and arrival rate does not change. Unless a different mechanism is specified, the units are served based on first-in-first-out. Arrivals are defined by the distribution of the time between arrivals and inter-arrival time. Service times are defined by a distribution. An entity leaves the system immediately after completing the service. Following the completion of a service, the first entity occupies the service. When the arrival rate is less than the service rate, then the system is defined as stable; otherwise, the system is called unstable and is one in which a queue will grow unbounded. The time durations that are necessary to model the system are shown in ■ Table 10.11. We will explain where those numbers come from and how to use them in the following sections.

■ **Table 10.11** Arrival, inter-arrival, and service times

Part number	Arrival time	Inter-arrival time	Service time
1	0.00	1.85	2.87
2	1.85	1.63	1.75
3	3.48	1.38	3.75
4	4.86	1.89	2.67
5	6.75	.	4.79
.	.	.	.
.	.	.	.
.	.	.	.
.	.	.	.

Source: Author's creation

10.3.6.4 Performance Measures

The milling center system is comprised of the service, parts (in waiting line or being served), and a simulation clock. The system state variables are the number of parts waiting in the queue, status of the milling-cutter (idle, busy), and the inter-arrival time of the parts. The events that would change the system status are the arrival or the departure of a part. The performance measures of such a queuing system can be summarized as follows:

- Total production number during the 25 minutes of operation
- The maximum waiting time in queue of parts
- Average total time in system of parts (cycle time)
- Average waiting time in queue of parts.

Let WQ_i be the waiting time in queue for part i and N the number of parts processed in 25 minutes. Then, the average waiting time per part in queue is

$$\frac{\sum_{i=1}^{20} WQ_i}{N} \tag{10.14}$$

- Time-average number of parts waiting in the queue
  Let $Q(t)$ be the number of parts in the queue at any time t, and then the average time of parts in the queue is the total area under the $Q(t)$ curve divided by the length of the run, 25 minutes.

$$\int_0^{25} \frac{Q(t)dt}{25} \tag{10.15}$$

- The utilization of milling-cutter
  Utilization is the proportion of time it is busy within the simulation.

$$B(t) = \begin{Bmatrix} 1 \text{ if the milling cutter is busy at time } t \\ 0 \text{ if the milling cutter is idle at time } t \end{Bmatrix} \tag{10.16}$$

  The utilization is the area under $B(t)$, divided by the length of the run.

$$\int_0^{25} \frac{B(t)dt}{25} \tag{10.17}$$

It is not always preferable for the value of utilization to be very high, that is, "close to 1." Utilization with value 1 shows that the system works at full capacity, but this can result in long queues and low throughputs. The values must be evaluated according to the system features.

10.3.7 Performing an Event-Driven Simulation

Before presenting the steps for performing a simulation analysis, the mechanism behind the time advancement in discrete models is clarified in detail. This section is devoted to the explanation of how the simulation clock changes over time, and performing a simulation analysis manually is explained in the second part.

10.3.7.1 Simulation Clock and Time Advancement Mechanism

Due to the structure of the DES, the clock for the simulation must be known at each step. Therefore, there should be a mechanism defined to advance the time from point to another. The clock for the simulation has nothing to do with the real computational time for the simulation. The most common approach for the advancement of time in DES is the next event time advancement. The steps for the next time advancement are given below:

- Step 1: Initialize simulation clock to zero.
  Next event time advance mechanism estimates the time of futuristic events that are going to happen on the basis of a list of events (in terms of arrival state or departure state). Under this approach, the mechanism is started along with locating the simulation clock at zero. The simulation clock is initialized at 0.
- Step 2: Determine the times of occurrences of future events.
  Times of all known future events are determined and placed in the future events list (FEL), ordered by time.
- Step 3: Advance clock to the most imminent event of the future event.
  The clock advances to the most imminent event, then to the next imminent event, etc.

- Step 4: Update system variables.
  At each event, the system state is updated. The system state can only change at event times. Nothing really happens between events. The simulation progresses by sequentially executing the most imminent event on the FEL. When the clock advances to the most imminent event, the system state is updated, depending on what type of event it is (arrival, departure, etc.). For example, if it is an arrival event, you may need to change an idle server to busy or add a new part to the queue if the server is already busy.
- Step 5: Update knowledge of the times for future events.
  The FEL is updated by inserting new events or deleting events. For instance, when a part arrives at a system, typical simulation programs will immediately spawn the next arrival time. We will place this new event in the FEL considering the ordered times. If this arrival time is later than the departure of the previous part, it is simply put at the "end" of the FEL. If the next arrival occurs before the previous part departure, then that next arrival has to be inserted in the interior of the FEL. An example of next time event advancement mechanism is given in ◘ Fig. 10.8 for three customers in a single server queue where

t_i : the arrival time for the entity $i\,(t = 0)$

$a_i : t_i - t_{i-1}$ interarrival time between two consecutive entities

s_i : the service time for entity i

d_i : the time in the queue for entity i

$c_i : t_i + d_i + s_i$, the time for the departure of entity i

e_i : an event occurring time

F_a : distribution for interarrival times

F_s : distribution for service times.

The service is idle at $e_0 = 0$. The first entity arrives at t_1, which is obtained by the random variable a_1 generated from F_a distribution function, $0 + a_1 = t_1$. The simulation clock is advanced from e_0 to e_1. The entity that arrived at t_1 finds the service idle. The wait-

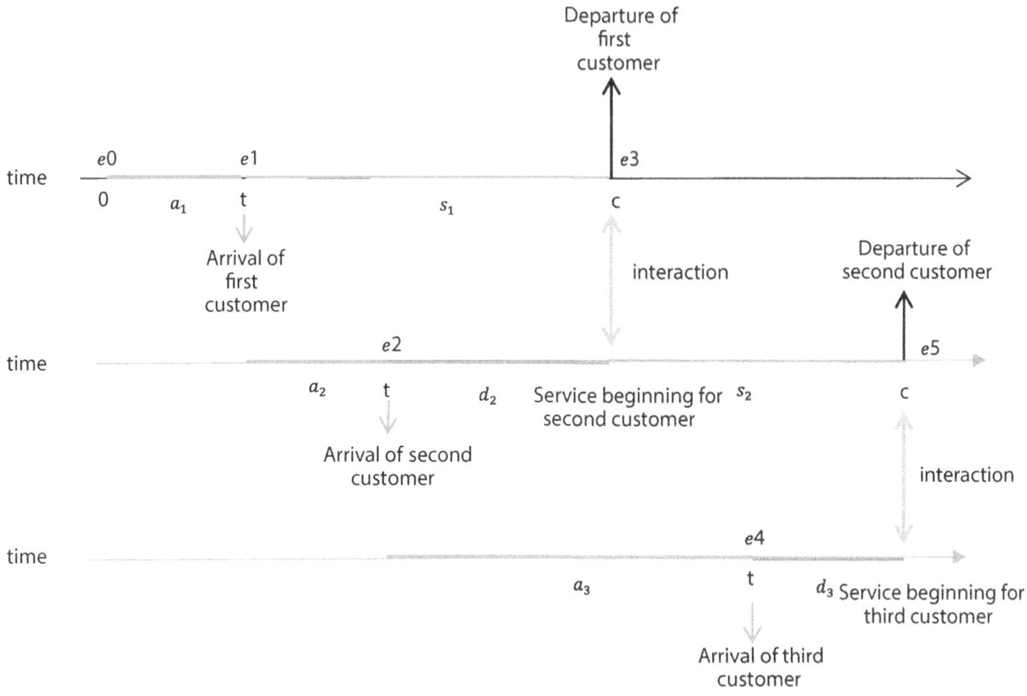

◘ **Fig. 10.8** Three customers' process interaction in a single server queue. (Source: Author's creation)

ing time in the queue for the first entity is 0 ($d_1 = 0$).The state of the service is turned to busy. The departure time for the first entity is $c_1 = t_1 + d_1 + s_1$. The s_1 is generated from the distribution function F_s. After the first entity arrives at a queue, typical simulation programs will immediately spawn the next arrival time, which is obtained by the random variable a_2 generated from F_a distribution function, $t_1 + a_2 = t_2$. When the new arrival time is generated, the simulation determines the most imminent event. If $t_2 < c_1$, then the simulation clock is advanced to $e_2 = t_2$. If $c_1 < t_2$, the simulation clock would advance to $e_2 = c_1$. In the example above, the imminent event is the arrival of the second customer, and therefore $e_2 = t_2$.

The second customer will find the server busy and waits in the queue until the first customer's departure, which is c_1 . Following the arrival of the second customer, the new arrival time for the third customer is generated, $t_2 + a_3 = t_3$. Now the simulation compares if $t_3 < c_1$. In the example above, since $c_1 < t_3$, the imminent event is c_1. Simulation clock is advanced from e_2 to e_3, where $e_3 = c_1$. At time c_1, the first customer departs, and the second customer receives the service. The waiting time for the second customer is $d_2 = c_1 - t_2$. When the second customer starts receiving the service, the service time s_2 is generated from F_s, and the departure time for the second customer is calculated $c_2 = c_1 + + s_2$. Since $t_3 < c_2$, the simulation clock advances to t_3, $e_4 = t_3$. The third customer waits until the second customer departs, and when the second customer departs, the third customer starts receiving the service, and simulation time advances to c_2, $e_5 = c_2$. The waiting time for the third customer is $d_3 = c_2 - t_3$. For this example, the simulation will run over the time interval [0, TE], where TE is the termination criteria specified for the simulation. For this example, TE is the time of the second service completion. It could also be a pre-specified time which corresponds to the departure of the second customer. The FEL generated at the end of the simulation will be $e_0 < t_1 < t_2 < c_1 < t_3 < c_2$.

10

The single server system has two types of events: arrivals and departures. To simulate using next event time advance, we need to keep track of events in chronological order and know how to model the events. The idea behind the modeling of the arrival and departure events is given in ◘ Figs. 10.9 and 10.10.

Arrival Event Whenever a new part arrives at the system, the next new part arrival time is scheduled and placed into the FEL chronologically. If the milling-cutter is idle when the part gets into the service, then the milling-cutter is changed to busy, and service time for this part is generated. On the other hand, if the milling-cutter is busy when the part enters in the queue, then the queue length is incremented. Events occur randomly depending on the inter-arrival times (a_i) defined by a distribution (F_a) and service times (s_i) defined according to a distribution (F_s).

Departure Event In the departure event, the part is finished with the milling-cutter and ready to leave the system. If queue is not empty, it begins servicing next part; otherwise milling-cutter will be idle. Events occur randomly depending on the inter-arrival times (a_i) defined by a distribution (F_a) and service times (s_i) defined according to a distribution (F_s).

10.3.7.2 Event-Driven Simulation by Hand

The idea behind event-driven algorithm is based on the next event time advancement mechanism described in the previous section. At each event occurrence (arrival or departure), the future event is generated and added into FEL. The basic way of implementing this algorithm is to use simulation tables, which make it easier to track the system over the time. The steps of performing an event-driven simulation manually are as follows:

- Step 1: Read input data and generate random data.
- Step 2: Build a simulation table specific to the problem.

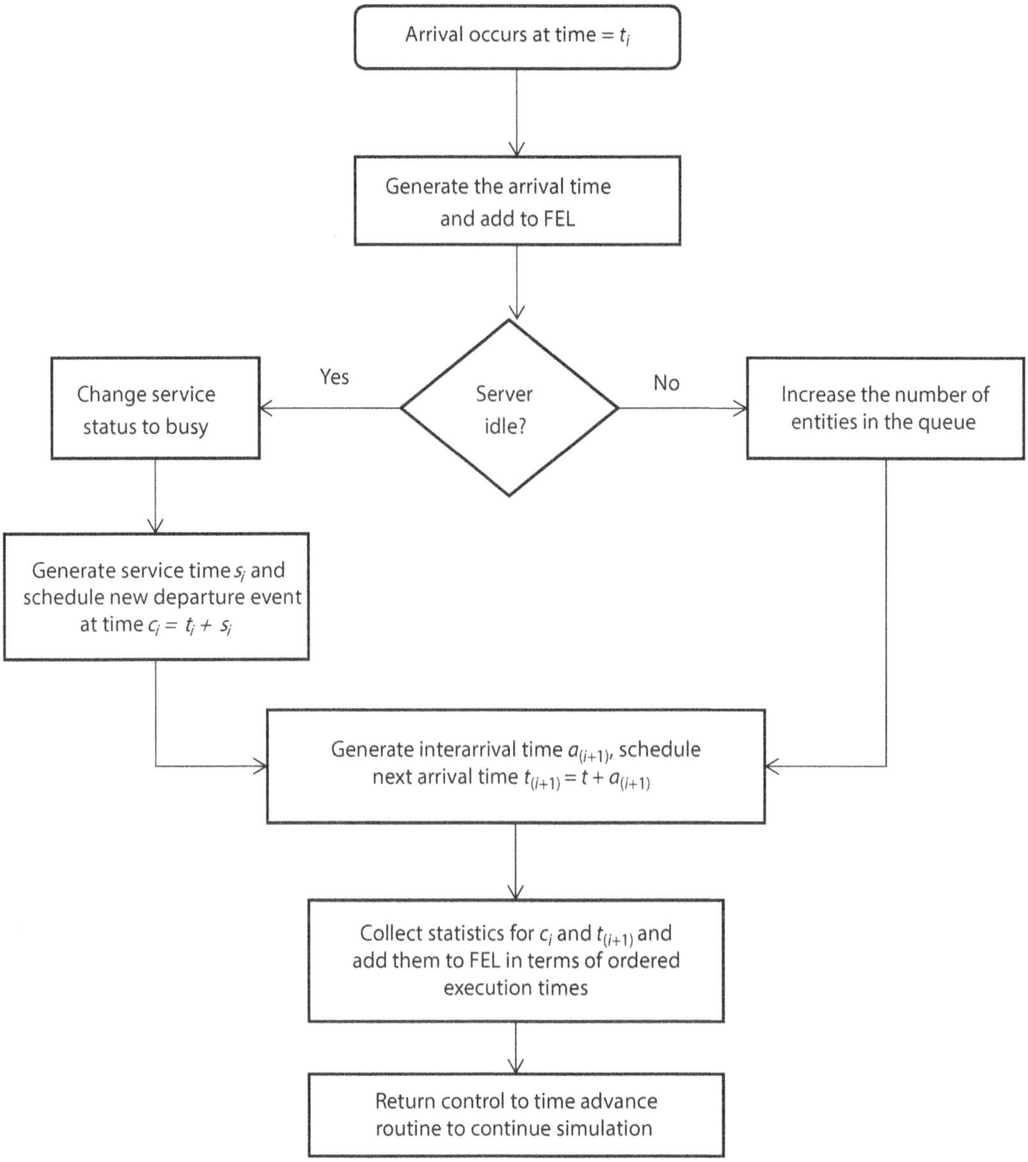

Fig. 10.9 Flow chart of an arrival event. (Source: Author's creation)

- Step 3: FEL updating process
  while (FEL is not empty),
  - determine the next event on FEL,
  - apply arrival or departure subroutine,
  - advance simulation clock to the tie of next event,
  - remove the occurring event from FEL, and
  - collect statistics from current state.
- Step 4: Statistical analysis.
- Step 5: Report and recommendations.

The steps are explained based on the example given below.

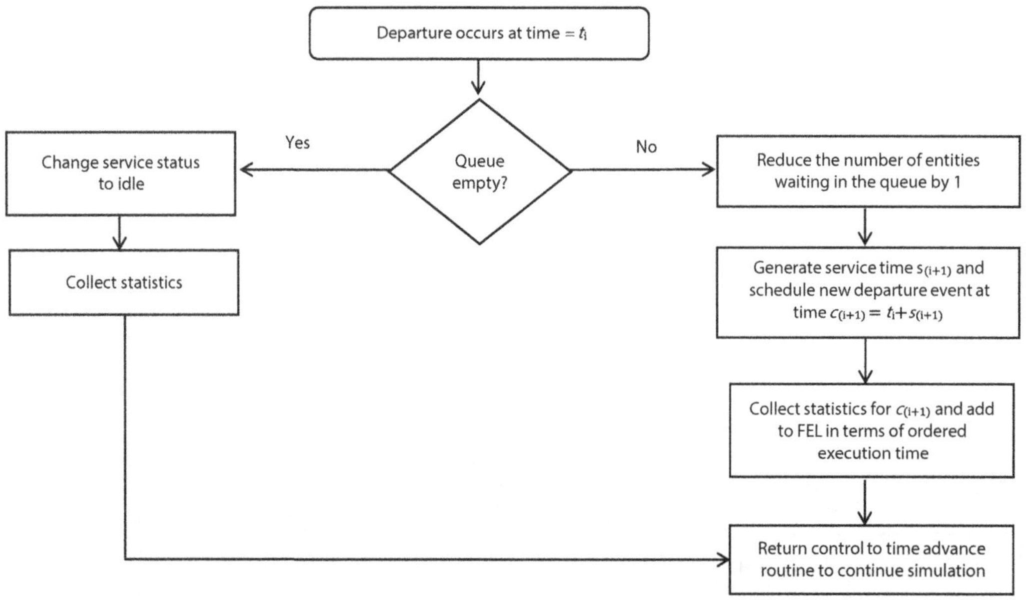

Fig. 10.10 Flow chart of a departure event. (Source: Author's creation)

▶ Example 7 Single-Service Queue Example

A small electronics store has only one cashier counter. Customers arrive at the cashier randomly 2–9 minutes apart, as shown in Table 10.11. Each inter-arrival time has the same probability of occurrence. Inter-arrival times vary from 2 to 9 minutes, with the probabilities shown in Table 10.12. The goal of the problem is to simulate the system for 15 customers and analyze the statistics. ◀

- Step 1: Read input data and generate random data.

  The first step is determining characteristic inputs by probability distributions, either continuous or discrete. The queuing systems described above can be modeled as discrete event simulation. In such a system, events are usually observed at random times, which corresponds to a probabilistic system classification. For example, it is not known when the next customer will arrive at the cashier counter (arrival times) or how long it will take for a customer to complete a payment (service times). In these cases, statistical analyses are used to determine the probability distributions of inter-arrival and service times. The system is observed, and required data are collected, e.g., the arrival times and service times of a cashier counter are recorded for certain periods. Once the data are collected and proper distribution functions are determined, the inter-arrival times and service times are generated from the distributions. For this example, generating random data from input data is as follows: It should be noted that we assume that the system starts with zero customers and an idle service. However, in real-life cases, the initial conditions of the systems could be different. Here, starting conditions are ignored to keep the calculations simple. The inter-arrival times for customers vary between 2 and 9 minutes with equal probability. Table 10.12 depicts the probability of each inter-arrival time and the random digit assignments obtained from cumulative probability values. The rightmost column data (random digit assignment) are obtained using the cumulative probability values. For instance, the second random digit assignment is 126–250 and is obtained from the cumulative probabilities of the first and second column

Table 10.12 Distribution of inter-arrival times and service times

Inter-arrival times (minutes)	Probability	Cumulative probability	Random-digit assignment
2	0.125	0.125	001-125
3	0.125	0.250	126-250
4	0.125	0.375	251-375
5	0.125	0.500	376-500
6	0.125	0.625	501-625
7	0.125	0.750	626-750
8	0.125	0.875	751-875
9	0.125	1.000	876-000
Service time (minutes)	**Probability**	**Cumulative probability**	**Random-digit assignment**
6	0.20	0.20	01-10
7	0.40	0.60	11-60
8	0.25	0.85	61-85
9	0.15	1.00	86-00

Source: Author's creation

(0.125 and 0.250). There are 125 possible random-digit assignments (126–250) assigned to the occurrence of inter-arrival times of 3 minutes. Table 10.12 is also a representation for service times.

When the distributions for the arrival and service times are obtained, the next step is generating the values from those distributions, that is, to generate a set of random numbers. A random number is a uniformly distributed value between 0 and 1 and independent from successive random numbers. Random digit tables can be used to form random numbers by selecting appropriate number of digits for each random number and placing a decimal point to the left of the value. The random numbers can also be generated in Excel using Excel RAND() command. In this example, we will use random digit tables to generate random numbers. Below is an example of two lines of random digits:

- Line 1: 95637 05007 16605 81194 14873 04197 85576 45195
- Line 2: 11169 15529 33241 83594 01727 86595 65723 82322

- The properties of random digits are as follows:
  - Each digit is equally likely to be any of the ten digits 0 through 9.
  - The digits are independent from each other.

The random number that could be generated from the upper random digits would be .95637, .05007, .16605, .81194...etc. In the example, the probabilities of the inter-arrival times are three digits. Therefore, three-place random digits would be enough to generate the random numbers. For inter-arrival probabilities given in Table 10.12, 14 random numbers are generated because the first customer is assumed to arrive at time zero, and 14 more arrivals have to be processed in order to simulate 15 customers. Similarly, two-place random digits will be sufficient for generating service times from distribution given in Table 10.12.

Table 10.13 Inter-arrival times determination

Customer	Random digits	Inter-arrival times (minutes)	Customer	Random digits	Inter-arrival times (minutes)
1	–	–	9	887	9
2	578	6	10	284	4
3	128	3	11	054	2
4	985	9	12	394	5
5	684	7	13	376	5
6	592	6	14	741	7
7	208	3	15	369	4
8	911	9			
Customer	**Random digits**	**Service times (minutes)**	**Customer**	**Random digits**	**Service times (minutes)**
1	66	8	9	51	7
2	12	7	10	80	8
3	64	8	11	76	8
4	89	9	12	24	7
5	48	7	13	30	7
6	04	6	14	45	7
7	22	7	15	60	7
8	86	9			

Source: Author's creation

10

The third and fourth columns of Table 10.12 are used to generate random inter-arrival and service times. The third column shows the cumulative distributions, and the fourth column shows the random digit assignments. The random values are generated by listing appropriate number of random digits and comparing them to the random digit assignments given in the last columns of the tables.

The inter-arrival times determination is shown in Table 10.13. The first random digits used in Table 10.13 are 578. These random digits fall into the [501-625] interval corresponding to 6 minutes in Table 10.12. Hence, the first inter-arrival time is generated as 6 minutes. The other inter-arrival times in Table 10.13 are determined in similar way.

Service times are generated similar to those determined in inter-arrival times in Table 10.13. For example, the first customer's service time is 8 minutes since the random digit 66 corresponds to the interval of [61–85] in Table 10.12.

- Step 2: Build a simulation table specific for the problem.
  Each simulation table could be different depending on the problem properties and the performance measures. The framework of a simulation table is given in Table 10.14. In that table x_{ij} represents the repetitions and inputs, where $i = 1,2,..,n$ is for repetitions and $j = 1,2,..,m$ is for inputs. For each repetition, there is one response denoted as y_i.
- Step3. FEL updating process.
  For each repetition i, the required input data determined in Step 1 is used, and FEL is generated. While FEL is not empty,

Table 10.14 An example for simulation table

	Inputs							Response
Repetitions	x_{i1}	x_{i2}	x_{i3}	x_{i4}	…………………	$x_{i(m-1)}$	x_{im}	y_i
1								
2								
3								
.								
.								
.								
.								
n								

Source: Author's creation

continue with determining the next event on FEL, applying arrival or departure subroutine, advancing simulation clock to the tie of next event, and removing the occurring event from FEL. Table 10.15 shows the implementation of this process for the cashier example. The first column (just finished event) corresponds to the event deleted from the FEL. For instance, the event [2,6,Arr] (which means the arrival of the customer 2 at time 6) will be deleted from FEL and written as a just finished event when it occurs. The variables are the number of customers in the queue ($Q(t)$) and the utilization of the cashier ($B(t)$) as given in the second main column in Table 10.15. Attributes, which are the properties of the customers (in this case arrival and service times), are given in two separate columns as arrival times in queue and in service in Table 10.15. Those attributes are recorded to calculate statistical accumulators, which will be explained in the following step. The last column in Table 10.15 is the FEL, where the future events are ordered chronologically. Below a brief explanation of the process is given:

- Customer 1 arrives at the system at time zero [1,0,Arr] and starts getting the service immediately. Customer 1 will not wait in the queue ($Q(t)$=0 and $B(t)$=1). As explained in ► Sect. 10.3.7.1 following an arrival event, the next arrival time will be produced and placed to FEL, comparing the ordered execution times. In this case, arrival of Customer 2 occurs before the departure of Customer 1. Therefore, the next event that will be deleted from FEL is [2,6,Arr]. Customer 2 will find the cashier busy and enter into the queue at time 6. Now, $Q(t)$=1 due to waiting Customer 2 at the queue and $B(t)$=1 since Customer 1 still is receiving service. Then, the most imminent event is [1,8,Dep], meaning that the departure event of Customer 1 will change the cashier to idle, and Customer 2 receives service immediately. Now, Customer 2 makes the cashier busy, and the arrival time of Customer 2 is stored in its attributes (not in parenthesis). The next event after [1,8,Dep] is [3,9,Arr], and Customer 3 waits until Customer 2 departs. The arrival of Customer 3 at time 9 results in increase in $Q(t)$, and the event is deleted from FEL. This process continues until Customer 15 departs the system as shown in Table 10.15.

- Step 4: Statistical analysis.
  The statistics for the example can be calculated using the simulation table generated manually. Some of the findings from the simulation in Table 10.15 are as follows:

Table 10.15 Record of the manual simulation

Just finished event			Variables		Attributes	Future event list	Statistics	
Customer number	Time (*t*)	Event type	Q(*t*)	B(*t*)	Arrival times (in queue) in service	FEL [cus. number, time, type]	Total time customer wait in queue*	Total time customer spend in system*
1	0	Arr	0	1	0	[2, 6,Arr] [1,8,Dep]		
2	6	Arr	1	1	(6) 0	[1,8,Dep] [3,9,Arr]		
1	8	Dep	0	1	() 6	[3,9,Arr] [2,15,Dep]	8–6 = 2 (#2)	8–0 = 8 (#1)
3	9	Arr	1	1	(9) 6	[2,15,Dep] [4,18, Arr]		
2	15	Dep	0	1	() 9	[4,18, Arr] [3,23,Dep]	15–9 = 6 (#3)	15–6 = 9 (#2)
4	18	Arr	1	1	(18) 9	[3,23,Dep] [5,25,Arr]		
3	23	Dep	0	1	() 18	[5,25,Arr] [4,32,Dep]	23–18 = 5 (#4)	23–9 = 14 (#3)
5	25	Arr	1	1	(25) 18	[6,31,Arr] [4,32,Dep]		
6	31	Arr	2	1	(25, 31) 18	[4,32,Dep] [7,34,Arr]		
4	32	Dep	1	1	(31) 25	[7,34,Arr] [5,39,Dep]	32–25 = 7 (#5)	32–18 = 14 (#4)
7	34	Arr	2	1	(31,34) 25	[5,39,Dep] [8,43,Arr]		
5	39	Dep	1	1	(34) 31	[8,43,Arr] [6,45,Dep]	39–31 = 8 (#6)	39–25 = 14 (#5)
8	43	Arr	2	1	(34,43) 31	[6,45,Dep] [9,52,Arr]		
6	45	Dep	1	1	(43) 34	[9,52,Arr] [7,52,Dep]	45–34 = 11 (#7)	45–31 = 14 (#6)
7	52	Dep	0	1	() 43	[9,52,Arr] [8,61,Dep]	52–43 = 9 (#8)	52–34 = 18 (#7)
9	52	Arr	1	1	(52) 43	[10,56,Arr] [8,61,Dep]		
10	56	Arr	2	1	(52,56) 43	[11,58,Arr] [8,61,Dep]		
11	58	Arr	3	1	(52,56,58) 43	[8,61,Dep] [12,63,Arr]		
8	61	Dep	2	1	(56,58) 52	[12,63,Arr] [9,68,Dep]	61–52 = 9 (#9)	61–43 = 18 (#8)
12	63	Arr	3	1	(56,58,63) 52	[9,68,Dep] [13,68,Arr]		
9	68	Dep	2	1	(58,63) 56	[13,68,Arr] [10,76,Dep]	68–56 = 12 (#10)	68–52 = 16 (#9)
13	68	Arr	3	1	(58,63,68) 56	[14,74,Arr] [10,76,Dep]		
14	74	Arr	4	1	(58,63,68,74) 56	[10,76,Dep] [14,78,Arr]		
10	76	Dep	3	1	(63,68,74) 58	[14,78,Arr] [11,84,Dep]	76–58 = 18 (#11)	76–56 = 20 (#10)
14	78	Arr	4	1	(63,68,74,78) 58	[15,82,Arr] [11,84,Dep]		
15	82	Arr	5	1	(63,68,74, 78,82) 58	[11,84,Dep]		
11	84	Dep	4	1	(68,74,78,82) 63	[12,91,Dep]	84–63 = 21 (#12)	84–58 = 26 (#11)
12	91	Dep	3	1	(74,78,82) 68	[13,98,Dep]	91–68 = 23 (#13)	91–63 = 28 (#12)
13	98	Dep	2	1	(78,82) 74	[14,105,Dep]	98–74 = 24 (#14)	98–68 = 30 (#13)
14	105	Dep	1	1	(82) 78	[15, 112, Dep]	105–78 = 27 (#15)	105–78 = 27 (#14) 112–82 = 30 (#15)

*The numbers in the parentheses demonstrate the number of the customer.
Source: Author's creation

1. The average waiting time for a customer

$$\text{Average waiting time(min.)} = \frac{\text{total time customers wait in queue(min.)}}{\text{total number of customers}} \tag{10.18}$$

Total number of customers waiting in queue can be found by summing each customer time spent in the queue. The time a customer spent in queue is obtained by subtracting the arrival time of the customer from the starting time of service for the customer. From ◘ Table 10.15, this value can be derived only from departure events as departure of a customer will be the start of next one to start getting service.

$$\text{Average waiting time(min.)} = \frac{\begin{array}{c}2+6+5+7+8+11+9+9+12+\\18+21+23+24+27\end{array}}{15} = 12.13\,\text{minutes}$$

2. The probability that a customer has to wait in the queue

The number of customers who wait is obtained from the last column of the ◘ Table 10.15.

The total number of customers waiting in the queue is the number of customers having waiting time greater than 0. In our example, 14 customers among 15 have to wait in the queue. The customer 1 finds the service line empty and therefore he/she does not have to wait.

$$\text{Probability(wait)} = \frac{\text{number of customer who wait}}{\text{total number of customers}} = \frac{14}{15} = 0.933 \tag{10.19}$$

3. The fraction of idle time of cashier

Idle time of the server can be obtained from the variable $B(t)$. In our example the value of $B(t)$ is always 1, meaning the server is always busy. Therefore, the idle time is 0. Total run of simulation is the time the 15th customer departs [15, 112, Dep].

$$\text{Probability of idle server} = \frac{\text{total idle time of server(min.)}}{\text{total run of simulation(min.)}} = \frac{0}{112} = 0 \tag{10.20}$$

4. The average service time

$$\text{Average service time(min.)} = \frac{\text{total service time(min.)}}{\text{total number of customers}} \tag{10.21}$$

$$\text{Average service time(min.)} = \frac{\begin{array}{c}8+7+8+9+7+6+7+9+\\7+8+8+7+7+7+7\end{array}}{15} = \frac{112}{15} = 7.466\,\text{minutes}$$

5. The average time a customer spends in the system

$$\text{Average time customer spends in system (min.)} = \frac{\text{total time customer spend in the system (min.)}}{\text{total number of customers}} \quad (10.22)$$

For a customer, the time spent in the system can be calculated by subtracting the arrival time from its departure time. Those values can be obtained for each customer from ■ Table 10.15, events list.

$$\text{Average time customer spends in system (min.)} = \frac{8+9+14+14+14+14+18+18+16+20+26+28+30+27+30}{15} = \frac{286}{15} = 19.666\,\text{minutes}$$

- Step 5: Report and recommendation.
  Simulation results can be summarized according to the purpose of the project, using the results obtained from the statistical analysis step. The overall analysis of the system performance measures based on the manual simulation is summarized as follows:
  - The average waiting time for a customer: 12.13 minutes
  - The probability that a customer has to wait in the queue: 0.933
  - The fraction of idle time of cashier: 0
  - The average service time: 7.466 minutes
  - The average time a customer spends in the system: 19.66 minutes.

  These performance measures are used to evaluate the existing system, and alternative systems can be proposed based on the results. For instance, the cashier works full time, and there is an increasing queue in front of the cashier. This results in average 12.13 minutes of queue waiting time for 15 customers. The Six Sigma team can propose alternative system designs (additional cashier, lower service times, etc.) and compare the system designs based on the performance measure values. The outputs of the simulation are random as the input variables are random. In such a situation, for higher accuracy in results (to obtain better confidence intervals), the number of replications is increased. The randomness of the simulation is explained more in detail in the following section.

10.3.7.3 Randomness in Simulation

As we explained in the previous sections, running the simulation model for one time would lead to less accuracy in the results due to random nature of the inputs and outputs. When performing a stochastic simulation, the analysts should increase the number of runs when making important decisions. Several independent runs are called replications in simulation terminology, and they are used to analyze how the results change in every replication. Every replication has to be identical, starting with the same conditions, stopping with the same termination criteria, and using the same parameter settings. When the replications are completed, we will have n independent and identically distributed results for a specific performance measure. The confidence interval for the performance measure results obtained by the replications can be calculated as follows:

$$\overline{X} \mp t_{n-1,1-\frac{\alpha}{2}} \frac{s}{\sqrt{n}} \tag{10.23}$$

where $\overline{X}$ is the mean of the performance measure values (sample mean), s is the standard deviation of the sample, n is the number of replications, and $t_{n-1,1-\frac{\alpha}{2}}$ is the value obtained from the Student's t distribution with $n-1$ degrees of freedom. To obtain smaller confidence intervals, the Six Sigma team should simply increase the number of replications. The required number of replications for desired confidence interval is calculated using statistical methods. Since Student's t distribution and confidence intervals are detailed in ▶ Chap. 7, we will not analyze confidence intervals in this section.

10.4 Lean Philosophy and Principles

Lean principles and philosophy are implemented in the Improve phase of DMAIC process. Lean principles were successfully implemented in a great variety of industries after Toyota Production System (TPS) originated lean in the automobile industry. Many industries benefited from TPS by reducing waste and increasing reliability, quality, and overall performance. Essentially, the norms and values of Japanese societal culture had an influence in shaping and fostering TPS. Lean philosophy focuses on minimizing waste (*muda*) throughout the system, increasing value for all stakeholders, eliminating variation and NVA activities in the processes (*mura*), minimizing poor work conditions, and continuously improving the overall performance. In lean, "doing the right thing at the first time" is a general rule where wastes are identified and eliminated throughout the system. It is also critical to identify "waste" and "value" terms in stakeholders', and specifically the customer's points of view.

Lean philosophy categorizes waste (*muda*) into seven categorizes: (1) unnecessary transportation (T), (2) excessive inventory (I), (3) unnecessary motion (M), (4) excessive waiting (W), (5) overprocessing (O), (6) overproduction (O), and (7) defective products or services (D). Lean philosophy was built upon two pillars in the TPS: just-in-time (*JIT*) and *Jidoka* (see ◘ Fig. 1.3 in ▶ Chap. 1). JIT systems emphasize delivering the right service or product, at the right place, at the right time. JIT systems are designed to supply or produce only what is needed and ordered by the customers, internally and externally. *Jidoka* is a Japanese term that means that the production systems automatically detect and identify defects and halt the processes until the issues are fixed. As an overarching mechanism, PDCA (Plan-Do-Check-Act) system triggers and drives all lean effort. The details of the lean philosophy are categorized into 4 sections and 14 principles as follows:

Philosophy – Long Term

1. Base your management decisions on a long-term philosophy, even at the expense of short-term financial goals.

Process – Promote Flow (Creating a Pull Production System That Has Continuous Flow and Balanced Workload)

2. Create a continuous process flow to bring problems to the surface.
3. Use pull systems to avoid overproduction.
4. Level out the workload (heijunka).
5. Build a culture of stopping to fix problems, to get quality right the first time.
6. Standardized tasks are the foundation for continuous improvement and employee empowerment.
7. Use visual control so no problems are hidden.
8. Use only reliable, thoroughly tested technology that serves your people and processes.

People – Respect and Development

9. Growing leaders who thoroughly understand the work, live the philosophy, and teach it to others.
10. Developing exceptional people and teams who follow your company's philosophy.
11. Respecting your extended network of partners and suppliers by challenging them and helping them improve.

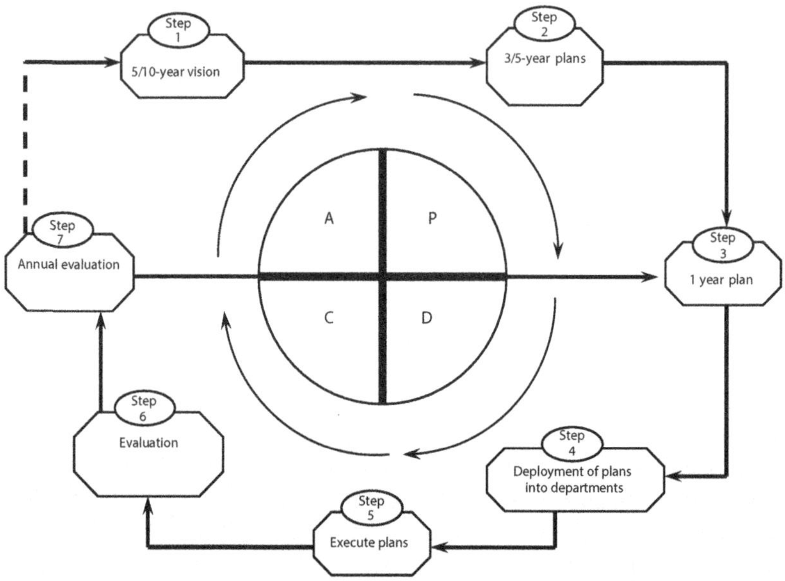

Fig. 10.11 *Hoshin Kanri* process. (Source: Adapted from King, (1989))

10

Problem-Solving – Continuous Improvement (Organize Their Continuous Improvement Activities)

12. Go and see for yourself to thoroughly understand the situation (genchi genbutsu).
13. Make decisions slowly by consensus, thoroughly considering all options; implement decisions rapidly.
14. Become a learning organization through relentless reflection (hansei) and continuous improvement (kaizen).

Each principle of lean philosophy is analyzed in the following sections. The first section of lean, *philosophy – long term*, includes the first principle of lean.

- *Principle 1: Base your management decisions on a long-term philosophy, even at the expense of short-term financial goals.*

Lean philosophy recommends organizations to focus on the long-term goals and objectives for creating a sustainable future in dynamic market conditions. Because of the nature of the Japanese societal culture in lean, managing the future of the organization is pivotal. Rather than focusing on short-term profits and financial advantages, the future of the organization should be built on a concrete organizational structure and system to achieve long-term goals and objectives. Generating value for all stakeholders, including customers, employees, and investors, is the main driving force behind lean philosophy. Principle 1 emphasizes that generating value requires management processes focusing on long-term goals and objectives. *Hoshin Kanri* (policy deployment) is the main strategic management approach implemented by Toyota and other successful companies to direct decision-making processes toward long-term philosophy. The structure of *Hoshin Kanri* is shown in Fig. 10.11.

Using a cascade system and long-term approach, organizations set up organization-wide long-term goals, the vision and mission, in the first step. Then, long-term goals are broken into 3–5-year plans and 1-year plans. Departmental goals and objectives aligned with 1-year plan are identified in the next step. Later on, each position in each department is linked with the departmental goals and objectives. In the cascade system, the "strategies" of a higher hierarchical level department are employed as "goals and objectives" by a lower-level department. The cascade system is based on catch-ball process through which all departments, units, and individual employees contribute to identify and achieve organization-wide goals and objectives. Employees' individual goals are generated from departmental goals and objectives using the employee performance management systems. AT&T, Danone, Florida Power & Light, Proctor & Gamble, and HP are some examples of Western organizations that implemented *Hoshin Kanri* into their long-term planning processes. We will not go further into the details. King (1989) can be used for further readings for *Hoshin Kanri*.

Stability, long-term growth, mutual benefits, creating lasting customer satisfaction, and investing in long-term technology are the driving forces of Principle 1. Six Sigma teams accept these points as their motto while running the projects. Six Sigma teams are likely to benefit from Principle 1 in the long run, as the overarching principle of lean.

The second section of lean, *process – promote flow*, includes seven principles. This section mainly focuses on production-related functions and requirements of lean.

- *Principle 2: Create a continuous process flow to bring problems to the surface.*

The continuous flow in processes is perceived as the most critical part of the lean philosophy to identify and minimize waste and increase value, quality, and reliability for the stakeholders. Providing a continuous flow in process enables organizations to deliver finished goods and services on time to the marketplace. The main idea of Principle 2 is to minimize lead time from raw material to finished goods/services. Continuous process flow with preventative maintenance and *jidoka* aims to bring any problems to the attention of those operating the processes to minimize potential wastes. Decreasing the level of inventory that hides some of the problems in manufacturing systems implements continuous process flow. Along with other performance indicators, the efficiency of the flow is measured using *takt time*. Takt time represents the pace of customer orders and production system. Takt time answers these questions:

1. How many finished goods should be produced each hour or day?
2. What is the average number of daily outputs for each assembly or production line?

Former President of Toyota Motor Manufacturing, North America, Teruyuki Minoura says:

> If some problem occurs in one-piece flow manufacturing, then the whole production line stops. In this sense it is a very bad system of manufacturing. But when production stops everyone is forced to solve the problem immediately. So, team members have to think, and through thinking team members grow and become better team members and people.

The flow of the processes is also constructed on *one-piece flow systems* in lean. Contrary to mass production system that relies on batches/lots, lean manufacturing emphasizes working on one piece in the whole process flow until all operations are done on one piece. In a one-piece flow system, the processes are lined up in the sequence that will produce the customers' order in the shortest time. If the batch/lot size is one, that one piece of the batch/lot will not wait for the next pieces that are processed in the other batches/lots. This is how cycle time and lead time are minimized using one-piece flow. Ideally, a batch/lot size in lean is one, since the one-piece flow system is integrated into continuous flow. Practically, lean leads to minimize batch/lot sizes from large numbers to very low levels.

VSMs help the Six Sigma team understand the general structure of any continuous flow system, as it pictures the flow of input and information. Lean production systems use small lots to make the flow of the material through processes as smooth as possible. U-shaped cells also enable continuous flow of information and inputs in lean (◘ Fig. 10.12). Each cell contains all required machines, equipment, raw material, components, operations, and multi-skilled employees. In ◘ Fig. 10.12, the cell has seven work steps and two employees. One piece is processed through seven work steps one by one in the U-shaped cell. Once the first piece completes the whole process, the cell starts working on the second piece. In mass production system, continuous flow is designed based on work stations. As the first batch/lot completes

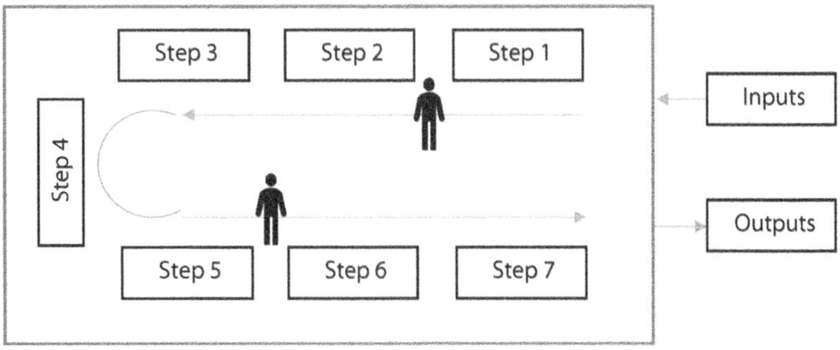

Fig. 10.12 Continuous flow in U-shaped cells in lean production systems. (Source: Author's creation)

10

Step 1 in station 1, the batch/lot is moved forward to Step 2 in station 2. Each station is specialized for a particular activity and task. Each batch/lot that contains multiple parts is processed by going through each work station. Continuous flow, one-piece flow, and takt time are easily used in high volume and repetitive manufacturing processes.

- *Principle 3: Use pull systems to avoid overproduction.*

One of the ways of minimizing overproduction in lean philosophy is to integrate pull systems into production systems, where customer orders are used to start the production process. The manufacturing process does not produce to stock but only when receiving a customer order. Customer orders pull through production functions that include *kanban*, supermarket, *heijunka*, and shipping subfunctions. Commonly, mass production systems are constructed based on push systems that deliver finished goods to marketplace through distributors and retailers before demand is known, creating stock or inventory. Lean systems are known as "make-to-order," while mass production systems are "make-to-stock."

Lean philosophy emphasizes that overproduction and excessive inventory enable wastes to be hidden. Ideally, lean philosophy aims to implement zero inventory practices. However, a great variety of factors may prevent organizations from implementing zero inventory. For example, uncertainties in suppliers or marketplace and fluctuations of input prices may influence organizations to keep safety stock. As the components of pull production system, *kanban* cards and supermarket inventory stocking systems are integrated into manufacturing systems in lean management. Supermarkets are buffers or work-in-process inventory systems. Depending on daily or hourly demand, each cell is expected to keep a certain amount of stock in cells and supermarket.

- *Principle 4: Level out the workload (heijunka).*

Supply is aligned with demand through *heijunka*, which helps level production planning in lean production systems. *Heijunka* is accepted as a fundamental system to eliminate *mura*, which means NVA activities in the processes. *Heijunka* levels production plans by implementing a mix-model production planning approach (◘ Image 10.14). Rather than producing the same amount of product in a day, mix-model production system enables to divide total number of products ordered into days and produce a variety of products daily to fulfill customer orders. ◘ Table 10.16 demonstrates how to implement a mix-model production system in a weekly production plan. The total number of products produced in a week does not change in option 1 and option 2. However, the product mix produced every day changes in option 2 in the level schedule, which gives manufacturers and customers flexibility. While option 1 produces only product A on Monday, option 2 produces each product in a way that by the end of Monday, the production system produces 3 for A, 4 for B, 2 for C, 2 for D, and 3 for E.

Heijunka helps reduce inventory and lead time, allows flexibility to meet customer's orders on time for each item, and minimizes the risk of increasing the company's stock of unsold products, increasing throughput level (volume of product) using the mix-model production system. The only time-consuming

◘ **Image 10.14** Heijunka box. (Source: Courtesy of Wiremold-Legrand)

step in *heijunka* is during the changeover or setup process. As a remedy for this potential problem, Single Minute of Exchange of Dies (SMED) keeps the changeover time to less than 10 minutes.

- *Principle 5: Build a culture of stopping to fix problems, to get quality right the first time.*

Lean suggests stopping the assembly and production lines as needed to prevent defects and errors. This approach is not accepted in mass production systems because the main goal of the system is increasing the volume of products. After the industrial revolution, the manufacturing industry focused on increasing throughput in the early 1900s. The quality of the outputs was not a concern during this time, unlike today. Rather than stopping the lines to prevent defects, increasing the volume of the outputs was the motto and the golden rule in manufacturing industry.

Taichi Ohno implemented a quality culture in TPS that allowed stopping the lines to correct defects. Instead of wasting resources by producing defective products and making errors, Ohno proposed stopping the activities until the problems are fixed. The customers do not accept nonconforming or defective products, and producers cannot tolerate the cost of defects, errors, rework, and defectives. Stopping the processes, *jidoka* or *autonomation*, is one of the two pillars of lean as presented in the House of Lean. Mistake-proofing (*poka-yoke*), *jidoka*, *andon*, and standardization are pivotal components of Principle 5. Mistake-proofing, or *poka-yoke*, refers to use of automatic systems that prevent or immediately identify any abnormality, defect, error, or problem. It performs a critical role in lean production to avoid simple human errors. When a *poka-yoke* system detects an error, it automatically shuts down the equipment and gives a warning as coded in *andon* systems (◘ Image 10.15). *Poka-yoke* is also seen as a way of preventing inadvertent errors. *Jidoka*, or *autonomotion*, refers to the ability of stopping production lines when problems and issues are detected by employees or automated systems. The process flow of *jidoka* starts with designing and implementing pro-

Table 10.16 Mix-model production system in lean

Customer orders						
Product 1 (A)	15 units					
Product 2 (B)	20 units					
Product 3 (C)	10 units					
Product 4 (D)	10 units					
Product 5 (E)	15 units					
Option 1: Mass production schedule						
Monday	**Tuesday**	**Wednesday**	**Thursday**	**Friday**	**Saturday**	**Total**
AAAAA AAAAA AAAAA	BBBBB BBBBB	BBBBB BBBBB	CCCCC CCCCC	DDDDD DDDDD	EEEEE EEEEE EEEEE	A = 15 B = 20 C = 10 D = 10 E = 15
Daily number of products						
A = 15 B = 0 C = 0 D = 0 E = 0	A = 0 B = 10 C = 0 D = 0 E = 0	A = 0 B = 10 C = 0 D = 0 E = 0	A = 0 B = 0 C = 10 D = 0 E = 0	A = 0 B = 0 C = 0 D = 10 E = 0	A = 0 B = 0 C = 0 D = 0 E = 15	
Option 2: Level schedule						
Monday	**Tuesday**	**Wednesday**	**Thursday**	**Friday**	**Saturday**	
AAABBBB CCDDEEE	AAABBBB CCDDEEE	AAABBB CCDDEEE	AABBB CCDEE	AABBB CCDEE	AABBB DDEE	
Daily number of products						
Monday	**Tuesday**	**Wednesday**	**Thursday**	**Friday**	**Saturday**	**Total**
A = 3 B = 4 C = 2 D = 2 E = 3	A = 3 B = 4 C = 2 D = 2 E = 3	A = 3 B = 3 C = 2 D = 2 E = 3	A = 2 B = 3 C = 2 D = 1 E = 2	A = 2 B = 3 C = 2 D = 1 E = 2	A = 2 B = 3 C = 0 D = 2 E = 2	A = 15 B = 20 C = 10 D = 10 E = 15

Source: Author's creation

duction tools with built-in sensors and ends with stopping the process if an immediate action is needed. *Jidoka* minimizes the probability of passing defective products, components, or parts to the next work stations, cells, or the marketplace.

After identifying issues and problems, employees focus on correcting those problems and getting the system back to the regular pace to catch the *takt time*. Employees are authorized to stop the lines, analyze the root causes of the problems, and solve those problems in lean production. Building things right the first time is the motto of Principle 5. Similar to *jidoka*, *andon* systems are used to alert employees, supervisors, team leaders, and other decision-makers about potential line stopping. Flags, colorful lights, or sounds are used in *andon* systems. Each flag, color, or sound is coded with a message to categorize the types of the potential issues and problems.

Image 10.15 Andon systems. (Source: Courtesy of Wiremold-Legrand)

- *Principle 6: Standardized tasks are the foundation for continuous improvement and employee empowerment.*

Standardization is the backbone of TPS since it eliminates NVA activities, errors, reworks, nonconforming products, and services. Standardization supports the PDCA cycle. The continuous quality improvement system in TPS is constructed based on standardized tasks and work instructions.

Standardization has a long historical background that goes back to Henry Ford's mass production system, Taylor's scientific management approach, and Frank and Lillian Gilbreth's time and motion studies in the early 1900s. Ohno's and Shingo's suggestions significantly contribute to standardization. TPS utilizes three systems to implement standardization in the processes: (1) takt time, (2) the sequence of activities and processes, and (3) standardized inventory level. Standardization at TPS is also the most effective enablers of zero defects. Global, national, and industry-oriented standards, standardized work instructions, VSMs, process flow charts, bills of material, and checklists are some ways to implement standardization.

- *Principle 7: Use visual control so no problems are hidden.*

Visual control in lean is analyzed both in the micro and macro level. At the micro level, lean philosophy emphasizes making problems visible to each employee in the workplace. In Principle 7, standardized work instructions, *kanban* cards, *poka-yoke* tools and devices, *andon* systems, display panels, and *jidoka* are well-structured methods used in lean. Visual control focuses on process, equipment, inventory, information, standards, and employees. One of the main goals of Principle 7 is to improve the continuous flow of the processes and items as detailed in Principle 2. Visual control methods are so naturally integrated into processes in lean that employees automatically implement those methods in their daily activities. Additionally, visual control in lean is built upon on 5S: Sort (*Seiri*), Set in Order (*Seiton*), Shine (*Seiso*), Standardize (*Seiketsu*), and Sustain (*Shitsuke*). Lean production systems do not tolerate messy work stations, cells, offices, warehouses, and working areas.

Visual control contains any kind of communication devices that tell us how work should be done or whether the work deviates from the standard. Similar to previous principles, Principle 7 aims to minimize waste, errors, defects, and injuries in the workplace. Visual controls also help answer the following questions for the items and tools used in the workplace:

1. *Where* does the item belong?
2. *How many* items should be in that space?
3. What are *the standard procedures* to store items and tools?

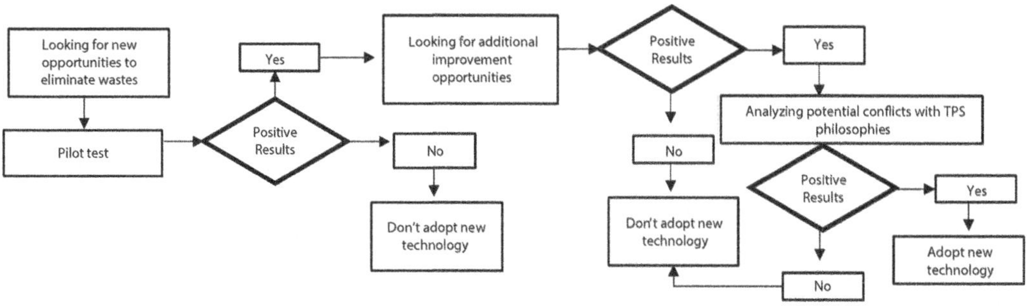

Fig. 10.13 Technology adaptation at TPS. (Source: Author's creation)

Shadow tool boards and boxes, designated areas, lines, and arrows for identifying places for specific items are used in visual controls.

At the macro level, visual control in lean implements the big room (*obeya*) strategy in project management. Project details such as work plans, timetables, Gantt charts, deliverables, and flow charts are displayed prominently and publicly on boards in *obeya*. During the meetings, the team members visualize the details of materials displayed in *obeya*. That strategy keeps members informed, improves communication, and allows better and more accurate decisions.

10

- *Principle 8: Use only reliable, thoroughly tested technology that serves your people and processes.*

Technology adaptation in TPS involves decision-making processes, as shown in Fig. 10.13. In lean, new technology requires a detailed approval process in which all relevant stakeholders have a critical role. In TPS, technology is perceived as an *input* that supports people, processes, and the overall system. This means that the process of analyzing and evaluating emerging technologies takes longer than in traditional Western organizations. The TPS does not implement a new technology without analyzing it comprehensively and considering many different angles of the new technology. The knowledge, skills, abilities, and experience of engineers at TPS make them qualified to decide what technology would be value-added for the system. The engineers focus not only on the capability of new technology but also on the long-run payback and on having a short return on investment ratio. Although analysis and testing processes go very slowly and safely, implementing new technology is done very quickly with all key stakeholders in TPS. In practice, Principle 8 has been ignored by lean practitioners in various industries, which is one of the root causes of failing lean implementation in Western organizations.

People – Respect and Development

Lean transformation requires a tremendous organizational and cultural change for Western organizations. This change is constructed based on the following three principles. These principles have been ignored by Western organizations over the last decades and are the other root causes of failing lean implementation experiences in Western organizations. The next sections will present the details of Principles 9, 10, and 11.

- *Principle 9: Growing leaders who thoroughly understand the work, live the philosophy, and teach it to others.*

TPS raises leaders within the lower levels of the hierarchy in the organization because this is a cultural value and norm. Employees work for companies by starting at entry level jobs. Then, they climb the hierarchical levels, being promoted based on their skills in years. Japanese management culture emphasizes the importance of seniority and being trained by the seniors in business organizations. More recently, TPS transferred leaders, specifically in subsidiaries abroad. However, in the TPS, it is preferable that the organization use the senior leaders to educate and train future leaders. Deming's consistency of purpose motto is aligned to this principle. A few leaders raised using TPS are Sakichi Toyoda, Kiichiro Toyoda, Eiji Toyoda, Taiichi Ohno, and Akio Toyota. The first foreign top executive raised through TPS was Gary Convis, the Chairman of Toyota Motor Manufacturing, Kentucky Inc. (TMMK), and Executive Vice President of Toyota Motor Engineering and Manufacturing North America Inc. (TEMA).

- *Principle 10: Develop exceptional people and teams who follow your company's philosophy.*

In general, respect for people means respect for all human resources' experience, knowledge, skills, and abilities. Mutual respect, trust, and understanding are the values that support Principle 10. TPS trains, educates, and prepares next generation of employees for the benefit of the future of the organization. In turn, employees commit and contribute to the vision of the organization and engage in designing of the future of the organization using *Hoshin Kanri* methodology.

Teamwork is one of the critical foundations of the lean production system, similar to Six Sigma. Continuous quality improvement projects are run by teams, and decisions are made by teams in lean and Six Sigma efforts. Employees' individual performance is also critical for lean success, not just their teamwork. This balance between individual excellence and effective teamwork is considered in Principle 10. Excellent individual performance elevates team's performance in the lean philosophy. Employee recruitment and hiring processes include various sub-processes, and it can take more than 10 months to make sure a candidate is the best fit for an organization using TPS. The approach taken in the recruitment and hiring process indicates the value of employees in TPS.

- *Principle 11: Respect your extended network of partners and suppliers by challenging them and helping them improve.*

The success of lean implementation depends on the performance of suppliers. The quality and reliability of supplies determine the likelihood of meeting *takt time* in the production processes, allowing the organization to fulfill customers' orders at the right time at the right place. Any delays or issues caused by suppliers cause delays in fulfilling customer orders. Therefore, creating effective and efficient partnerships with suppliers is critical in lean success.

Lean organizations also train their suppliers. To minimize reject and rework rates of the supplies delivered by the suppliers, collaborating with suppliers based on long-term contracts, transferring knowledge, and training them provide a huge return on investment and performance improvement in the main organization. Overall, the core competencies of TPS and the degree of self-reliance determine the level of outsourcing. The high standards and expectations set by lean organizations push suppliers toward complying. A win-win policy, with mutual benefits and trust, is the driving force of Principle 11.

Problem-Solving – Continuous Improvement

- *Principle 12: Go and see for yourself to thoroughly understand the situation (Genchi Genbutsu).*

Genchi Genbutsu means go and see in lean literature. When a problem is discovered in an input, process, or output, the stakeholders examine and analyze *gemba*, that is, the place where the problem is identified. Seeing the place firsthand may help analyze and identify the root causes of the problem more accurately. Taiichi Ohno recommends visiting and observing areas without preconceptions. Lean philosophy emphasizes going and seeing the problematic area firsthand by the parties taking a role in the problem investigated, rather than analyzing the problems in meeting rooms with reports. This approach automatically integrates all decision-makers into problem-solving processes. An Ohno Circle requires decision-makers to spend at least 30 minutes to carefully observe and analyze the workplace, including the flow of the process, standardized work instructions, and the way employees work. In essence, lean philosophy wants leaders to get their hands dirty.

- *Principle 13: Make decisions slowly by consensus, thoroughly considering all options; implement rapidly.*

Lean organizations make their decisions slowly and collaboratively at different hierarchical levels. Their decision-making processes effectively utilize data and implement a great variety of data analysis tools and techniques as presented in the Measure and Analyze steps in the DMAIC process in previous chapters. As a traditional Japanese decision-making approach, *ringi* is implemented in decision-making processes in lean. *Ringi* system starts with a memorandum that is shared with all relevant employees and departments in the organization. To provide consensus, the memorandum is articulated by an authority and presents the details of the decision to all parties. When the *ringi* is accepted by everyone, the authority who prepared the *ringi* implements the decision. The only disadvantage of *ringi* in Japanese management is that it is a time-consuming process. After getting approval from all parties, the implementation phase is more efficient, however.

- *Principle 14: Become a learning organization through relentless reflection (hansei) and continuous improvement (Kaizen)*

Lean philosophy implements a continuous improvement and continuous learning process throughout the organization. *Kaizen* and breakthrough improvement practices create the basis for continuous quality improvement. Using either explicit or tacit knowledge, business organizations are expected to become a learning organization in their lean journey. The previous 13 principles are strongly supported by Principle 14. Since continuous quality improvement has been discussed previously, it will not be detailed in this section.

10.5 Failure Modes and Effects Analysis

Failure Modes and Effects Analysis (FMEA) is a preventive action to prevent failures in design and business processes, services, products, and parts. The main goal of FMEA is to identify and eliminate potential failures and risks before they occur. FMEA is a step-by-step process through which all possible failures are identified and quantified in design, manufacturing, assembly, or service delivery processes. FMEA is taken as proactive problem-solving and risk assessment approach to prioritize potential risks. As a predictive and proactive approach, risk is a vital concept to maximize the performance of the product. A new or current product or system is assessed to determine where a failure may potentially occur within the processes. After identifying potential failures, each failure is analyzed in terms of its potential effects on the entire system. FMEA contributes to increase customer satisfaction, market share, and eventually overall organizational performance. FMEA was first developed and used in the aerospace industry and the US military in the 1940s. Military standard MIL-P-1629 Military Standard Procedures for Performing a Failure Mode Effects and Criticality Analysis (FMECA) was developed in the 1940s. The aerospace industry, and particularly NASA, implemented FMEA in design processes during the 1960s and 1970s. The Ford Motor Co. started using FMEA during the 1970s, and FMEA became popular throughout the automotive industry. The British Standards Institute published BS 5760–5: 1991 (Reliability of systems, equipment and components. Guide to failure modes, effects and criticality analysis (FMEA and FMECA)) in 1991. Later, BS EN IEC 60812:2018 (Failure modes and effects analysis – FMEA and FMECA) was released in 2018. Since then, FMEA has also functioned as a risk assessment and analysis tool in various industries, including the automotive, marine, and healthcare industries.

Failure refers to a state of inability of performing a function as designed. Not only are the failures detected before delivery and shipment, but also the possible failures in the entire life cycle of the process are considered. Failure modes are ways in which a product or service, process, or function may fail to show expected performance. Defectives, errors, and nonconformities are also considered to be failure. Failure is a term that needs to be identified by the FMEA team prior to starting a FMEA process, either potential or actual failures.

Error is defined as "the concept of an individual performing a task that is not completed as intended or an individual making a decision that does not produce the desired results" (Stewart and Grout 2001, p. 442). *Defective* and *nonconformity* refer to outcomes of a process that do not meet technical specifications and requirements. In the FMEA process, each potential failure is ranked by severity level, detectability status, and occurrence.

The types of FMEA are:

1. System FMEA

2. Design FMEA
3. Process FMEA
4. Service FMEA.

FMEA follows a straightforward process. The following section details that process.

- Step 1: Establish an FMEA team.
  First and foremost, FMEA requires effective teamwork that involves multi-disciplines and functions that may be influential on the topic analyzed. Depending on the complexity of the topic, all relevant departments or functions are represented in a cross-functional FMEA team. Team dynamics should be taken into consideration prior to establishing the team. Team members also need to be trained in FMEA before beginning the process.
- Step 2: Identify the scope of the FMEA.
  After establishing an FMEA team, the team focuses on identifying the scope of the subject that is analyzed through the FMEA process. Since the outcomes of FMEA are expected to identify and eliminate potential/actual failures, the scope needs to be identified using a broad perspective to include all relevant external or internal stakeholders. Also, some functions such as marketing, design, production, quality, delivery, purchasing, supply chain, and sales may be included in FMEA. Customer complaints and returns, quality and process control results, design outcomes, R&D drawings, and prototype test results are analyzed to formalize the scope of the FMEA and the FMEA team. Depending on the scope, the type of the FMEA is decided in this step as well. The following questions should be answered in this step:
  1. What is the purpose of this system, design, product, process, or service?
  2. What do our customers expect it to do?
  The scope is broken into separate subsystems, items, parts, assemblies, or process steps. As another critical requirement in this step, functions of each system, design, product, process, or service are pictured using appropriate tools and methods to visualize the details of the topic. VSMs, process flow charts, SIPOC diagrams, engineering drawings, and design outcomes are used in this step.
  For example, if the FMEA team works on minimizing mean failure time in accelerated life tests of finished goods in a dishwasher producer, the team would focus on particular parts or components of the finished product in the accelerated life test process. Considering the complexity of the product, a limited scope may be more beneficial to accomplish the mission of the FMEA team.
- Step 3: Determine the process.
  The scope of the FMEA addresses relevant process/es. If there is more than one related process to be analyzed in FMEA, the team may consider splitting into multiple teams. Process determination is considered the core part of FMEA, since the probability of identifying and eliminating potential or actual failures depends on the structure of the process.
- Step 4: Draw process flow.
  A picture of the actual process is needed in FMEA to demonstrate the actual process flows. To identify the current state of the process, process maps, process flow charts, VSMs, swim lanes, SIPOC diagrams, or spaghetti diagrams are used in this step. The details of those tools were presented in the previous chapters (see ▶ Chap. 4 for process analysis and documentation tools). Potential or current bottlenecks, obstacles, and issues are identified by picturing the flow of the process.
- Step 5: Identify potential failure modes.
  After analyzing the process and drawing the process flow, FMEA team can identify potential failure modes that are identified in the relevant processes. The failure mode addresses the reason for an inspector rejecting a part in incoming/process control, an auditor suspending a process, or a customer returning a product. Process flow charts, VSMs, swim lanes, spaghetti diagrams, and SIPOC diagrams identify potential failure modes and where these potential failures may occur. The question "How can the process go wrong?" helps the team identify potential failure modes. In this step, parts/

functions and failure modes are matched to identify potential failure modes in part/function and failure mode matrix. The relationships between the parts of the product and expected functions directly refer to potential failure modes. The questions to be asked while developing part/function matrix are as follows:

1. What is this part supposed to do?
2. What is this part supposed not to do?
3. What are the related function/s of this part?

The other questions asked by FMEA team to identify the failure modes are:

1. Why are the parts rejected or what parts are likely be rejected?
2. Why does the part not meet the technical specifications?
3. What scenario is unacceptable for an end user?

- Step 6: Identify potential effects of failures.

10

In light of the relationships between functions and parts, the potential effects of failures are listed by the FMEA team. Design outcomes, engineering drawings, process flow charts, experiment results, and part/function matrixes are helpful in this step. Effect is the result of a failure. Potential effects of failures are used to understand issues that may occur because of potential failures while the system, processes, products, or services are functioning. Some of the questions that need to be asked in this step are:

1. What does the internal customer experience because of this failure?
2. What do the external customers and end users experience because of this failure? and
3. What happens in the system when this failure occurs?

The answers to these questions help identify where, how, and when the failures may occur. Identifying potential effects of those failures may lead the FMEA team to (1) identify potential risks and (2) determine the factors contributing to those risks and (3) the solutions and improvement ways to minimize and eliminate the risks and their effects.

- Step 7: Rate the severity of effects.

In this step, the FMEA team focuses on rating the severity of all potential effects of the potential failure modes. *Severity* is a rating that indicates the seriousness of the effect of the potential process failure mode. Generally, severity level is ranked on a scale varying between 1 and 10. If the effect is identified as critical, then the severity level is as high level. If the effect is not critical, the severity is low. The expert knowledge of the team members and technical reports help accurately rate how serious each effect is to the stakeholders. An example of severity levels and their explanations is given below:

- Hazardous – without warning: Very high severity ranking. A potential failure mode affects safety of the operation and all relevant parties without warning. Noncompliance with regulations (10 points).
- Hazardous – with warning: Very high severity ranking. A potential failure mode affects safety of the operation and all relevant parties with warning. Noncompliance with regulations (9 points).
- Very high: Product/item loses primary function or 100% of product may have to be scrapped. Repair time >1 h (8 points).
- High: Product/item functions at reduced level of performance. Customer dissatisfaction. Product/item may have to be sorted and less than 100% scrapped. 1 h>Repair time> 0.5 (7 points).
- Moderate: Product/item functions, but no comfort/convenience is present. Customer dissatisfaction. A portion of the product may have to be scrapped. Repair time <0.5 h (6 points).
- Low: Product/item functions, but comfort/convenience is at a reduced level. Or less than 100% of the product may have to be reworked. Repairing off-line (5 points).
- Very Low: Certain items do not conform. Defect noticed by greater than 75% of customers (4 points).
- Minor: Certain items do not conform. Defect noticed by 50% of the customers (3 points).

- Very minor: Certain items do not conform. Defect noticed by less than 25% of customers (2 points).
- None: No negative effect (1 point).

To rate the severity of the failure modes, deterministic and stochastic approaches are used by FMEA teams. The example given above represents a deterministic approach since each effect has a certain severity. When uncertainty arises, FMEA teams may use stochastic (probabilistic) and fuzzy approaches. If severity is related to more than one factor, multi-criteria decision-making methods are implemented in severity rating processes. Research shows that advanced decision-making methods have been implemented in this step of FMEA (e.g., Dağsuyu et al. 2016; Fattahi and Khalilzadeh 2018; Liu 2016).

- Step 8: Identify potential causes of the failures.

Shingo (1985) indicates that defects can be eliminated by identifying the causes. To prevent potential failures, the potential root causes of the failures are identified by the FMEA team in this step. Similar to the importance of discovering the root causes in regular problem-solving and continuous quality improvement processes, Step 8 focuses on digging out the details of the causes of potential failures systems, processes, and products. Root cause analysis and process mapping tools are utilized in this step. The questions to ask in this step are:

1. "In what way can this system fail to perform its intended function?"
2. "What circumstances could cause the failure?"
3. "Why or how can the part fail to meet its specifications?"

Potential and actual causes are identified using problem-solving tools such as fishbone diagrams, 5 Whys analysis, affinity diagrams, or process mapping tools (see ► Chap. 8 for other data analysis tools).

- Step 9: Rate the detectability of failures.

ASQ[1] states that this rating is an estimate of how well the controls detect either a specific root cause or its failure mode after they have happened *but* before the customer is affected. The assumption here is that a failure has already occurred. The detectability variable ranges between 1 and 10, where 1 means that the control is definitely capable of detecting the problem and 10 means that the control is not capable of detecting it. An example of detectability scale is given as follows:

- Almost impossible: Absolute certainty of non-detection (10 points).
- Very rare: Controls will probably not detect (9 points).
- Rare: Controls have low chance of detection (8 points).
- Very low: Controls have low chance of detection (7 points).
- Low: Controls may detect (6 points).
- Moderate: Controls may detect (5 points).
- Moderately high: Controls have a high chance to detect (4 points).
- High: Controls have a high chance to detect (3 points).
- Very high: Controls almost certain to detect (2 points).
- Almost certain: Controls certain to detect (1 point).

The technical complexity of the process or product and accuracy of the control methods will affect the decisions of Six Sigma teams as far as the detectability rate in this step. To rate the detectability of the failure modes, deterministic and stochastic approaches are used by FMEA teams. The example given above represents a deterministic approach since each effect has a certain detectability. Similar to the severity rate, when uncertainty arises in detectability, FMEA teams may prefer to use stochastic (probabilistic) and fuzzy approaches. If detectability is related to more than one factor, multi-criteria decision-making methods are implemented in detectability rating processes.

- Step 10: Rate the occurrence of failures.

Occurrence is the probability that potential failures occur in the processes. Each cause is assigned with an occurrence rate that varies between 1 and 10, where

1 ► http://asq.org/learn-about-quality/process-analysis-tools/overview/fmea.html

1 means that that cause is *least likely* to happen and 10 means that cause is *most likely* to happen. The rate may be assigned by the team members based on statistical reports and frequencies. Historical data and occurrences may give a better idea of more accurate rates for failures. Sigma levels and process capability indices, such as *Cpk* and *Cp*, can also be utilized in this step (see ▶ Chap. 5 for Six Sigma measurements and ▶ Chap. 7 for process capability analysis). Occurrence ratings based on historical data can be assigned as follows:

  - Very high: 1 in 2 or ≥100 per thousand pieces (10 points)
  - Very high: 1 in 3 or 50 per thousand pieces (9 points)
  - High: 1 in 8 or 20 per thousand pieces (8 points)
  - High: 1 in 20 or 10 per thousand pieces (7 points)
  - Moderate: 1 in 80 or 5 per thousand pieces (6 points)
  - Moderate: 1 in 400 or 2 per thousand pieces (5 points)
  - Moderate: 1 in 2000 or 1 per thousand pieces (4 points)
  - Low: 1 in 15,000 or 0.5 per thousand pieces (3 points)
  - Low: 1 in 150,000 or 0.1 per thousand pieces (2 points)
  - Remote: < 1 in 500 M or ≤0.01 per thousand pieces (1 point).

  Companies can develop their own occurrence ratings based on the industry dynamics and company-oriented needs and requirements. To rate the occurrence of the failure modes, deterministic and stochastic approaches are used by FMEA teams. The example given above represents a deterministic approach, since each effect has a certain occurrence level. Similar to severity and detectability rates, when uncertainty arises in occurrence, FMEA teams may prefer to use stochastic (probabilistic) and fuzzy approaches. If occurrence is related to more than one factor, multi-criteria decision-making methods are implemented.

- Step 11: Calculate Risk Priority Number (RPN).

  RPN is used to prioritize the failure mode/s in terms of preventing the occurrence of possible failure modes. *RPN* is calculated as the product of severity (*S*), detectability (*D*), and occurrence (*O*) as shown in Eq. 10.24.

$$RPN = S * D * O \qquad (10.24)$$

  The higher the *RPN*, the higher the risk. An *RPN* less than 40 refers to a situation where there is no risk, whereas an *RPN* between 40 and 100 indicates relatively lower risk. An *RPN* higher than 100 refers to a definite risk.

- Step 12: Determine the highest risky potential cause.

  Higher *RPN*s point out more serious and risky potential causes. The potential cause/s that have the highest *RPN*s should be identified as the highest risk cause/s. An FMEA team focuses on the highest *RPN*s and analyzes those processes and potential causes in detail to minimize severity, detectability, and occurrence of the highest potential risk cause/s. Preventing those high-risk cause/s from happening is the main goal of an FMEA team in this step.

- Step 13: Identify improvement activities.

  Considering potential causes of the failure modes, the FMEA team identifies potential improvement activities to reduce the *RPN* calculated for the relevant cause/s in the previous steps. Accurate, effective, and efficient improvement activities increase the probability of eliminating the potential failures. The FMEA team can implement the simulation method to generate and assess improvement alternatives in this step. The quality of improvement activities developed by the FMEA team depends on the budget of the team and team members' expert knowledge and determines the effectiveness of FMEA outcomes.

- Step 14: Implement activities.

  Improvement activities proposed and approved by the FMEA team are implemented in the system, design, product, process, or service to analyze the effectiveness of improvement activities. All planned activities are executed and integrated into the systems, and the *RPN* number is expected to decrease, or, in other words, the risk of occurrence of the potential failures is lessened.

- Step 15: Recalculate RPN.
  After implementation of improvement activities, the *RPN* is recalculated and is expected to be significantly reduced. The probability of reducing the *RPN* depends on the quality of improvement activities. If quality improvement activities are effective and efficient, RPNs are likely to be reduced.

▶ Example 8

In light of the patient satisfaction and the Joint Commission reports, the Patient Safety Committee at Eastern Connecticut Hospital wants to analyze potential strategies and tactics for improving patient safety. Assume that you are the Director of the Quality and Safety Department, and the hospital already implemented Six Sigma methodology several months ago. The committee decides to implement FMEA in the project and go through the process of FMEA step by step. Outline the FMEA process and identify what needs to be done in each step. ◀

Solution

To help them with FMEA process, let's follow steps as given below:

- Step 1: Establish an FMEA team.
  The FMEA team at the hospital includes four members: J. Doe from Medical Operations, L. Smith from Hospital Operations, B. Jones from Patient Outreach, and Z. Thomas from Nursing. The team demonstrates all steps in the FMEA form as shown in ◘ Table 10.17. The full version of ◘ Table 10.17 is presented in the data files of the textbook in web site.
- Step 2: Identify the scope of the FMEA.
  The scope of the project is identified as "patient safety" by the Patient Safety Committee. Patient satisfaction and the Joint Commission reports identify the scope of the project. Let's answer the questions given in the flow above.
  1. What is the purpose of this system, design, product, process, or service? The purpose of the hospital system in this example is to improve patient safety and eventually increase patient satisfaction.
  2. What do our customers expect it to do? The customers of the hospital, that is, patients, expect to experience safe healthcare services at the Eastern Connecticut Hospital.
- Step 3: Determine the process.
  The scope of the FMEA addresses that the scope is patient safety in this example. Considering the size of the hospital, the team may want to focus on a few functions/departments or the entire system of the hospital. Our FMEA form, shown in ◘ Table 10.17, demonstrates that the team preferred to go with the second option. Because of the nature of patient safety, the concept covers many different processes, departments, and functions.
- Step 4: Draw process flow.
  Since the scope of patient safety covers and affect many processes, we will not see other detailed processes. In this example, the process is articulated as "patient safety."
- Step 5: Identify potential failure modes.
  The team identifies several potential failure modes of patient safety as follows:
  1. Wait time
  2. Interactions with other patients
  3. Cleanliness and hygiene
  4. Failing in following standardized medical procedures
  5. Unstable patients not being properly monitored
  6. Feeling threat from environment.

  When the question "How can the process go wrong?" is asked by FMEA team members, let's assume that these six potential failure modes were identified. The other question that can help team members accurately identify potential failure modes is "what scenario is unacceptable for an end user?" That question focuses on patient safety and helps generate identification of those failures.
- Step 6: Identify potential effects of failures.
  The team identifies several potential effects of failures as shown in ◘ Table 10.17. While some potential failures have only one effect, the others have multiple effects. For example, the team identifies wait time as the first potential failure mode. The potential effect of wait

time is anticipated as too long wait time. The third potential failure mode, cleanliness and hygiene, is linked with four different but interrelated potential effects of failures: unclean waiting area, unclean restrooms, unclean patient rooms, and hand hygiene.

- Step 7: Rate the severity of effects.
  The severity rates are assigned by the FMEA team for each potential effect of failure. The S column in ◘ Table 10.17 shows those rates, and the team may use or modify severity scales shown above to rate each potential effect.
- Step 8: Identify potential causes of the failures.
  Potential causes of failures are analyzed in depth by the team and listed in the column named "Potential cause(s) of failure" in ◘ Table 10.17. In this step, we assume that the team already systematically worked on root cause analysis and answered relevant questions given in Step 8 above to identify accurate causes.
- Step 9: Rate the detectability of failures.
  The detectability rates are assigned by the FMEA team for each potential effect of failure. The D column in ◘ Table 10.17 shows those rates, and the team may use or modify detectability scale shown above to rate each potential effect.
- Step 10: Rate the occurrence of failures.
  The occurrence rates are assigned by the FMEA team for each potential effect of failure. The O column in ◘ Table 10.17 shows those rates, and the team may use or modify occurrence scales shown above to rate each potential effect.
- Step 11: Calculate Risk Priority Number (RPN).
  Using Eq. 10.24, the *RPN* is calculated for each potential cause of failure as shown in ◘ Table 10.17. For example, "feel unsafe in hospital" is assigned with two for severity, five for occurrence, and four for detectability by the FMEA team. As the product of these three variables, the *RPN* is 40.
- Step 12: Determine the highest risky potential failure mode.
  Among all *RPN*s, "unstable patients not being properly monitored" has the highest *RPN* with 280 points. In the next step, this potential failure mode should be analyzed, and use root cause analysis to minimize the potential effects on stakeholders.
- Step 13: Identify improvement activities.
  The FMEA team particularly focuses on "unstable patients not being properly monitored" failure mode, aims to develop multiple improvement alternatives, and selects the most beneficial one/s. In this step, the team identifies "implement individual patient monitoring" as the best improvement activity, as shown in ◘ Table 10.17.
- Step 14: Implement activities.
  In this step, we assume that improvement activities proposed by the team were implemented in the relevant processes. For example, "individual patient monitoring" practice was implemented for "unstable patients not being properly monitored" potential failure mode in this example.
- Step 15: Recalculate RPN.
  The *RPN*s are recalculated in the last column in ◘ Table 10.17. For example, after the improvement actions, the failure mode that has the highest *RPN* (280) decreases to 24. A similar trend is seen in the other failure modes and shows the effectiveness of improvement actions.

Key Concepts

Improve phase, simulation, simulation model, static-dynamic models, deterministic-stochastic models, discrete-continuous models, queueing theory, Design of Experiment (DOE), lean implementation, and Failure Modes and Effects Analysis (FMEA); Go and No-Go decision, multi-criteria decision (MCDM), Value Stream Map (VSM), factors, levels of the factors, response variable (*Y* variable), number of experiments (runs), controllable and uncontrollable factors, single factor experiments, one-variable-at-a-time, two-factor factorial designs, full factorial experiments, fractional factorial experiments, screening experiments, and response surface methods; Taguchi designs, decision tree of experiment plan,

10

Table 10.17 An example of FMEA form

Topic: Patient safety		FMEA number: 123456				Prepared by: Core team			Status: Initial		Start date:				
Team members: J. Doe (Medical), J. Smith (Operations), B. Jones (Patient Outreach), Z. Thomas (Nursing) Rev.: 01											Finish date:				
Process function	Potential failure mode	Potential effect(s) of failure	S	Potential cause(s) of failure	O	Current process controls	D	RPN	Improvement actions	Roles and dues	Action results				
											Actions taken	S	O	D	RPN
Safety of patient	Wait time	Too long wait time	2	Feel unsafe in hospital	5	Schedule for how often to see new patients	4	40	Increase staff in peak times to see more patients	J. Smith 11/10/20		1	2	3	6
			3	Feel sicker while waiting	4	Check patients in waiting room	7	84	Identify contributors of long wait times (root cause analysis)	B. Jones 11/8/20		2	2	4	16
	Cleanliness and hygiene	Unclean waiting area	5	Feel unsafe in terms of their health	6	Frequent cleaning and sanitizing of waiting area	2	60	Implement daily waiting area cleaning schedule	J. Smith 11/7/20		2	2	1	4
	Safety from themselves	Unstable patients not being properly monitored	8	Causing harm to themselves or others	7	Monitoring unstable patients/ use of security guards	5	280	Monitor patients more closely	B. Jones 11/6/20		4	2	3	24
	Perceived safety	Not feeling safe in a safe environment	4	Desire to leave hospital without receiving necessary treatment	5	Role of authority used to maintain feelings of safety	7	140	Ask patients how they are feeling/get their opinions on hospital procedures	B. Jones 11/9/20		1	1	2	2

Analysis of Variance (ANOVA), Two-way ANOVA, main effects plot, interaction plot, lean principles, Toyota Production System (TPS), *muda*, *mura*, JIT, *jidoka*, Andon systems, Poka-yoke, 5S, PDCA, takt time, one-piece flow, U-shaped cells, *heijunka*, zero inventory, mix-model production system, Single Minute of Exchange of Dies (SMED), "big room" (*obeya*), *Hoshin Kanri*, Ohno Circle, Ringi, *Kaizen*, part/function and failure mode matrix, Risk Priority Number, severity (S), detectability (D), and occurrence (O).

Summary

The goal of the Improve phase is to identify and implement the best solutions for the topic analyzed in the project charter in a Six Sigma team. To determine the best solution, qualitative and quantitative methods are used by the team. Some of these methods are simulation, lean philosophy, DOE, and FMEA. Using these methods, teams can develop various alternative solutions and analyze the effectiveness of those solutions for the system analyzed. Using a six-step process in the Improve phase as presented in this chapter, Six Sigma teams fulfill the requirements of the Improve phase of the DMAIC process.

Practice and Discussion Questions

1. Identify the steps of the Improve phase in DMAIC.
2. What does "optimal solution" mean in the Improve phase?
3. Using cost-benefit matrix, when should "Go" and "No-Go" decisions be made?
4. Describe why MCDM should be integrated into optimal solution selection process in step 2 in the Improve phase.
5. Describe potential benefits of creating future-state map in step 3 in the Improve phase.
6. Discuss why Six Sigma teams need DOE.
7. What are the variables that need to be considered at the planning stage in DOE?
8. Describe the types of DOE.
9. Describe the steps involved in the process of DOE.
10. Describe when single factor experiments are preferred in DOE.
11. Describe when full factorial experiments are preferred in DOE.
12. Describe when two-factor factorial designs are preferred in DOE.
13. What are the components of total variation (SST) in two-factor factorial designs?
14. Describe the functions of (−) and (+) signs and notations of (1), a, b, and ab in two-factor factorial designs.
15. Describe three tests of hypotheses conducted in two-way ANOVA in two-factor factorial designs.
16. Describe how main effects of factors are analyzed in two-factor factorial designs.
17. Describe how interactions between factors are analyzed in two-factor factorial designs.
18. Describe how potential number of combinations are calculated in full factorial designs.
19. What does simulation mean in quality improvement and Six Sigma projects?
20. What are the expected outcomes of a simulation model?
21. What are the types of simulation?
22. What are the main steps of building a simulation model?
23. In the hand simulation example, suppose that queue discipline is changed and when the cashier gets idle, instead of serving the first entity in the queue, it serves the entity with shortest processing time. Re-do the hand simulation, and compare the results in terms of performance measures.
24. In the hand simulation example, suppose that two cashiers serve the entities in parallel. When an entity arrives, it is served with the idle cashier. If both cashiers are idle, entity is served by the first cashier. If both cashiers are busy, entity waits in the queue until one

cashier gets idle. In such a case, you will need to define service times for the second cashier. Suppose that the service time of the second cashier equals to the ones obtained for the first cashier. Re-do the hand simulation, and compare the results in terms of performance measures.

25. Discuss the role of lean philosophy in Six Sigma.
26. Discuss the concepts of *mura* and *muri*.
27. What are types of waste in lean philosophy?
28. Discuss two pillars of lean philosophy.
29. What are the four sections of lean philosophy?
30. Discuss the details of each principle of lean philosophy.
31. Discuss why Six Sigma teams need FMEA.
32. Discuss why Six Sigma teams need simulation.
33. Describe the types of FMEA.
34. Describe the types of simulation.
35. Describe the steps involved in the process of FMEA.
36. Using the severity, occurrence, and detectability scales, conduct an FMEA analysis for a food preparation section in a fast food restaurant.

References

Ahmed, A., Page, J., and J. Olsen. (2017). Process improvement based on an integrated approach of DMAIC and multi-method simulation. The 22nd international congress on modelling and simulation, Hobart, Tasmania, 3–8thDecember 2017.

Altiparmak, F., Dengiz, B., & Bulgak, A. A. (2002). Optimization of buffer sizes in assembly systems using intelligent techniques. In *Proceedings of the Winter Simulation Conference* (Vol. 2, pp. 1157–1162). IEEE.

Dağsuyu, C., Göçmen, E., Narlı, M., & Kokangül, A. (2016). Classical and fuzzy FMEA risk analysis in a sterilization unit. *Computers & Industrial Engineering, 101*, 286–294.

Fattahi, R., & Khalilzadeh, M. (2018). Risk evaluation using a novel hybrid method based on FMEA, extended MULTIMOORA, and AHP methods under fuzzy environment. *Safety Science, 102*, 290–300.

Hussein, N. A., Abdelmaguid, T. F., Tawfik, B. S., & Ahmed, N. G. (2017). Mitigating overcrowding in emergency departments using Six Sigma and simulation: A case study in Egypt. *Operations Research for Health Care, 15*, 1–12.

ISO 3534-3. *Statistics — Vocabulary and symbols — Part 3: Design of experiments*. ISO.

Karnon, J., Stahl, J., Brennan, A., Caro, J. J., Mar, J., & Möller, J. (2012). Modeling using discrete event simulation: A report of the ISPOR-SMDM modeling good research practices task force–4. *Medical Decision Making, 32*(5), 701–711.

King, B. (1989). *Hoshin planning: The developmental approach*. Salem: GOAL/QPC.

Liu, H. C. (2016). FMEA using uncertainty theories and MCDM methods. In *FMEA using uncertainty theories and MCDM methods* (pp. 13–27). Singapore: Springer.

Montgomery, D. C. (2013). *Introduction to statistical quality*. New York: Wiley/NYC.

Ricki G. I. (2008). Introduction to simulation. Proceedings of the 2008 Winter Simulation Conference.

Shannon, R. E. (1975). *Systems simulation – The art and science*. Englewood Cliffs: Prentice-Hall.

Shingo, S. (1985). *Zero quality control: Source inspection and the Poka-Yoke System*. Massachusetts: Productivity Press.

Stewart, D. M., & Grout, J. R. (2001). The human side of mistake-proofing. *Production and Operations Management, 10*(4), 440–459.

Taneja, M., & Manchanda, A. (2013). Six sigma an approach to improve productivity in manufacturing industry. *International Journal of Engineering Trends and Technology (IJETT), 5*(6), 281–286.

Ungureanu, D., Sisak, F., Kristaly, D. M., & Moraru, S. (2005). Simulation modeling. Input data collection and analysis. In The 14th international scientific and applied science conference electronics ET, Sozopol, Bulgaria: 43–50.

Further Readings

Greasley, A. (2017). *Simulation modelling for business*. London, UK: Routledge.

Monden, Y. (1998). *Toyota production system, an integrated approach to just-in-time*. Norcross: Engineering and Management Press.

Montgomery, D. C. (2017). *Design and analysis of experiments*. New York: Wiley.

Morgan, B. J. (2018). *Elements of simulation*. Milton: Routledge.

Myers, R. H., Montgomery, D. C., & Anderson-Cook, C. M. (2016). *Response surface methodology: Process and product optimization using designed experiments*. Hoboken: Wiley.

Control Phase: C Is for Control

Contents

F. Pakdil, *Six Sigma for Students*, https://doi.org/10.1007/978-3-030-40709-4_11

Learning objectives

- Understand how to conduct the Control phase in DMAIC
- Understand how to use statistical process control in the Control phase
- Explain when and how to adapt successful project outputs across the organization
- Establish a control plan
- Implement standardization
- Organize and implement project closure.

11.1 Introduction

The goals of the "Control" phase in DMAIC are to:

1. Make sure that all activities planned in the Improve phase are successfully implemented, standardized, and sustained in the processes
2. Document all phases of Six Sigma project properly
3. Make sure that proper documents used in the processes are revised and revised documents are in-use
4. Make sure that Six Sigma team achieved the goals of the team
5. Make sure achievements are not lost in the near future
6. Adapt successful examples across the organization
7. Properly close the project.

The Control phase focuses on determining whether improvement activities produced expected outputs and improvements in the previous steps of DMAIC. The Control phase also aims to identify what else should be done to achieve the goals in the project, if the expected outputs were not provided by the improvement activities. If the goals were already achieved in the project, the Control phase focuses on identifying whether any additional improvement action is needed or whether permanent success is guaranteed in the process. The team needs to ensure that revisions and changes developed in the previous phases in DMAIC were successfully implemented in the processes. Six Sigma practitioners and decision-makers understand whether interventions created the intended and expected impact in this phase. Afterward, they can decide whether the improvement is likely to be sustainable after the project is done.

Deming's PDCA cycle (◘ Fig. 1.2 in ► Chap. 1) helps understand how the Control phase of DMAIC contributes to Six Sigma project performance. As presented in the previous chapters, in the PDCA cycle, plans are developed and executed, actual performance is compared with target performance, and, if necessary, corrective actions are carried out in a continuous quality improvement cycle. PDCA is also a process where new standards are established, revised, and replaced with newer and more improved standards (Imai 1986). The Control in PDCA is the step where actual performance is compared with plans to make sure that the details of the plans have been executed properly. Similarly, the Control phase in DMAIC aims to ensure that the Six Sigma team achieves the goals and sustains the success in the processes.

Feigenbaum (1983: 10) states that four steps are involved in control functions: (1) setting standards, (2) appraising conformance, (3) acting when necessary, and (4) planning for improvements. In his approach, standards refer to expected improvement outputs and goals in Six Sigma projects. Appraising conformance refers to making sure that the goals are achieved and performance is improved. The performance of the process across time before and after improvement implementation is compared by the team. If any unexpected gaps occur between goals and actual performance, the Six Sigma team begins to set up corrective actions. Then, the Six Sigma team implements corrective actions to improve performance of the project.

Gitlow et al. (2015) itemize the relevant activities in the Control phase as follows:

1. Reduce the impact of negative consequences on the processes analyzed in the Six Sigma project.
2. Standardize improvements implemented in the Improve phase.
3. Develop a control plan for the process.

4. Identify costs and benefits of the project.
5. Add the project details in the Six Sigma database.
6. Deploy the improvements in the organization.

To evaluate and monitor the performance of the previous phases of DMAIC, the performance of the process under study is analyzed using various methods and performance indicators in the Control phase. The statistical process control (SPC) that includes run charts, control charts, and process capability analysis is considered one of the most effective tools used in the Control phase. The other effective tools of the Control phase are Failure Mode and Effects Analysis (FMEA) and control plans. Since FMEA is discussed in ► Chap. 10, we will not detail FMEA in this chapter. The control plans developed in this phase help monitor and sustain planned and implemented solutions. After analyzing steps of the Control phase in the next section, following sections will discuss SPC and control plans.

11.2 Steps in the Control Phase

The Control phase is designed to implement, document, measure, and maintain the improvements. In the Control phase, several steps are followed by the Six Sigma team. These steps are:

- Step 1: Implementing ongoing measurements
- Step 2: Standardization of the improvement
- Step 3: Monitoring the improvement
- Step 4: Project closure.

11.2.1 Implementing Ongoing Measurements

After implementing improvements, the Six Sigma team identifies what will be measured and who will measure and monitor the process. The team determines how progress of the process is measured and monitored. This step plans performance measurement and monitoring activities to be done after implementing the improvement. Some of the questions to be asked in this step are:

- How will the performance of the new process be measured?
- How will the data be collected after the improvement implementation?
- How will the process owner make sure the improvements become a routine part of the process?
- How will the process owner make sure the performance is improving after implementation?

By answering these questions, the Six Sigma team seeks to identify potential risks. To minimize these risks, the Six Sigma team identifies performance measurement and monitoring points in the process in which improvements have been embedded.

11.2.2 Standardization of the Solutions

The Six Sigma team needs to ensure that the solutions and improvements, new procedures, and control methods are in place and will be sustained in both the short and long runs. When a change takes place in a business process, the system may tend to revert to the previous settings. Therefore, the Six Sigma team is expected to design methods to prevent the system from returning to the old version. Similar to other improvement-oriented practices, such as 5S and total productive maintenance (TPM), the standardization step is pivotal to maintain and sustain the improvements. If standardization is not maintained in the processes, it is likely that expected goals cannot be achieved in the team and the system will revert to the previous activities.

Revising work instructions, policies, procedures, flow charts, process maps, and checklists is the first requirement of standardization. Once the document revision is done, employees will run the revised processes and implement improvements. The Six Sigma team needs to make sure that the process owner and other responsible parties revised these documents and put them in place for use. Using visual tools and methods to teach employees the revised process is an effective way to maintain the new process.

As a critical component of standardization, Six Sigma teams and process owners ensure that employees are properly trained on the revised process to minimize failures and nonconformities. In this step, employees are trained, and supervisors make sure that employees follow the revised procedures, policies, work instructions, and plans.

If the improvements identified in the project are applicable to other processes, the Six Sigma team needs to make sure that these improvements are transferred and implemented in those processes.

11.2.3 Monitoring the Improvements

While monitoring and assessing the performance of the improvements, Six Sigma team compares the performance before and after the improvements. A comparative analysis helps the team determine whether improvements result in any gains in the system. Particularly, Six Sigma measures such as DPU, DPO, DPMO, and PPM are used to make such comparative analysis. The main output of this step is a before and after improvement comparison using Six Sigma measures. To monitor the processes and improvements implemented in the Improve phase, SPC and control plans are extensively utilized by Six Sigma team. SPC and control plans are detailed in the next sections in this chapter.

11.2.4 Project Closure

After going through all the phases of the DMAIC process, the Six Sigma team systematically documents and reports outputs of each phase. Well-documented Six Sigma projects can guide the next projects and the executive steering committee in organization-wide Six Sigma effort. In project closure, benefits and costs of the project are itemized by the Six Sigma team. Master black belt and black belt members systematically report overall benefits and costs of the project to the executive steering committee. The benefits that have been recognized so far and other potential benefits that could be seen in the short and long term are itemized and detailed, including all operational, financial, and functional benefits of the project. All cost items that are related to the project are reported. Additionally, a comparative report that includes "what-if" analyses can be presented by the Six Sigma team. For example, the Six Sigma team may compare (1) the actual benefits of the improvements implemented in the project and (2) the potential loss that would have been experienced if the project did not perform that way.

The champion, master black belt, or black belt of the project enters comprehensive information about the completed project into the organization's Six Sigma database. To keep organizational memory alive and guide the next teams, Six Sigma organizations keep such databases up-to-date. Lessons learned are shared with the next teams for future projects. These lessons learned and improvement outcomes can be adapted to some other similar projects across the organization. Therefore, proper documentation and record keeping systems have a critical role in Six Sigma projects. The documents for each phase of DMAIC can be listed as follows:

Define
- Scope of the project, project plan, process documentation, stakeholder analysis, and project charter

Measure
- Data collection plan, baseline Sigma level, and Gage R&R analysis results

Analyze
- Data analysis results

Improve
- Pilot test results, meeting minutes, and after improvement Sigma level

Control
- Control plans, control charts, and before and after comparison results.

When the Control phase is finalized, the Six Sigma team makes sure that improvements are already implemented in the processes and the revised process/product/service is successfully handed over to the process owner for ongoing maintenance. Afterward, the Six Sigma team may celebrate the project outcomes and think about topics for the next project. The success of the Control phase is likely to impact the next Six Sigma projects. Celebrating Six Sigma teams and recognizing their success are considered pivotal in Six Sigma methodology to encourage other teams and employees.

11.3 Tools in Control Phase

The tools used in the Control phase focus on monitoring improvements and sustaining the success of the improvements. Some of these tools are FMEA, key performance indicators (KPI) trees, control plans, standardized processes, 5S, SPC, TPM, mistake proofing (poka-yoke), standard operating procedures, Six Sigma measures, and Six Sigma project team reports. This section analyzes some of these tools.

11.3.1 Statistical Process Control

SPC is a branch of statistics that combines time series analysis methods with graphical presentation of data (Benneyan et al. 2003). SPC and statistical quality control are used interchangeably both in the literature and practice. "The SPC strategy incorporates the concepts of an analytic study, process thinking, prevention, stratification, stability, capability, and prediction. SPC incorporates measurement, data collection methods, and planned experimentation" (Carey 2003: xviii). SPC is also known as graphical and statistical analysis of process data to allow process stability, understanding, monitoring, and improve capability by reducing the variability in the processes (Benneyan 1998; Montgomery 2009). Process controls are placed throughout the operations (Juran 1992). SPC reduces variability and assists to maintain consistency in the processes. Amsden et al. (1998: 2) state that the following rules dominate SPC studies in manufacturing settings:

1. No two things are exactly alike.
2. Variation in a product/process can be measured.
3. Things vary according to a definite pattern.
4. Whenever things of the same kind are measured, a large group of the measurements will tend to cluster around the middle.
5. It is possible to determine the shape of the distribution curve for parts produced by any process.
6. Variations due to assignable causes tend to distort the normal distribution curve.

SPC has been extensively utilized in process control activities since the 1920s. The tools used in SPC significantly contributed to effective process controls and quality management, and many supervisors and employees were trained in SPC. SPC is a solid theoretical and statistical foundation and a highly robust method for analyzing process performance (Thor et al. 2007).

SPC is not used as a tool to solve problems. Instead, it identifies causes of variation and monitors the processes. The ultimate goal of SPC is to identify assignable causes of the variability, start corrective actions before variation affects the quality of the outputs, and eliminate all current and potential variability of the process. One of the advantages of SPC is that the data are not "time static" (Benneyan et al. 2003). Instead, the data, specifically in control charts, show the time order. Data collection and measurement activities are fundamental in SPC (Carey 2003). Also, changes and trends occurring in the process can be detected by SPC.

After the Critical-to-Quality (CTQ) characteristics are identified and the relationship among the CTQ characteristics and process outputs are quantified, SPC can be implemented with considerable effectiveness for monitoring the processes (Montgomery 2013). Hart and Hart (2002) suggest that SPC may be used to learn how to incorporate quantitative performance measurement into daily activities to build an improvement-oriented culture throughout the organization.

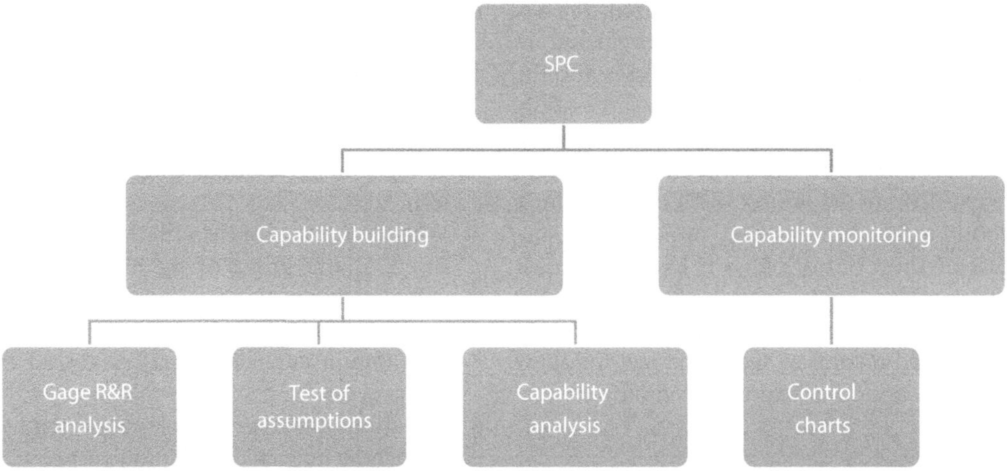

Fig. 11.1 The components of SPC. (Source: Author's creation)

11

As an *online* quality control component, SPC methodology is comprised of two main phases: *capability building* and *capability monitoring* (Fig. 11.1). In the *capability building* phase, the Six Sigma team analyzes the capability of measurement system using Gage R&R studies (as presented in ► Chap. 6) and examines process capability ratios (as presented in process capability section in ► Chap. 7). In the *capability monitoring* phase, CTQ characteristics are monitored through control charts (as presented in ► Chap. 9). In the *capability building* phase in SPC, (1) Gage R&R analysis, (2) testing of assumptions, and (3) capability analysis studies are performed, respectively. Three assumptions of testing are:

1. The data used in process capability analysis using process potential capability (C_p) and actual capability (C_{pk}) are normally distributed.
2. The process is stable.
3. The process mean is centered between the *LSL* and *USL* in the two-sided specifications (Montgomery 2009).

11.3.2 Control Plans

Organizational systems do not easily accept changes, and resistance to organizational change is common among organizational members. Any kind of revision, change, or improvement in individual processes, functions, departments, or systems brought by Six Sigma teams may face significant resistance. The size and impact of that resistance vary depending on the magnitude of the proposed improvements and changes. Considering this resistance, Six Sigma teams need to work on sustaining proposed improvements in the processes.

As identified by Pyzdek and Keller (2018), in the Control phase, Six Sigma teams can use several methods to make sure that proposed improvements are successfully implemented and sustained. For instance, modifying relevant policies and procedures, training associated employees on the modifications, implementing and compliance with new standards, and revising quality audit requirements, production plans, engineering drawings, and relevant business functions are examples of tasks in the Control phase.

Variation is considered to be an enemy in all of the quality literature. To minimize variation in the improvement implementation process and ensure that improvements are successfully implemented in the processes, the Six Sigma team may develop a control plan. Using control plans, process owners and employees (1) articulate new standards and (2) observe and monitor the performance of CTQ characteristics and independent variables (*X*s). It is similar to the quality plans developed for the CTQ characteristics of the products/services in the Define phase previously. The variables needed in control plans are:

- CTQ characteristics
- X variables
- Unit of measure
- Target value of the CTQ characteristic
- Specifications of the CTQ characteristic
- Control and record-keeping methods
- Frequency of measurement
- Sample size
- Decision-making criteria
- Measurement tool
- Measurement operator.

An example of a control plan is presented in ◘ Table 11.1. The product exemplified in ◘ Table 11.1 is a round metal part produced in a manufacturing firm. There are five CTQ characteristics identified by the Six Sigma team to monitor in this phase: diameter, weight, volume, color tone, and strength. The variables listed above are in the control plan for each CTQ characteristic. When the process owner needs to ensure that the improvements are successfully integrated into manufacturing process, control plans may function as a base. Additionally, control plans function as a guide for employee training and new employee orientation processes. Moreover, employees learn the details of the CTQ characteristics from control plans, rather than learning from other employees, which can introduce variability. When the operators inspect products/services or parts, they can easily see the details of CTQ characteristics to be inspected.

Key Concepts
DMAIC, control phase, control plans, PDCA, statistical process control (SPC), capability building, capability monitoring, organizational change, organizational memory, standardization, Six Sigma measures, lessons learned, and project closure.

Summary
The Control phase in DMAIC focuses on whether all activities planned in the Improve phase are successfully implemented, standardized, and sustained in the processes. Documenting all phases of DMAIC, revising the systems and documents, and controlling the processes and improvements using tools such as SPC and control plans are some of the main tasks of the Control phase. The steps followed in the Control phase are (1) implementing ongoing measurements, (2) standardization of the improvement, (3) monitoring the improvement, and (4) project closure.

Practice and discussion questions

1. What are the goals of the Control phase in DMAIC?
2. Discuss similarities between PDCA cycle and the Control phase.
3. Discuss each step of the Control phase.
4. What questions are answered while implementing ongoing measurements in the Control phase?
5. Why is it critical for Six Sigma teams to implement ongoing measurements in the Control phase?
6. What is the importance of standardizing the solutions for Six Sigma projects?
7. Discuss "standardization of the solutions" in the Control phase.
8. Discuss "monitoring the improvements" in the Control phase.
9. Discuss "project closure" in the Control phase.
10. What documents are kept in project closure in the Control phase?
11. What are the tools used in the Control phase in DMAIC?
12. Discuss how to use SPC in the Control phase in DMAIC.

Table 11.1 An example of control plan

CONTROL PLAN									
Product/service: Round metal part			**Prepared by: Six Sigma team**		**Revision date & number: 2/1/2020 & 003**		**Process owner: Product management**		**Approved: ☐**
CTQ characteristic	**Unit of measure**	**Target**	**Specifications**	**Control & record-keeping methods**	**Frequency of measurement**	**Sample size**	**Decision-making criteria**	**Measurement tool**	**Measurement operator**
Diameter	cm	0.5 cm	±0.01 cm USL = 0.51 cm LSL = 0.49 cm	Instruction (INS) 001	Each batch per line	10/ batch	If 1/10 is rejected, reject the batch	Digital gauge	Quality control (QC) technician 1
Weight	gram	3.5 g	±0.009 gr USL = 3.509 gr LSL = 3.491 gr	Instruction (INS) 002	Each batch per line	10/ batch	If 1/10 is rejected, reject the batch	Digital gauge	Quality control technician 2
Volume	cm^3	2.5 cm^3	±0.0004 cm^3 USL = 2.5004 cm^3 LSL = 2.4996 cm^3	Instruction (INS) 003	Each batch per line	10/ batch	If 1/10 is rejected, reject the batch	Digital gauge	Quality control technician 3
Color tone	Color palette	Gray	0	Instruction (INS) 004	Each batch per line	10/ batch	If no defects found, passes	Color pallet	Quality control technician 4
Strength	Kg	75 kg min	0 kg	Instruction (INS) 005	Each batch per line	10/ batch	If no defects found, passes	Visual count	Quality control technician 5

Source: Author's creation

11

13. What are the two components of SPC? Discuss the details of these components.
14. Discuss the functions of "control plans" in the Control phase in DMAIC.
15. What variables are used in "control plans"?

References

Amsden, D. M., Amsden, R. T., & Butler, H. E. (1998). *SPC simplified workbook: Practical steps to quality*. New York: SteinerBooks.

Benneyan, J. C. (1998). Statistical quality control methods in infection control and hospital epidemiology, part I introduction and basic theory. *Infection Control and Hospital Epidemiology, 19*(3), 194–214.

Benneyan, J. C., Lloyd, R. C., & Plsek, P. E. (2003). Statistical process control as a tool for research and healthcare improvement. *BMJ Quality and Safety, 12*(6), 458–464.

Carey, R. G. (2003). *Improving healthcare with control charts: Basics and advanced SPC methods and case studies*. Milwaukee: ASQ Quality Press.

Feigenbaum, A. V. (1983). *Total quality control*. New York: McGraw-Hill.

Gitlow, H. S., Melnyck, R. J., & Levine, D. M. (2015). *A guide to six sigma and process improvement for practitioners and students* (2nd ed.). Upper Saddle River, NJ: Pearson.

Hart, M. K., & Hart, R. F. (2002). *Statistical process control for health care*. Duxbury: Thomson Learning.

Imai, K. (1986). *Kaizen: The key to Japan's competitive success*. New York: The Kaizen Institute.

Juran, J. M. (1992). *Juran on quality by design: The new steps for planning quality into goods and services*. New York: The Free Press.

Montgomery, D. C. (2009). *Introduction to statistical quality control* (6th ed.). New York: John Wiley & Sons.

Montgomery, D. C. (2013). *Introduction to statistical quality control*. New York: Wiley.

Pyzdek, T., & Keller, P. A. (2018). *The six sigma handbook* (5th ed.). New York: Mc-Graw Hill.

Thor, J., Lundberg, J., Ask, J., Olsson, J., Carli, C., Härenstam, K. P., & Brommels, M. (2007). Application of statistical process control in healthcare improvement: Systematic review. *BMJ Quality and Safety, 16*(5), 387–399.

Supplementary Information

F. Pakdil, *Six Sigma for Students*, https://doi.org/10.1007/978-3-030-40709-4

Appendix

Table A.1 Common probability distributions

Name of the distribution	Probability distribution function	Mean of the distribution	Variance of the distribution
	Discrete distributions		
Bernoulli	$f(x)=\begin{cases} p, & x=1 \\ 1-p, & x=0 \\ 0, & \text{otherwise} \end{cases}$ $0 \le p \le 1$	p	$p(1\text{-}p)$
Binomial	$f(x)=\binom{n}{i}p^i(1-p)^{n-i}=\frac{n!}{i!(n-i)!}p^i(1-p)^{n-i}$ $x = 0, 1, 2, \ldots, n$, and $0 \le p \le 1$.	np	$np(1-p)$
Hypergeometric	$f(x)=P\{X=k\}=\frac{\binom{K}{k}\binom{N-K}{n-k}}{\binom{N}{n}}$ $x = 0, 1, 2, \ldots, n$, $\min(K,n)$, $K \le N$, $n \le N$	np where $p=\frac{K}{N}$	$np(1-p)\left(\frac{N-n}{N-1}\right)$
Geometric	$f(x) = P\{X = n\} = (1-p)^{n-1}p$ $X = 1, 2, \ldots$ and $0 \le p \le 1$	$\frac{1}{p}$	$\frac{(1-p)}{p^2}$
Poisson	$f(x)=P\{X=i\}=\frac{e^{-\lambda}\lambda^i}{i!}$ $i = 0, 1, 2, \ldots, \lambda > 0$	λ	λ

	Continuous distributions		
Uniform	$f(x)=\begin{cases}\dfrac{1}{b-a}, & a<x<b\\ 0, & \text{otherwise}\end{cases}$	$\dfrac{(b+a)}{2}$	$\dfrac{(b-a)^2}{12}$
Exponential	$f(x)=\begin{cases}\lambda e^{-\lambda x}, & x\geq 0\\ 0, & x<0\end{cases}$	$\dfrac{1}{\lambda}$	$\dfrac{1}{\lambda^2}$
Triangular	$\begin{cases}0, & x<a\\ \dfrac{2(x-a)}{(b-a)(c-a)}, & a\leq x\leq c\\ \dfrac{2(b-x)}{(b-a)(b-c)}, & c\leq x\leq b\\ 0, & x>b\end{cases}$	$\dfrac{(a+b+c)}{3}$	$\dfrac{(a^2+b^2+c^2-ab-ac-bc)}{18}$
Normal	$f(x)=\dfrac{1}{\sqrt{2\pi}\sigma}e^{-(x-\mu)^2/2\sigma^2}, -\infty<x<\infty$ $-\infty<\mu<\infty, \sigma>0$	μ	σ^2
Weibull	$f(x)=\begin{cases}\dfrac{k}{\lambda}\left(\dfrac{x}{\lambda}\right)^{k-1}e^{-\left(\frac{x}{\lambda}\right)^k}, & x\geq 0\\ 0, & x<0\end{cases}$ $k>0, \lambda>0, x>0$	$\lambda\Gamma\left(1+\dfrac{1}{k}\right)$	$\lambda^2\Gamma\left(1+\dfrac{2}{k}\right)-\lambda^2\left[\Gamma\left(1+\dfrac{1}{k}\right)\right]^2$

Source: Author's creation

Table A.2 The standard normal distribution table (the values represent left side of the Z score)

Z	0.00	0.01	0.02	0.03	0.04	0.05	0.06	0.07	0.08	0.09
0.0	0.50000	0.50399	0.50798	0.51197	0.51595	0.51994	0.52392	0.5279	0.53188	0.53586
0.1	0.53983	0.54380	0.54776	0.55172	0.55567	0.55962	0.56356	0.56749	0.57142	0.57535
0.2	0.57926	0.58317	0.58706	0.59095	0.59483	0.59871	0.60257	0.60642	0.61026	0.61409
0.3	0.61791	0.62172	0.62552	0.62930	0.63307	0.63683	0.64058	0.64431	0.64803	0.65173
0.4	0.65542	0.65910	0.66276	0.66640	0.67003	0.67364	0.67724	0.68082	0.68439	0.68793
0.5	0.69146	0.69497	0.69847	0.70194	0.70540	0.70884	0.71226	0.71566	0.71904	0.72240
0.6	0.72575	0.72907	0.73237	0.73565	0.73891	0.74215	0.74537	0.74857	0.75175	0.75490
0.7	0.75804	0.76115	0.76424	0.76730	0.77035	0.77337	0.77637	0.77935	0.78230	0.78524
0.8	0.78814	0.79103	0.79389	0.79673	0.79955	0.80234	0.80511	0.80785	0.81057	0.81327
0.9	0.81594	0.81859	0.82121	0.82381	0.82639	0.82894	0.83147	0.83398	0.83646	0.83891
1.0	0.84134	0.84375	0.84614	0.84849	0.85083	0.85314	0.85543	0.85769	0.85993	0.86214
1.1	0.86433	0.86650	0.86864	0.87076	0.87286	0.87493	0.87698	0.87900	0.88100	0.88298
1.2	0.88493	0.88686	0.88877	0.89065	0.89251	0.89435	0.89617	0.89796	0.89973	0.90147
1.3	0.90320	0.90490	0.90658	0.90824	0.90988	0.91149	0.91309	0.91466	0.91621	0.91774
1.4	0.91924	0.92073	0.92220	0.92364	0.92507	0.92647	0.92785	0.92922	0.93056	0.93189
1.5	0.93319	0.93448	0.93574	0.93699	0.93822	0.93943	0.94062	0.94179	0.94295	0.94408
1.6	0.94520	0.94630	0.94738	0.94845	0.94950	0.95053	0.95154	0.95254	0.95352	0.95449
1.7	0.95543	0.95637	0.95728	0.95818	0.95907	0.95994	0.96080	0.96164	0.96246	0.96327
1.8	0.96407	0.96485	0.96562	0.96638	0.96712	0.96784	0.96856	0.96926	0.96995	0.97062
1.9	0.97128	0.97193	0.97257	0.97320	0.97381	0.97441	0.97500	0.97558	0.97615	0.97670

2.0	0.97725	0.97778	0.97831	0.97882	0.97932	0.97982	0.98030	0.98077	0.98124	0.98169
2.1	0.98214	0.98257	0.98300	0.98341	0.98382	0.98422	0.98461	0.98500	0.98537	0.98574
2.2	0.98610	0.98645	0.98679	0.98713	0.98745	0.98778	0.98809	0.98840	0.98870	0.98899
2.3	0.98928	0.98956	0.98983	0.99010	0.99036	0.99061	0.99086	0.99111	0.99134	0.99158
2.4	0.99180	0.99202	0.99224	0.99245	0.99266	0.99286	0.99305	0.99324	0.99343	0.99361
2.5	0.99379	0.99396	0.99413	0.99430	0.99446	0.99461	0.99477	0.99492	0.99506	0.99520
2.6	0.99534	0.99547	0.99560	0.99573	0.99585	0.99598	0.99609	0.99621	0.99632	0.99643
2.7	0.99653	0.99664	0.99674	0.99683	0.99693	0.99702	0.99711	0.99720	0.99728	0.99736
2.8	0.99744	0.99752	0.99760	0.99767	0.99774	0.99781	0.99788	0.99795	0.99801	0.99807
2.9	0.99813	0.99819	0.99825	0.99831	0.99836	0.99841	0.99846	0.99851	0.99856	0.99861
3.0	0.99865	0.99869	0.99874	0.99878	0.99882	0.99886	0.99889	0.99893	0.99896	0.99900
3.1	0.99903	0.99906	0.99910	0.99913	0.99916	0.99918	0.99921	0.99924	0.99926	0.99929
3.2	0.99931	0.99934	0.99936	0.99938	0.99940	0.99942	0.99944	0.99946	0.99948	0.99950
3.3	0.99952	0.99953	0.99955	0.99957	0.99958	0.99960	0.99961	0.99962	0.99964	0.99965
3.4	0.99966	0.99968	0.99969	0.99970	0.99971	0.99972	0.99973	0.99974	0.99975	0.99976
3.5	0.99977	0.99978	0.99978	0.99979	0.99980	0.99981	0.99981	0.99982	0.99983	0.99983
3.6	0.99984	0.99985	0.99985	0.99986	0.99986	0.99987	0.99987	0.99988	0.99988	0.99989
3.7	0.99989	0.99990	0.99990	0.99990	0.99991	0.99991	0.99992	0.99992	0.99992	0.99992
3.8	0.99993	0.99993	0.99993	0.99994	0.99994	0.99994	0.99994	0.99995	0.99995	0.99995
3.9	0.99995	0.99995	0.99996	0.99996	0.99996	0.99996	0.99996	0.99996	0.99997	0.99997
−3.9	0.00005	0.00005	0.00004	0.00004	0.00004	0.00004	0.00004	0.00004	0.00003	0.00003

(continued)

Table A.2 (continued)

Z	0.00	0.01	0.02	0.03	0.04	0.05	0.06	0.07	0.08	0.09
−3.8	0.00007	0.00007	0.00007	0.00006	0.00006	0.00006	0.00006	0.00005	0.00005	0.00005
−3.7	0.00011	0.00010	0.00010	0.00010	0.00009	0.00009	0.00008	0.00008	0.00008	0.00008
−3.6	0.00016	0.00015	0.00015	0.00014	0.00014	0.00013	0.00013	0.00012	0.00012	0.00011
−3.5	0.00023	0.00022	0.00022	0.00021	0.00020	0.00019	0.00019	0.00018	0.00017	0.00017
−3.4	0.00034	0.00032	0.00031	0.00030	0.00029	0.00028	0.00027	0.00026	0.00025	0.00024
−3.3	0.00048	0.00047	0.00045	0.00043	0.00042	0.00040	0.00039	0.00038	0.00036	0.00035
−3.2	0.00069	0.00066	0.00064	0.00062	0.00060	0.00058	0.00056	0.00054	0.00052	0.00050
−3.1	0.00097	0.00094	0.00090	0.00087	0.00084	0.00082	0.00079	0.00076	0.00074	0.00071
−3.0	0.00135	0.00131	0.00126	0.00122	0.00118	0.00114	0.00111	0.00107	0.00104	0.00100
−2.9	0.00187	0.00181	0.00175	0.00169	0.00164	0.00159	0.00154	0.00149	0.00144	0.00139
−2.8	0.00256	0.00248	0.00240	0.00233	0.00226	0.00219	0.00212	0.00205	0.00199	0.00193
−2.7	0.00347	0.00336	0.00326	0.00317	0.00307	0.00298	0.00289	0.00280	0.00272	0.00264
−2.6	0.00466	0.00453	0.00440	0.00427	0.00415	0.00402	0.00391	0.00379	0.00368	0.00357
−2.5	0.00621	0.00604	0.00587	0.00570	0.00554	0.00539	0.00523	0.00508	0.00494	0.0048
−2.4	0.00820	0.00798	0.00776	0.00755	0.00734	0.00714	0.00695	0.00676	0.00657	0.00639
−2.3	0.01072	0.01044	0.01017	0.00990	0.00964	0.00939	0.00914	0.00889	0.00866	0.00842
−2.2	0.01390	0.01355	0.01321	0.01287	0.01255	0.01222	0.01191	0.01160	0.01130	0.01101
−2.1	0.01786	0.01743	0.01700	0.01659	0.01618	0.01578	0.01539	0.01500	0.01463	0.01426
−2.0	0.02275	0.02222	0.02169	0.02118	0.02068	0.02018	0.01970	0.01923	0.01876	0.01831
−1.9	0.02872	0.02807	0.02743	0.02680	0.02619	0.02559	0.02500	0.02442	0.02385	0.02330

−1.8	0.03593	0.03515	0.03438	0.03362	0.03288	0.03216	0.03144	0.03074	0.03005	0.02938
−1.7	0.04457	0.04363	0.04272	0.04182	0.04093	0.04006	0.03920	0.03836	0.03754	0.03673
−1.6	0.05480	0.05370	0.05262	0.05155	0.05050	0.04947	0.04846	0.04746	0.04648	0.04551
−1.5	0.06681	0.06552	0.06426	0.06301	0.06178	0.06057	0.05938	0.05821	0.05705	0.05592
−1.4	0.08076	0.07927	0.07780	0.07636	0.07493	0.07353	0.07215	0.07078	0.06944	0.06811
−1.3	0.09680	0.09510	0.09342	0.09176	0.09012	0.08851	0.08691	0.08534	0.08379	0.08226
−1.2	0.11507	0.11314	0.11123	0.10935	0.10749	0.10565	0.10383	0.10204	0.10027	0.09853
−1.1	0.13567	0.13350	0.13136	0.12924	0.12714	0.12507	0.12302	0.12100	0.11900	0.11702
−1.0	0.15866	0.15625	0.15386	0.15151	0.14917	0.14686	0.14457	0.14231	0.14007	0.13786
−0.9	0.18406	0.18141	0.17879	0.17619	0.17361	0.17106	0.16853	0.16602	0.16354	0.16109
−0.8	0.21186	0.20897	0.20611	0.20327	0.20045	0.19766	0.19489	0.19215	0.18943	0.18673
−0.7	0.24196	0.23885	0.23576	0.23270	0.22965	0.22663	0.22363	0.22065	0.21770	0.21476
−0.6	0.27425	0.27093	0.26763	0.26435	0.26109	0.25785	0.25463	0.25143	0.24825	0.24510
−0.5	0.30854	0.30503	0.30153	0.29806	0.29460	0.29116	0.28774	0.28434	0.28096	0.27760
−0.4	0.34458	0.34090	0.33724	0.33360	0.32997	0.32636	0.32276	0.31918	0.31561	0.31207
−0.3	0.38209	0.37828	0.37448	0.37070	0.36693	0.36317	0.35942	0.35569	0.35197	0.34827
−0.2	0.42074	0.41683	0.41294	0.40905	0.40517	0.40129	0.39743	0.39358	0.38974	0.38591
−0.1	0.46017	0.45620	0.45224	0.44828	0.44433	0.44038	0.43644	0.43251	0.42858	0.42465
0	0.50000	0.49601	0.49202	0.48803	0.48405	0.48006	0.47608	0.47210	0.46812	0.46414

Source: Author's creation

Table A.3 Percentage points of the χ^2 distribution

	α								
ν	**0.995**	**0.99**	**0.975**	**0.95**	**0.50**	**0.05**	**0.025**	**0.010**	**0.005**
1	0.00	0.00	0.00	0.00	0.45	3.84	5.02	6.63	7.88
2	0.01	0.02	0.05	0.10	1.39	5.99	7.38	9.21	10.60
3	0.07	0.11	0.22	0.35	2.37	7.81	9.35	11.34	12.84
4	0.21	0.30	0.48	0.71	3.36	9.49	11.14	13.28	14.86
5	0.41	0.55	0.83	1.15	4.35	11.07	12.38	15.09	16.75
6	0.68	0.87	1.24	1.64	5.35	12.59	14.45	16.81	18.55
7	0.99	1.24	1.69	2.17	6.35	14.07	16.01	18.48	20.28
8	1.34	1.65	2.18	2.73	7.34	15.51	17.53	20.09	21.96
9	1.73	2.09	2.70	3.33	8.34	16.92	19.02	21.67	23.59
10	2.16	2.56	3.25	3.94	9.34	18.31	20.48	23.21	25.19
11	2.60	3.05	3.82	4.57	10.34	19.68	21.92	24.72	26.76
12	3.07	3.57	4.40	5.23	11.34	21.03	23.34	26.22	28.30
13	3.57	4.11	5.01	5.89	12.34	22.36	24.74	27.69	29.82
14	4.07	4.66	5.63	6.57	13.34	23.68	26.12	29.14	31.32
15	4.60	5.23	6.27	7.26	14.34	25.00	27.49	30.58	32.80
16	5.14	5.81	6.91	7.96	15.34	26.30	28.85	32.00	34.27
17	5.70	6.41	7.56	8.67	16.34	27.59	30.19	33.41	35.72

18	6.26	7.01	8.23	9.39	17.34	28.87	31.53	34.81	37.16
19	6.88	7.63	8.91	10.12	18.34	30.14	32.85	36.19	38.58
20	7.43	8.26	9.59	10.85	19.34	31.41	34.17	37.57	40.00
25	10.52	11.52	13.12	14.61	24.34	37.65	40.65	44.31	46.93
30	13.79	14.95	16.79	18.49	29.34	43.77	46.98	50.89	53.67
40	20.71	22.16	24.43	26.51	39.34	55.76	59.34	63.69	66.77
50	27.99	29.71	32.36	34.76	49.33	67.50	71.42	76.15	79.49
60	35.53	37.48	40.48	43.19	59.33	79.08	83.30	88.38	91.95
70	43.28	45.44	48.76	51.74	69.33	90.53	95.02	100.42	104.22
80	51.17	53.54	57.15	60.39	79.33	101.88	106.63	112.33	116.32
90	59.20	61.75	65.65	69.13	89.33	113.14	118.14	124.12	128.30
100	67.33	70.06	74.22	77.93	99.33	124.34	129.56	135.81	140.17

Source: The entries in this table were computed by the author

Table A.4 Percentage points of the t distribution

	Level of significance (α)				
ν	**0.100**	**0.05**	**0.025**	**0.010**	**0.005**
1	3.078	6.314	12.706	31.821	63.657
2	1.886	2.920	4.303	6.965	9.925
3	1.638	2.353	3.182	4.541	5.841
4	1.533	2.132	2.776	3.747	4.604
5	1.476	2.015	2.571	3.365	4.032
6	1.440	1.943	2.447	3.143	3.707
7	1.415	1.895	2.365	2.998	3.499
8	1.397	1.860	2.306	2.896	3.355
9	1.383	1.833	2.262	2.821	3.250
10	1.372	1.812	2.228	2.764	3.169
11	1.363	1.796	2.201	2.718	3.106
12	1.356	1.782	2.179	2.681	3.055
13	1.350	1.771	2.160	2.650	3.012
14	1.345	1.761	2.145	2.624	2.977
15	1.341	1.753	2.131	2.602	2.947
16	1.337	1.746	2.120	2.583	2.921
17	1.333	1.740	2.110	2.567	2.898
18	1.330	1.734	2.101	2.552	2.878
19	1.328	1.729	2.093	2.539	2.861
20	1.325	1.725	2.086	2.528	2.845
21	1.323	1.721	2.080	2.518	2.831
22	1.321	1.717	2.074	2.508	2.819
23	1.319	1.714	2.069	2.500	2.807
24	1.318	1.711	2.064	2.492	2.797
25	1.316	1.708	2.060	2.485	2.787
26	1.315	1.706	2.056	2.479	2.779
27	1.314	1.703	2.052	2.473	2.771

Table A.4 (continued)

	Level of significance (α)				
v	**0.100**	**0.05**	**0.025**	**0.010**	**0.005**
28	1.313	1.701	2.048	2.467	2.763
29	1.311	1.699	2.045	2.462	2.756
30	1.310	1.697	2.042	2.457	2.750
35	1.306	1.690	2.030	2.438	2.724
40	1.303	1.684	2.021	2.423	2.704
45	1.301	1.679	2.014	2.412	2.690
50	1.299	1.676	2.009	2.403	2.678
100	1.290	1.660	1.984	2.364	2.626
200	1.286	1.653	1.972	2.345	2.601
400	1.284	1.649	1.966	2.336	2.588
999	1.282	1.646	1.962	2.330	2.581
∞	1.282	1.645	1.960	2.326	2.576

Source: The entries in this table were computed by the author

Table A.5 Percentage points of the F distribution ($\alpha = 1\ \%$)

$F_{0.01\ (v1,\ v2)}$

v1=degrees of freedom for the numerator

	v1	**1**	**2**	**3**	**4**	**5**	**6**	**7**	**8**	**9**	**10**	**12**	**15**	**20**	**24**	**30**	**40**	**60**	**120**	**∞**
v2	1	4052	4999.5	5403	5625	5764	5859	5928	5982	6022	6056	6106	6157	6209	6235	6261	6287	6313	6339	6366
v2=degrees of freedom for the denominator	2	98.5	99	99.17	99.25	99.3	99.33	99.36	99.37	99.39	99.4	99.42	99.43	99.45	99.46	99.47	99.47	99.48	99.49	99.5
	3	34.12	30.82	29.46	28.71	28.24	27.91	27.67	27.49	27.35	27.23	27.05	26.87	26.69	26.6	26.5	26.41	26.32	26.22	26.13
	4	21.2	18	16.69	15.98	15.52	15.21	14.98	14.8	14.66	14.55	14.37	14.2	14.02	13.93	13.84	13.75	13.65	13.56	13.46
	5	16.26	13.27	12.06	11.39	10.97	10.67	10.46	10.29	10.16	10.05	9.89	9.72	9.55	9.47	9.38	9.29	9.2	9.11	9.02
	6	13.75	10.92	9.78	9.15	8.75	8.47	8.26	8.1	7.98	7.87	7.72	7.56	7.4	7.31	7.23	7.14	7.06	6.97	6.88
	7	12.25	9.55	8.45	7.85	7.46	7.19	6.99	6.84	6.72	6.62	6.47	6.31	6.16	6.07	5.99	5.91	5.82	5.74	5.65
	8	11.26	8.65	7.59	7.01	6.63	6.37	6.18	6.03	5.91	5.81	5.67	5.52	5.36	5.28	5.2	5.12	5.03	4.95	4.86
	9	10.56	8.02	6.99	6.42	6.06	5.8	5.61	5.47	5.35	5.26	5.11	4.96	4.81	4.73	4.65	4.57	4.48	4.4	4.31
	10	10.04	7.56	6.55	5.99	5.64	5.39	5.2	5.06	4.94	4.85	4.71	4.56	4.41	4.33	4.25	4.17	4.08	4	3.91
	11	9.65	7.21	6.22	5.67	5.32	5.07	4.89	4.74	4.63	4.54	4.4	4.25	4.1	4.02	3.94	3.86	3.78	3.69	3.6
	12	9.33	6.93	5.95	5.41	5.06	4.82	4.64	4.5	4.39	4.3	4.16	4.01	3.86	3.78	3.7	3.62	3.54	3.45	3.36
	13	9.07	6.7	5.74	5.21	4.86	4.62	4.44	4.3	4.19	4.1	3.96	3.82	3.66	3.59	3.51	3.43	3.34	3.25	3.17
	14	8.86	6.51	5.56	5.04	4.69	4.46	4.28	4.14	4.03	3.94	3.8	3.66	3.51	3.43	3.35	3.27	3.18	3.09	3
	15	8.68	6.36	5.42	4.89	4.56	4.32	4.14	4	3.89	3.8	3.67	3.52	3.37	3.29	3.21	3.13	3.05	2.96	2.87
	16	8.53	6.23	5.29	4.77	4.44	4.2	4.03	3.89	3.78	3.69	3.55	3.41	3.26	3.18	3.1	3.02	2.93	2.84	2.75
	17	8.4	6.11	5.18	4.67	4.34	4.1	3.93	3.79	3.68	3.59	3.46	3.31	3.16	3.08	3	2.92	2.83	2.75	2.65
	18	8.29	6.01	5.09	4.58	4.25	4.01	3.84	3.71	3.6	3.51	3.37	3.23	3.08	3	2.92	2.84	2.75	2.66	2.57

19	8.18	5.93	5.01	4.5	4.17	3.94	3.77	3.63	3.52	3.43	3.3	3.15	3	2.92	2.84	2.76	2.67	2.58	2.49
20	8.1	5.85	4.94	4.43	4.1	3.87	3.7	3.56	3.46	3.37	3.23	3.09	2.94	2.86	2.78	2.69	2.61	2.52	2.42
21	8.02	5.78	4.87	4.37	4.04	3.81	3.64	3.51	3.4	3.31	3.17	3.03	2.88	2.8	2.72	2.64	2.55	2.46	2.36
22	7.95	5.72	4.82	4.31	3.99	3.76	3.59	3.45	3.35	3.26	3.12	2.98	2.83	2.75	2.67	2.58	2.5	2.4	2.31
23	7.88	5.66	4.76	4.26	3.94	3.71	3.54	3.41	3.3	3.21	3.07	2.93	2.78	2.7	2.62	2.54	2.45	2.35	2.26
24	7.82	5.61	4.72	4.22	3.9	3.67	3.5	3.36	3.26	3.17	3.03	2.89	2.74	2.66	2.58	2.49	2.4	2.31	2.21
25	7.77	5.57	4.68	4.18	3.85	3.63	3.46	3.32	3.22	3.13	2.99	2.85	2.7	2.62	2.54	2.45	2.36	2.27	2.17
26	7.72	5.53	4.64	4.14	3.82	3.59	3.42	3.29	3.18	3.09	2.96	2.81	2.66	2.58	2.5	2.42	2.33	2.23	2.13
27	7.68	5.49	4.6	4.11	3.78	3.56	3.39	3.26	3.15	3.06	2.93	2.78	2.63	2.55	2.47	2.38	2.29	2.2	2.1
28	7.64	5.45	4.57	4.07	3.75	3.53	3.36	3.23	3.12	3.03	2.9	2.75	2.6	2.52	2.44	2.35	2.26	2.17	2.06
29	7.6	5.42	4.54	4.04	3.73	3.5	3.33	3.2	3.09	3	2.87	2.73	2.57	2.49	2.41	2.33	2.23	2.14	2.03
30	7.56	5.39	4.51	4.02	3.7	3.47	3.3	3.17	3.07	2.98	2.84	2.7	2.55	2.47	2.39	2.3	2.21	2.11	2.01
40	7.31	5.18	4.31	3.83	3.51	3.29	3.12	2.99	2.89	2.8	2.66	2.52	2.37	2.29	2.2	2.11	2.02	1.92	1.8
60	7.08	4.98	4.13	3.65	3.34	3.12	2.95	2.82	2.72	2.63	2.5	2.35	2.2	2.12	2.03	1.94	1.84	1.73	1.6
120	6.85	4.79	3.95	3.48	3.17	2.96	2.79	2.66	2.56	2.47	2.34	2.19	2.03	1.95	1.86	1.76	1.66	1.53	1.38
∞	6.63	4.61	3.78	3.32	3.02	2.8	2.64	2.51	2.41	2.32	2.18	2.04	1.88	1.79	1.7	1.59	1.47	1.32	1

Source: The entries in this table were computed by the author

Table A.6 Percentage points of the *F* distribution ($\alpha = 2.5\ \%$)

$F_{0.025\,(v1,\,v2)}$

***v1*=degrees of freedom for the numerator**

	v1	1	2	3	4	5	6	7	8	9	10	12	15	20	24	30	40	60	120	∞
v2	1	647.79	799.5	864.16	899.58	921.85	937.11	948.22	956.66	963.28	968.63	976.71	984.87	993.1	997.25	1001.41	1005.6	1009.8	1014.02	1018.26
v2=	2	38.51	39	39.17	39.25	39.3	39.33	39.36	39.37	39.39	39.4	39.41	39.43	39.45	39.46	39.46	39.47	39.48	39.49	39.5
degrees of	3	17.44	16.04	15.44	15.1	14.88	14.73	14.62	14.54	14.47	14.42	14.34	14.25	14.17	14.12	14.08	14.04	13.99	13.95	13.9
freedom	4	12.22	10.65	9.98	9.6	9.36	9.2	9.07	8.98	8.9	8.84	8.75	8.66	8.56	8.51	8.46	8.41	8.36	8.31	8.26
for the denomi-	5	10.01	8.43	7.76	7.39	7.15	6.98	6.85	6.76	6.68	6.62	6.52	6.43	6.33	6.28	6.23	6.18	6.12	6.07	6.02
nator	6	8.81	7.26	6.6	6.23	5.99	5.82	5.7	5.6	5.52	5.46	5.37	5.27	5.17	5.12	5.07	5.01	4.96	4.9	4.85
	7	8.07	6.54	5.89	5.52	5.29	5.12	4.99	4.9	4.82	4.76	4.67	4.57	4.47	4.41	4.36	4.31	4.25	4.2	4.14
	8	7.57	6.06	5.42	5.05	4.82	4.65	4.53	4.43	4.36	4.3	4.2	4.1	4	3.95	3.89	3.84	3.78	3.73	3.67
	9	7.21	5.71	5.08	4.72	4.48	4.32	4.2	4.1	4.03	3.96	3.87	3.77	3.67	3.61	3.56	3.51	3.45	3.39	3.33
	10	6.94	5.46	4.83	4.47	4.24	4.07	3.95	3.85	3.78	3.72	3.62	3.52	3.42	3.37	3.31	3.26	3.2	3.14	3.08
	11	6.72	5.26	4.63	4.28	4.04	3.88	3.76	3.66	3.59	3.53	3.43	3.33	3.23	3.17	3.12	3.06	3	2.94	2.88
	12	6.55	5.1	4.47	4.12	3.89	3.73	3.61	3.51	3.44	3.37	3.28	3.18	3.07	3.02	2.96	2.91	2.85	2.79	2.72
	13	6.41	4.97	4.35	4	3.77	3.6	3.48	3.39	3.31	3.25	3.15	3.05	2.95	2.89	2.84	2.78	2.72	2.66	2.6
	14	6.3	4.86	4.24	3.89	3.66	3.5	3.38	3.29	3.21	3.15	3.05	2.95	2.84	2.79	2.73	2.67	2.61	2.55	2.49
	15	6.2	4.77	4.15	3.8	3.58	3.41	3.29	3.2	3.12	3.06	2.96	2.86	2.76	2.7	2.64	2.59	2.52	2.46	2.4
	16	6.12	4.69	4.08	3.73	3.5	3.34	3.22	3.12	3.05	2.99	2.89	2.79	2.68	2.63	2.57	2.51	2.45	2.38	2.32
	17	6.04	4.62	4.01	3.66	3.44	3.28	3.16	3.06	2.98	2.92	2.82	2.72	2.62	2.56	2.5	2.44	2.38	2.32	2.25

18	5.98	4.56	3.95	3.61	3.38	3.22	3.1	3.01	2.93	2.87	2.77	2.67	2.56	2.5	2.44	2.38	2.32	2.26	2.19
19	5.92	4.51	3.9	3.56	3.33	3.17	3.05	2.96	2.88	2.82	2.72	2.62	2.51	2.45	2.39	2.33	2.27	2.2	2.13
20	5.87	4.46	3.86	3.51	3.29	3.13	3.01	2.91	2.84	2.77	2.68	2.57	2.46	2.41	2.35	2.29	2.22	2.16	2.09
21	5.83	4.42	3.82	3.48	3.25	3.09	2.97	2.87	2.8	2.73	2.64	2.53	2.42	2.37	2.31	2.25	2.18	2.11	2.04
22	5.79	4.38	3.78	3.44	3.22	3.05	2.93	2.84	2.76	2.7	2.6	2.5	2.39	2.33	2.27	2.21	2.14	2.08	2
23	5.75	4.35	3.75	3.41	3.18	3.02	2.9	2.81	2.73	2.67	2.57	2.47	2.36	2.3	2.24	2.18	2.11	2.04	1.97
24	5.72	4.32	3.72	3.38	3.15	2.99	2.87	2.78	2.7	2.64	2.54	2.44	2.33	2.27	2.21	2.15	2.08	2.01	1.94
25	5.69	4.29	3.69	3.35	3.13	2.97	2.85	2.75	2.68	2.61	2.51	2.41	2.3	2.24	2.18	2.12	2.05	1.98	1.91
26	5.66	4.27	3.67	3.33	3.1	2.94	2.82	2.73	2.65	2.59	2.49	2.39	2.28	2.22	2.16	2.09	2.03	1.95	1.88
27	5.63	4.24	3.65	3.31	3.08	2.92	2.8	2.71	2.63	2.57	2.47	2.36	2.25	2.19	2.13	2.07	2	1.93	1.85
28	5.61	4.22	3.63	3.29	3.06	2.9	2.78	2.69	2.61	2.55	2.45	2.34	2.23	2.17	2.11	2.05	1.98	1.91	1.83
29	5.59	4.2	3.61	3.27	3.04	2.88	2.76	2.67	2.59	2.53	2.43	2.32	2.21	2.15	2.09	2.03	1.96	1.89	1.81
30	5.57	4.18	3.59	3.25	3.03	2.87	2.75	2.65	2.57	2.51	2.41	2.31	2.2	2.14	2.07	2.01	1.94	1.87	1.79
40	5.42	4.05	3.46	3.13	2.9	2.74	2.62	2.53	2.45	2.39	2.29	2.18	2.07	2.01	1.94	1.88	1.8	1.72	1.64
60	5.29	3.93	3.34	3.01	2.79	2.63	2.51	2.41	2.33	2.27	2.17	2.06	1.94	1.88	1.82	1.74	1.67	1.58	1.48
120	5.15	3.8	3.23	2.89	2.67	2.52	2.39	2.3	2.22	2.16	2.05	1.94	1.82	1.76	1.69	1.61	1.53	1.43	1.31
∞	5.02	3.69	3.12	2.79	2.57	2.41	2.29	2.19	2.11	2.05	1.94	1.83	1.71	1.64	1.57	1.48	1.39	1.27	1

Source: The entries in this table were computed by the author

Table A.7 Percentage points of the *F* distribution ($\alpha = 5$ %)

$F_{0.05\,(v1,\,v2)}$

***v1*=degrees of freedom for the numerator**

	v1	1	2	3	4	5	6	7	8	9	10	12	15	20	24	30	40	60	120	∞
v2	1	161.45	199.5	215.71	224.58	230.16	233.99	236.77	238.88	240.54	241.88	243.91	245.95	248.01	249.05	250.1	251.14	252.2	253.25	254.31
v2=	2	18.51	19	19.16	19.25	19.3	19.33	19.35	19.37	19.38	19.4	19.41	19.43	19.45	19.45	19.46	19.47	19.48	19.49	19.5
degrees of	3	10.13	9.55	9.28	9.12	9.01	8.94	8.89	8.85	8.81	8.79	8.74	8.7	8.66	8.64	8.62	8.59	8.57	8.55	8.53
freedom	4	7.71	6.94	6.59	6.39	6.26	6.16	6.09	6.04	6	5.96	5.91	5.86	5.8	5.77	5.75	5.72	5.69	5.66	5.63
for the denomi-	5	6.61	5.79	5.41	5.19	5.05	4.95	4.88	4.82	4.77	4.74	4.68	4.62	4.56	4.53	4.5	4.46	4.43	4.4	4.36
nator	6	5.99	5.14	4.76	4.53	4.39	4.28	4.21	4.15	4.1	4.06	4	3.94	3.87	3.84	3.81	3.77	3.74	3.7	3.67
	7	5.59	4.74	4.35	4.12	3.97	3.87	3.79	3.73	3.68	3.64	3.57	3.51	3.44	3.41	3.38	3.34	3.3	3.27	3.23
	8	5.32	4.46	4.07	3.84	3.69	3.58	3.5	3.44	3.39	3.35	3.28	3.22	3.15	3.12	3.08	3.04	3.01	2.97	2.93
	9	5.12	4.26	3.86	3.63	3.48	3.37	3.29	3.23	3.18	3.14	3.07	3.01	2.94	2.9	2.86	2.83	2.79	2.75	2.71
	10	4.96	4.1	3.71	3.48	3.33	3.22	3.14	3.07	3.02	2.98	2.91	2.85	2.77	2.74	2.7	2.66	2.62	2.58	2.54
	11	4.84	3.98	3.59	3.36	3.2	3.09	3.01	2.95	2.9	2.85	2.79	2.72	2.65	2.61	2.57	2.53	2.49	2.45	2.4
	12	4.75	3.89	3.49	3.26	3.11	3	2.91	2.85	2.8	2.75	2.69	2.62	2.54	2.51	2.47	2.43	2.38	2.34	2.3
	13	4.67	3.81	3.41	3.18	3.03	2.92	2.83	2.77	2.71	2.67	2.6	2.53	2.46	2.42	2.38	2.34	2.3	2.25	2.21
	14	4.6	3.74	3.34	3.11	2.96	2.85	2.76	2.7	2.65	2.6	2.53	2.46	2.39	2.35	2.31	2.27	2.22	2.18	2.13
	15	4.54	3.68	3.29	3.06	2.9	2.79	2.71	2.64	2.59	2.54	2.48	2.4	2.33	2.29	2.25	2.2	2.16	2.11	2.07
	16	4.49	3.63	3.24	3.01	2.85	2.74	2.66	2.59	2.54	2.49	2.42	2.35	2.28	2.24	2.19	2.15	2.11	2.06	2.01
	17	4.45	3.59	3.2	2.96	2.81	2.7	2.61	2.55	2.49	2.45	2.38	2.31	2.23	2.19	2.15	2.1	2.06	2.01	1.96
	18	4.41	3.55	3.16	2.93	2.77	2.66	2.58	2.51	2.46	2.41	2.34	2.27	2.19	2.15	2.11	2.06	2.02	1.97	1.92

19	4.38	3.52	3.13	2.9	2.74	2.63	2.54	2.48	2.42	2.38	2.31	2.23	2.16	2.11	2.07	2.03	1.98	1.93	1.88
20	4.35	3.49	3.1	2.87	2.71	2.6	2.51	2.45	2.39	2.35	2.28	2.2	2.12	2.08	2.04	1.99	1.95	1.9	1.84
21	4.32	3.47	3.07	2.84	2.68	2.57	2.49	2.42	2.37	2.32	2.25	2.18	2.1	2.05	2.01	1.96	1.92	1.87	1.81
22	4.3	3.44	3.05	2.82	2.66	2.55	2.46	2.4	2.34	2.3	2.23	2.15	2.07	2.03	1.98	1.94	1.89	1.84	1.78
23	4.28	3.42	3.03	2.8	2.64	2.53	2.44	2.37	2.32	2.27	2.2	2.13	2.05	2.01	1.96	1.91	1.86	1.81	1.76
24	4.26	3.4	3.01	2.78	2.62	2.51	2.42	2.36	2.3	2.25	2.18	2.11	2.03	1.98	1.94	1.89	1.84	1.79	1.73
25	4.24	3.39	2.99	2.76	2.6	2.49	2.4	2.34	2.28	2.24	2.16	2.09	2.01	1.96	1.92	1.87	1.82	1.77	1.71
26	4.23	3.37	2.98	2.74	2.59	2.47	2.39	2.32	2.27	2.22	2.15	2.07	1.99	1.95	1.9	1.85	1.8	1.75	1.69
27	4.21	3.35	2.96	2.73	2.57	2.46	2.37	2.31	2.25	2.2	2.13	2.06	1.97	1.93	1.88	1.84	1.79	1.73	1.67
28	4.2	3.34	2.95	2.71	2.56	2.45	2.36	2.29	2.24	2.19	2.12	2.04	1.96	1.91	1.87	1.82	1.77	1.71	1.65
29	4.18	3.33	2.93	2.7	2.55	2.43	2.35	2.28	2.22	2.18	2.1	2.03	1.94	1.9	1.85	1.81	1.75	1.7	1.64
30	4.17	3.32	2.92	2.69	2.53	2.42	2.33	2.27	2.21	2.16	2.09	2.01	1.93	1.89	1.84	1.79	1.74	1.68	1.62
40	4.08	3.23	2.84	2.61	2.45	2.34	2.25	2.18	2.12	2.08	2	1.92	1.84	1.79	1.74	1.69	1.64	1.58	1.51
60	4	3.15	2.76	2.53	2.37	2.25	2.17	2.1	2.04	1.99	1.92	1.84	1.75	1.7	1.65	1.59	1.53	1.47	1.39
120	3.92	3.07	2.68	2.45	2.29	2.18	2.09	2.02	1.96	1.91	1.83	1.75	1.66	1.61	1.55	1.5	1.43	1.35	1.25
∞	3.84	3	2.6	2.37	2.21	2.1	2.01	1.94	1.88	1.83	1.75	1.67	1.57	1.52	1.46	1.39	1.32	1.22	1

Source: The entries in this table were computed by the author

Table A.8 Percentage points of the F distribution ($\alpha = 10\,\%$)

$F_{0.10\,(v1,\,v2)}$

***v1*=degrees of freedom for the numerator**

	v1	1	2	3	4	5	6	7	8	9	10	12	15	20	24	30	40	60	120	∞
v2	1	39.86	49.5	53.59	55.83	57.24	58.2	58.91	59.44	59.86	60.19	60.71	61.22	61.74	62	62.26	62.53	62.79	63.06	63.33
v2= degrees of freedom for the denominator	2	8.53	9	9.16	9.24	9.29	9.33	9.35	9.37	9.38	9.39	9.41	9.42	9.44	9.45	9.46	9.47	9.47	9.48	9.49
	3	5.54	5.46	5.39	5.34	5.31	5.28	5.27	5.25	5.24	5.23	5.22	5.2	5.18	5.18	5.17	5.16	5.15	5.14	5.13
	4	4.54	4.32	4.19	4.11	4.05	4.01	3.98	3.95	3.94	3.92	3.9	3.87	3.84	3.83	3.82	3.8	3.79	3.78	3.76
	5	4.06	3.78	3.62	3.52	3.45	3.4	3.37	3.34	3.32	3.3	3.27	3.24	3.21	3.19	3.17	3.16	3.14	3.12	3.1
	6	3.78	3.46	3.29	3.18	3.11	3.05	3.01	2.98	2.96	2.94	2.9	2.87	2.84	2.82	2.8	2.78	2.76	2.74	2.72
	7	3.59	3.26	3.07	2.96	2.88	2.83	2.78	2.75	2.72	2.7	2.67	2.63	2.59	2.58	2.56	2.54	2.51	2.49	2.47
	8	3.46	3.11	2.92	2.81	2.73	2.67	2.62	2.59	2.56	2.54	2.5	2.46	2.42	2.4	2.38	2.36	2.34	2.32	2.29
	9	3.36	3.01	2.81	2.69	2.61	2.55	2.51	2.47	2.44	2.42	2.38	2.34	2.3	2.28	2.25	2.23	2.21	2.18	2.16
	10	3.29	2.92	2.73	2.61	2.52	2.46	2.41	2.38	2.35	2.32	2.28	2.24	2.2	2.18	2.16	2.13	2.11	2.08	2.06
	11	3.23	2.86	2.66	2.54	2.45	2.39	2.34	2.3	2.27	2.25	2.21	2.17	2.12	2.1	2.08	2.05	2.03	2	1.97
	12	3.18	2.81	2.61	2.48	2.39	2.33	2.28	2.24	2.21	2.19	2.15	2.1	2.06	2.04	2.01	1.99	1.96	1.93	1.9
	13	3.14	2.76	2.56	2.43	2.35	2.28	2.23	2.2	2.16	2.14	2.1	2.05	2.01	1.98	1.96	1.93	1.9	1.88	1.85
	14	3.1	2.73	2.52	2.39	2.31	2.24	2.19	2.15	2.12	2.1	2.05	2.01	1.96	1.94	1.91	1.89	1.86	1.83	1.8
	15	3.07	2.7	2.49	2.36	2.27	2.21	2.16	2.12	2.09	2.06	2.02	1.97	1.92	1.9	1.87	1.85	1.82	1.79	1.76
	16	3.05	2.67	2.46	2.33	2.24	2.18	2.13	2.09	2.06	2.03	1.99	1.94	1.89	1.87	1.84	1.81	1.78	1.75	1.72
	17	3.03	2.64	2.44	2.31	2.22	2.15	2.1	2.06	2.03	2	1.96	1.91	1.86	1.84	1.81	1.78	1.75	1.72	1.69

18	3.01	2.62	2.42	2.29	2.2	2.13	2.08	2.04	2	1.98	1.93	1.89	1.84	1.81	1.78	1.75	1.72	1.69	1.66
19	2.99	2.61	2.4	2.27	2.18	2.11	2.06	2.02	1.98	1.96	1.91	1.86	1.81	1.79	1.76	1.73	1.7	1.67	1.63
20	2.97	2.59	2.38	2.25	2.16	2.09	2.04	2	1.96	1.94	1.89	1.84	1.79	1.77	1.74	1.71	1.68	1.64	1.61
21	2.96	2.57	2.36	2.23	2.14	2.08	2.02	1.98	1.95	1.92	1.87	1.83	1.78	1.75	1.72	1.69	1.66	1.62	1.59
22	2.95	2.56	2.35	2.22	2.13	2.06	2.01	1.97	1.93	1.9	1.86	1.81	1.76	1.73	1.7	1.67	1.64	1.6	1.57
23	2.94	2.55	2.34	2.21	2.11	2.05	1.99	1.95	1.92	1.89	1.84	1.8	1.74	1.72	1.69	1.66	1.62	1.59	1.55
24	2.93	2.54	2.33	2.19	2.1	2.04	1.98	1.94	1.91	1.88	1.83	1.78	1.73	1.7	1.67	1.64	1.61	1.57	1.53
25	2.92	2.53	2.32	2.18	2.09	2.02	1.97	1.93	1.89	1.87	1.82	1.77	1.72	1.69	1.66	1.63	1.59	1.56	1.52
26	2.91	2.52	2.31	2.17	2.08	2.01	1.96	1.92	1.88	1.86	1.81	1.76	1.71	1.68	1.65	1.61	1.58	1.54	1.5
27	2.9	2.51	2.3	2.17	2.07	2	1.95	1.91	1.87	1.85	1.8	1.75	1.7	1.67	1.64	1.6	1.57	1.53	1.49
28	2.89	2.5	2.29	2.16	2.06	2	1.94	1.9	1.87	1.84	1.79	1.74	1.69	1.66	1.63	1.59	1.56	1.52	1.48
29	2.89	2.5	2.28	2.15	2.06	1.99	1.93	1.89	1.86	1.83	1.78	1.73	1.68	1.65	1.62	1.58	1.55	1.51	1.47
30	2.88	2.49	2.28	2.14	2.03	1.98	1.93	1.88	1.85	1.82	1.77	1.72	1.67	1.64	1.61	1.57	1.54	1.5	1.46
40	2.84	2.44	2.23	2.09	2	1.93	1.87	1.83	1.79	1.76	1.71	1.66	1.61	1.57	1.54	1.51	1.47	1.42	1.38
60	2.79	2.39	2.18	2.04	1.95	1.87	1.82	1.77	1.74	1.71	1.66	1.6	1.54	1.51	1.48	1.44	1.4	1.35	1.29
120	2.75	2.35	2.13	1.99	1.9	1.82	1.77	1.72	1.68	1.65	1.6	1.55	1.48	1.45	1.41	1.37	1.32	1.26	1.19
∞	2.71	2.3	2.08	1.94	1.85	1.77	1.72	1.67	1.63	1.6	1.55	1.49	1.42	1.38	1.34	1.3	1.24	1.17	1

Source: The entries in this table were computed by the author

Table A.9 Percentage points of the *F* distribution ($\alpha = 25\ \%$)

$F_{0.25\,(v1,\,v2)}$

v1=degrees of freedom for the numerator																			
	v1	**1**	**2**	**3**	**4**	**5**	**6**	**7**	**8**	**9**	**10**	**12**	**15**	**20**	**24**	**30**	**40**	**60**	**120**	**∞**
v2	1	5.83	7.5	8.2	8.58	8.82	8.98	9.1	9.19	9.26	9.32	9.41	9.49	9.58	9.63	9.67	9.71	9.76	9.8	9.85
***v2*= degrees of freedom for the denominator**	2	2.57	3	3.15	3.23	3.28	3.31	3.34	3.35	3.37	3.38	3.39	3.41	3.43	3.43	3.44	3.45	3.46	3.47	3.48
	3	2.02	2.28	2.36	2.39	2.41	2.42	2.43	2.44	2.44	2.44	2.45	2.46	2.46	2.46	2.47	2.47	2.47	2.47	2.47
	4	1.81	2	2.05	2.06	2.07	2.08	2.08	2.08	2.08	2.08	2.08	2.08	2.08	2.08	2.08	2.08	2.08	2.08	2.08
	5	1.69	1.85	1.88	1.89	1.89	1.89	1.89	1.89	1.89	1.89	1.89	1.89	1.88	1.88	1.88	1.88	1.87	1.87	1.87
	6	1.62	1.76	1.78	1.79	1.79	1.78	1.78	1.78	1.77	1.77	1.77	1.76	1.76	1.75	1.75	1.75	1.74	1.74	1.74
	7	1.57	1.7	1.72	1.72	1.71	1.71	1.7	1.7	1.7	1.69	1.68	1.68	1.67	1.67	1.66	1.66	1.65	1.65	1.65
	8	1.54	1.66	1.67	1.66	1.66	1.65	1.64	1.64	1.63	1.63	1.62	1.62	1.61	1.6	1.6	1.59	1.59	1.58	1.58
	9	1.51	1.62	1.63	1.63	1.62	1.61	1.6	1.6	1.59	1.59	1.58	1.57	1.56	1.56	1.55	1.54	1.54	1.53	1.53
	10	1.49	1.6	1.6	1.59	1.59	1.58	1.57	1.56	1.56	1.55	1.54	1.53	1.52	1.52	1.51	1.51	1.5	1.49	1.48
	11	1.47	1.58	1.58	1.57	1.56	1.55	1.54	1.53	1.53	1.52	1.51	1.5	1.49	1.49	1.48	1.47	1.47	1.46	1.45
	12	1.46	1.56	1.56	1.55	1.54	1.53	1.52	1.51	1.51	1.5	1.49	1.48	1.47	1.46	1.45	1.45	1.44	1.43	1.42
	13	1.45	1.55	1.55	1.53	1.52	1.51	1.5	1.49	1.49	1.48	1.47	1.46	1.45	1.44	1.43	1.42	1.42	1.41	1.4
	14	1.44	1.53	1.53	1.52	1.51	1.5	1.49	1.48	1.47	1.46	1.45	1.44	1.43	1.42	1.41	1.41	1.4	1.39	1.38
	15	1.43	1.52	1.52	1.51	1.49	1.48	1.47	1.46	1.46	1.45	1.44	1.43	1.41	1.41	1.4	1.39	1.38	1.37	1.36
	16	1.42	1.51	1.51	1.5	1.48	1.47	1.46	1.45	1.44	1.44	1.43	1.41	1.4	1.39	1.38	1.37	1.36	1.35	1.34
	17	1.42	1.51	1.5	1.49	1.47	1.46	1.45	1.44	1.43	1.43	1.41	1.4	1.39	1.38	1.37	1.36	1.35	1.34	1.33
	18	1.41	1.5	1.49	1.48	1.46	1.45	1.44	1.43	1.42	1.42	1.4	1.39	1.38	1.37	1.36	1.35	1.34	1.33	1.32

19	1.41	1.49	1.49	1.47	1.46	1.44	1.43	1.42	1.41	1.41	1.4	1.38	1.37	1.36	1.35	1.34	1.33	1.32	1.3
20	1.4	1.49	1.48	1.47	1.45	1.44	1.43	1.42	1.41	1.4	1.39	1.37	1.36	1.35	1.34	1.33	1.32	1.31	1.29
21	1.4	1.48	1.48	1.46	1.44	1.43	1.42	1.41	1.4	1.39	1.38	1.37	1.35	1.34	1.33	1.32	1.31	1.3	1.28
22	1.4	1.48	1.47	1.45	1.44	1.42	1.41	1.4	1.39	1.39	1.37	1.36	1.34	1.33	1.32	1.31	1.3	1.29	1.28
23	1.39	1.47	1.47	1.45	1.43	1.42	1.41	1.4	1.39	1.38	1.37	1.35	1.34	1.33	1.32	1.31	1.3	1.28	1.27
24	1.39	1.47	1.46	1.44	1.43	1.41	1.4	1.39	1.38	1.38	1.36	1.35	1.33	1.32	1.31	1.3	1.29	1.28	1.26
25	1.39	1.47	1.46	1.44	1.42	1.41	1.4	1.39	1.38	1.37	1.36	1.34	1.33	1.32	1.31	1.29	1.28	1.27	1.25
26	1.38	1.46	1.45	1.44	1.42	1.41	1.39	1.38	1.37	1.37	1.35	1.34	1.32	1.31	1.3	1.29	1.28	1.26	1.25
27	1.38	1.46	1.45	1.43	1.42	1.4	1.39	1.38	1.37	1.36	1.35	1.33	1.32	1.31	1.3	1.28	1.27	1.26	1.24
28	1.38	1.46	1.45	1.43	1.41	1.4	1.39	1.38	1.37	1.36	1.34	1.33	1.31	1.3	1.29	1.28	1.27	1.25	1.24
29	1.38	1.45	1.45	1.43	1.41	1.4	1.38	1.37	1.36	1.35	1.34	1.32	1.31	1.3	1.29	1.27	1.26	1.25	1.23
30	1.38	1.45	1.44	1.42	1.41	1.39	1.38	1.37	1.36	1.35	1.34	1.32	1.3	1.29	1.28	1.27	1.26	1.24	1.23
40	1.36	1.44	1.42	1.4	1.39	1.37	1.36	1.35	1.34	1.33	1.31	1.3	1.28	1.26	1.25	1.24	1.22	1.21	1.19
60	1.35	1.42	1.41	1.38	1.37	1.35	1.33	1.32	1.31	1.3	1.29	1.27	1.25	1.24	1.22	1.21	1.19	1.17	1.15
120	1.34	1.4	1.39	1.37	1.35	1.33	1.31	1.3	1.29	1.28	1.26	1.24	1.22	1.21	1.19	1.18	1.16	1.13	1.1
∞	1.32	1.39	1.37	1.35	1.33	1.31	1.29	1.28	1.27	1.25	1.24	1.22	1.19	1.18	1.16	1.14	1.12	1.08	1

Source: The entries in this table were computed by the author

Table A.10 Poisson cumulative distribution

	λ=	0.1	0.2	0.3	0.4	0.5	0.6	0.7	0.8	0.9	1	1.2	1.4	1.6	1.8	2	2.2	2.4	2.6	2.8	3	3.2	3.4	3.6	3.8
$x=$	0	0.9048	0.8187	0.7408	0.6703	0.6065	0.5488	0.4966	0.4493	0.4066	0.3679	0.3012	0.2466	0.2019	0.1653	0.1353	0.1108	0.0907	0.0743	0.0608	0.0498	0.0408	0.0334	0.0273	0.0224
	1	0.9953	0.9825	0.9631	0.9384	0.9098	0.8781	0.8442	0.8088	0.7725	0.7358	0.6626	0.5918	0.5249	0.4628	0.406	0.3546	0.3084	0.2674	0.2311	0.1991	0.1712	0.1468	0.1257	0.1074
	2	0.9998	0.9989	0.9964	0.9921	0.9856	0.9769	0.9659	0.9526	0.9371	0.9197	0.8795	0.8335	0.7834	0.7306	0.6767	0.6227	0.5697	0.5184	0.4695	0.4232	0.3799	0.3397	0.3027	0.2689
	3	1	0.9999	0.9997	0.9992	0.9982	0.9966	0.9942	0.9909	0.9865	0.981	0.9662	0.9463	0.9212	0.8913	0.8571	0.8194	0.7787	0.736	0.6919	0.6472	0.6025	0.5584	0.5152	0.4735
	4	1	1	1	0.9999	0.9998	0.9996	0.9992	0.9986	0.9977	0.9963	0.9923	0.9857	0.9763	0.9636	0.9473	0.9275	0.9041	0.8774	0.8477	0.8153	0.7806	0.7442	0.7064	0.6678
	5	1	1	1	1	1	1	0.9999	0.9998	0.9997	0.9994	0.9985	0.9968	0.994	0.9896	0.9834	0.9751	0.9643	0.951	0.9349	0.9161	0.8946	0.8705	0.8441	0.8156
	6	1	1	1	1	1	1	1	1	1	0.9999	0.9997	0.9994	0.9987	0.9974	0.9955	0.9925	0.9884	0.9828	0.9756	0.9665	0.9554	0.9421	0.9267	0.9091
	7	1	1	1	1	1	1	1	1	1	1	1	0.9999	0.9997	0.9994	0.9989	0.998	0.9967	0.9947	0.9919	0.9881	0.9832	0.9769	0.9692	0.9599
	8	1	1	1	1	1	1	1	1	1	1	1	1	1	0.9999	0.9998	0.9995	0.9991	0.9985	0.9976	0.9962	0.9943	0.9917	0.9883	0.984
	9	1	1	1	1	1	1	1	1	1	1	1	1	1	1	1	0.9999	0.9998	0.9996	0.9993	0.9989	0.9982	0.9973	0.996	0.9942
	10															1	1	1	0.9999	0.9998	0.9997	0.9995	0.9992	0.9987	0.9981
	11															1	1	1	1	1	0.9999	0.9999	0.9998	0.9996	0.9994
	12															1	1	1	1	1	1	1	0.9999	0.9999	0.9998
	13															1	1	1	1	1	1	1	1	1	1
	14															1	1	1	1	1	1	1	1	1	1
	15															1	1	1	1	1	1	1	1	1	1
	16															1	1	1	1	1	1	1	1	1	1
	17															1	1	1	1	1	1	1	1	1	1

	λ=	4	4.5	5	5.5	6	6.5	7	7.5	8	8.5	9	9.5	10	11	10	12	14	15
x=	0	0.0183	0.0111	0.0067	0.0041	0.0025	0.0015	0.0009	0.0006	0.0003	0.0002	0.0001	0.0001	0	0	0	0	0	0
	1	0.0916	0.0611	0.0404	0.0266	0.0174	0.0113	0.0073	0.0047	0.003	0.0019	0.0012	0.0008	0.0005	0.0002	0.0005	0.0001	0	0
	2	0.2381	0.1736	0.1247	0.0884	0.062	0.043	0.0296	0.0203	0.0138	0.0093	0.0062	0.0042	0.0028	0.0012	0.0028	0.0005	0.0001	0
	3	0.4335	0.3423	0.265	0.2017	0.1512	0.1118	0.0818	0.0591	0.0424	0.0301	0.0212	0.0149	0.0103	0.0049	0.0103	0.0023	0.0005	0.0002
	4	0.6288	0.5321	0.4405	0.3575	0.2851	0.2237	0.173	0.1321	0.0996	0.0744	0.055	0.0403	0.0293	0.0151	0.0293	0.0076	0.0018	0.0009
	5	0.7851	0.7029	0.616	0.5289	0.4457	0.369	0.3007	0.2414	0.1912	0.1496	0.1157	0.0885	0.0671	0.0375	0.0671	0.0203	0.0055	0.0028
	6	0.8893	0.8311	0.7622	0.686	0.6063	0.5265	0.4497	0.3782	0.3134	0.2562	0.2068	0.1649	0.1301	0.0786	0.1301	0.0458	0.0142	0.0076
	7	0.9489	0.9134	0.8666	0.8095	0.744	0.6728	0.5987	0.5246	0.453	0.3856	0.3239	0.2687	0.2202	0.1432	0.2202	0.0895	0.0316	0.018
	8	0.9786	0.9597	0.9319	0.8944	0.8472	0.7916	0.7291	0.662	0.5925	0.5231	0.4557	0.3918	0.3328	0.232	0.3328	0.155	0.0621	0.0374
	9	0.9919	0.9829	0.9682	0.9462	0.9161	0.8774	0.8305	0.7764	0.7166	0.653	0.5874	0.5218	0.4579	0.3405	0.4579	0.2424	0.1094	0.0699
	10	0.9972	0.9933	0.9863	0.9747	0.9574	0.9332	0.9015	0.8622	0.8159	0.7634	0.706	0.6453	0.583	0.4599	0.583	0.3472	0.1757	0.1185
	11	0.9991	0.9976	0.9945	0.989	0.9799	0.9661	0.9467	0.9208	0.8881	0.8487	0.803	0.752	0.6968	0.5793	0.6968	0.4616	0.26	0.1848
	12	0.9997	0.9992	0.998	0.9955	0.9912	0.984	0.973	0.9573	0.9362	0.9091	0.8758	0.8364	0.7916	0.6887	0.7916	0.576	0.3585	0.2676
	13	0.9999	0.9997	0.9993	0.9983	0.9964	0.9929	0.9872	0.9784	0.9658	0.9486	0.9261	0.8981	0.8645	0.7813	0.8645	0.6815	0.4644	0.3632
	14	1	0.9999	0.9998	0.9994	0.9986	0.997	0.9943	0.9897	0.9827	0.9726	0.9585	0.94	0.9165	0.854	0.9165	0.772	0.5704	0.4657
	15	1	1	0.9999	0.9998	0.9995	0.9988	0.9976	0.9954	0.9918	0.9862	0.978	0.9665	0.9513	0.9074	0.9513	0.8444	0.6694	0.5681
	16	1	1	1	0.9999	0.9998	0.9996	0.999	0.998	0.9963	0.9934	0.9889	0.9823	0.973	0.9441	0.973	0.8987	0.7559	0.6641
	17	1	1	1	1	0.9999	0.9998	0.9996	0.9992	0.9984	0.997	0.9947	0.9911	0.9857	0.9678	0.9857	0.937	0.8272	0.7489

(continued)

Table A.10 (continued)

λ=	4	4.5	5	5.5	6	6.5	7	7.5	8	8.5	9	9.5	10	11	10	12	14	15
18					1	0.9999	0.9999	0.9997	0.9993	0.9987	0.9976	0.9957	0.9928	0.9823	0.9928	0.9626	0.8826	0.8195
19					1	1	1	0.9999	0.9997	0.9995	0.9989	0.998	0.9965	0.9907	0.9965	0.9787	0.9235	0.8752
20					1	1	1	1	0.9999	0.9998	0.9996	0.9991	0.9984	0.9953	0.9984	0.9884	0.9521	0.917
21					1	1	1	1	1	0.9999	0.9998	0.9996	0.9993	0.9977	0.9993	0.9939	0.9712	0.9469
22					1	1	1	1	1	1	0.9999	0.9999	0.9997	0.999	0.9997	0.997	0.9833	0.9673
23					1	1	1	1	1	1	1	0.9999	0.9999	0.9995	0.9999	0.9985	0.9907	0.9805
24					1	1	1	1	1	1	1	1	1	0.9998	1	0.9993	0.995	0.9888
25					1	1	1	1	1	1	1	1	1	0.9999	1	0.9997	0.9974	0.9938
26					1	1	1	1	1	1	1	1	1	1	1	0.9999	0.9987	0.9967
27					1	1	1	1	1	1	1	1	1	1	1	0.9999	0.9994	0.9983
28					1	1	1	1	1	1	1	1	1	1	1	1	0.9997	0.9991
29					1	1	1	1	1	1	1	1	1	1	1	1	0.9999	0.9996
30					1	1	1	1	1	1	1	1	1	1	1	1	0.9999	0.9998
31					1	1	1	1	1	1	1	1	1	1	1	1	1	0.9999
32					1	1	1	1	1	1	1	1	1	1	1	1	1	1

Source: The entries in this table were computed by the author

Table A.11 Coefficients for control charts

Observation in batch	d_2	c_4	A_2	D_3	D_4	A_3	B_3	B_4
2	1.128	0.798	1.880	0	3.267	2.659	0	3.267
3	1.693	0.886	1.023	0	2.575	1.954	0	2.568
4	2.059	0.921	0.729	0	2.282	1.628	0	2.266
5	2.326	0.940	0.577	0	2.115	1.427	0	2.089
6	2.534	0.952	0.483	0	2.004	1.287	0.030	1.970
7	2.704	0.959	0.419	0.076	1.924	1.182	0.118	1.882
8	2.847	0.965	0.373	0.136	1.864	1.099	0.185	1.815
9	2.970	0.969	0.337	0.184	1.816	1.032	0.239	1.761
10	3.078	0.973	0.308	0.223	1.777	0.975	0.284	1.716
11	3.173	0.975	0.285	0.256	1.744	0.927	0.321	1.679
12	3.258	0.978	0.266	0.283	1.717	0.886	0.354	1.646
13	3.336	0.979	0.249	0.307	1.693	0.850	0.382	1.618
14	3.407	0.981	0.235	0.328	1.672	0.817	0.406	1.594
15	3.472	0.982	0.223	0.347	1.653	0.789	0.428	1.572
16	3.532	0.984	0.212	0.363	1.637	0.763	0.448	1.552
17	3.588	0.985	0.203	0.378	1.622	0.739	0.466	1.534
18	3.640	0.985	0.194	0.391	1.608	0.718	0.482	1.518
19	3.689	0.986	0.187	0.403	1.597	0.698	0.497	1.503
20	3.735	0.987	0.180	0.415	1.585	0.680	0.510	1.490
21	3.778	0.988	0.173	0.425	1.575	0.663	0.523	1.477
22	3.819	0.988	0.167	0.434	1.566	0.647	0.534	1.466
23	3.858	0.989	0.162	0.443	1.557	0.633	0.545	1.455
24	3.895	0.989	0.157	0.451	1.548	0.619	0.555	1.445
25	3.931	0.990	0.153	0.459	1.541	0.606	0.565	1.435

Source: The entries in this table were computed by the author

Table A.12 Table K for tolerance factors for normal distribution

Values of *k* for two-sided intervals									
	Confidence level								
	0.90			**0.95**			**0.99**		
	Probability of coverage								
Sample size	**0.90**	**0.95**	**0.99**	**0.90**	**0.95**	**0.99**	**0.90**	**0.95**	**0.99**
2	15.978	18.800	24.167	32.019	37.674	48.430	160.193	188.491	242.300
3	5.847	6.919	8.974	8.380	9.916	12.861	18.930	22.401	29.055
4	4.166	4.943	6.440	5.369	6.370	8.299	9.398	11.150	14.527
5	3.949	4.152	5.423	4.275	5.079	6.634	6.612	7.855	10.260
6	3.131	3.723	4.870	3.712	4.414	5.775	5.337	6.345	8.301
7	2.902	3.452	4.521	3.369	4.007	5.248	4.613	5.488	7.187
8	2.743	3.264	4.278	3.136	3.732	4.891	4.147	4.936	6.468
9	2.626	3.125	4.098	2.967	3.532	4.631	3.822	4.550	5.966
10	2.535	3.018	3.959	2.839	3.379	4.433	3.582	4.265	5.594
11	2.463	2.933	3.849	2.737	3.259	4.277	3.397	4.045	5.308
12	2.404	2.863	3.758	2.655	3.162	4.150	3.250	3.870	5.079
13	2.355	2.805	3.682	2.587	3.081	4.044	3.130	3.727	4.893
14	2.314	2.756	3.618	2.529	3.012	3.955	3.029	3.608	4.737
15	2.278	2.713	3.562	2.480	2.954	3.878	2.945	3.507	4.605
16	2.246	2.676	3.514	2.437	2.903	3.812	2.872	3.421	4.492
17	2.219	2.643	3.471	2.400	2.858	3.754	2.808	3.345	4.393
18	2.194	2.614	3.433	2.366	2.819	3.702	2.753	3.279	4.307
19	2.172	2.588	3.399	2.337	2.784	3.656	2.703	3.221	4.230
20	2.152	2.564	3.368	2.310	2.752	3.615	2.659	3.168	4.161
21	2.135	2.543	3.340	2.286	2.723	3.577	2.620	3.121	4.100
22	2.118	2.524	3.315	2.264	2.697	3.543	2.584	3.078	4.044
23	2.103	2.506	3.292	2.244	2.673	3.512	2.551	3.040	3.993
24	2.089	2.489	3.270	2.225	2.651	3.483	2.522	3.004	3.947
25	2.077	2.474	3.251	2.208	2.631	3.457	2.494	2.972	3.904
30	2.025	2.413	3.170	2.140	2.529	3.350	2.385	2.841	3.733
40	1.959	2.334	3.066	2.052	2.445	3.213	2.247	2.677	3.518
50	1.916	2.284	3.001	1.996	2.379	3.126	2.162	2.576	3.385
60	1.887	2.248	2.955	1.958	2.333	3.066	2.103	2.506	3.293
70	1.865	2.222	2.920	1.929	2.299	3.021	2.060	2.454	3.225
80	1.848	2.202	2.894	1.907	2.272	2.986	2.026	2.414	3.173
90	1.834	2.185	2.872	1.889	2.251	2.958	1.999	2.382	3.130
100	1.822	2.172	2.854	1.874	2.233	2.934	1.977	2.355	3.096

Source: The entries in this table were computed by the author

Table A.13 Table K for tolerance factors for normal distribution

Values of *k* for one-sided intervals									
	Confidence level								
	0.90			**0.95**			**0.99**		
	Probability of coverage								
Sample size	**0.90**	**0.95**	**0.99**	**0.90**	**0.95**	**0.99**	**0.90**	**0.95**	**0.99**
2	10.253	13.090	18.500	20.581	26.260	37.094	103.029	131.426	185.617
3	4.258	5.311	7.340	6.155	7.656	10.553	13.995	17.370	23.896
4	3.188	3.957	5.438	4.162	5.144	7.042	7.380	9.083	12.387
5	2.742	3.400	4.666	3.407	4.203	5.741	5.362	6.578	8.939
6	2.494	3.092	4.243	3.006	3.708	5.062	4.411	5.406	7.335
7	2.333	2.894	3.972	2.755	3.399	4.642	3.859	4.728	6.412
8	2.219	2.754	3.783	2.582	3.187	4.354	3.497	4.285	5.812
9	2.133	2.650	3.641	2.454	3.031	4.143	3.240	3.972	5.389
10	2.066	2.568	3.532	2.355	2.911	3.981	3.048	3.738	5.074
11	2.011	2.503	3.443	2.275	2.815	3.852	2.898	3.556	4.829
12	1.966	2.448	3.371	2.210	2.736	3.747	2.777	3.410	4.633
13	1.928	2.402	3.309	2.155	2.671	3.659	2.677	3.290	4.472
14	1.895	2.363	3.257	2.109	2.614	3.585	2.593	3.189	4.337
15	1.867	2.329	3.212	2.068	2.566	3.520	2.521	3.102	4.222
16	1.842	2.299	3.172	2.033	2.524	3.464	2.459	3.028	4.123
17	1.819	2.272	3.137	2.002	2.486	3.414	2.405	2.963	4.037
18	1.800	2.249	3.105	1.974	2.453	3.370	2.357	2.905	3.960
19	1.782	2.227	3.077	1.949	2.423	3.331	2.314	2.854	3.892
20	1.765	2.028	3.052	1.926	2.396	3.295	2.276	2.808	3.832
21	1.750	2.190	3.028	1.905	2.371	3.263	2.241	2.766	3.777
22	1.737	2.174	3.007	1.886	2.349	3.233	2.209	2.729	3.727
23	1.724	2.159	2.987	1.869	2.328	3.206	2.180	2.694	3.681
24	1.712	2.145	2.969	1.853	2.309	3.181	2.154	2.662	3.640
25	1.702	2.132	2.952	1.838	2.292	3.158	2.129	2.633	3.601
30	1.657	2.080	2.884	1.777	2.220	3.064	2.030	2.515	3.447
40	1.598	2.010	2.793	1.697	2.125	2.941	1.902	2.364	3.249
50	1.559	1.965	2.735	1.646	2.065	2.862	1.821	2.269	3.125
60	1.532	1.933	2.694	1.609	2.022	2.807	1.764	2.202	3.038
70	1.511	1.909	2.662	1.581	1.990	2.765	1.722	2.153	2.974
80	1.495	1.890	2.638	1.559	1.964	2.733	1.688	2.114	2.924
90	1.481	1.874	2.618	1.542	1.944	2.706	1.661	2.082	2.883
100	1.470	1.861	2.601	1.527	1.927	2.684	1.639	2.056	2.850

Source: The entries in this table were computed by the author

Index

D

E

F

G

H

I

O

P

Q

R

S

T

U

V

W

X

Y

Z

The manufacturer's authorised representative in the EU is Springer Nature Customer Service Centre GmbH, Europaplatz 3, 69115 Heidelberg, Germany. If you have any concerns regarding our products, please contact ProductSafety@springernature.com

Printed and bound by CPI Group (UK) Ltd, Croydon, CR0 4YY
30/09/2026
02225601-0020